# Searching for Sustainability

*Interdisciplinary Essays in the Philosophy of Conservation Biology*

This book examines from several disciplinary viewpoints the question of what we mean – what we should mean – by setting sustainability as a goal for environmental management. The author, trained as a philosopher of science and language, searches through multiple disciplines for insights necessary to develop a comprehensive and integrated concept of sustainable living. This book explores ways to break down the disciplinary barriers to communication and deliberation about environmental policy, and to integrate science and evaluations into a more effective environmental strategy. Choosing sustainability as the key concept of environmental policy, the author explores what we can learn about sustainable living from the philosophy of pragmatism, from ecology, from economics, from planning, from conservation biology, and from related disciplines. The idea of adaptive, or experimental, management provides the context, while insights from various disciplines are integrated into a comprehensive philosophy of environmental management.

The book will appeal to students and professionals in the fields of environmental policy and ethics, conservation biology, and philosophy of science.

Bryan G. Norton is Professor of Philosophy, Science, and Technology at the School of Public Policy, Georgia Institute of Technology.

# CAMBRIDGE STUDIES IN PHILOSOPHY AND BIOLOGY

*General Editor*
Michael Ruse   *Florida State University*

*Advisory Board*
Michael Donoghue   *Yale University*
Jean Gayon   *University of Paris*
Jonathan Hodge   *University of Leeds*
Jane Maienschein   *Arizona State University*
Jesús Mosterín   *Instituto de Filosofía (Spanish Research Council)*
Elliott Sober   *University of Wisconsin*

Alfred I. Tauber   *The Immune Self: Theory or Metaphor?*

Elliott Sober   *From a Biological Point of View*

Robert Brandon   *Concepts and Methods in Evolutionary Biology*

Peter Godfrey-Smith   *Complexity and the Function of Mind in Nature*

William A. Rottschaefer   *The Biology and Psychology of Moral Agency*

Sahotra Sarkar   *Genetics and Reductionism*

Jean Gayon   *Darwinism's Struggle for Survival*

Jane Maienschein and Michael Ruse (eds.)   *Biology and the Foundation of Ethics*

Jack Wilson   *Biological Individuality*

Richard Creath and Jane Maienschein (eds.)   *Biology and Epistemology*

Alexander Rosenberg   *Darwinism in Philosophy, Social Science, and Policy*

Peter Beurton, Raphael Falk, and Hans-Jörg Rheinberger (eds.)   *The Concept of the Gene in Development and Evolution*

David Hull   *Science and Selection*

James G. Lennox   *Aristotle's Philosophy of Biology*

Marc Ereshefsky   *The Poverty of the Linnaean Hierarchy*

Kim Sterelny   *The Evolution of Agency and Other Essays*

William S. Cooper   *The Evolution of Reason*

Peter McLaughlin   *What Functions Explain*

Steven Hecht Orzack and Elliott Sober (eds.)   *Adaptationism and Optimality*

# Searching for Sustainability

## Interdisciplinary Essays in the Philosophy of Conservation Biology

BRYAN G. NORTON

*Georgia Institute of Technology*

CAMBRIDGE
UNIVERSITY PRESS

PUBLISHED BY THE PRESS SYNDICATE OF THE UNIVERSITY OF CAMBRIDGE
The Pitt Building, Trumpington Street, Cambridge, United Kingdom

CAMBRIDGE UNIVERSITY PRESS
The Edinburgh Building, Cambridge CB2 2RU, UK
40 West 20th Street, New York, NY 10011-4211, USA
477 Williamstown Road, Port Melbourne, VIC 3207, Australia
Ruiz de Alarcón 13, 28014 Madrid, Spain
Dock House, The Waterfront, Cape Town 8001, South Africa

http://www.cambridge.org

First published 2003

Printed in the United Kingdom at the University Press, Cambridge

*Typeface* Times Roman 10.25/13 pt.     *System* LATEX 2$_\varepsilon$   [TB]

*A catalog record for this book is available from the British Library.*

*Library of Congress Cataloging in Publication Data*

Norton, Bryan G.
Searching for sustainability : interdisciplinary essays in the philosophy of conservation
biology / Bryan G. Norton.
p.   cm. – (Cambridge studies in philosophy and biology)
Includes bibliographical references (p.).
ISBN 0-521-80990-8 (hb) – ISBN 0-521-00778-X (pb)
1. Conservation biology – Philosophy.   I. Title.   II. Series.
QH75 .N66   2002
333.95′16 – dc21          2002017404

ISBN 0 521 80990 8 hardback
ISBN 0 521 00778 X paperback

# Contents

# Searching for Sustainability

*Interdisciplinary Essays in the Philosophy of Conservation Biology*

# General Introduction

## An Interdisciplinary Experiment

In the late 1980s, I was in the process of finishing a book in which I had argued – to my satisfaction, at least – that environmental philosophers should adopt a new role in the process of environmental policy development, that they should reduce their appeal to abstractions and arguments regarding universal principles, and become more pluralistic and problem-oriented. My goal was to encourage philosophers to contribute to the larger policy process from within democratic decision procedures and to venture out of the insulated atmosphere of academic departments of philosophy. While I doubt that I deserve much credit for the trend, I am happy to say that a number of very bright and talented young philosophers have begun to work in a more problem-based and process-oriented mode, and today I believe environmental ethics stands at the threshold of a new era, an era in which "environmental ethics," defined narrowly as the search for justifications of general moral principles such as biocentrism or ecocentrism, will give way to a new era of "environmental philosophy," and I predict that there will be an efflorescence of new ideas and practical suggestions for responding – rationally and democratically – to specific, place-based environmental problems.

During this same time, I undertook a sort of intellectual experiment on my own, an experiment that resulted in the writing of a series of papers, some of which are collected here. I will introduce these papers by explaining my experiment and how it evolved over the decade of the 1990s, and then I will point forward a bit by summarizing some of the main lessons that can be drawn from my experiment, admitting, of course, that one person's experience can only be suggestive and never definitive. I hope that my recounting of my experiment and what I learned from it will contribute to a better understanding of the complex process by which environmental policy is proposed, modified, and implemented in modern democratic societies.

1

The purpose of my experiment was to determine whether philosophers' tools can be used to reduce differences among disputing parties. At first, the methods to be used in such an experiment were unclear even to me; I only knew that I wanted to experiment with various ways I could use my own philosophical skills – which tend toward semantics and pragmatics of scientific languages, philosophy of science, epistemology, and the philosophy of ecology and nature – to contribute to policy formation.

From my vantage point in a school of public policy, constantly involved in research on case studies with colleagues and collaborators, I learned that a social-problem orientation to analyzing proposed environmental policy encouraged testing, in practice, of my initially implicit hypothesis that the skills of a philosopher might prove useful "in place," useful in clarifying the public discourse about how to solve particular environmental problems in particular communities that face them. In fact, my expanded explorations went in two directions. In one direction, I became involved with Environmental protection Agency (EPA) policy panels and, for example, became embroiled in some controversies about how to define "ecological significance" and "ecological risk" as a framework for expanding the EPA's docket to include ecological concerns as well as human health concerns.

This book, however, while shaped by those policy wars, traces a second, more academic, line of intellectual experimentation. I set out, as a philosopher of science with an interest in language and communication, to try to build some bridges to connect the various disciplinary islands composing the archipelago referred to as "environmental science.' At about this same time, building on an excellent opportunity to serve as Gilbert White Fellow at Resources for the Future (the respected economic think tank on the economics of environmental policy) in the mid-1980s, I had begun to interact professionally less and less with philosophers and more and more with scientists, including social scientists and natural scientists – a tendency that was made possible by my academic appointment in a school of public policy. I began accepting invitations to scientific- and management-oriented conferences, conferences that dealt in some way with environmental values and policy; and, as a result of my experiences and contacts at these conferences, I began writing for an audience of scientists and environmental management practitioners.

I cannot claim that the experiment had a definite hypothesis or a clear methodology when I began, so a big part of the process has been to rethink what I was doing and why. I can say, in other words, that my own plunge into the maelstrom has given me far more reason for humility than for hubris. I am sure that my tentative interactions with other disciplines taught me far more than I was able to teach to others; all I can offer is my intention to recount my adventures as accurately as possible and to present my lessons with the humility of one who has been often, and justifiably, corrected while speaking with too much confidence in fields where I had hardly achieved the status of shameless dilettante. Fortunately, in cases where I perceived I was in over my head, I was able to rely upon coauthors with stronger disciplinary credentials to correct the worst excesses of my dilettantism. I thank these long-suffering colleagues from the various sciences for their guidance, and for the permission of some of them to publish our joint efforts here. By the end of the 1990s, I had written dozens of papers, publishing them in both prominent and obscure outlets in a half dozen disciplines.

Looking at the situation from the viewpoint of policy and practice, and recognizing the need for a unifying concept to anchor normative theories of environmental protection, it seemed to me that the most promising candidate was the idea of sustainability. Here was a concept that seemed noncontroversial in many political contexts – perhaps because of its vagueness – but the concept seemed also to fit into a lot of conversations and even to act as a rallying point for diverse interests among those seeking a more environmentally sound way forward in environmental policy. I began examining the possibility of forming management goals around the general concept of sustainability, using this concept as a way of bridging the often isolated discourses of the various disciplines, with the goal of creating more interdisciplinary communication among the disciplines that study environmental science and policy.

What unifies these papers, then, is that they all stem from this same evolving interdisciplinary experiment. Each paper, taken individually, tells one aspect of the story of sustainability from a given disciplinary viewpoint, clarifying value issues as they arise within the context of specific policy-relevant scientific disagreements that emerge within management conflicts. It is hoped that, taken as a whole, these multiple, disciplinary threads will form an intelligible interdisciplinary tapestry and suggest a richer multidisciplinary approach to living sustainably.

In presenting these papers I had two complementary – though not easily integrated – goals in mind. The unifying goal, as just noted, was my struggle to understand sustainability as a policy goal, independent of the multiple disciplinary perspectives that can be taken on the topic. The complementary goal was to show how philosophical discourse and argumentation, carried on within scientific and management contexts, can result in new insights and in changes in philosophical views. This second goal had to be pursued by showing how my own ideas evolved aross time as I tried to justify them using more and more disciplinary vocabularies. Demonstrating this evolution, then required a temporal dimension, a way to reveal the ways my thinking evolved over a decade of intense interdisciplinary and policy-oriented research.

The organization of the book is thus a bit of a compromise, a compromise that represents my best attempt at a thematic treatment of the topic of sustainability as it can be viewed from several disciplinary perspectives, balanced against a chronological organization of my papers written from such perspectives. The book is therefore organized into themes that, roughly speaking, represent somewhat different disciplinary perspectives on the problem of sustainability. My strategy for unifying the book has been to choose six themes that run through my work and to treat each of these as a thread in a larger tapestry, a tapestry that tells the story of one academic's search for an interdisciplinary theory of sustainability. Balancing this thematic treatment, I try also to convey how my thinking developed on each of these themes – what I learned over time – by arranging papers on them in a chronological order within sections.

The six concurrent themes each trace one perspective on the subject of sustainability across at least a decade of writings. In part I, "Pragmatism as an Environmental Philosophy," five papers exhibit my changing approach to philosophical problems, which began with my discovery of unquestionable pragmatist influences on Leopold in the mid-1980s and progressed to a more aggressive pragmatism that places philosophical work in the trenches of practical policy debate. During the late 1980s and early 1990s, I moved further and further from the ideal of environmental philosophy as metaethics and became more and more interested in recasting environmental ethics as a pragmatic philosophy of policy discourse. The goal of this new approach was to forgo general arguments regarding the general nature of environmental value, and to strive to improve communication and cooperation through improved problem formulation in the search for more sustainable policies in particular situations. Accordingly, my philosophical approach became more and more interdisciplinary and more and more pragmatic.

Part II, "Science, Policy, and Policy Science," consists of four papers that examine problems of cross-disciplinary communications that hinder environmental policy discourse and decision making, denying decision makers the integrated science they so desperately need when the time for judgment and decisions is at hand. Part III, "Economics and Environmental Sustainability," develops several stages in the argument that, while economic analysis certainly has a role to play in the process of environmental evaluation, economics cannot provide a comprehensive evaluation process for examining and deciding environmental problems and priorities. In part IV, "Scaling Sustainability: Ecology as if Humans Mattered," a series of papers on scaling, taken together,

4

provide a means – hierarchy theory – to represent environmental problems in multiscaled ecological models. These papers suggest that ecology's most important theoretical contribution to environmental policy may be to provide insight regarding how to array environmental problems on multiscaled systems, using a family of models that are defined hierarchically, in order to sort out temporal and spatial scales that are important for social values. Accordingly, it is also argued that ecologists must recognize more clearly that human concerns cannot be avoided in their choices of scales of nature to study. Part IV ends with a call for a new role for environmental modeling, that of creating "demand-driven" models useful in dispute resolution. Part V, "Some Elements of a Philosophy of Sustainable Living," returns to philosophical themes, examining again – in light of insights from hierarchy theory and systems theory – what we can say about the environment in which we must define and address environmental problems, and search for sustainable policies. Here, the focus is mainly on how to understand, philosophically, the long-term obligations we feel toward the future. Arguing that these obligations cannot be counted in terms of comparisons of welfare across generations – the usual approach – I insist that ecologically scaled values play an essential role in managing the environment for public goods, especially the public good of protecting the well-being of the future through adoption of sustainable policies. Finally, in part VI, "Valuing Sustainability: toward a more comprehensive approach to environmental evaluation," the various multidisciplinary and multiscalar threads of argument are pulled together in a series of papers that turn back to a question of action: How should we, in seeking ecological polices for sustainable living, put together what the various disciplines offer us and define a new approach to environmental evaluation?

Having introduced these papers by mentioning their origins, and by briefly stating their scope and main concerns, I complete this general introduction to the volume by surveying some of the lessons I learned at the synoptic and transdisciplinary levels at which the book as a whole is directed. The lessons I have learned by my forays into special disciplines of environmental science – and my evolving view of my own discipline of environmental philosophy – will be explored briefly in the part introductions.

First, one of my expectations – that I would encounter important perspectival differences among the members of the various disciplines that study environmental policy – was strongly reinforced. As I wandered through the literatures of various disciplines – from ecology to planning and from environmental ethics to decision theory – I saw that certain assumptions shared by authors within one discipline created a distinct perspective on the environment in general and on the meaning and measurement of sustainability more particularly. These assumptions, because they are apparently settled and beyond discussion within the discipline in question, are little discussed by its practitioners. As a result, they do not receive adequate critical appraisal within the discipline; further, because they are not much discussed, they are often invisible, buried in methodological norms, and difficult for outsiders to identify. I rediscovered, in the environmental sciences, what Thomas Kuhn had described for science more generally. Practitioners of a mature science operate within a "paradigm" – a constellation of assumptions, norms, and accepted practices that are usually unquestioned and that give meaning and substance to the explanatory models developed by the science.

At first, I saw this as a problem, and certainly the multiple paradigms used by practitioners of the various sciences bearing upon environmental policy have created miscommunications, confusions, and impasses, both intellectually and practically. Recognition of these problems, however – recognizing the variety of perspectives represented in discourse about environmental policy – can be turned into an opportunity. If we become reflective, it becomes possible to learn from these differing viewpoints, allowing us to create understanding collage-style, recognizing the strengths and weaknesses of various types of analyses, sifting and integrating insights from specific sciences. On this view of science, multiplicity and diversity of perspective are the driving forces for creating new and more broadly applicable models, models that prove useful in the public discussion of what to do to solve environmental problems.

What is needed is a public discourse that is broad and flexible enough to encourage both specialized learning and transdisciplinary integration. I have learned that the best means to achieve integration is from within the maelstrom of policy formation and management decision making; so I have learned to emphasize a problem-oriented, locally based dialogue as the place where integration can occur. People from multiple disciplines, if allowed to speak in abstractions to each other, will talk past each other because of the assumptions that shape their disciplinary perspective. If, however, these same individuals, coming from multiple disciplines, focus their shared attention on a real problem or crisis – how best to characterize it, what causes it, and what they should do about it – the multiple perspectives become multiple resources for envisioning new models and new solutions. So the recognition that scientists see problems from multiple perspectives may be turned from a problem into an opportunity, creating the possibility of interdisciplinary science organized in response to real environmental problems, but in which there is also an integrative dialogue across participating disciplines. To this end, I have begun to emphasize that the environmental policy process must be seen as an iterative process. I suggest, as a useful fiction, that policy choice should be viewed as embodying two "phases," an action phase (in which we consider what to do, given adopted goals, current rules and laws, and current knowledge) and a reflective phase (in which we reconsider goals, reconsider indicators and monitoring practices, and consider evidence from recent management interventions).

This view of environmental policy and the process by which it is proposed, implemented, reconsidered, and revised opens up a new role for environmental philosophers who, by encouraging the dialogue in the reflective phase, keep pressure on to create and use new concepts to evaluate and integrate new information. From the analysis of the first lesson, then, I have drawn a second, applicable to my own discipline: environmental ethicists and environmental philosophers do better when they enter public discourse, offering conceptual clarification and value analysis from within ordinary discourse. This allows philosophers to be integrators – learning from other disciplines and thinking hard about how to integrate specialized knowledge into a rational decision process. This task is best undertaken in the ordinary discourse of politics and decision making, not in the specialized language of any particular discipline.

For this reason, I disagree with my colleagues in environmental ethics who argue that we should create a distinctive language and subject matter – the intrinsic value of nature – as the subject of environmental ethics. If environmental philosophers are to fulfill the integrative function proposed here, they must speak the language of public discourse, paying attention to how language functions to convey information and also to how our languages fail to convey important information. This process will involve theory-building, but it is theory-building in the service of developing and broadening consensus, not for the sake of theory itself. The ideas and concepts of environmental philosophers must, if they are to serve an integrative function, eschew jargon and specialized concepts not explicable to decision makers and the public. When environmental ethicists create a separate, distinct discipline with its own terms and assumptions, they disqualify themselves as the integrative contributors they could be if their discipline were built in public discourse, the rich, highly textured language upon which we fall back when we are faced with a crisis or problem and must decide collectively what to do about it.

Third, I learned that we need a whole new approach to the philosophy of science; what I learned in graduate school, based on the "ideal' of modern physics, does not help very much when one thinks about science within the maelstrom of policy formation. In order to enjoy the advantages of the problem-oriented approach, and the alternation between action and reflection it makes possible, the relevant science – and the relevant paradigm for philosophers to study – is the science of environmental management. The

7

special disciplines, devoted to the truth as they understand it, are not well suited for integration. But public crises and forced action have a way of identifying working hypotheses and actions that have wide support. So we need a new way to understand science, a way that respects the importance of objectivity and minimizes bias, but a way that addresses these problems through action – what has been called "mission-oriented" science. Mission-oriented science must be an open science. It must be at least capable of translation into a public vernacular, because it will ultimately be judged openly, within public discourse, by participants in management discussions and controversies.

To summarize, what I learned from my experiment is the importance of addressing conceptual and value problems – philosophical problems – within the context of concrete environmental problems faced by real communities. When one does this, disciplinary assumptions, no matter how well hidden, will eventually be identified and called into question; if environmental policy discourse and implementation were thought of as an ongoing iterative process with a reflective phase, in which philosophers can make a case for new goals with new justifications, in alteration with an action phase in which management experiments are undertaken to achieve stated goals and to reduce uncertainty about the impacts of our actions.

The common insight here is that it is useful to recognize that our current academic and intellectual practice of forming disciplines and developing distinct paradigms makes sense, ultimately, only if these disciplines are understood as useful and temporary outposts at the frontiers of knowledge and ignorance. Disciplinary boundaries represent our current imperfect understanding; the specialized assumptions and languages that constitute special disciplines and their paradigms are only temporarily useful to expand understanding. If that expanded understanding is to affect environmental policy – if it can be used to justify and explain a management policy in the nonspecialized language of public policy and discourse – it must eventually be integrated in a more inclusive transdisciplinary language from the perspective of a wise manager, not from the perspective of any disciplinary specialist. If this collection of essays is successful, the reader will come away with a strong sense of the importance of consciously occupying – and consciously improving – the transdisciplinary, ordinary language discourse in which scientific knowledge and social evaluation must be integrated if we are to find a viable environmental morality.

# I

# Pragmatism as an Environmental Philosophy

Intellectual histories, no less than histories of societies, are often shaped – and sometimes misshaped – by accidents of their birth. In my view, the discipline of environmental philosophy was in fact misshaped by a confluence of small accidents, beginning in 1967 with the provocative comment by the historian Lynn White, Jr., that our environmental crisis results from the "anthropocentric" nature of Christianity, the dominant religious tradition in the West. Then, when professional philosophers began asking, in the early 1970s, what philosophers could contribute to environmental thought and action, they responded to White's highly ambiguous, even offhand, statement with a particular analysis and a particular solution.

First, they interpreted White as having associated Christianity with a particular substantive *theory* about moral value, that all and only human individuals have "intrinsic/inherent value" and are morally considerable. Nonhuman organisms, on this interpretation of Christianity, are at best of "instrumental" value to the needs and whims of humans. Second, without considering other possible interpretations of White's comments, they responded with an alternative theory, which denies the anthropocentric theory by attributing intrinsic value and moral considerability to some nonhuman elements of nature. As a result, most philosophical discussion of environmental issues has centered on the question of whether natural objects other than humans have intrinsic or inherent value.

As noted in the General Introduction to this volume, I have adopted an experimental attitude, even toward philosophical ideas, and I gave the idea of intrinsic value of nonhumans a chance early in my career as an environmental policy advocate. I was living and working in Washington, D.C., just after beginning my work as an environmental philosopher in the early 1980s. I joined a small working group formed to monitor and improve the nation's policy toward biological diversity. At one of our brown bag lunch meetings, we were talking about the importance of various actions and policies and, taking the philosophical plunge, I tried to engage my new friends and colleagues in a discussion of whether the species and ecosystems we were trying to protect had intrinsic value. I still remember – better than the conversation that preceded my query – the profound silence that followed it. This group of activists could not see any point in discussing the question of *how* we should value nature and natural objects. They took it as obvious that these objects have value; to them, the question was not how to value them but how to save them.

At first, the profundity of the silence engendered by my philosophical question bothered me; surely, I thought, if I could just think of the right entree to the conversation, my activist friends would see the interest and value of the question of whether intrinsic value exists in nature, and my commitment to better understanding the metaethical nature of the moral obligation would prove useful. Later, after having had several such experiences as I moved back and forth between academia and activism, I eventually stopped asking "What's wrong with activists? Why don't they care about the philosophical bases for their action?" and I began asking "What's wrong with environmental ethicists, who seem to provide answers to questions no activists find pressing?"

The papers in this part are chosen to illuminate my gradual, fits-and-starts transformation from a regular, well-behaved environmental ethicist, deeply concerned to develop a metaphysic – or at least a metaethic – of environmental value, into a rather crotchety, ill-mannered critic of the field's initial trajectory across intellectual space. Following the initial shock of learning that activists had no interest in my philosophical wares, I lapsed into silence and was forced to become a listener. I gradually learned to listen in context. I began to hear, that is, the language of environmental activists as a language of action, a language that was already embedded in practical, and highly political, activities and projects. Having studied, early in my philosophical career, the linguistic turn in twentieth-century philosophy, this new way of listening suddenly made sense. The almost universal lesson of the various strains of twentieth-century philosophy was a recognition that language is important, that language holds the key to deeper understanding of our world. Another persistent lesson of the twentieth century was the rediscovery that the most important language is the public language we have evolved in order to *do things*. Logical positivism, which sought certain foundations on which to rebuild the world through a "logically perfect language," as it failed philosophically, reverted to a more linguistically sophisticated version of the pragmatism of Charles Sanders Peirce and the nineteenth century. W. V. O. Quine and Rudolf Carnap worked at midcentury to reconfigure analytic philosophy into a working pragmatism.

The lesson I had learned from my study of twentieth-century philosophy now, having been forced to listen to activists in the process of acting, took on real meaning: philosophy, I found, works best when it engages language and meaning in problematic situations. Philosophy was driven by the pragmatists out of the language of pure reason and logical perfection into the maelstrom of action, politics, and decision making. My listening to activists, originally enforced by the deadening effect of my early attempts to inject philosophical jargon into policy discussions, now began to feel like a conscious method; watching my activist friends act and listening to them explain why they did what they did opened my eyes to a philosophical laboratory. As a pragmatist, I had a ready explanation of the method that was imposed by necessity; language and its meanings are best understood when they are observed at work, in problematic situations.

The first paper in this part, "The Constancy of Leopold's Land Ethic," represents my first attempt to establish a historical thread connecting the important conservationist Aldo Leopold to American pragmatism. The first step was to track down the explicit 1923 reference by Leopold to the pragmatist definition of "truth," which he attributed to "Hadley." Who was Hadley? By answering this question, this 1988 essay represents my first clear endorsement of pragmatism as a historically valid, and philosophically promising, support for central conservationist themes. Pragmatism proved especially useful as a way of understanding Leopold's fascination with temporal scales, an idea that is explored in "Thoreau and Leopold on Science and Values." By emphasizing the tradition of dynamism and its emphasis on change ushered in by Darwinism, the American naturalist tradition, culminating in American pragmatism, provides the basis for understanding Leopold's value system as a naturalistic and science-based ethic. In this paper, I argue that both Thoreau and Leopold, in good pragmatist fashion, transcended the fact–value dichotomy in a way that creatively integrated science and human values in the search for better environmental policies.

The pivotal paper in this part, "Integration or Reduction," represents a wide-ranging argument that begins by rejecting the usual questions asked by environmental ethicists – embodied in a critique of J. Baird Callicott's approach to ethics and policy – and ends by showing how pragmatism and naturalism provide a more adequate philosophical basis for sustainability and for an adaptive approach to environmental management. A central tenet of my pragmatism has been what I have called the "convergence hypothesis," the view that, if reasonably interpreted and translated into appropriate policies, a nonanthropocentric ethic will advocate the same policies as a suitably broad and long-sighted anthropocentrism. A few years ago, I had the opportunity to put this empirical hypothesis about environmental policies to the test when the editors of a festshrift volume for the famous Norwegian nonanthropocentrist Arne Naess asked me to comment on two papers by Naess, one of them coauthored with a biologist, Ivar Mysterud, on wolf and bear policies. The question was whether I, who reject nonanthropocentrism, would agree or disagree with Naess's policies, as they are based on his qualified nonanthropocentrism. What I found was that Naess's policies would be different in no important ways from the policy adopted by someone committed to protecting nature as the natural heritage of future generations, providing corroboration for the hypothesis of convergence.

The final paper in this part, "Pragmatism, Adaptive Management, and Sustainability," is something of a promissory note for the future. It puts forward the hypothesis that there is an illuminating analogy between the pragmatists' approach to truth – which locates truth in the future, a result of human intelligence, with a method operating on unlimited experience – with the search for a sustainable lifestyle. Both are forward-looking and both require learning along the way, with learning understood not as accumulations of bits of a single total truth, but rather as an unfolding and ongoing social process requiring community and communication, a process that advances by gradually correcting errors through the use of intelligence in the activity of problem-solving. This paper picks up the theme that pragmatism can be useful in environmental philosophy and projects that theme forward, speculating that the future of environmental philosophy will be closely entwined with the future of pragmatism.

# 1

## The Constancy of Leopold's Land Ethic

In 1920 Aldo Leopold enthusiastically described his predator eradication program. He had formed a coalition of sportsmen and stockmen to eliminate wolves, mountain lions, and other large predators from Arizona and New Mexico: "But the last one must be caught before the job can be called fully successful," he said (Flader 1974: 3). Twenty-four years later Leopold repented his war on wolves in a graceful and humble essay, "Thinking Like a Mountain," which was drafted in 1944 and published in *A Sand County Almanac and Sketches Here and There*. What happened in the meantime?

It is tempting to believe that, during this period, Leopold discovered his revolutionary land ethic, that his thinking underwent a profound religious-metaphysical-moral change, and that his about-face on predator control programs was a direct result of this profound philosophical conversion. Since Leopold was acting, in 1920, as a representative of the U.S. Forest Service, which remained under the philosophical domination of Gifford Pinchot's humanistic utilitarianism, this interpretation sees Leopold as later rejecting utilitarian management because he came to espouse "a right to exist" for all members of the land community (*see*, for example, Petulla 1980: 16, 20).

This essay seeks to show that Leopold's intellectual odyssey during this period was more complex than this straightforward account would suggest. In particular, Leopold had embraced the main philosophical elements of his land ethic early in his career, even while he was advocating predator eradication. These main elements include important influences, hitherto unnoticed, derived from American pragmatism, a philosophical approach that Leopold borrowed from Arthur Twining Hadley, who was president of Yale University when Leopold was a student there. Leopold never abandoned the

From *Conservation Biology* 2 (1988): 93–102. Reprinted with permission.

main elements of his early philosophy, and I will argue that his shift from predator eradication to predator protection was motivated, not by a shift in religious, metaphysical, or moral views, but rather by a recognition that scientific knowledge is inadequate to guide gross manipulations of ecosystems and by an increasingly pessimistic view of the prospects of environmental management. I conclude that, while Leopold was fascinated by organicism and its metaphysical and moral implications, these abstract views had little direct impact on his managerial style.

I

In 1923 Leopold drafted an essay, "Some Fundamentals of Conservation in the Southwest" (Leopold 1979). He included, as a final section, some brief remarks that he called "Conservation as a Moral Issue." This essay remained unpublished until 30 years after Leopold's death; commentators have treated the final section as an immature draft of Leopold's conservation ethic, and some have suggested that Leopold later abandoned significant elements of the philosophy expressed there (Flader 1979: 143–144, Callicott 1987, Rolston 1987, note 3).

It cannot be denied that Leopold's brief 1923 discussion of conservation morality is confusing. In only three pages Leopold stated that "economic determinism" is insufficient to understand land conservation; invoked the prophet Ezekiel; considered the Russian organicist philosopher P. D. Ouspensky's view that the world is a "living thing" with "a soul, or consciousness"; questioned whether the world "was made for man's use, or has man merely the privilege of temporarily possessing an earth made for inscrutable purposes"; approvingly quoted John Muir on the rights of rattlesnakes but decided "I will not dispute the point"; and finally concluded that we have an obligation to future generations to prove ourselves "capable of inhabiting the earth without defiling it." Along the way, he admitted that most scientists and laymen hold "anthropomorphic" views.[1] He also considered the effect that Ouspensky's organicism would have on "most men of affairs," and observed that for them "this reason is too intangible to either accept or reject as a guide to human conduct" (Leopold 1979: 138–141).

It is difficult to see, at first glance, a unifying principle in this densely packed presentation of so many grand ideas. It seems clear that, after introducing nonanthropocentric ideas, Leopold opted in the end for a conservation ethic based on our obligations to future generations of humans – a forward-looking anthropocentrism. But the reasoning by which he shifted from his prior discussion of organicism and nonanthropocentrism to longsighted anthropocentrism is so compact as to nearly defy understanding. Fortunately, a clue to Leopold's thinking appears in a parenthetical comment, which is embedded in his discussion of our obligations to future generations: "How happy a definition is that one of Hadley's which states, "Truth is that which prevails in the long run."

Since the essay was never published. Leopold prepared no notes or list of references. This definition, however, clearly derives from the American pragmatists (see, for example, Peirce 1978: 288). Arthur Twining Hadley, a child prodigy in Greek, graduated at the head of his class at Yale in 1876. He studied political economy at the University of Berlin and returned to become a tutor and later a professor at his undergraduate institution. Noted for the breadth of his knowledge, his classes on economics and political ethics were extremely popular. Hadley became the first lay president of Yale in 1899 (M. Hadley 1948). He described himself as a "thoroughgoing pragmatist" (M. Hadley 1948: 197) and generally quoted William James's work as representative of modern philosophical thinking.

In his most thoroughly philosophical book, *Some Influences in Modern Philosophic Thought*, Hadley said: "The criterion which shows whether a thing is right or wrong is its permanence. Survival is not merely the characteristic of right; it is the test of right."[2] These views he characterized as the views of pragmatists and in the next paragraph he discussed James's view, which he stated as "We hold the beliefs which have preserved our fathers" and accepted this view while changing its emphasis somewhat:

> I do not mean that we should consciously adopt a belief because it is useful to us, as James seems to imply. I would rather take the ground that we hold the belief that has preserved our fathers as an intuition and act on it as an instinct. (A. T. Hadley 1913: 73)

Leopold began "Conservation as a Moral Issue" with a quotation from Ezekiel:

> Seemeth it a small thing unto you to have fed upon good pasture, but you must tread down the residue of your pasture? And to have drunk of the clear water, but ye must foul the residue with your feet? (Leopold, 1979; 138)

When Leopold invoked Ezekiel, he was invoking the "beliefs that have preserved our fathers." As understood by Hadley, the pragmatists' notion of truth amounted to a recommendation that we respect the wisdom of our ancestors. Hadley said:

15

> The moral and religious instincts that bind the group together, which some men, not so many years ago, were condemning as outworn prejudices, count for even more than individual intelligence. In our practical philosophy, of politics and of life, we are reverting to the words of Edmund Burke: "We are afraid to put men to live and trade each on his own private stock of reason, because we suspect that this stock in each man is small, and that the individuals would do better to avail themselves of the general bank and capital of nations and of ages."
> (Hadley 1913: 75)

These ideas, due to Hadley and the pragmatists, provide the key to understanding an important passage in Leopold's "Conservation as a Moral Issue." Leopold said, "Possibly, in our intuitive perceptions, which may be truer than our science and less impeded by our words than our philosophies, we realize the indivisibility of the earth." This brief passage shows a connection between three important and related ideas in Leopold's thought. He was referring to Hadley's conception of "truth" or "rightness" of social practices interpreted as the "intuitive perceptions" we have inherited from Ezekiel and others who have counseled protection of resources. Second, he implied that our science is fallible, and perhaps less reliable than these intuitive perceptions. Finally, he suggested that our intuitions are more helpful than philosophies, because the latter are "impeded by language."

In a discussion of Darwin's influence on history and political science, Hadley explained that Darwin's idea of survival of the fittest was readily accepted by historians, who had long recognized that survival of a culture depended on the adaptability of its institutions to situation (Hadley 1913: 121–126). He criticized Herbert Spencer and others for trying to apply the Darwinian criterion over too brief time periods and to individual behaviors (Hadley 1913: 130). Hadley concluded: "It is the institution even more than the man that has been marked out for survival by the process of natural selection" (Hadley 1913: 127).

I have found no evidence that Leopold read or studied the American pragmatists in detail, although he no doubt knew about the publication of Hadley's book, as it was given a major review in the *Yale Review* (Sherman 1913). Apparently, Leopold read this magazine regularly.[3] Further, Leopold was clearly familiar with, and impressed by, Hadley's conception of truth, because he referred to it on several occasions in the journals he kept during this early period of his career.[4] It may, at this point, be impossible to determine the extent of Leopold's study of Hadley, but it is clear that Leopold absorbed and applied the basic elements of Hadley's approach to evaluating cultures.

When Leopold mentioned intuitive perceptions, Ezekiel's admonition to treat the land with respect, he was invoking Hadley's intuitions of our fathers.

In this same passage Leopold compared this intuition favorably with both science and philosophy. Since Leopold was here striving for a very large-scale understanding, a kind of worldview, to support his conservation goals, it is important to examine the treatment he gave both philosophy and science in this large perspective.

Leopold is obviously wary of philosophical pronouncements. In only three pages "Conservation as a Moral Issue" contains no less than five cautions against the "pitfalls of language," and three of them are related to doubts about the efficacy of philosophical ideas. While on first reading these passages may appear as expressions of humility and little more, I believe they also provide a clue to the philosophical approach underlying Leopold's conservation ethic. These references to language derive from the linguistic pluralism of Ouspensky's *Tertium Organum*, which Leopold here blended with American pragmatism as a justification for a longsighted anthropocentrism to support his conservation ethic.

In his discussion of Ouspensky, Leopold characterized two "conceptions of the earth." There is a "mechanistic conception of the earth as our physical provider and abiding place." This conception was opposed to another: the world is a living organism and the "soil, mountains, rivers, atmosphere, etc. – [are] organs, or parts of organs, of a coordinated whole, each part with a definite function." On this view of the world, which "many of the world's most penetrating minds" (he cited Ouspensky) have found compelling, the earth has "a soul or consciousness." It begins even to make sense to respect it as a living thing and to relate to it morally. In this context of competing worldviews, Leopold's references to the importance of language appear more significant:

> There is not much discrepancy, except in language, between this conception of a living earth, and the conception of a dead earth, with enormously slow, intricate, and interrelated functions among its parts, as given by physics, chemistry, and geology." (Leopold 1979; 139–140)

A similar form of conceptual pluralism, also linked with cautions about the inadequacy of language, is a major theme of Ouspensky's *Tertium Organum* (Ouspensky 1968: 222f).

I am suggesting that, in "Conservation as a Moral Issue," Leopold amalgamated ideas from the American pragmatists with the organicism of Ouspensky, concluding that the most effective conservation ethic represents a concern that we pass on, to future generations, a world not despoiled by our current activities. The relationships between these ideas become more explicit if one distinguishes several levels of discourse. When Leopold compared different conceptions of the earth, he was speaking of what might be called second-order beliefs. The facts about the world around us, the facts of physics, chemistry, and geology, are first-order beliefs about the way the world is. Organicism and mechanism, two alternative conceptions of the world in Leopold's terminology, are second-order beliefs about how to interpret the first-order facts of the particular sciences. Leopold was arguing that organicism and mechanism can accept the same first-order facts about the world and that the choice between these two interpretations is mainly a difference of language. "The essential thing for present purposes is that both admit the interdependent functions of the elements" (Leopold 1979: 139–140). Thus, the choice of a second-order interpretation of scientific facts, the choice between organicism and mechanism as alternative conceptions of the world, is essentially a linguistic choice of how to conceptualize this interrelatedness.

Leopold placed so much emphasis on the pitfalls of language because he believed that empirical data from the natural sciences will never determine our second-order, linguistic choices as to how to conceptualize that data. These questions depend on human perception, not reality: "The very words *living thing* have an inherited and arbitrary meaning derived not from reality, but from human perceptions of human affairs" (Leopold 1979: 139). Further, he recognized that the choice of a conception of the world, an interpretation of the data, will be closely related to the way we think and act. Here we see the amalgamation of Ouspensky's organicism with Hadley's pragmatism. It is a central idea of American pragmatism that linguistic forms depend upon perception and perception depends in turn upon human affairs. What we do determines what we say and think as much as vice versa. Leopold combined pragmatism with Ouspensky's organicism and arrived at a form of homegrown perspectivism, a view that metaphysical conceptions of the world are projections of human perceptions which depend, in turn, upon cultural practices.

From this amalgamation, Leopold concluded that different cultures with radically different practices will have different characteristic vocabularies and therefore different conceptions of the world. But since these different conceptions of the world amount only to different conceptualizations of the same hard, scientific data, Leopold was cautious whenever he compared ideas from one conception with those from another. Leopold thus shied away from metaphysical and theological pronouncements: "It is just barely possible that God himself likes to hear birds sing and see flowers grow. But here again we encounter the insufficiency of words as symbols for reality." Similarly, his concern that philosophical and theological ideas are artifacts of conceptions of the world led Leopold to express a deep distrust of pronouncements about the "truth" of anthropocentrism or nonanthropocentrism:

> Probably many of us who have neither the time nor the ability to reason out conclusions on such matters by logical processes have felt intuitively that there existed between man and the earth a closer and deeper relation than would follow the mechanistic conception of the earth.... Of course, in discussing such matters we are beset on all sides with the pitfalls of language. (Leopold 1979: 139)

So far, we have seen that Leopold understood Ouspensky's organicism and modern, atomistic science as two alternative conceptions of the world, the choice between which is a second-order choice among vocabularies for presenting first-order data, and unlikely to be fully decided by that data. At the same time, he recognized that the choice between these conceptions has profound consequences for the way we treat the earth. On a mechanistic worldview, the earth is dead and we would have no moral concern for it. This viewpoint has led to "economic determinism," which Leopold disparagingly characterized as the "language of compound interest." On the organicist view, the earth is alive and worthy of our respect and moral concern. But the choice between the two conceptions of the world is mainly a linguistic choice, underdetermined by the scientific data available.

This combination of views, without supplementation, would result in a radical relativism: we either speak as mechanistic scientists or as organicists and, depending on this arbitrary choice, we will either be economic determinists or we will react morally to the land. It is at this crucial juncture that Leopold introduced Hadley's definition of truth as that which prevails in the long run. Following Hadley, Leopold chose not to dispute the metaphysical and theological issues of anthropocentrism, but rather to rely on the intuitive perceptions of the tradition – Ezekiel. Or, to put the point in more Hadleian terms, the test of rightness of cultural practices is their long-term survivability. Hadley's pragmatic definition of truth therefore functioned, in Leopold's early philosophy, as a third-order principle, as a means to judge second-order conceptions of the world and to provide a criterion for distinguishing acceptable cultural practices from unacceptable ones.

This was surely an attractive idea to Leopold for several reasons. First, it provided a unification of his philosophical thought with his biological belief in the Darwinian principle of "survival of the fittest." Second, it allowed him to relate, in a larger perspective, scientific knowledge with his speculations about Ouspensky's organicism – these are two differing linguistic approaches to the same "earth." Third, it provided him an ideal platform from which to denounce the practices, described in detail in the first two sections of "Some Fundamentals of Conservation in the Southwest," which he believed were destroying the land. Our treatment of the land is wrong because it is not sustainable, as Ezekiel said.

Immediately after he decided not to dispute anthropocentrism, Leopold granted "that the earth is for man." He then said, "there is still a question: what man? The cliff dwellers, the Pueblos, the Spaniards, and now the Americans all believed the earth was their possession." But the prior cultures "left the earth alive, undamaged." If we are "logically anthropomorphic," he said, we must consider what the next civilization will say of us. If there is, indeed, "a special nobility inherent in the human race – a special cosmic value, distinctive from and superior to all other life," then it must manifest itself in "a society decently respectful of its own and all other life, capable of inhabiting the earth without defiling it. "If we do not manifest that nobility, we shall "be judged in the derisive silence of eternity" (Leopold 1979: 141). Leopold was, ultimately, basing his argument for conservation on the fact that the land in the Southwest was going through a series of less and less productive cycles owing mainly to overgrazing. This factual basis, outlined in detail in earlier sections of the draft article, seemed to him an adequate basis for conservation when combined with Hadley's pragmatic definition of truth as survivability. This argument could be made independent of appeals to nonanthropocentrism, which is a second-order belief not possible to establish conclusively. By applying Hadley's definition of truth to the cultural practices that were causing deterioration of the land in the Southwest, Leopold could sidestep the issue of anthropocentrism and declare those practices "false." He relied not upon Ouspensky (in spite of his obviously deep attraction to organicism) but upon Ezekiel buttressed by Hadley's definition of truth.

Leopold's philosophical pragmatism also provides an explanation of his comment that "most men of affairs" will find organicism and other forms of nonanthropocentrism "too intangible to either accept or reject as a guide to human conduct" (Leopold 1979: 138–141). Leopold recognized that both modern science and traditional Judaeo-Christian religion are anthropocentric in their conception of the world and, given his admission that conceptions of the world involve unresolvable differences in linguistic forms, he recognized that he would be unsuccessful in preaching nonanthropocentrism to them, at least in the short run. He resolved, instead, to argue for a longsighted anthropocentrism based on the intuitive perceptions of Ezekiel: we ought not to "tread down the residue of [our] pasture." Leopold therefore resolved, early in his career, to enter the policy arena armed only with arguments based on longsighted anthropocentrism, rather than basing his moral strictures on nonanthropocentrism.

II

Leopold's approach to environmental management underwent a profound change between 1920 and 1944. The question I am posing is whether this change resulted from a change in his religious-metaphysical-moral views, or whether the change was motivated by expanding scientific information and hands-on management experience. Having sketched a unified philosophy underlying Leopold's thinking in 1923, it is now possible to ask whether he changed this philosophy in subsequent years. How significantly do Leopold's views, as expressed in the final version of "The Land Ethic" (which dates from 1947), differ from his early views, as stated in "Conservation as a Moral Issue"?

"The Land Ethic" begins with the story of Odysseus' arbitrary hanging of a dozen slave-girls on mere suspicion of misbehavior, a historical example of the changing nature of moral judgments. This example, which was first introduced in 1933 (Leopold 1933a), evokes his earlier view that moral ideas are tied closely to changing patterns of behavior and to the conceptions of thought associated with them. These, in turn, are likely to differ across epochs. Subsequent developments have seen the extension of moral concepts to human individuals previously treated as mere property. This much-discussed example stands as an analogy for the eventual emergence of a full-blown land ethic: "The land-relation is still strictly economic, entailing privileges but not obligations" (Leopold 1949: 201–203). The emergence of a land ethic

> is actually a process in ecological evolution. . . . An ethic, ecologically, is a limitation on freedom of action in the struggle for existence. An ethic, philosophically, is a differentiation of social from anti-social conduct. These are two definitions of the same thing.

Here we see a later representation of Leopold's view that there are multiple conceptions of the world (here represented as two different fields of study, philosophy and ecology) and that these conceptions, which are associated with differing practices, are winnowed by the forces of competition and evolution. An ethic "has its origin in the tendency of interdependent individuals or groups to evolve modes of cooperation. . . . [T]he original free-for-all competition has been replaced, in part, by cooperative mechanisms with an ethical content" (Leopold 1949: 202).

In 1923 Leopold was drawn to organicism and the ethic associated with it, but he was wary of trying to establish these views rationally because an ethic is so tied up with its characteristic vocabulary and world conceptions. An ecologically sensitive conception of the world will emerge only gradually as we learn, in practice, the extent of our interdependence. In the meantime we must rely on the intuitive wisdom of Ezekiel and our forefathers. In the 1947 version, Leopold was looking forward, predicting that, in the face of greater recognition of our mutual dependencies with other species, we will eventually develop an ethic consonant with the ecological conception of the world. Assuming we survive at all, we will have discovered a world view that is adapted to the modern world, a new set of Hadleian intuitions that will promote survival: "Ethics are possibly a kind of community instinct-in-the making" (Leopold 1949: 203). Leopold's temporary reliance on the traditional strictures of Ezekiel, as well as his faith that a new ethical age will dawn (if we do not destroy ourselves first), are both manifestations of the pluralistic view of world conceptions and the pragmatic conception of truth as survivability.

Near the end of "The Land Ethic" Leopold discussed "Land-Health and the A-B Cleavage":

> Conservationists are notorious for their dissensions. Superficially these seem to add up to mere confusion but a more careful scrutiny reveals a single plane of cleavage common to many specialized fields. In each field one group (A) regards the land as soil, and its function as commodity-production; another group (B) regards the land as a biota, and its function as something broader.

Here, again, we see the conception of a "dead earth" contrasted with that of a "living earth," and Leopold understood the dissensions among conservationists as based, ultimately, on the acceptance of one or the other of these conceptions. References to Ouspensky do not occur, but Leopold's words indicate that he was still fascinated with second-order systems that he called in 1923 "conceptions of the earth." "In all of these cleavages, we see repeated the same basic paradoxes: man the conqueror *versus* man the biotic citizen: science the sharpener of his sword *versus* science the search-light on his universe; land the slave and servant *versus* land the collective organism" (Leopold 1949: 221–223).

"The Land Ethic" differs from "Conservation as a Moral Issue" in lacking the latter's numerous cautions about language and its pitfalls. This is indicative of a deeper shift in Leopold's strategy – he apparently decided not to emphasize philosophical theory (which he saw as raising issues that can be settled only in the slow-heating crucible of evolutionary selection), but to rely instead on his vast experience as an environmental manager. But the pragmatism and conceptual relativity are still there in muted form, as evidenced by the references, quoted above, to different fields (ethics and ecology) as using alternative vocabularies to describe similar processes and in the discussion of the A-B cleavage.

Thus, while Leopold's philosophical scaffolding is less explicit in his later work, traces of his theoretical commitments remain. The strategic decision to emphasize his experiences as a forester and wildlife manager is deeply pragmatic in spirit. Theory, according to the pragmatist, must ultimately be tested against experience.

Leopold was admittedly tentative in his 1923 pronouncements, perhaps because he had to rely on the philosophical and linguistically relative concepts that he cautioned against. As he matured, gaining experience and replacing philosophical speculation with knowledge of ecological science and the consequences of management strategies, he relied more confidently on his managerial experience (Flader 1974: 18). If, therefore, we use the term "philosophy" to refer to a basic worldview including a metaphysics and value system, we can conclude that Leopold acted, throughout this period, against the backdrop of a consistent and unified philosophical approach.

III

If, however, we use the term "philosophy" more broadly, in a sense in which one might say that Leopold had a "philosophy of environmental management," his philosophy did indeed change. Why did Leopold, acting against a backdrop of unchanging metaphysical and moral worldviews, change his approach to predator control and other management strategies? How could Leopold, in the early twenties, have approved organicism while eradicating wolves and mountain lions?

There is a ready explanation for Leopold's early attempts at predator eradication: trained in the Pinchot mold of utilitarian forestry, Leopold set out to maximize resources for human use (Flader 1974: 25). When he became interested in game management in 1915, he simply transferred forestry management practices to a new resource, fish and game. If a little game is valuable, more is correspondingly so. Predators compete for game, so he set out to eliminate them, as well as to enforce hunting laws and to stock streams – all were management methodologies designed to maximize sport resources.

What requires an explanation is not Leopold's use of these management practices (which were assumed as part of his job), but his doing so while approving organicism and questioning the adequacy of "economic determinism." Leopold already believed in this early stage of his career that the health of the land community is important; but he had not yet realized that all species are necessary to promote that health. He believed that resource managers, provided they are scientific in their management practices, can manipulate populations and he seems to have believed that predator control would enhance the overall health and productivity of biological systems.

With this view as a starting point, Leopold's views of management began a slow evolution toward less radical interventions. By 1925 he believed that wolves and mountain lions contributed to the diversity of an area and he retreated from a goal of eradication to one of control (Flader 1974: 154). When Leopold met Charles Elton, who had initiated the transformation of ecology from a purely descriptive to a more functionally oriented science with the publication of *Animal Ecology* in 1927, he integrated these ideas into his own 1933 text, *Game Management* (Flader 1974: 24–25). Defining "management" as "the coordination of science and use," Leopold stated that "the central thesis of game management is this: game can be restored by the *creative use* of the same tools which have heretofore destroyed it – axe, plow, cow, fire, and gun" (Flader 1974: 25, quoted from Leopold 1933b).

In the 1933 essay "The Conservation Ethic" Leopold noted that there was emerging a new and more positive approach to conservation of the land and living things. The means to the goal of protection was biological research:

> The duty of the individual is to apply its findings to the land. . . . [because] the soil and plant succession are recognized as the basic variables which determine plant and animal life. . . . and likewise the quality of human satisfactions. . . . Leopold 1933a: 641)

Leopold applied this strategy to species preservation, arguing that species become rare and extinct because their habitats have shrunk. He asks:

> Can such shrinkage be controlled? Yes, once the specifications are known. How known? Through ecological research. How controlled? By modifying the environment with those same tools and skills already used in agriculture and forestry. (Leopold 1933a: 641)

In 1933, then, Leopold optimistically believed that ecological research would usher in a new era of plenty based on a positive program of environmental management:

> Given, then, the knowledge and the desire, this idea of controlled wild culture
> or 'management' can be applied not only to quail and trout, but to *any living
> thing* from bloodroots to Bell's vireos." (Leopold 1933a: 641)

He as yet saw no inherent contradiction between conservation and intense
management for production of resources. Given ecological knowledge, populations could be managed for the good of humans. The reduction of predators
would be unproblematic: human hunters would simply absorb the ecological
function of wolves and mountain lions. While accepting organicism, he did
not yet see that organicism implied a goal of saving all species. He believed
that, given sufficient knowledge and sensitivity in management, the living
organism (the land) could be kept thriving, while some of its less desirable
organs were removed. Faith in ecological technique, therefore, shielded him
from the conclusion that destruction of wolves and mountain lions would
cause serious illness in the organic system of nature.

By 1939, however, Leopold's view of the role of ecology had changed
drastically. He stili believed that ecology "is the new fusion point for all the
natural sciences," but its results were not those he had hoped for:

> The emergence of ecology has placed the economic biologist in a peculiar
> dilemma: with one hand he points to the accumulated findings of his search for
> utility, or lack of utility, in this or that species; with the other he lifts the veil from
> a biota so complex, so conditioned by interwoven cooperations and competi-
> tions, that no man can say where utility begins or ends...the old categories
> of 'useful' and 'harmful' have validity only as conditioned by time, place,
> circumstance. The only sure conclusion is that the biota as a whole is useful,
> and biota includes not only plants and animals, but soils and waters as well.
> (Leopold 1939: 727)

Leopold concluded that human management activities do not successfully
mimic nature in the creation of habitats because "evolutionary changes are
slow and local," while human use of tools "has enabled him to make changes of
unprecedented violence, rapidity, and scope." Unforeseen changes in species
composition result as "larger predators are lopped off the cap of the pyramid."
The effects of such changes "are seldom foreseen; they represent unpredicted
and often untraceable readjustments in the structure" (Leopold 1939: 728).

Leopold was now regretting his war on wolves. His predator control program and the restrictions on hunting he began to enforce in 1915 in the
Southwest had resulted in huge, but starving, herds of deer. The deer over-
browsed the land and caused yet one more downward turn in the cycle of
succession. Weedy, brushy species replaced more useful trees and shrubs,
and the diversity of the area diminished (Flader 1974: 117).

The 1939 essay "A Biotic View of Land" was in effect a retrospective view of what Leopold had learned as a forester and wildlife manager. He referred to the German experience in tree farming: "Thus the Germans, who taught the world to plant trees like cabbages, have scrapped their own teachings and gone back to mixed woods of native species...." (Leopold 1939: 730; see also Flader 1974: 139f). Similarly, he came out definitively against predator control as a "highly artificial (i.e., violent)" method of management (Leopold 1939: 729). Additionally, he had observed the extent and seriousness of the dust bowl phenomenon, of what pervasive "economic management" could do to fragile ecosystems.

Leopold therefore changed his views of management because:

> In short, economic biology assumed that the biotic function and economic utility of a species was partly known and the rest could shortly be found out. That assumption no longer holds good; the process of finding out added new questions faster than new answers. The function of species is largely inscrutable, and may remain so. (Leopold 1939: 727)

We need not posit any shift in Leopold's metaphysical or moral views to explain the changes in his views on wildlife management. He learned through practice that "violent" methods of management and control are inappropriate because they also cause unforeseen effects and damage the biotic community. This is an insight that was implicit in his belief in the importance of ecology; but it was obscured by his initial faith that ecology would teach us enough about ecological interactions among species to allow manipulation of populations for utilitarian purposes. He underestimated the complexity of systems and overestimated our ability to control them; he consequently failed to see that predator protection was one of the principles implied by the holistic approach that he advocated in opposition to the economic determinism he rejected. In the face of practical evidence, the pest problems of monocultural forestry, and deer starving on overgrazed reserves, Leopold eventually adopted a less violent and disruptive approach toward management.

IV

Should Leopold be classed as an anthropocentrist? Yes and No.

Yes, in the sense that he believed that, for better or worse, humans must and should manage the natural world. Given that, and current attitudes in society, arguments based on the good of the human species will carry more weight in policy debates. After summarizing the structure of the land pyramid and describing the land as an "energy circuit," Leopold summarized his land ethic in three basic ideas: "(1) That land is not merely soil. (2) That the native plants and animals kept the energy circuit open; others may not. (3) That man-made changes are of a different order than evolutionary changes, and have effects more comprehensive than is intended or foreseen. These ideas, collectively, raise two basic issues: Can the land adjust itself to the new order? Can the desired alterations be accomplished with less violence?" (Leopold 1949: 218). This is anthropocentrism of sorts. Leopold accepted that humans will alter the biota. Their management will be successful if they protect life and if the human race survives. It will be a failure if the human race, "like . . . John Burrough's potato bug, which exterminated the potato, . . . exterminates itself" (Leopold 1979: 141). Leopold never questioned the right of humans to alter nature, provided these alterations were consistent with ecological knowledge and would protect, in the long run, human life and the living land on which it depends.

But Leopold regarded both anthropocentrism and its denial as representing only human conceptions, as artifacts of human perceptions rather than reality. To try to determine the truth of these speculative pronouncements without reference to the systems in which they are embodied is to go beyond the possibilities of language. And yet there is a legitimate sense in which Leopold was a nonanthropocentrist. He saw organicism as an alternative to mechanism, one that would carry with it a deeper, even moral, reaction to the land. This view, while it can hardly be expressed in our current vocabulary, is true in the pragmatic sense – it has survival value. Leopold's dream that someday our culture will evolve a more sensitive reaction to the land explains the central role he always placed, as an environmental professional, on developing public perception. He believed that, as Americans become more aware of their interdependence with the rest of the biotic world, they will gradually develop a new conception of the world, including a moral reaction to the community of life. This development will, he thought, improve the survival chances of our culture and he therefore devoted his career to improving the perception of the American public.[5]

Leopold's actual target was not anthropocentrism, however. He concluded that nonanthropocentrism raises issues too intractable to make it useful in management discussions. Instead, he attacked shortsighted economic reasoning that ignores the scientific evidence that intense management often leads to gradual decline in productive systems.[6] Leopold recognized, in the degeneration of vegetative systems in the Southwest, in German forestry, and in the dust bowl phenomenon, the inadequacy of management practices based solely on Pinchot's utilitarian criterion. The search for profit, "economic determinism," leads inevitably to an undervaluing of future resources. But shortsighted, destructive practices are wrong even if we are "logically" anthropocentric. Anthropocentrism itself should imply a concern for future generations.

Leopold acted upon what I call the "convergence hypothesis" (Norton, 1991). According to that hypothesis, the interests of humans and the interests of nature differ only in the short run. If we recognize the extent to which the human species is an integral part of the community of life, long-term human interests coincide with the "interests" of nature. To protect the fullness of life is to protect the far-distant future of the human species and its evolutionary successors, and vice versa. Since the survival of our culture depends upon the survival of the ecosystems on which we, in turn, depend, the conception of the world one adopts is less important than the longsightedness with which it is applied in environmental management.

## NOTES

1. Leopold used the term "anthropomorphic" as we currently use "anthropocentric" – to refer to a value system that bases all value in human motives. Except in quotations of Leopold, I will follow current practice and use "anthropocentric."
2. Pragmatists described their conception of truth in several ways. James tended to emphasize the "effectiveness" of true ideas (see, for example, James 1948: 162). Peirce emphasized that truth is that which will last through an indefinite number of experiments and actions. He said that truth is "the predestined result to which sufficient inquiry *would* lead" (Peirce 1978: 288). The reader should not be disconcerted by Leopold's shift from Hadley's discussion of "right" to a definition of "truth" – the pragmatists drew no sharp distinction between facts and values and therefore treated "true" and "right" as largely interchangeable. Leopold's application of "truth" to cultural practices would be acceptable to pragmatists such as Hadley.
3. Curt Meine, who has just completed a biography of Leopold based on the collection of Leopold's papers, informs me that he found evidence that Leopold read the *Yale Review* regularly.
4. Again, Meine is my source of information here.
5. For a fuller discussion of the central role of "transformative values" in environmental ethics, see Norton (1987, Chapter 10).
6. For a discussion of Leopold's use of the term "economic," see Norton (1986: 208).

ACKNOWLEDGMENTS. I gratefully acknowledge the help of Curt Meine, J. Baird Callicott, Sara Ebenreck, and two very helpful anonymous reviewers who read and commented on earlier drafts of this paper.

LITERATURE CITED

Callicott, J. B., 1987. The conceptual foundations of the land ethic. In J. B. Callicott, editor. *A Companion to a Sand County Almanace: Interpretive and Critical Essays.* University of Wisconsin Press, Madison, Wisconsin.

Flader, S. L. 1974. *Thinking Like a Mountain.* University of Nebraska Press, Lincoln, Nebraska.

_____. 1979. Leopold's some fundamentals of conservation: a commentary. *Environmental Ethics* 1:143–148.

Hadley, A. T. 1913. *Some Influences on Modern Philosophy.* Yale University Press, New Haven, Connecticut.

Hadley, M. 1948. *Arthur Twining Hadley.* Yale University Press, New Haven, Connecticut.

James, W. 1948. Pragmatism's conception of truth. Pages 159–176 in A. Castell, editor. *Essays in Pragmatism.* Hafner Publishing Company, New York, New York.

Leopold, A. 1933a. The conservation ethic. *Journal of Forestry* 31:634–643.

_____. 1933b. *Game Management.* Scribner Publishing, New York, New York.

_____. 1939. A biotic view of land. *Journal of Forestry* 37:727–730.

_____. 1949. A *Sand County Almanac and Sketches Here and There.* Oxford University Press, Oxford.

_____. 1979. Some fundamentals of conservation in the Southwest. *Environmental Ethics* 8:195–220.

Norton, B. G. 1986. Conservation and preservation: a conceptual rehabilitation. *Environmental Ethics* 8:195–220.

_____. 1987. *Why Preserve Natural Variety?* Princeton University Press, Princeton, New Jersey.

_____. 1991. *Toward Unity among Environmentalists.* Oxford University Press, New York.

Ouspensky, P. D. 1968. *Tertium Organum.* Alfred Knopf. New York, New York.

Peirce, C. S. 1978. Pragmatism in retrospect: a last formulation. Page 269–289 in J. Buchler, editor. *The Philosophy of Peirce.* AMS Press, Inc., New York.

Petulla, J. 1980. *American Environmentalism: Values, Tactics, Briorities.* Texas A&M University Press, College Station, Texas.

Rolston. H., III. 1987. Duties to ecosystems. In J. B. Callicott, editor. *A Companion to a Sand County Almanac: Interpretive and Critical Essays.* University of Wisconsin Press, Madison, Wisconsin, USA.

Sherman, S. P. 1913. Review of some influences in modern philosophic thought. *Yale Review* 3:383–385.

# 2

# Thoreau and Leopold on Science and Values

When we ask, "What is the value of biodiversity?" we can expect that respondents, assuming that they answer the question at all, will answer in one of two quite different ways. Let us sketch these alternatives.

Some answers are mainly economic, emphasizing the actual and potential *uses* of living species. To this group, the value of biodiversity will be stated in quantifiable terms (Randall, 1988). This approach is utilitarian and anthropocentric. It measures value as contributions to human welfare. And it is "reductionistic" in the sense that it reduces to dollars all of the apparently disparate values and uses associated with wild species. Reductionists discuss the value of biodiversity by trying to put fair prices on its uses; they are most comfortable with the language of mainstream, neoclassical microeconomics. Natural objects, on this approach, are simply "resources" for human use and enjoyment. One characteristic of this approach, which makes it attractive in decision processes, is that it promises an aggregation of values: the contribution of nature to human welfare is made commensurable and interchangeable with other human benefits. This approach, therefore, holds open the possibility of a bottom-line figure that tells us what we should do in complex policy decisions; we should have exactly as much preservation of biodiversity as society is willing to pay for, given competing social needs.

Other approaches employ moral terminology and insist that we have an obligation to protect all species, an obligation that transcends economic reasoning and trumps our mere interests in using nature for our own welfare (Ehrenfeld, 1978; Rolston, 1988). These moralists limit human activities using nature by appeal to obligations that are independent of human welfare.

From Ke Chung Kim and Robert D. Weaver, eds., *Biodiversity and Landscapes: A Paradox of Humanity* (New York: Cambridge University Press, 1994). Reprinted with the permission of Cambridge University Press.

Moralists do not believe our obligations to protect nature can be traded off against other obligations. Their language is a moral and sometimes a theological one. Moralists, who believe that wild species have "rights" or "intrinsic value" – value independent of human interests and consciousness – recognize our obligations to protect other species as prima facie commands; they posit at least a strong presumption against trading them off against values based in human welfare. John Muir, first president of the Sierra Club and a passionate advocate of preserving nature in its multiple forms, once said: "The battle we have fought and are still fighting for the forests is a part of the eternal conflict between right and wrong, and we cannot expect to see the end of it" (Fox, 1981, p. 107).

I call the question of the value of biodiversity, when posed as a choice between these approaches, the "Environmentalists' Dilemma" (Norton, 1991; Ehrenfeld, 1978) because most commentators have assumed that we should give one answer or the other – either our obligation is to save natural resources *for* future consumption, or we should save nature *from* consumption and for its own sake (see, for example, Passmore, 1974). I argue below that this is a *false* dilemma – the works of Henry David Thoreau show a way between the horns of the Environmentalists' Dilemma, though specific application of Thoreau's moral insight did not emerge until it was given expression, by Aldo Leopold, in the vernacular of community ecology. I will describe and advocate a system of values that follows in the well-chosen footsteps of Thoreau and Leopold.

## THOREAU'S TRANSFORMATIVE VALUES

Thoreau's *Walden*, as well as his other writings, is sprinkled with analogies and metaphors drawn from wild species and applied to human life. I will begin by citing two passages in which Thoreau uses insect analogies to make points about people. First, in the most explicitly philosophical chapter of *Walden*, "Higher Laws," Thoreau notes that entomologists of his day had recognized that some insects in their "perfect state" (after transformation from the larval into the winged state) are "furnished with organs of feeding, [but] make no use of them." More generally, he noted, all insects eat much less in their perfect state. Thoreau applies this to human society: "The abdomen under the wing of the butterfly still represents the larva. . . . The gross feeder is a man in the larval state; and there are whole nations in that condition, nations without fancy or imagination, whose vast abdomens betray them" (Thoreau, 1960 [1854], p. 146).

This analogy illustrates Thoreau's peculiarly *dualistic* conception of human nature. Ostensively, Thoreau is explaining that hunting, and carnivorous habits more generally, are appropriate for the young, that these are a necessary *stage* in the individual's evolution (and in a culture's evolution), and that the urges to indulge in these practices will give way to higher sentiments and the abandonment of killing and meat-eating.[1] But it is obvious that carnivorous habits symbolize "gross feeding" more generally, and that Thoreau's intent is to portray materialistic consumerism as an immature developmental stage of the person.

Thoreau was a dualist not in the Cartesian sense, whereby two substances, mind and body, coexist in tandem through time, but in a dynamic or emergent sense. There exists a prior, primitive instinct that tempted Thoreau, upon seeing a woodchuck cross his path, "to seize and devour him raw." But Thoreau sensed in himself and in others also an instinct "toward a higher, or as it is named, spiritual life" (p. 143). Dynamic dualism, as Thoreau describes it, sees human nature as tensionally stretched between an older, primitive, "rank and savage" self and an emergent, higher, spiritual self in which the person's relationship with his surroundings changes from a consumptive one to a contemplative one. His economics – the first chapter of *Walden* – explains how one can retain and enhance one's creativity by living a simple, nonmaterialistic life.

Thoreau embraced both aspects of his humanity ("I love the wild not less than the good"), but he leaves no doubt that the emergent instinct toward a spiritual and perceptual relationship with nature is a "higher" instinct than the consumptive one, which is based in our animal nature: "The voracious caterpillar when transformed into a butterfly and the gluttonous maggot when become a fly content themselves with a drop or two of honey or some other sweet liquid" (p. 143). Note that, in this and other passages, Thoreau casually mixes moralism with description.

Thoreau confidently espouses a distinction between "higher" and "lower" satisfactions, much as John Stuart Mill advocated in his argument that it is "better to be Socrates dissatisfied than a fool satisfied" (Mill, 1957, Chapter II). Thoreau, however, is more specific than Mill; he advocates a life of simplicity and contemplation, of freedom from dependence on material needs. Thoreau therefore goes beyond Mill in proposing a substantive characterization of "higher" satisfactions. The important point for our present purposes is that Thoreau describes the benefits of the transformation to higher values in terms of human maturation and fulfillment of potential, as improvements *within* human consciousness, not in terms of obligations *to* nature and *extrinsic* to human consciousness.

Thoreau's dualism regarding human nature has three rather unusual aspects: (1) It is a *dynamic* dualism. The emergence of the spiritual aspect of the person represents a transformation from a "lower," primitive state to a "higher," more spiritual one. (2) It involves a comprehensive shift in perception and consciousness. In the immature, "larval" state, the person relates to nature physically, as a consumer; nature is thereby perceived in this immature state of the person as mere physical stuff, raw material, resources. In the second, perfect state, the person has undergone a perceptual transformation and now relates to nature nonconsumptively. Physical needs are minimized; the result is a life of "fancy and imagination." The world of nature is no longer seen as mere raw material for consumption; it is now seen as alive, soulful, and inspirational. (3) Thoreau explains this dynamic emergence of the higher self with an insect analogy, which illustrates the power of natural objects to teach us about human nature and simultaneously introduces the Aristotelian idea that the higher self is implicit within the lower state just as the butterfly is implicit in the caterpillar. In using organic analogies, Thoreau is emphasizing the power of nature study – observation – to hasten an inherent, systematic change in perceptual relations between the person and the natural world. Observation of dynamically unfolding natural processes, according to this view, hastens the evolution of a higher consciousness which is a potential of the human spirit. This perceptual and conceptual shift coincides with a deeper and more important shift in metaphysical assumptions, as well as a changing pattern of specific needs and values. Thoreau's dualism is therefore perceptual and psychological, but it is also behavioral. Post-transformational individuals will consume less.

In the penultimate paragraph of *Walden*, Thoreau again uses an insect analogy. He tells an anecdote of a strong and beautiful bug which gnawed its way out of an old table, "hatched perchance by the heat of an urn." It grew, he noted, from an egg that had been deposited many years before in the living apple tree. Thoreau explains the analogy explicitly, saying that the egg represents the human potential to live a free life of beauty, a potential which can only be released after the spirit of independence gnaws through "many concentric layers of woodenness in the dead dry life of society." So Thoreau returns to the dualistic idea at the end of *Walden*. The urn, I think, represents *Walden* itself – it is intended to provide the catalyst for truly human individuals to "hatch," to achieve the potential for individual freedom implicit in each person, and to escape the life of "quiet desperation" Thoreau saw as the plight of his neighbors. The book encourages the reader to begin gnawing through layers of social custom and expectations of material success toward individual freedom (p. 221).

Thoreau stood at a crossroads in American thought in the sense that he can be aptly described as both a transcendentalist and as a naturalist (Miller, 1968). Looking backward, he owed clear debts both to Puritanism and to Emerson's idealistic pantheism. But his boyhood days were filled with "nature studies" long before he knew of Emerson's philosophical glorification of nature. Thoreau's transcendentalism therefore represented from the start an uneasy compromise between the earthly love of natural events and a heady preoccupation with intuitions of transcendence.

Whether or not Thoreau rejected idealistic transcendence explicitly, I am arguing that he developed a theory of perceptual and psychological (world-view) change that could stand independently of idealism – the only transcendence involved is the transition from one state of consciousness to another. This shift away from commitments to "pure" insight unsullied by preconceptions and presuppositions puts him more in the tradition of contemporary naturalism than in the tradition of pure idealism. The validity of the change in world-view will be demonstrated not by pure reason but by improved experience of those who have undergone the transformation.

Thoreau's program was to describe living nature in such a way as to evoke analogies that will set in motion a shift to a higher, less materialistic and consumptive set of needs and style of life. And these analogies were essential to his ambitious project of reforming his contemporaries, of freeing them from slavery to commercialism and removing the "quiet desperation" from their lives.

## THE DYNAMICS OF NATURE AND THE DYNAMICS OF CONSCIOUSNESS

Thoreau ultimately rejected Emerson's Platonism because he could not understand nature in terms of fixed essences. Thoreau thus chose Heraclitus over Plato: "All is in flux." All things in nature, including human beings and the values they live by, are constantly changing in a great interrelated whole. Thoreau's dynamic dualism required the rejection of Emerson's world of fixed forms and real essences – human nature itself shifts in response to its changing environment.

Heraclitean dynamism was abundantly confirmed by Thoreau's experience: *Walden* is a spring → summer → fall → winter guide to the dynamic transformations of nature, and it ends with the second spring. The return of spring is symbolic of the individual's resurrection from the "death" of social conformity and of rebirth into spirituality. In response to the beautiful bug that gnawed its way out of the table, Thoreau says: "Who does not feel his faith in a resurrection and immortality strengthened by hearing this?" (p. 221).

Thoreau thought he saw a way out of the trap of materialistic consumerism and its concomitant understanding of nature as raw resources, and hence the symbolic "urn," *Walden* is Thoreau's spiritual legacy, a catalytic agent to unlock human potential. Human beings have an individual genius, a potential to be good, to live perceptually rich and consumptively simple lives; consciousness and imagination represent for Thoreau the chance humanity has to achieve transcendence, to become the butterfly free of addiction to consumption. But these instincts to a "higher" life of freedom are crushed within "the dead dry layers of society" (p. 221), the indoctrination in consumerism that Thoreau foresaw so clearly would become the fate of moderns.

And here Thoreau turns moralist and social philosopher by uniting his theory of world-view change with his social criticism. We experience *wonder* in observing nature, by being still to let other living things provide lessons, in the form of analogies. Thoreau believed we can learn how to live by observing wild species. Careful attention to natural analogies can make us wise.

Thoreau's dynamism, and the multiple layers of symbols it offered, was at its heart organicist. By this I mean that the world-view he advocated was governed on all levels by a dynamic, organic metaphor. In describing the break-up of the ice on Walden Pond in spring, he said: "Who would have suspected so large and cold and thick-skinned a thing could be so sensitive?... [T]he earth is all alive and covered with papillae. The largest pond is as sensitive to atmospheric changes as the globule of mercury in its tube" (p. 201).

The change in perception that accompanies the understanding of the natural world as alive and soulful, a shift in governing metaphors and world-view, also encourages the perceiver to experience a change in values – natural objects are transformed, within the new world-view, into objects of contemplation and inspiration, not mere objects of exploitation. Post-transformational consciousness experiences nature within an organicist world-view, a world of dynamic change and development, a world in which natural objects have a spiritual as well as a material value.

Thoreau's analogical method is therefore blatantly moralistic; one might even say "naive."[2] Thoreau chose observation as knowledge in the service of wonder – it is wonder at nature's intricacy, complexity, and economy that drives us to a changed understanding of ourselves and our place in nature. The plausibility of Thoreau's transformational theory therefore depends on his frank acceptance of the value-ladenness of facts – "Our whole life is startlingly moral" (p. 148).[3] Accordingly, he was not surprised to find significance in facts on the eating habits of insects. The sense of wonder inspired by these value-charged facts leads to deeper insights than those of botany or zoology. Because nature's facts have this significance, they can build moral character.

But Thoreau was not simply naive; he went far beyond quaint proverbs – he recognized that the adoption of a new, organic world-view would represent a shift toward more holistic thinking. Thoreau thought that his anecdotes and analogies could act as a catalyst for world-view transformation in his readers. But he also recognized that the change results more directly from the rediscovery of a sense of wonder. Thoreau thought that if he could induce people to develop their perception through patient and sensitive observation of nature, he could also induce them to reject mechanism and with it materialistic consumerism. Once the new world-view was embraced, he believed a new perceptual and evaluative relationship with nature would also emerge. Thoreau saw the conversion to the new world-view as a transformation in attitudes, values, desires, and demands as well as beliefs. The vehicle of this transformation is a change in the form of perception that emerges within the new world-view. Careful and patient observation is the trigger that initiates the life-long process of seeking beauty and truth in the dynamic natural world.

It is tempting to ridicule Thoreau's experiments in moral reasoning as native-sounding analogies, comparing them to Aesop's fables. The temptation is increased by the writings of naturalists, such as Annie Dillard (1974) and David Quammen (1985), who have undermined the naive use of individual analogies by exhibiting the horrors and perversities (when viewed from a human perspective) that coexist with beauty in nature. Thoreau's simple analogies therefore sound quaint today. A sympathetic reader could nevertheless credit him with recognizing the holistic nature of world-view change, coupling this with a clear understanding that economics is as much about "managing" our preferences as it is about fulfilling them. The process that Thoreau undertook at Walden Pond was action, catalyzed by an intuitive wonder at the observation of nature, in service of liberation. Thoreau saw clearly that ethical development will be a dynamic process that will reach to the deepest assumptions of the modern world-view.

Thoreau did not cast his learn-from-nature analogies into the story of his experiences at Walden Pond without offering a supporting theory. He argued cogently that nature's analogies are illuminating on all levels, that the analogies from insects are just a part of a larger transformation in world-view, and that an organic, holistic metaphor is demanded if we are to understand both nature and human consciousness. Holism, the view that the whole is more than the sum of its parts, applies to understanding no less than to physical systems. A change in world-view will require an intuitive leap, an integrative act of creation catalyzed by the sense of wonder; it follows that the new consciousness cannot be expressed, much less justified, in the immature consciousness. The transformation is in this sense nondeterministic and requires an intuitive spark as much as observation and logic.

Thus, while he emasculated Emerson's theory of intuition regarding timeless essences, Thoreau remained true to his transcendentalist past in an important sense: the intuitive leap to a new, holistic world-view cannot be understood in a mechanistic system of psychology or logic. What Thoreau learned from nature was dynamism, the view that becoming is more fundamental than being, which entails that the system of nature is not deterministic; its most basic law is the law of creative activity. The striving to create goes beyond what can be known by a deductive process. Thoreau was an indeterminist who believed that the whole, both in nature and consciousness, is more than the sum of the parts.

Because the shift to the new consciousness cannot be described in the mechanistic models of nature popular in the modern age, Thoreau could not promise rational proof, but only catalyze a process. He never achieved, and despaired of achieving, an "objective" proof of the truth as it is experienced in evolving systems; and therefore he recognized the crucial role of an aesthetic, extralogical leap into a new world-view with more expressive concepts.

Thoreau also foresaw with remarkable clarity the conclusions that have emerged in contemporary physical theory and in contemporary philosophy of science. The breakdown of the logical positivists' program of a unified, reductionistic science, philosophical analyses that explain and justify this breakdown (Quine, 1969; Sellars, 1956), and conclusions in the new physics (Prigogine and Stengers, 1984), have all converged on the conclusion that there exists no complete and consistent theory that can represent the world on all levels simultaneously. The truth, it follows, cannot all be of the form of deductions from pure facts. Thoreau therefore anticipated the epistemological problems of postmodernism – he struggled to explain how one can explain and justify a change in world-view if there is no single, correct "description" of the physical world.

The problem of justification is, of course, tied to the problem of objectivity. Thoreau recognized that a demonstration of the inadequacy of mechanism would require a richer vocabulary than the one of deterministic science. Hence his experiments with analogies as ways to enlist scientific description in the larger, intuitive task of recognizing our proper "place" in the larger systems of nature. Thoreau's understanding of world-view change could not be explained in the objectivist tradition of Descartes and Newton; he needed the language of aesthetic creativity as much as the language of descriptive science. Thoreau recognized that dynamism and change is nature's most profound lesson and, by applying that insight to human consciousness and cultural development, Thoreau provided a plausible model of world-view change. He avoided serious mystical commitments by treating intuitions of holism as intuitive leaps that provide insight about our place in natural systems. He therefore reduced intuition to wonder, to a catalytic process that sets in motion a systematic rethinking of facts, and avoided commitment to intuitively justified know-ledge – scientific or moral – of the physical world.

## THE ENVIRONMENTALISTS' DILEMMA REVISITED

By combining the religious idea of a moral transformation with the more mundane idea of consumption trimmed to fit context, Thoreau avoided the Environmentalists' Dilemma. Thoreau's commitment to simplicity was *both* a call to live in harmony with nature economically *and* a response to the alienation evident in the quiet desperation characteristic of industrial society. The analogy of hunting as appropriate for young individuals and for "young" societies, but inappropriate for mature individuals and mature societies, represents a linkage of psychological maturation with social maturation. Nature thus holds more than economic value to humans – it is a sacred talisman, the honored symbol and guide leading humanity toward spiritual and material freedom. Nature exhibits a "higher" value than satisfying consumptive wants and needs. This does not mean that natural objects will no longer be used; but they will be "used" appropriately, with respect, and with a sense of awe at their sacred powers to transform consciousness and values.

Thoreau, like Darwin, recognized that the key to following nature is to strive to fit our needs to nature's demands, rather than to alter nature to serve our unexamined demands. An ethic that takes into account our "place" in nature, a larger whole of which we are parts, will also be more satisfying. In this sense, "cultural survival" is determined, in a dynamic world, by appropriateness to the larger context that sets the conditions for survival. Morality is not determined by fixed moral principles, knowable by pure intuition; morality is determined, rather, by situation, and requires analogical insight based on careful observation of constantly changing situations.

Thoreau escaped the Environmentalists' Dilemma by insisting that nature has didactic value, value that stands outside the aggregated demands expressed by individuals in our "immature" society, but which is represented in world-view changes that will reshape those very demands. If individuals achieve freedom through transformation, they will also adopt a new, nonconsumptive lifestyle appropriate to the postmodern world. Sensitive observation of nature and an appropriate sense of wonder at nature's complex organization sensitize us to the "whole" of which we are a part. Thoreau therefore anticipated the insight of Carl Jung, who once said that he never succeeded in helping any patient who was not convinced he was a part of some larger whole.

Nature's value is manifest within human consciousness and experience, and implies no commitment to values that are defined outside human consciousness. Thoreau said: "I am not interested in mere phenomena, though it were an explosion of a planet, only as it may have lain in the experience of a human being" (Richardson, 1986, p. 309). Observation, understanding, and appreciation of nature are inseparable parts of a process of change by which our lives are illuminated and seen in a new way. The process represents a constant transformation to a new form of perception and action, perception and action that seek harmony with a larger, dynamic whole, of which we are a part.

Thoreau left Walden Pond when he discovered that, in two-and-a-half years, his "spontaneous" walks were cutting pathways in the woods, imposing patterns on nature, rather than finding them there. And so he left Walden Pond "for as good a reason as [he] went there." He had gone there to observe and react to stimuli he sensed within nature, rather than to manipulate nature and its resources to serve his own preferences. Thoreau's paths, radiating outward from his cabin, showed him that this project of living within nature is a worthy ideal, but one that cannot be fully achieved. Humans inevitably reconstruct their habitat from their own perspective. Even the "pure" experiencer, Thoreau in his cabin, altered the natural context.

Thoreau, the optimist, did not despair: "I learned this, at least, by my experiment: that if one advances confidently in the direction of his dreams, and endeavors to live the life which he has imagined, he will meet with a success unexpected in common hours" (p. 214–215).

THOREAU'S SCIENCE

While Thoreau showed a way through the Environmentalists' Dilemma by recognizing the transformational role of observation in value shifts, he fell short, I think, of providing a clear connection between science and values. To say that scientific observations of living things suggest analogies which, in turn, "catalyze" a world-view change is to leave the role of science, especially theoretical science, as a mysterious black box in Thoreau's account.

Assessments of Thoreau's scientific acuity differ and are currently undergoing reassessment (Scholfield, 1992). Some critics have found little more than idiosyncratic description in Thoreau's journals (Richardson, 1986), but recent scholarship has revealed that Thoreau was actively working on a pioneering book in protoecology in the last years of his life. This book has been recently reassembled and published (Thoreau, 1993), and Gary Nabhan argues on its basis that "More than any botanist of his time. Thoreau moved past the mere naming of trees – the nouns of the forest – to track its verbs: the birds, rodents, and insects that pollinate flowers or disperse seeds, and all the other agents that shape the forest's structure" (Nabhan, 1993).

Apparently, on the basis of this newly assembled and rethought evidence, Thoreau quite explicitly recognized that the forest, a dynamic system, had a "language" of its own, and that the transition from the immature state was both literary and scientific. Thoreau argued that the important place to look for insight from wild species is in their natural habitat, not on a dissecting table (Thoreau, 1960). He saw that one learns more important things by relating an organism to its environment than by dissecting an organism into parts. This indicates that Thoreau was on the right track, seeking the secret of life and its organization in the larger systems in which species live. Especially, he thought we learn more important things about human behavior, and the evaluation of it, by observing organisms in environments. He believed that if he could unlock the code of nature's language, it would provide the key to a new, dynamic and scientific understanding of nature. The key prerequisite for this change to a more contemplative consciousness was development of a new "language" of human values based on analogies from the "language" of nature. Nabhan asserts that, on this basis, we can conclude that Thoreau never gave up his attempt to become a romantic poet: "Instead of turning his back on these literary traditions, Thoreau tried to incorporate them into his search for a language more difficult but more enduring: the language of the forest itself" (Nabhan, 1993).

Thoreau saw that humans are analogous to other animals in the levels of organic nature, and he recommended sympathy for the hare "which holds its life by the same tenure" as a human person (Thoreau, 1960, p. 144). Understanding the roles of animals, including ourselves, in larger systems can therefore teach us a new code of behavior. The core change, the heart of the new worldview, is the adoption of an organic metaphor for understanding nature.

Thoreau's early death, unfortunately, prevented him from completing his project, which would surely have developed in new directions as he reacted to Darwin's *On the Origin of Species*, which he first read only two years before his death. It is interesting to speculate whether he would have developed an epistemology not unlike [that of] the American pragmatists such as John Dewey or Charles Peirce. In a dynamic system, truth must be dynamic and adaptive or become irrelevant. Natural selection of world-views for adaptability and contribution to cultural survivability would be the missing piece in Thoreau's theory, the link to tie his observations of natural events to his moralistic analogies. It might have guided him toward a full-blown evolutionary epistemology and an evolutionary ethic. We learn from observation because our role in nature is functional in a larger system. Analogies – Thoreau's preferred moral method – would be relevant because all species, including humans, must adapt to a constantly changing environment, even while functioning as an element in the systems that compose that environment. The ultimate lesson learned from nature is therefore a re-organization of thought that expresses atomistic facts as parts of an integrated understanding of a dynamic whole.

## ALDO LEOPOLD AND SCIENTIFIC CONTEXTUALISM

It was left to Aldo Leopold, born in 1882, twenty years after Thoreau's early death, to recognize and explain the importance of ecology and evolutionary theory to perceptual and ethical transformations. Leopold, who described ecology as the biological science that runs at right angles to evolutionary biology, chose cranes as his illustration: "our appreciation of the crane grows with the slow unraveling of earthly history. His tribe, we now know, stems out of the remote Eocene. . . . When we hear his call we hear no mere bird. We hear the trumpet in the orchestra of evolution. He is the symbol of our untamable past, of that incredible sweep of millennia which underlies and conditions the daily affairs of birds and men" (Leopold, 1949, p. 96). The essay "Marshland Elegy," which is one of Leopold's finest, begins on the edge of a crane marsh, with the narrator hearing "[o]ut of some far recesses of the sky a tinkling of little bells," and carefully describes in dragging prose how a human experiences the spectacle of the arrival of the cranes. He describes how there are periods of silence and periods of growing clamor until at last, "[o]n motionless wing they emerge from the lifting mists, sweep a final arc of sky, and settle in clangorous descending spirals to their feeding grounds" (Leopold, 1949, p. 95). Note that Leopold avoids *direct* moralizing from cranes to humans (which would be similar to Thoreau's use of insect analogies), arguing that cranes teach us about evolution and our role in those processes. The emphasis is on broadening perception, not on providing moral maxims.

Over the next six pages, Leopold guides the reader back and forth among many scales of time. He moves abruptly from human, perceptual time to place

41

the marsh in geological time: "A sense of time lies thick and heavy on such a place. Yearly since the ice age it has awakened each spring to the clangor of cranes" (Leopold, 1949, p. 96). After stepping out of time altogether to discuss the aesthetics of time, he stresses the ability of scientifically enlightened perception to transform the arrival of the cranes into a semireligious experience, and, simultaneously, to explain why the destruction of crane marshes is, if not a sin, at least a tragedy.

Leopold then plunges back into geological time, tracing the path of the last glacier, describing the formation of the pond, and then braking the time machine down to the pace of ecological time and describing the development of ecological conditions that allowed the cranes to find a niche in Wisconsin. He recognizes that they have survived many earlier, gradual transformations of their habitat, and then laments how, in so many marshes, they had succumbed to human alteration of their habitat in just a few generations. Initially, in an arcadian time, farmers and cranes cohabited harmoniously. But technology, avarice, and inappropriate land use destroyed crane habitat even as it impoverished the human inhabitants. The downward spiral, looked at in ecological time, represented a deterioration of both crane and human habitats. But humans, unaccustomed to think like a mountain, unable to perceive the value of wolves or cranes, harm themselves, both economically and spiritually, by failing to see the difference between nature's gradual changes and the accelerated pace of change in larger systems attendant upon the technology augmented economic activities of modern humans.

Leopold's aesthetic explanation: "Our ability to perceive quality in nature begins, as in art, with the pretty. It expands through successive stages of the beautiful to values as yet uncaptured by language. The quality of cranes lies, I think, in this higher gamut, as yet beyond the reach of words" (Leopold, 1949, p. 96). The cranes, linking as they do the various scales of our history, unlock our understanding of time and our origins, and exhibit to us our "place" in nature. Leopold concludes that "The ultimate value in these marshes is wildness, and the crane is wildness incarnate" (Leopold, 1949, p. 101). The value of wildness to us is that it illustrates the multilevelled complexity, the dynamically stable system that has enough constancy to allow organization through evolution, and yet enough change to foster nature's creativity. The cranes act as living metaphors that locate our own species and its cultures in its larger context; they contribute to the larger-scale change in our world-view, a change that can be described metaphorically as embracing an organicist world-view, but experienced and made truly meaningful only in a crane marsh or some other such wild place. Here, Leopold follows Thoreau in his emphasis on the ways in which observation can catalyze changes in perception and in world-view. The moralizing, here, is indirect: a change in perception, aided by ecology, helps us to place ourselves in a larger dynamic, as evolved animals, and we consequently see that our destruction of the crane habitat is wrong – it cuts us loose from our evolutionary and cultural history.

When Leopold introduces his land ethic near the end of *A Sand County Almanac*, he returns to these potent evolutionary and ecological themes. "The extension of ethics to [land and to the animals and plants that live there] is, if I read the evidence correctly, an evolutionary possibility and an ecological necessity" (Leopold, 1949, p. 203). These passages hark back to passages written in 1923 in which Leopold expressly argued that a culture will be judged according to its treatment of the land it lives upon and that a society unable to sustain itself on its land will "be judged in 'the derisive silence of eternity'." Expressing this idea in the terms of an evolutionary epistemology, Leopold explicitly linked cultural survival with truth: "Truth is that which prevails in the long run" (Leopold, 1979, p. 141; paper 1, this volume).

Leopold's land ethic represents the fruition of Thoreau's breakthrough in moral theory. Our society, even as it has tamed the wilderness and damaged natural systems, has created a society so productive that it can consciously forbear from further destruction. Change to a new world-view and a new form of life, one that is harmonious with its context, one that uses nature even as it recognizes the higher values embodied in the complex systems that form our ecological context, is now an "evolutionary possibility." Understanding of our ecological role clarifies our "place" in life's larger enterprise and establishes our moral bearings in the changing world we face.

In this way, ecology is the key to a new morality. To understand our role in the biological world is to reject hubris: "a land ethic changes the role of *Homo sapiens* from conqueror of the land-community to plain member and citizen of it" (Leopold, 1949, p. 204). The goal, after the transformation to a new perspective, is to understand the role of our cultures in the ecological communities they have evolved within.

The central idea of Leopold's land ethic is that the land is a complex *system composed of many levels and subsystems that change according to many rates of speed* (Norton, 1991). Humans armed with conscious goals and powerful technologies can disrupt that system. But the same dynamic system of causes that created consciousness can, in the face of changing conditions, create a *new* consciousness. In the tradition of Thoreau, Leopold thought that the best antidote to disruptive behavior is a transformation in human consciousness and a new style of perception, perception that is informed by ecology and by evolution, and perception that recognizes, according to these sciences, the true role of the human species in the natural order. These changes, in turn, result in a very different conception of environmental "management." This new conception, informed by ecological science and tempered by the humility appropriate not to conquerors but to "plain citizens" of ecological communities, can be aptly called "scientific contextualism."

Scientific contextualism is a sort of holism. It is holism because it understands human activities, world-views, and ethics as a part of the evolving systems of nature. Science thus informs ethics by describing the appropriate role of humans in the system. Observation, as Thoreau so clearly recognized, unlocks treasures of understanding and self-understanding. But Leopold, who saw more clearly the role of ecological science, was able to conceptualize the metaphor of organicism as a systems approach to management, and while he often anthropomorphised animals he also "animalized" humans by insisting that we are, like every other species, extruded into a niche, and therefore cannot destroy that niche – our natural context – with impunity. Human activities, including economic ones, represent subsystems within a larger ecological and physical whole, or context.

But contextualism is a *limited* holism; it reifies no single model as reality and need assume no superorganism that intentionally organizes the complex systems of nature (Wallace and Norton, 1992). According to scientific contextualism, many different models, which are no more than technical analogies, are useful in differing contexts. The difficult part is to choose analogies/models; in particular, it is difficult to conceive environmental problems on the correct scale and from the right perspective. Leopold's brilliant simile, "thinking like a mountain," is an exercise in choosing the correct *scale* for analysis and management. He first thought, while supporting predator eradication programs, that deer/wolves/hunters formed an equilibrium system and that if he removed wolves, hunters would increase their take and create a new system with greater human utility. But he learned from experience that the larger ecological community, the mountain ecosystem, which changes slowly over millennia, was thrown into an accelerated pace of change by his actions. The vegetative cover – the skeleton and skin of the mountain "organism" – was destroyed and erosion began to set in. The key, Leopold concluded, is to see our activities as changing subsystems that function within a larger whole (Leopold, 1949).

Contextualism is a sort of "poor man's systems theory." It proclaims no single model that can capture and relate all phenomena on all levels to all others. Models are seen more modestly as tools of the understanding, and there is an implicit recognition that systems of different scales will be chosen to deal with different problems. Choice of the proper model will depend *both* upon social purposes (values) *and* scientific understanding, and will involve experiments, both social and ecological. But the experiments must be conducted with great care. Their purpose is to learn what nature is telling us, not merely to manipulate nature for human uses.

The American naturalist tradition, if not burdened by unreasonable epistemological requirements – such as strict adherence to a separation of science and ethics, or a belief in intuited, timeless moral principles – provides an interesting and plausible moral theory. This moral theory most centrally involves a commitment to world-view transformations that are only partially objective, and which evolve organically in reaction to changing values, concepts, and theoretical beliefs. But the criterion by which to judge such transformations must be improved human experience, not some abstract and timeless principle. The need for a transformation in consciousness is evident both in the illness and alienation of modern society and in the illness and deterioration of the ecological context. As Thoreau saw, the two problems have a common solution in the phenomenon of world-view change triggered by a sense of wonder at nature's complexity.[4]

## NOTES

1. Thoreau stressed a point that has been a recurring theme of naturalists since – sport hunting is appropriate for a young man, but a more mature individual will become a less consumptive naturalist. See, for example, Leopold, 1949, pp. 168–176.
2. As did Holmes Rolston III upon reading an earlier version of this paper.
3. I have discussed transformational theories and arguments in more detail in Norton 1987, especially chapters 10 and 11.
4. Portions of this paper appeared in "Thoreau's Insect Analogies" in *Environmental Ethics*, 13, 235–251.

## REFERENCES

Dillard, A. (1974). *Pilgrim at Tinker Creek*. New York: Harper & Row.

Ehrenfeld, D. (1978). *The Arrogance of Humanism*. New York: Oxford University Press.

Fox, S. (1981). *John Muir and His Legacy*. Boston: Little, Brown and Company.

Krutch, J. W. (1948). *Henry David Thoreau*. New York: William Sloane Associates.

Leopold, A. (1949). *A Sand County Almanac*. Oxford: Oxford University Press.

_____. (1979). Some fundamentals of conservation in the Southwest. *Environmental Ethics*, 1, 131–148.

Mill, J. S. (1957) (originally published 1861). *Utilitarianism*. Indianapolis, IN: The Bobbs-Merrill Company, Inc.

Miller, P. (1968). Thoreau in the Context of International Romanticism. In *Twentieth Century Interpretations of Walden*, ed. R. Ruland. New York: Prentice-Hall.

Nabhan, G. P. (1993). Learning the Language of Field and Forest (Foreword to *Faith in a* Seed). Island Press, Washington, DC.

Norton, B. G. (1987). *Why Preserve Natural Variety?* Princeton: Princeton University Press.

———. (1991). *Toward Unity among Environmentalists*. New York: Oxford University Press.

Passmore, John (1974). *Man's Responsibility for Nature*. New York: Charles Scribner's Sons.

Prigogine, I. & Stengers, I. (1984). *Order Out of Chaos: Man's New Dialogue with Nature*. New York: Bantam.

Quammen, D. (1985). *Natural Acts*. New York: Schocken.

Quine, W. V. (1969). Epistemology Naturalized. In *Ontological Relativity and Other Essays*. New York: Columbia University Press.

Randall, A. (1988). What Mainstream Economists Have to Say about the Value of Biodiversity. In *Biodiversity*, ed. E. O. Wilson. National Academy Press, Washington, DC.

Richardson, R. (1986). *Henry Thoreau: A Life of the Mind*. Berkeley: University of California Press.

Rolston, H. (1988). *Environmental Ethics: Duties to and Values in the Natural World*. Philadelphia: Temple University Press.

Scholfield, E. (1992). *A Natural Legacy: Thoreau's World and Ours*. Golden, CO: Fulcrum Publishing.

Sellars, W. (1956). Empiricism and the Philosophy of Mind. *Minnesota Studies in the Philosophy of Science, Vol. 1*. Minneapolis: University of Minnesota Press.

Thoreau, H. D. (1960). *Walden and "Civil Disobedience."* New York: NAL Penguin Inc. (Originally published 1854 and 1848.)

———. (1993). *Faith in a Seed: The Dispersion of Seeds*, ed. B. Dean. Island Press, Washington, DC.

Wallace, R. & Norton, B. G. (1992). The policy implications of Gaian theory, *Ecological Economics*, 6, 103–118.

# 3

## Integration or Reduction

### Two Approaches to Environmental Values

INTRODUCTION: THE ROLE OF ENVIRONMENTAL ETHICISTS
IN POLICY PROCESS

Environmental ethics has been dominated in its first twenty years by questions of axiology, as practitioners have mainly searched for a small set of coherent principles to guide environmental action. In axiological studies, a premium is placed on the systematization of moral intuitions, which is achieved when all moral judgments are shown to be derivable from a few central principles. The goal of these studies is to propose and defend a set of first principles that is (1) *complete* in the sense that this small set of principles can generate a single correct answer for every moral quandary and (2) *jointly justifiable* in the sense that, once the principles are warranted, then every particular moral directive derived from the principles must also be warranted.

The limiting case of axiological simplification is *moral monism*, the view that a single principle suffices to support a uniquely correct moral judgment in every situation.[1] Monism represents to some philosophers an ideal because, provided the adopted principle is self-consistent, problems of coherence and consistency are resolved once and for all – there is no need to worry about what to do if two principles imply differing actions in a given situation, no worry that there will be irresoluble conflicts among competing and equally worthy moral claims. This reasoning motivates the drive towards unification.[2] The goal of environmental ethics as a discipline, in keeping with this ideal, has most centrally been to offer a unified and monistic account of our moral

From A. Light and E. Katz, eds., *Environmental Pregmatism* (London: Routledge, 1995). Reprinted with permission.

obligations. The adoption of this goal is what has given environmental ethics its axiological character.

What is curious is that both sides in what has become a polarized debate: neoclassical welfare economists – who believe that all value is expressible in units of individual, human welfare – and advocates of attributing inherent value to non-humans – who argue that the moral force of environmental principles derives from the moral considerability of natural objects – are unyieldingly monistic in their approaches. The adoption of the monistic viewpoint and the associated goal of developing a universal moral theory applicable in all cases are inevitably "reductionistic." Because all values, which are experienced in multiple modes and contexts, must on the monistic approach be accounted for under a single theory, the basic strategy must be to reduce all moral concerns to a unified analytic vernacular in which solutions to specific moral quandaries are generated, by unavoidable inferences, from a single theory.

This shared assumption of monism has, I believe, locked environmental ethicists into a paralyzing dilemma, a dilemma that lies at the heart of most discussions of environmental values. Most participants in these discussions have subscribed to a crucial alternation in the theory of environmental valuation: either the value of nature is entirely instrumental to human objectives, or elements of nature have a "good of their own" – value not dependent on human valuations.[3] Could it be that the polarized thinking that paralyzes environmental policy today results from false alternatives forced upon us by the assumption, unquestioned by neoclassical economists and by most of their opponents among environmental ethicists alike, that whatever the units of environmental value turn out to be, there will be only one kind of them?

The thesis of this paper is that the goal of seeking a unified, monistic theory of environmental ethics represents a misguided mission, a mission that was formulated under a set of epistemological and moral assumptions that harks back to Descartes and Newton. An assessment of the contribution of environmental ethics to environmental policy in its first two decades is accordingly bleak. The search for a "Holy Grail" of unified theory in environmental values has not progressed towards any consensus regarding what inherent value in nature is, what objects have it, or what it means to have such value. Nor have environmental ethicists been able to offer useful practical advice by providing clear management directives regarding difficult and controversial problems in environmental planning and management.[4] One very practical effect of the monistic assumption is that the range of topics open for discussion in environmental ethics has been narrowed, and opportunities for building bridges with other, more practice-oriented disciplines have been lost.

Another effect has been to define an often unhelpful role for environmental ethicists in environmental policy debates.

In order to emphasize these practical implications of the issues raised in this paper, I complete this introduction by drawing a distinction between "applied" and "practical" philosophy as representing two somewhat different roles for environmental ethicists in the process of developing and implementing environmental policy. After this practice-oriented introduction, the remainder of the paper falls into two parts, one destructive and critical and the other positive and speculative. Part 1 illustrates the problems of formulating a monistic environmental ethic by exploring the evolution of the monistic, ecocentric theory of J. Baird Callicott. This retrospective of Callicott's position, and a criticism of his current position, provide reasons to be very skeptical both of Callicott's specific monistic theory and also of his mission as he understands it. As counterpoint to this negative argument, I briefly present a pluralist conception of the role and possible content of an environmental ethic in Part 2. Influenced by a pragmatist attitude towards social problems, I will sketch an environmental ethic that applies multiple principles, but one which seeks integration of these principles in a way that is sensitive to place orientation and to temporal and spatial scales.

Let me explain my basic methodology and advocate the importance of practice by reference to a distinction between two kinds of non-theoretical philosophy: I call them "applied" philosophy and "practical" philosophy, despite the fact that these terms are sometimes used interchangeably. I use them here to correspond to two somewhat different roles for philosophers in the process of public policy formation. Applied philosophy refers to the application of general philosophical principles in adjudications among policy goals and options. Applied philosophy's method is usually to develop very general and abstract principles and then to illustrate their use by discussing a few carefully circumscribed hypothetical cases. This conception of the role of environmental ethicists has encouraged the confinement of philosophers, in their day-to-day work, within their traditional academic roles of teaching and writing. The actual applications of these principles is usually left to others such as environmental managers or environmental groups.[5] Moral monism and applied philosophy are naturally complementary – a single principle agreed upon by all disputants provides just the sort of moral guidance that applied philosophers would like to give. They want to furnish a universal principle from which actual decision-makers can derive moral directives and then apply them to the cases they encounter in the day-to-day process of setting policy. Since the universal principle functions as an essential premise in an argument that one or another policy is justified, agreement on a policy option will emerge only if the general principle is accepted by all parties to the dispute.[6] Philosophers' contributions, given the role envisaged by applied philosophers, can only be as strong and decisive as the case for one universal principle. If some disputants do not accept the unitary moral principle proposed by applied philosophers, or if applied philosophers cannot agree among themselves regarding the formulation of the universal principle, they must retreat to theoretical arguments and attempt to establish more definitively the universal, monistic principle/premise before returning to applications. I therefore turn to a discussion of the assumptions that shape environmental ethicists' view of what they can offer in the policy process.

Practical philosophy, as I am defining it here in contrast to applied philosophy, is more problem-oriented; its chief characteristic is an emphasis on theories as tools of the understanding, tools that are developed to resolve specific policy controversies. It shares with applied philosophy the goal of contributing to problem solution; but practical philosophy does not assume that useful theoretical principles will be developed and established independent of the policy process and then applied within that process. It works towards theoretical principles by struggling with real cases, appealing to less sweeping rules of thumb that can be argued to be appropriate in a particular context, rather than establishing a universal theory and "applying" it to real cases. Practice is prior to theory in the sense that principles are ultimately generated from practice, not vice versa.

I do not mean to claim that theorizing is worthless; on the contrary, theory-building that addresses real-world problems, in the spirit of John Dewey and Aldo Leopold – the forester-philosopher – is absolutely essential if the environmental movement is to develop a vision for the future.[7] In the meantime, however, theoretical differences often need not impede progress in developing current policy; if all disputants agree on central management principles, even without agreeing on ultimate values, management can proceed on these principles.[8] And philosophers have a lot to offer policy-makers in specific complex situations in which they face many conflicting moral directives, even though it has proved impossible for them to deliver the Holy Grail of monism as promised.

What sets practical philosophy in contrast to applied philosophy is the differing practices and impacts they envision for philosophers in the processes of policy articulation, evaluation and implementation. Not surprisingly, such deep differences in conceiving the role of environmental ethics are associated with differences of philosophical theory. This paper explores the philosophical beliefs and assumptions that shape the thinking of avowed applied philosophers.

PART 1: MONISM AND THE MISSION OF ENVIRONMENTAL ETHICS

Having noted that moral monism and applied philosophy tend to go together, I now explain the reasons that have convinced me that the search for a monistic ethic is intellectually, as well as practically, misguided in a profound sense. Monists are not simply wrong in that they have not yet proposed the correct universal principle, or because they have not quite successfully specified the precise boundaries of moral considerability in nature. No, I believe that the entire project of shoehorning all of our obligations regarding other humans and nature into a "monistic" system of analysis is the wrong strategy at the wrong time, given that it allows decisive intervention in public policy formation only after a single unified moral principle is articulated and agreed upon, an outcome that seems unlikely in the foreseeable future.

As noted above, monism is embraced by both environmental ethicists and economic theorists – both are equally "reductionistic" in this sense. In this paper I will focus on the dominant form of monism in environmental ethics – the large and diverse collection of theories that assert that nature has value independent of humans in some sense.[9] It may not at first be clear how such theories achieve monism, so I begin by tracing briefly the development of the idea of human-independent value in the writings of J. Baird Callicott. Callicott provides an excellent case study for several reasons. First, his position claims less than other non-anthropocentric theories in the sense that Callicott does not assert that human-independent values in nature are independent of human consciousness; he claims only that value in nature is independent of human *valuations*. So criticisms of this moderate view may apply equally to theorists who defend a more Radical independence of values in nature.[10] Second, Callicott has experimented with several different versions, or at least formulations, over the past fifteen years, and has readily noted shifts in his own thinking. By tracing Callicott's changing formulations of inherent value we can better understand the dynamic of a complex argument. Finally, Callicott has explicitly embraced monism, explained explicitly the sense in which he considers himself a monist and criticized pluralistic alternatives, providing us with greater insight into the nature and implications of monism as a moral mission.[11]

As a preliminary to this brief historical examination, it must be noted that Callicott and his monistic colleagues never question an underlying conceptual assumption, an assumption that lies at the heart of the assumed mission of applied environmental ethics: success in the axiology of environmental ethics must include as a centerpiece an answer to the question of *moral standing*: "What beings are morally considerable?" Given the project of applied philosophy, it is not surprising that non-anthropocentrists believe that, whatever monistic principle or theory turns out to be the correct one, this principle will fulfill two conditions: (1) The principle/theory must specify what objects in nature are morally considerable. Interestingly, success in this specification has been identified with the task of identifying which objects in nature "own" their own inherent worth, of which more below. (2) The principle/ theory must also provide some *motivation* for moral beings to protect natural objects. The universal underlying principle is that moral individuals act to protect inherent value, wherever it is determined to reside. Condition (1) ensures that environmental activists can identify which objects deserve moral consideration in any given situation. Condition (2) ensures that goals to protect inherent value are invested with moral gravity.[12] Any morally committed environmentalist ought always to act so as to maximize the protection of inherent value, wherever it occurs. Monistic, non-anthropocentric theory can on these conditions rival economists in universalism. This monistic principle is attractive to philosophers who hope to resolve environmental problems by throwing fully formed, general principles over the edge of the ivory tower, to be used as intellectual armaments by the currently outgunned environmental activists, to aid them against the economic philistines in the political street wars that determine the fate of natural environments.

## Callicott's Dilemma

This heroic version of applied philosophy's role in the policy process can only be realized if environmental ethicists, laboring in the tower, can agree on which principles to throw down to the streetfighters. And, if these principles are to exert moral force to protect the environment, they must be "objectively" supportable.

The measure of objectivity, on Callicott's view, is the extent to which the central theory of environmental values succeeds in attributing human-independent value to natural objects themselves. In the words of Callicott, the blue whale and the Bridger wilderness "may therefore, be said in a quite definite, straightforward sense to own inherent value, that is to be valued for *themselves*." Callicott goes on in the same paragraph to state that the institution of a "genuine" environmental ethic – one that recognizes inherent value in nature – provides the only defensible basis for the environmentalists' platform of social reforms: "Environmental policy decisions, because they may thus be based upon a genuine environmental ethic, may thus be rescued from reduction to cost-benefit analyses in which valued natural aesthetic, religious, and epistemic experiences are shadow priced and weighed against the usually overwhelming material and economic benefits of development and exploitation."[13] In this passage, Callicott commits himself to a good, old-fashioned realist interpretation of the problem of objectivity: "the very sense of the hypothesis that inherent or intrinsic value [sic: exists?] in nature seems to be that value *inheres* in natural objects as an intrinsic characteristic. To assert that something is inherently or intrinsically valuable seems, indeed, to entail that its value is objective."[14] What is interesting is that Callicott, having formulated the problem of objectivity in terms of representational realism, immediately retreats from asserting an objectivist solution. Instead he argues that his own Humean subjectivist solution asserts as much objectivity for claims that objects in nature own their inherent values as exists for scientific claims. So, Callicott's "ownership" theory of inherent value, which attributes to ecosystems their own inherent value, is offered to environmental activists as the fruits of his search for the Holy Grail of monistic ecocentrism.

Three comments are necessary. First, the general principle of "ecocentrism," so defined, hardly resolves the question of what beings in nature are proper owners of inherent value. The Bridger wilderness and the blue whale are given as examples, but they themselves represent different scales on the biological hierarchy, and Callicott owes his readers an account of the breadth to which he would generalize these examples. Second, as long as the first comment remains unanswered, Callicott cannot claim to have provided any definitive policy direction to activists because they can only know what they are obliged under the universal principles of non-anthropocentrism to protect after they know what particular entities in nature have inherent value. Third, Callicott, who wishes to interpret the land ethic as a moral theory, betrays an underlying commitment to moral individualism. He interprets Leopold's holism as attributing inherent value to ecosystems *as individuals* who can "own" their own goodness. But this conclusion only brings us to the heart of the matter – Callicott's original assumption that the land ethic is to be interpreted as monistic and holistic. It will therefore be necessary to look briefly at Callicott's changing definitions of ecocentric holism, and to question whether they express the kernel ideas of Leopold's land ethic.

In an important 1980 article Callicott established himself as the leading interpreter and proponent of Aldo Leopold's land ethic, and also caused an important alteration in the intellectual terrain on which the principles of environmental ethics were to be debated in the subsequent decade.[15] Callicott showed that, if one took Leopold's holistic pronouncements and arguments seriously, the land ethic was logically incompatible with extensionist ethics of animal rights advocates whose individualistic ethics are based in utilitarianism or rights theory.

Callicott's initial interpretation of Leopold's land ethic was therefore boldly holistic and strongly non-anthropocentric. He argued that ecological communities, not individuals, are the real locus of values in nature (as we have learned by drawing out the metaphysical implications of ecology) and that individuals have value in so far as they contribute to ecosystematic processes that support the community, leaving the clear implication that protecting inherent value in nature might involve sacrificing individual specimens of *any* species – including, presumably, human individuals – if those individuals threaten the "ecological integrity" of the biotic community. Not surprisingly, his position was attacked as brutal towards individual animals and apparently misanthropic. In particular, it was pointed out by critics that this reduction of all individual value to functional value in a larger whole smacks of fascism.[16]

Subsequently, admitting that he had left his holism unqualified in order to be provocative, Callicott offered a much more conventional view of our moral obligations to human individuals, other species, and ecosystems.[17] But Callicott revised the apparent implications of the bold holism of the 1980 paper without so much as questioning his earlier conclusions that the land ethic establishes the whole biotic community as a morally considerable being. What Callicott did instead was simply to specify how we should rank our obligations to various objects that have inherent value, theoretically and practically, in accord with the "communitarian" principles of the land ethic, taking communitarianism to imply that humans and other elements of nature make up one moral community among others. Nonhuman elements of nature, including species and biotic communities, have inherent value because they are morally considerable "owners" of their own value as members, along with humans, in the land community. He therefore explains why we may give precedence to obligations to family members or our human community over obligations to ecosystems: "we are members of nested communities each of which has a different structure and therefore different moral requirements . . . I have obligations to my fellow citizens which I do not have to human beings in general *and* I have obligations to human beings in general which I do not have towards animals in general."[18] Building on this gradation of obligations, Callicott extricates himself from charges of fascism thus: "our holistic environmental obligations are not preemptive. We are still subject to all the other more particular and individually oriented duties to the members of our various more circumscribed and intimate communities. And since they are closer to home, they come first."[19]

It is significant that Callicott shifts the grounds of the debate over fascism from obligations flowing from attributions of inherent value – the central source of normative obligation in his monistic theory – to the origins of special obligations that emerge in specific communities in biology, culture and ecology; he differentiates the obligations according to the intimacy of the community. There can be multiple criteria of right action – one stringent criterion of protectionism applicable in cases of parents to children and a less stringent moral criterion of protectionism that applies to the broader, ecological community. The commitment to moral monism recedes into the theoretical background as these special, community-based (and presumably not universal) obligations do the hard work of resolving conflicts that were introduced by generalizations of inherent value to species and ecosystems as well as individuals in human communities.

Callicott's 1980 formulation, which alarmed readers who wondered if persons and individual animals would be sacrificed to a single principle that apparently followed from his theory that all value originates in ecosystems, has given way more recently to an endorsement of theoretical monism rather than a monism of principles (as explained in note 1). Callicott therefore adopted a qualified version of holism that recognizes a plurality of rules applicable according to specific circumstances, explaining that pluralism on the level of principles of action is not inconsistent with monism on a more general, theoretical level in which various practical rules are unified and related to a single moral ontology.

Callicott's version of monism allows, he thinks, *both* unification under a single theory of value *and* flexibility in the formulation of rules. He claims a unified theory because he relates all obligations to a moral ontology in inherent value. Recognizing the subjective source of inherent value in human consciousness and in cultural ideas and institutions, however, the theory can nevertheless be elaborated in ways that are appropriate, depending on the special circumstances of the communities in which the obligations arose. But this answer to the charge of fascism – and one assumes Callicott must answer this charge if his theory is to be taken seriously – surely taxes the semantic elasticity of the concept of "inherent" value. Inherent value of natural objects, on this account, is due to "virtual" characteristics of objects of value, even though the specific, practical implications of the evaluations of these characteristics is ultimately determined by individual actors, according to the moral sensibilities of particular, independent, moral communities.

Ignoring these semantic difficulties, the upshot is that Callicott advocates allegiance to monistic inherentism in theory, but recognizes that the more intimate obligations of kinship and culture will usually outrank obligations to protect species and ecosystems. If this seems a capitulation to business as usual in environmental affairs, with inherent value reduced to a meaningless slogan, it must in fairness be said that Callicott faces a difficult and apparently destructive theoretical dilemma, for which he has thus far offered no resolution.[20] If moral inherentism is to provide the unified foundation for an environmental ethic, inherent value must be protected wherever it occurs. This apparently implies that any conflicts between a person's obligations to protect her children, for example, and her obligations to protect some inherently valuable biotic community, should strictly be determined by the obligation to maximize the protection of inherent value. But this would leave the theory open to charges of fascism, unless Callicott can prove that our obligations to persons could never, in principle, conflict with our obligations to ecosystems. Such a proof seems highly unlikely, so he chooses the other horn of the dilemma. While we have obligations to ecosystems as owners of inherent value and obligations to persons as owners of inherent value, the latter can override the former because of the special circumstances of the moral agent within the specific community in which those obligations arose.

But what are we to make of these auxiliary rules that resolve disputes when the interests of inherently valuable and morally considerable entities conflict? Can they, or can they not, be derived from the central, monistic theory of ecocentrism? If they can, it would seem that inherent value must come in grades, providing objective resolution of conflicts in interest among inherently valuable entities. But Callicott has never, to my knowledge, offered even a sketch of the required theory of gradations of inherent value or an explanation of how such gradations should inform our choices of what protectionist goals to give priority in action. If, on the other hand, these auxiliary rules cannot be derived from the central theory, we apparently have uncovered a most important class of moral quandaries that require that we step outside Callicott's complete and unified theory, negating any claim to monism and universality. Until this dilemma has been resolved, it appears that Callicott's modified monistic ecocentrism can tell us nothing about what we are morally obligated to protect.

While this criticism of Callicott's adventures in ecocentric holism has proceeded on a theoretical level, it is important to note that Callicott's decision to interpret the land ethic as monistic and ecocentric has had at least two important practical consequences for the development of environmental ethics. Callicott's early comments on the land ethic established the nonanthropocentric interpretation of the land ethic as the standard interpretation of Leopold's mature thought. The criterion has not, accordingly, been operationalized because, on this standard interpretation it embodies all of the ambiguities of non-anthropocentrism, as just listed; worse, the criterion has been used as a shibboleth by one side in the polarized situation described above, rather than as the powerful practical guide it could be, because this interpretation identifies the criterion with the idiosyncratic views of a small subset of scientists and the public. Second, Callicott and other non-anthropocentrists have used this interpretation to support a highly tendentious, and narrowing, definition of the field of study of environmental ethics. These two issues are discussed in the following two subsections.

### Non-Anthropocentrism and the Land Ethic

Callicott has interpreted what is perhaps the most important passage in the history of conservation thought, Leopold's famous "criterion" of right management, as monistic, holistic and non-anthropocentric in its philosophical commitments. Leopold said: "A thing is right when it tends to preserve the integrity, stability and beauty of the biotic community. It is wrong when it tends otherwise."[21] Callicott and the Deep Ecologists have taken the two sentences of Leopold's criterion to imply that the community or ecosystem is the *object of value* which conservationists should be attempting to protect. They have assumed, accordingly, that the ecosystem/community must, for Leopold, be *an object of value independent of human values*. This passage is read as an endorsement of the view that the ecological community has "integrity" understood as wholeness, *and therefore that ecosystems are moral subjects*. To be a moral patient, however, requires, given the objectivity requirement, non-human ownership. And ownership requires a moral subject. Under the influence of his commitment to ownership as the basis of moral considerability, and implicitly unliberated from the subject–object dichotomies of Cartesian dualism, Callicott reifies biotic communities as "moral subjects" who can own human-independent value. Callicott and his followers therefore interpret this passage as Leopold's definitive statement that communities themselves are loci of inherent, human-independent value that can be considered in competition with human values. To the extent that this has become the standard interpretation of Leopold's land ethic, environmental ethicists have encouraged environmentalists to understand this passage as an assertion of non-anthropocentrism.[22]

While I agree with Callicott's identification of "integrity" as the key concept of environmental ethics and management, we nevertheless differ strongly regarding how to interpret this conceptual centerpiece of the land ethic. Callicott believes that, by attributing integrity to the biotic community, *taken as a whole*, Leopold stepped across the line to non-anthropocentrism and declared his moral allegiance to the hypothesis that nature has inherent value. Our obligations to protect this integrity are "objective" in the sense that they originate in the integrity of whole agents/objects which are morally

57

considerable owners of their own value. Being constantly tempted to think of ecosystems as persons by the requirement that they are "owners" of their own value, it is natural also to think of them as objects capable of strategizing and having "interests" and "strategies" of their own.[23] Philosophers and ecologists, under the influence of this misguided, morally driven holism, have unfortunately failed to confront another ancient philosophical problem – how an organ can behave relatively independently, as an object in its own right, while at the same time being an organ that functions as a part of a larger organism. In philosophy, it can be called the problem of parts and wholes. In ecology, it can be called the problem of scalar dynamics. It is a problem that cannot get a hearing on the current assumption that the objects of moral attention in nature are necessarily "wholes."

Consider an alternative interpretation of Leopold's famous remark; consider it as a *practical* remark on the *proper focus of conservation management*, not as a philosophical statement of what objects in nature are of ultimate value. On this view, Leopold is in this passage summarizing his wisdom, drawing a broad, inductive generalization from the experiences of his long and varied career in environmental management, rather than asserting a moral "first principle." On this reading Leopold was making the ontologically less committed but none the less insightful point that, because of the complexity of the interrelationships in nature, and because there are so many different values exemplified in nature, the only way to manage to protect *all* of these diverse and pluralistic values is to protect the integrity of community processes (which support and sustain the individuals and species of which they are composed). The latter two guidelines – stability and beauty – are then interpreted as glosses on or specific criteria to employ – in our search for the sometimes elusive analogy of integrity. On this interpretation Leopold is not telling us *what to value* in nature, but rather telling us *what to protect* in our practical environmental management (given the diversity of values and scales involved).

On this view, then, Leopold is proposing an approach to management: he is defining right action in environmental use and management, rather than addressing the problem of standing. Further, he is strongly endorsing an integrative, systems approach to environmental management as the only way to encompass multifarious human goals as we manage a many-level, complex system which is our habitat. An advantage of this interpretation is that it unites the land ethic with Leopold's seminal managerial metaphor of "thinking like a mountain."[24] As I have argued elsewhere, the key idea behind the admonition to think like a mountain is the recognition of multi-scalar relationships in time and space and a prescription that citizens and environmental managers in a technological age must pay attention to longer-scale values embodied in the structures and processes of slow-changing systems, as well as the immediate and short-term values of economics.[25] But on my interpretation of Leopold, he is emphatically *not* committing himself to a moral ontology of inherent value; nor is he expressing fealty to a single, monistic principle or theory of value. I see Leopold as a moral pluralist who was struggling to integrate multiple values rather than as an axiomatic deducer of applications from some universal theory.

My interpretation, it can be argued, fits much better into the context in which the criterion is found in the essay "The Land Ethic." In the section immediately preceding the one in which the criterion is stated, Leopold outlined a systematic practical difference, cutting across all fields of resource management, between Group A and Group B conservationists.[26] Group A conservationists are primarily concerned with commodity production, whereas Group B conservationists, among whom Leopold counted himself, regard "the land as a biota, and its function as something broader. How much broader is admittedly in a state of doubt and confusion." Now Leopold often used the term "broader" to apply to the philosophical aspects of a problem, so I interpret this passage as explicitly choosing *not* to embrace any particular moral ideology. Leopold confidently made the separation between the two activist groups but, faced with an opportunity to make a pronouncement on the philosophical principles of the Group B movement, Leopold deferred and turned humble, admitting that the question of what their principles meant philosophically is "in a state of doubt and confusion." It would be rather odd, would it not, if he had changed moods in the very next section of the essay and explicitly endorsed a full-blown theory of inherent moral goodness which implies that ecological communities are morally considerable beings who own their own good?

## The Scope of Environmental Ethics

Non-anthropocentrists such as Callicott and Tom Regan, though they disagree strongly on what non-anthropocentric principle to apply, and which beings/subjects have it, agree that, if environmental ethics is to have a distinctive subject matter, the field must embrace some form of non-anthropocentrism. Regan asserts that, lacking a commitment to biocentrism, environmental ethics "collapses into an ethic for managing the environment [for human purposes].[27] Remarkably, this definition excludes problems of environmental justice within generations and problems of intergenerational fairness from the discipline of environmental ethics. Callicott nevertheless quotes Regan approvingly and, seeing non-anthropocentrism as the only route to rationally defensible values, he accepts Regan's tendentious definition of environmental ethics as the search for a non-anthropocentric ethic.[28] One suspects that Callicott's belief that the land ethic is non-anthropocentric must be at work in predisposing him to accept Regan's otherwise implausible definitional narrowing; one also suspects that this definition is encouraged by the unquestioned goal of monistic theory-building and by the mission of applied philosophy as it is understood by Callicott and other monists. Non-anthropocentrism is not, apparently, an empirical hypothesis; it is a principle that is established "philosophically" – independent of experience, because that is what is required if philosophers are to operate independently of real environmental problems and constraints to establish moral principles independent of management science.

## Ontology and Epistemology

Having examined a number of characteristics of applied philosophy by examining the ideas and practices of a leading proponent of that approach, and generalizing from them, I conclude this first part with one last look at the standard problematic of environmental ethics since its beginning as a recognized sub-discipline having a "distinctive" subject matter. Because of his prior commitment to monism, Callicott assumed that the discipline of environmental ethics must achieve two goals with a single theory. There is, of course, the problem of determining who, or what, has *moral standing*, as noted above. And there is the quite distinct problem of the *warranted assertibility* of environmentalists' pronouncements that certain policies should be instituted because they are morally required and trump mere preferences of consumers. His solution is to provide a realistic moral ontology in which there exist moral objects in nature as well as among human individuals, subjects capable of "owning" inherent value. But whereas the first problem might (although I doubt it) be usefully addressed by an ontological theory of moral monism, the second problem is essentially a question of warranted assertibility of environmentalists' claims to priority in certain cases. This is an *epistemological* problem; we require no *ontological* solution to it. Once we avoid Callicott's conflation and problematize monism, there is no reason to consider it obligatory that our theory of value will be (1) realistic and (2) designed to provide *both* a theory of value *and* an epistemological warrant for assertions of priority.

The single ontological solution to two distinct problems seems plausible to Callicott, I submit, because he is operating on a fundamentally Cartesian conception of knowledge and reality, a conception in which epistemological justification requires location of a causal antecedent of perception and knowledge that can be located in reality, independent of human perception. He accordingly attempts to establish the independent existence of environmental values by reifying ecosystems and making them owners of value. His single solution to the two problems is to recognize Cartesian moral subjects in nature as the owners of objective moral characteristics.

It is especially puzzling that Callicott attempts to reify ecosystems in this way, given that he clearly rejects the description of ecosystems as tending towards an inevitable and stable context. Callicott brilliantly shows that the lesson from ecology is that ecosystems are multi-layered, dynamic processes, not self-like, organismic wholes that seek a stable equilibrium.[29] Is it even plausible to say that multi-layered dynamic processes are owners of inherent value?

We can now see why Callicott's ontological, monistic interpretation of the integrity criterion and of the land ethic has led to ambiguities and paralysis. If the land ethic must be morally monistic, then to fulfill its unifying role in the non-anthropocentrists' moral ontology, it must consider ecosystems analogous to human organisms or at least Cartesian subjects who can "own" inherent value. But to answer the epistemological demand of activists for warranted assertibility, Callicott must claim that these owned values must have their existence independently of human valuation – they must be "objective" and exist independently of human evaluations. The theoretical dragon, "inherent value in nature," is necessarily two-headed, because it was created to slay two logically distinct monsters: ontological pluralism in value theory and moral skepticism. But the ontological solution of positing independent moral subjects in nature has unfortunately encouraged the organicist interpretation of ecological assemblages and led to the fascist tendencies of reified holism.

The difficulty in all of this, of course, is that organicism does have an important point to make. Organicists are correct that mechanistic models do not explain the ability of ecological processes to create, to sustain and to heal themselves. Ecological management requires that we accept two elements of organicism – the idea that the whole is more than the sum of its parts and the related idea that relationships among multi-scalar processes – not the static characteristics of objects – provide the key to understanding ecosystems. The problem is to express this idea in a way that does not carry us all the way to teleology and personalism. We must emphasize the creative nature of environmental processes, and the key role of energy flows in those processes, without personalizing them.[30]

My concern in this paper is not so much with epistemology[31] as with the ontology that should unify the underlying theory of environmental ethics.

Callicott's overall position represents a curious mixture of Cartesian modernism and postmodernism. He attributes to Leopold, and seems to accept himself, a Darwinian, dynamic, particularistic and postmodern viewpoint in *ethics*, while at the same time addressing the problem of warranted assertibility within a distinctly modern epistemology. Despite direct evidence that Leopold employed a pragmatist theory of truth,[32] Callicott assumes Leopold would have given a *realist* answer to the question: how can we justify our pursuit of environmental protection? I, on the other hand, see Leopold's many cautions about specifying the *purposes* behind creation and his numerous remarks that the exact interpretation of reality is "beyond language" as indicating that Leopold had at least the glimmerings of a postmodern conception of knowledge and objectivity, as well as morals.[33] Had Callicott placed Leopold's remarks regarding good environmental management in the context of Darwinian epistemology as well as Darwinian ethics, he would have conceived the "objectivity" problem very differently.[34]

If we focus for a moment on the problem of warranted assertibility of environmentalists' goals, it seems likely that environmentalists will achieve more by appealing to the relatively non-controversial and intuitive idea that the use of natural resources implies an obligation to protect them for future users[35] – a sustainability theory based in intergenerational equity – rather than exotic appeals to hitherto unnoticed inherent values in nature. Callicott argues that it is an advantage of intrinsic value in nature that, if it can be shown to exist, then it would shift the burden of proof in environmental arguments from environmental protectors to the despoilers.[36] But from the fact that such value might be *sufficient* to shift the burden of proof, it does not follow that it is the only, nor the best, means available to environmentalists to shift the burden.

The epistemological problem is that environmentalists need to be able to enter the public arena armed with genuine and defensible moral principles so that they can assert the priority of their goals over the mere preferences of the consumer society. As long as we can assert *other* morally binding obligations – such as an obligation to sustain the integrity and health of the ecological systems we are now damaging so that future generations can enjoy the bounties of intact ecological communities – we have a basis warrantedly to assert obligations to protect biodiversity over many generations. But these obligations are anthropocentric and cannot, apparently, be comprehended in a monistic non-anthropocentrism, even though abiding by these less controversial obligations would lead to most of the environmental protections favored by inherent value theorists.

PART 2: AN ALTERNATIVE TO ONTOLOGY

In our search for an environmental ethic we will never, I submit, find any environmental values or goals more defensible than the sustainability principle, which asserts that each generation has an obligation to protect productive ecological and physical processes necessary to support options necessary for future human freedom and welfare. The normative force supporting the protection of the environment for future generations should be based on a commitment to building just, well-adapted and sustainable human communities. Accepting responsibility for our expanding numbers and for the power of our technologies follows simply from the recognition that we now affect the productivity of the human habitat and the very survival of the human community. This responsibility becomes less and less escapable as we learn the many consequences, expected and unexpected, of our increasingly violent and pervasive alteration of natural systems.[37] This principle is consistent with a Darwinian emphasis on survival and complements a pragmatic conception of truth. The acceptance of both the facts of human impacts and the associated *moral responsibility* to protect the integrity of ecological communities as repositories of many human options and values in the future is destined, in the terms of Peirce, to be adopted as the conclusion of all rational inquirers, as they struggle through many experiments to make coherent sense of human experience. I believe that *both* the descriptive problem of understanding the impacts of our actions on future generations *and* our resulting responsibilities as moral beings must be addressed within processes of inquiry constitutive of the Peircean community of inquirers/actors. For example, considering species threatened with extinction to represent "books" of information – information that may be essential to future generations in their struggle to understand and act within a changing environment – seems to entail that the obligation to contribute to the process of inquiry requires protection of the sources of information and knowledge for future inquirers.

While the sustainability principle can give a certain unity to environmental action, this unity represents an open-ended direction, an appeal to learning and small-scale adaptation, community-building, and experimentation, not a slavish commitment to a priori principles of value such as monism and anthropocentrism. Thus this principle is unifying without being monistic – it sees the task of living sustainably as a problem of social learning, guided by a method that is socially open and scientifically experimental. This means that diverse stakeholder groups will be encouraged to assert and defend their own values and interests and to participate in social experiments in search of solutions that allow diverse users to fulfill their diverse needs with minimal disruption of the interests of other groups. The goal of value and policy discussions in a democracy should be inclusive, recognizing that the diversity of values humans and communities find in nature is only the first step towards an integrative policy that preserves differences; in this sense it differs from monisms, which demand that diverse groups formulate their values in a single vernacular or have their considerations excluded from discussion and possible consensus.

I propose a new beginning for environmental philosophy – both for environmental ethics and environmental epistemology – a beginning based broadly in the pragmatic epistemology of Charles Sanders Peirce, who replaced the failed project of representational and foundational realism with a constructivist method that recognizes that the correctibility of scientific inquiry must be fully characterized within human experience, not by reference to "external objects" that exist beyond experience.[38] Truth and objectivity must be sought in the specific characteristics of specific situations in which action is required. If environmental values can justifiably be asserted to have priority in some situations, the mark of this will be their eventual emergence in a complex process of inquiry, including diverse groups with diverse interests and viewpoints, that will submit both values and scientific hypotheses to discussion and testing. The relevant intellectual community is not philosophers with a distinctive subject matter, but the activist community that is committed to human survival and improvement. Knowledge and moral discussion must be understood as a part of the struggle to determine adaptable policies, rather than as a distinct "field" of theoretical morality. The inclusion of values as well as information in the process of inquiry can be traced to the experimental approach to social activism of John Dewey, who clearly rejected as folly the search for certainty and deductivism in moral and social matters.[39]

I therefore place my work in the tradition of "adaptive management,"[40] first introduced by C. S. Holling and developed by his colleagues in the Pacific Northwest and in Northern Canada. Adaptive management has recently been given a political formulation by Kai Lee, who explicitly and concretely bases his political analysis in the philosophy of John Dewey.[41] Lee also shows that the Deweyan approach is usefully complemented by the creative work of policy analysts on "bounded rationality," who have recognized that arriving at improved policies is often a matter of "muddling through" rather than a matter of establishing idealistic goals and instituting decisive and abrupt changes to achieve those goals.[42] The pragmatic approach recognizes that there is great uncertainty in both human knowledge and human valuations and attempts to nurture processes and institutions that seek compromise and incremental change and improvement of understanding and goals. In the process, both information and values will be adjusted to become more appropriate and adaptive to particular situations.

## *Nature as a Multi-Scalar, Open System*

Modeling a natural system is not a simple matter of choosing the hierarchy which best represents natural processes because, for any given ecological system, there are many – no doubt an infinite number of – models which will correctly *describe* some aspect of ecosystem functioning.[43] Further, because ecological systems are irreducibly complex, it follows that these models are not reducible to a single model without loss of descriptive content. I represent this essential complexity by using *hierarchy theory* (HT) – a theoretical approach to multi-scalar systems of analysis developed by theoretical ecologists such as Holling, T. F. H. Allen, and Robert O'Neill. Hierarchy theory represents a scalar application of general systems theory.[44] Systems are best understood as open, multi-scaled processes in which each system is viewed also as a sub-system on a more encompassing level of the system. Multi-scalar analysis has just as important a positive role in understanding environmental values and management goals as it does in describing natural systems; it is the scientific cornerstone of the adaptive management approach. Multi-scalar analysis is based on the neutral, model-constituting assumption that smaller sub-systems function on faster dynamics, and that all models that capture the multi-scaled nature of ecological and physical systems should embody this spatio-temporal correlation between greater extent in physical space and slower dynamics of change. The central idea of a multi-scalar, contextual approach to analysis and management is therefore the very simple idea that human activities, economic and otherwise, take place within, and affect, larger systems that should be interpreted as multi-scaled ecological processes. As noted above and elsewhere, this multi-scalar approach represents a more precise and operational formalization of Aldo Leopold's insight that we must learn to "think like a mountain," and therefore provides a connection backward to Leopold's seminal managerial insights and also points forward to new directions for scientific and policy research that proceed by organizing information and value studies into a spatio-temporally structured system of analysis.[45]

Most uses of HT so far have been purely descriptive. T. F. H. Allen and his co-authors, for example, argue that the only type of reason that can be advanced to prefer one scalar model of an ecological phenomenon over another is that the preferred model *enhances our understanding of the system* more extensively than do less preferred models. While they refer to this justification as "utilitarian," it is clear from the context that utility is understood by Allen and Starr as "scientific" or methodological utility, not general social utility.[46] By contrast, I also apply a normative filter of *usefulness in understanding and protecting social values* as well as a descriptive filter, reducing drastically the class of descriptive models that must be considered as relevant to policy. The goal should be to develop scalar models that improve our understanding in the specific sense that they illuminate environmental problems and allow us to focus on those natural dynamics that are causally related to important social values. Hopefully, these models also help us to define our management goals clearly and to measure success and failure in attempts to achieve those goals.

Conservation biology – and also the "pure" disciplines of zoology, botany, etc. – must adopt the twin goals of understanding *and protecting* ecological systems just as human and veterinary medicine have adopted the twin, descriptive-and-normative goals of understanding *and healing* their patients. Commitment to the Peircean ideal of eventually understanding nature apparently carries with it the obligation to protect the sources of biological knowledge, living organisms, especially from irreversible losses such as species extinctions. The descriptive models chosen in normative sciences must therefore pass a double criterion. They must help us to understand nature, but they must also encourage us to understand nature in a way that will help us to formulate and measure environmental goals effectively and to propose and implement policies to achieve those goals. The problem in linking biological science, social science and value theory into an adequate plan for environmental management is to choose from among this multitude of *descriptively* adequate models a few models that truly help us to understand *and to integrate* human activities into the landscape.

This *multi-scalar and biogeographic* approach to environmental values assumes at the outset that management will proceed from a human perspective and also that human values quite legitimately shape the modeling decisions of ecological and physical scientists. This latter point is deserving of further explanation: the decisions of biological and physical scientists have an unavoidable normative component. The point is not to *purge* science of those values, which is both impossible and undesirable; the point is to *understand and justify* those values in specific contexts requiring action, and to attempt to adjust them through public discussion and education when they become maladaptive.

A successful integrative ethic for the environment must be morally pluralistic, but it must also be contextual, rather than either objectivist or subjectivist.

Good environmental decisions are ones that take into account likely impacts on a number of spatio-temporal scales in specific contexts. As the world becomes more full of humans and as technology becomes more powerful, there will be more and more cases in which there will be spill-over impacts from one level of hierarchical organization to another, especially from our expanding economic and social systems to the natural systems that form their ecological context. Environmental policy and action must do more than enhance values in one dynamic, such as the dynamic driving the economic decision of individual farmers; it is necessary also to examine the impacts on the larger and usually slower-changing dynamic that determines the structure and diversity of the landscape. Here the focus of moral analysis turns to multiple generations and to the landscape scale.

The goal of an integrative ethic should be to sort the many and various values that humans derive from their environment and to associate these variables with real dynamic processes unfolding on the various levels and scales of the physical and ecological context of our activities. Environmental problems are in this sense essentially scalar problems and I seek to define models that illuminate the dynamics which support human values.

## A Tri-Scalar Model

Since the practical philosopher and the adaptive manager set as a goal of all model-building that the models inform environmental decision-making, and since we firmly believe that environmental decision-making must be democratic, any model for this purpose must be fairly simple in structure. The model must be a simple enough representation of multi-scaled natural processes to serve as an aid in public discussion of the goals of a forest management plan or a plan for ecological restoration of a river system. Prescriptive, multi-scalar models must therefore provide a publicly useful vocabulary for discussing environmental goals; they must shape our models of management by associating them with the temporal and spatial scales of the natural dynamics that generate the values guiding our choice of goals. The goal is to develop a spatio-temporally organized and ecologically informed phenomenology of the moral space in which individuals formulate and pursue personal and environmental values.[47] The problem, on this pragmatic approach, is to characterize and categorize pluralistic values in a way that is sensitive to spatio-temporal features – what I call the "scale" – of human interactions with nature. It is to design a method of inquiry, which includes both a descriptive and an evaluative component so organized as to increase the likelihood of both achieving truth (defined in a broadly Darwinian sense) and sustainability of fair and equitable communities, and of traditional values and cultural diversity for the benefit of humans today and in the future.

To initiate discussion, let me suggest three basic scales, each of which corresponds to a temporally distinct *policy* horizon:

1. Locally developed values that express the preferences of individuals, given the established limits and "rules" – laws, physical laws, governmental laws, and market conditions, for example – within which individual transactions take place;
2. A longer and larger community-oriented scale on which we hope to protect and contribute to our community which might be taken to include the entire *ecological* community;
3. A global scale with essentially indefinite time scales on which humans express a hope that their own species, even beyond current cultures, will survive and thrive. (See Figure 3.1.)

On the first scale, which unfolds in the relatively short term and local space in which individuals make economic choices, an economics of cost-benefit, if supplemented with a sense of individual justice and equity, can provide useful decision models. The middle scale, on which we feel concern for our cultural connection to the past and future, is especially important for two reasons. Viewed socially, this multiple-generational level is the one on which we protect, develop and nurture our sense of who we are as a culture. It is on this level that we "decide" what kind of a society we want to be. These "decisions" are expressed in our art, in our religion and spirituality, and in our governing political institutions such as the Constitution. It is on this scale that we feel concern about the culture's interaction with the ecological communities that form its context. This second scale is doubly important because it corresponds

| Temporal Horizon of Human Concern | Time Scales | Temporal Dynamics in Nature |
|---|---|---|
| Individual/economic | 0–5 years | Human economies |
| Community, intergenerational bequests | up to 200 years | Ecological dynamics/ Interaction of species in communities |
| Species survival and our genetic successors | indefinite time | Global physical systems |

Figure 3.1. Correlation of human concerns and natural system dynamics at different temporal scales.

roughly to the ecological time scale on which multiple generations of human individuals, organized into communities, must relate to populations of other species that share their habitat. Thinking intergenerationally apparently requires that we pay special attention to large-scale aspects of the landscape.

Modern ecological knowledge has forced upon us the conclusion that we must act as members of the natural as well as the human social community; it follows also that we must pay attention to the context in which our values are formulated and acted upon, and that context is the interaction between a culture and its habitat that is described in the "natural history" of a place. That natural history must reach back into time and project itself in a creative way into the future.[48] According to the multi-scalar, contextual view, humans necessarily understand their world from a given local perspective. A preference for localism is really a preference over preferences – a favoring of values that emerge from experience of one's home place. This home place locates the perspective from which one understands and values elements and processes in the natural context of our actions. Localism is represented in the proposed theory of environmental values by an endorsement of the importance of *sense of place values*. I therefore advocate many different locally originating sustainability ethics, each of which is anchored to a particular place by a strong sense of the history and the future of the place. In particular, a sense of place manifests itself, for example, in the defense of local determination and in an inclination of citizens to fight and defeat solutions imposed by centralized and authoritarian institutions.[49]

But a full-blown sense of place must include also a sense of the space around that place.[50] And here we invoke our multi-scalar phenomenology of environmental concerns as the evaluative space in which each citizen, as an individual, as a member of an ongoing cultural community, and as a member of the global community, must seek local solutions which reduce impacts on larger systems.

We can represent action-decision aspects of the phenomenological models I am describing as in Figure 3.2. Imagine, conceptually, that the surface of the earth is represented as many points or individual perspectives, each of which is tied by a cultural history to a human community and by a "natural" history to the land community. These individuals are understood as representative individuals who live in a community – their personal identity is therefore associated with that community. They are therefore not selfish only – they seek to project their culture by affecting future generations, though they unavoidably conceive the world from their own local perspective. On these assumptions, our individuals must view themselves also as members of a community of plants and animals, as well as a community of humans. They

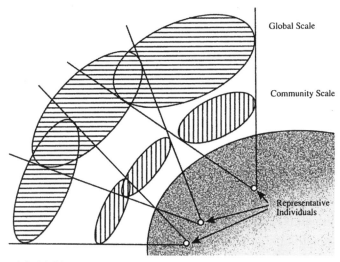

Figure 3.2. Multi-scalar relationship of individual, community, and global scales.

therefore experience, articulate and defend environmental values from a local perspective and from the present point in time. But impacts on the larger and long-term, intergenerational scale can also impact local and personal values if the ecological context changes so rapidly that traditional values and practices become meaningless in a few years. While individuals perceive values from a local perspective, those values are also shaped within a larger space in which there can be impacts on larger physical systems, which in turn constrain future choices. Note that greater population density and expansions in the scale of a community increase the likelihood of impacts from one community to another horizontally, across communities, and also vertically, across generations.

An implication of the multi-scalar approach outlined here is that the search for an acceptable environmental policy will not be a search for the policy that maximizes benefits to costs as measured in present dollars. A good environmental policy will be one that has positive implications for values associated with the various scales on which humans *are in fact* concerned, and also on the scales on which environmentalists think we *should* be concerned if we accept responsibility for the impacts of our current activities on the life prospects and options – the "freedom" of future generations.[51]

In a situation that recognizes multiple human values and associates these with various natural dynamics, it is possible to conceive, describe and seek policies that will protect or even enhance the processes maintaining ecological structures and processes that are crucial in future interactions of human and natural communities. The goal of policy is not to analyze and rank various policies with respect to how well they score on a single criterion, but rather to devise win-win policies that are robust enough to score highly on a number of relevant criteria of good management. Win-win policies, on the approach proposed here, are policies that have positive (or, failing that, neutral) results on all three "scales" of human concern – the individual welfare level, the community level and on the emerging values of the global community.

Consider an example which fulfills this robustness condition. In many deforested developing countries poor families must expend considerable effort to gather firewood for cooking and heating, as the scarcity of firewood can cause extended searches consuming many person-days per week per household. In such areas a creative environmental policy would institute many locally based tree-planting programs. Successful programs begin with small loans to private entrepreneurs who use the loan to purchase seedlings and for other start-up costs. Full payment to the entrepreneurs will occur when the trees reach a certain age or height, encouraging the entrepreneurs' clan to protect the trees so that they can reach the pay-off goal. If the trees are planted close together, they will provide increases to economic welfare within a few years as culled trees provide firewood nearer home. Meanwhile, local ecological processes should become more healthy as eroding land is replaced by forests, or at least small and diverse tree farms, improving water retention and improving streamwater quality. Finally, on the largest scale, the impact on current choices on the functioning of global physical systems, we can expect that birth rates will be reduced because peasant families will have less incentive to have children to help with household chores.

Figure 3.2 helps us to conceptualize a new criterion for acceptable and appropriate environmental action. I refer to the proposed criterion as advocating actions and policies that conform to the *scalar Pareto criterion*. The scalar Pareto criterion represents a multi-scalar application of the Pareto optimality criterion, which was originally stated on the individual level as the requirement that all actions have a positive impact on some individuals and negative impacts on nobody. The scalar application of the Pareto criterion is stated as follows: choose policies that, from the viewpoint of a representative individual in each community, will have positive (or at least non-negative) impacts on goals formulated by that person on the individual level, on the community level, and on the global level. While the scalar Pareto approach retains an individualist perspective (it is human individuals who formulate, discuss and defend values on all levels), it does not seek reduction of all values to economic preferences or to some generic form of "inherent value." It is *pluralistic* in the sense that the value of ecosystems is understood on a community, not an individual, scale and no reduction of community-level values

71

to individual values is attempted. But the pluralistic ethic is also *integrative* in the sense that we seek actions that will have positive (or at least non-negative) impacts on the relatively distinct dynamics that produce and support human values that are expressed on multiple scales (here hypothesized as three).

## CONCLUSION

Thus ends my explanation of, and plea for, a practical environmental ethic that seeks to *integrate* pluralistic principles across multiple levels/dynamics. Rather than *reducing* pluralistic principles by relating them to an underlying value theory that recognizes only economic preferences or "inherent" value as the ontological stuff that unifies all moral judgments, I have sought integration of multiple values on three irreducible scales of human concern and valuation, choosing pluralism over monism, and attempting to integrate values within an ecologically informed, multi-scalar model of the human habitat. I believe that the non-ontological, pluralistic approach to values can better express the inductively based values and management approach of Leopold's land ethic, which can be seen as a precursor to the tradition of adaptive management. And, if the problem of environmentalism is the need to support rationally the goals of environmental protection – the problem Callicott misconceived as the need for a realist moral ontology to establish the "objectivity" of environmental goals – then I endorse the broadly Darwinian approach to both epistemology and morals proposed by the American pragmatists. The environmental community *is* the community of inquirers; it is the community of inquirers that, for better or worse, must struggle, immediately as individuals and indefinitely as a community, both to survive and to know. In this struggle useful knowledge will be information about how to survive in a rapidly evolving culture and habitat. It is in this sense that human actors are a part of multi-layered nature; our actions have impacts on multiple dynamics and multiple scales. We humans will understand our moral responsibilities only if we understand the consequences of our action as they unfold on multiple scales; and the human community will survive to further evolve and adapt only if we learn to achieve individual welfare and justice in the present in ways that are less disruptive of the processes, evolving on larger spatio-temporal scales, essential to human and ecological communities.[52]

ACKNOWLEDGMENT. A slightly altered version of the first part of this paper appears in *Environmental Ethics* Vol. 17, under the title "Why I Am Not a Nonanthropocentrist."

NOTES

1. It is interesting, and perhaps not accidental, that the power and importance of the assumption that environmental ethics should be monistic in form were pointed out not by an environmental ethicist or a philosopher, but by a legal scholar. See Christopher Stone, *Earth and Other Ethics* (New York: Harper and Row, 1987).

   More recently, it has been noted that monisms come in several versions. See Peter Wenz, "Minimal, Moderate, and Extreme Moral Pluralism," *Environmental Ethics* 15 (1993): 61–74. Following J. Baird Callicott, I will restrict my discussion to "principles monism" and "theoretical monism," as explained in Callicott's "The Case Against Moral Pluralism," *Environmental Ethics* 12 (1990): 99–124. According to principles monism, there is a single principle that covers all moral quandaries, with principle understood as a moral standard sufficiently practical to imply a single correct action in every situation. Theoretical monism, by contrast, might employ more than one principle in different situations, but achieves monism on a theoretical level by providing an over-arching theory that explains and unifies the use of divergent principles in terms of a monistic theory. It seems reasonable to consider principles monism to be a special case of theoretical monism because, as in the case of some simple versions of utilitarianism, a single theory of value justifies a single principle applicable to all cases. This paper is directed at both principle and theoretical versions because of the reductionistic tendencies they share. For simplicity, I will refer only to the general form, theoretical monism, because arguments applicable to it will apply also to the special case of principles monism as well.

2. See Callicott, "The Case Against Moral Pluralism," op. cit., for a discussion of the strengths of monism as opposed to pluralism in environmental ethics.

3. I have discussed the "environmentalists' dilemma" in detail in *Toward Unity Among Environmentalists* (New York: Oxford University Press, 1991).

4. See B. Norton, "Applied Philosophy vs. Practical Philosophy: Toward an Environmental Policy Integrated According to Scale," in Donald Marietta and Lester Embree, eds, *Environmental Philosophy and Environmental Activism* (Totowa, NJ: Rowman and Littlefield, 1995), for a more detailed discussion of the failure of non-anthropocentrism to address real management problems.

5. See J. Baird Callicott, "Environmental Philosophy Is Environmental Activism: The Most Radical and Effective Kind," in Marietta and Embree, *Environmental Philosophy and Environmental Activism*, for a brief exposition and aggressive defense of applied philosophy and the role he sees for it.

6. Interestingly, Callicott asserts (ibid., p. 25) that applied philosophy does "rather little deducing of specific rules of conduct." Rather, he sees the role of philosophy as one of "articulating and thus helping to effect ... a radical change in outlook. The specific ethical norms of environmental conduct remain for the most part implicit – a project postponed to the future or something left for ecologically informed people to work out for themselves." Despite this well-placed skepticism regarding whether environmental ethicists can get much mileage out of abstract philosophical theories, Callicott never doubts that his monistic theories of value are important in effecting a change in consciousness of citizens and capable of bringing about a new environmental era: "Therefore, since human actions are carried out and

find their meaning and significance in a cultural ambience of ideas, we speculative environmental philosophers are inescapably environmental activists" (p. 33). So, whether deduction is involved or not, the applied philosopher's contribution is to provide thoughtful persons with reasons or some form of motivation to adopt a new "cultural ambience," which includes his particular monistic principle. Only after persons (at some time in the future) agree on the existence and meaning of non-anthropocentric monistic value can the principle begin to provide guidance in specific situations.

7. This point shows how badly Callicott misunderstands the anti-monistic position, which he refers to as "anti-philosophical." "Environmental Philosophy Is Environmental Activism" (p. 5). I, at least, am not anti-philosophical; I am against philosophical theory which is developed independently of real-world problems and I reject the role for theory assumed by Callicott and other monists and applied philosophers. Callicott mistakes an attack on *outmoded, modernist theory* for an attack on all theorizing.

8. See Bryan G. Norton, *Toward Unity Among Environmentalists* (New York: Oxford University Press, 1991).

9. See Bryan G. Norton, Papers 11 and 12, this volume, and "Thoreau's Insect Analogies: Or, Why Environmentalists Hate Mainstream Economists," *Environmental Ethics* 13 (1991): 235–261 for criticism of welfare economics as a similarly inadequate monistic theory of environmental valuation.

10. Such as Holmes Rolston III, *Environmental Ethics: Duties to and Values in the Natural World* (Philadelphia: Temple University Press, 1988), and Paul Taylor, *Respect for Nature* (Princeton, NJ: Princeton University Press, 1986).

11. Mainly in "The Case Against Moral Pluralism" and "Moral Monism in Environmental Ethics Defended," *Journal of Philosophical Research* XIX (1994): 51–60.

12. I use this vague phrase, "moral gravity," because Callicott has recently admitted that, because of the subjectivist foundations of his philosophy in David Hume, he cannot claim to offer a theory that exerts "moral force" (the more usual formulation), but only a "moral dimension." See "Can a Theory of Moral Sentiments Support a Genuinely Normative Environmental Ethic?" *Inquiry* 35 (1992): 183–198.

13. J. Baird Callicott, *In Defense of the Land Ethic* (Albany, NY: State University of New York Press, 1989), p. 163.

14. Ibid., p. 160.

15. J. Baird Callicott, "Animal Liberation: A Triangular Affair," *Environmental Ethics* 2 (1980): 311–338. Reprinted in Callicott, *In Defense.*

16. Tom Regan, *The Case for Animal Rights* (Berkeley, CA: University of California Press, 1983), p. 362.

17. Callicott, *In Defense*, pp. 55–59; 93–94.

18. Ibid., pp. 55–56.

19. Ibid., p. 58.

20. I first pointed out the difficulty of this dilemma in "Review of *In Defense of the Land Ethic*," *Environmental Ethics* 13 (1991): 185. As far as I know, Callicott has not responded in print or verbally to this dilemma, which apparently requires some response if we are to understand what is meant by monistic inherentism.

21. Aldo Leopold, *A Sand County Almanac and Sketches Here and There* (London: Oxford University Press, 1949), pp. 224–225.

22. There have, however, been dissenting opinions regarding the non-anthropocentrism of the land ethic. See Scott Lehmann, "Do Wildernesses Have Rights? *"Environmental Ethics* 3 (1981): 129–146; paper 1, this volume; and Norton, *Toward Unity.*

23. See, for example, Eugene Odum, "The Strategy of Ecosystem Development," *Science* 164 (1969): 262–270. I have criticized this form of strong organicism in "Should Environmentalists Be Organicists?" *Topoi* 12 (1991): 21–30.

24. Leopold, *A Sand County Almanac*, pp. 129–133.

25. See Bryan G. Norton, "Context and Hierarchy in Aldo Leopold's Theory of Environmental Management," *Ecological Economics* 2 (1990): 119–127; Norton, *Toward Unity.*

26. Leopold, *A Sand County Almanac*, pp. 221–222.

27. Tom Regan, "The Nature and Possibility of an Environmental Ethic," *Environmental Ethics* 3 (1981): 20.

28. Callicott, *In Defense*, p. 157; J. Baird Callicott, "Rolston on Intrinsic Value: A Deconstruction," *Environmental Ethics* 14 (1992): 130–131.

29. Callicott, *In* Defense, 107–112.

30. See Norton, "Should Environmentalists Be Organicists?"

31. See Norton, "Epistemology and Environmental Values," *Monist* 75 (1992): 208–226, for a detailed criticism of the epistemology of intrinsic value theory.

32. See Norton, paper 1, this volume for an argument that the land ethic has explicit roots in pragmatism.

33. See A. Leopold, "Some Fundamentals of Conservation in the Southwest," *Environmental Ethics* 1 (1979): 131–141; also see paper 1, this volume.

34. See Michael Ruse, *Taking Darwin Seriously: A Naturalist Approach to Philosophy* (Oxford: Blackwell, 1987), for a survey and convincing defense of Darwinian epistemology.

35. See E. B. Weiss, *In Fairness to Future Generations* (Tokyo, Japan, and Dobbs Ferry, NY: The United Nations University and Transnational Publishers, Inc., 1989). Weiss shows that every major world religion, and many minor ones as well, assert that the use of resources gives current generations an obligation to pass the resources on to future generations.

36. Callicott, "Environmental Philosophy Is Environmental Activism."

37. See A. Leopold, "A Biotic View of Land," *Journal of Forestry* 37 (1939): 727–730.

38. C. S. Peirce, *The Philosophical Writings of Peirce* (New York: Dover, 1955), especially pp. 21, 39.

39. See, especially, John Dewey, *The Public and Its Problems*, in *John Dewey: The Later Works, Volume 2: 1925–1927*, edited by Jo Ann Boydston (Carbondale, IL: Southern Illinois University Press, 1984), pp. 235–372.

40. The adaptive management approach was introduced by C. S. Holling, ed., *Adaptive Environmental Assessment and Management* (New York: John Wiley & Sons, 1978). Also see Carl J. Walters, *Adaptive Management of Renewable Resources* (New York: Macmillan, 1986), and, especially, Kai N. Lee, *Compass and Gyroscope: Integrating Science and Politics for the Environment* (Covelo, CA: Island Press, 1993).

41. Lee, *Compass and Gyroscope*, pp. 91–115.

42. Herbert Simon, *Administrative Behavior* (New York, Macmillan, 1954), and Charles Lindblom, "The Science of Muddling Through," *Public Administration*

*Review* (1959): 79–88. The present volume at last provides a discussion of the philosophical principles of pragmatism in connection with environmental policies and values.

43. See Simon Levin, "The Problem of Pattern and Scale in Ecology," *Ecology* 73(6): 1943–1967, for the scientific argument for this conclusion.

44. T. F. H. Allen and T. B. Starr, *Hierarchy: Perspectives for Ecological Complexity* (Chicago: University of Chicago Press, 1982), and R. V. O'Neill, D. L. DeAngelis, J. B. Waide and T. F. H. Allen, *A Hierarchical Concept of Ecosystems* (Princeton, NJ: Princeton University Press, 1986).

    Despite connotations sometimes associated with the term "hierarchy," it is important to understand that there is no implication that higher levels of the system dominate, or should dominate, lower levels. In fact, the processes described in hierarchical systems analysis exhibit communication both upward and downward through the hierarchy, and for this reason we prefer the more neutral terminology "multi-scalar analysis."

45. This use of hierarchy theory differs dramatically from recent applications of the theory by environmental ethicists Cheney and Warren, who have used the theory negatively to attack Callicott's uses of ecological theory. They argue that, because hierarchy theory does not claim ontological priority for any of the hierarchical models it proposes, the efforts of Callicott and others to use ecological theories to support "moral ontologies" is a doomed project. See Karen J. Warren and Jim Cheney, "Ecosystem Ecology and Metaphysical Ecology: A Case Study," *Environmental Ethics* 15 (1993): 99–116. But I believe hierarchy has much stronger potential as a useful method for dealing with troubling scalar issues in the discussion of environmental information, values and goals positively. See Norton, "Scale and Hierarchy in Aldo Leopold's Land Ethic."

46. See Allen and Starr, *Hierarchy: Perspectives for Ecological Complexity*, op. cit., p. 6. See also T. F. H. Allen and Thomas W. Hoekstra, *Toward a Unified Ecology* (New York: Columbia University Press, 1992), pp. 31–35.

47. B. Norton and Bruce Hannon, "Environmental Values: A Place-Based Theory," *Environmental Ethics* 19 (1997): 227–245.

48. Rolston, *Environmental Ethics*.

49. Paper 19, this volume.

50. Yi-Fu Tuan, *Space and Place: The Perspective of Experience* (Minneapolis, MN: University of Minnesota Press, 1977); Tuan, "Man and Nature," Commission on College Geography, Resource Paper 10, Association of American Geographers, Washington, DC, 1971; and Norton and Hannon, "Democracy and Sense of Place Values."

51. See Allen and Starr, *Hierarchy: Perspectives for Ecological Complexity*, p. 15, for a definition of "freedom" in this sense, a sense that can be understood within hierarchical models.

52. I gratefully acknowledge that my understanding of these problems has been improved by countless discussions, in formal and informal situations, with Baird Callicott. I have learned much from our discussions, and have concluded that Callicott and I agree on most practical issues of management, but that we cannot agree regarding the theoretical foundations of environmental ethics. I do not, of course, expect that Callicott will agree with my arguments or my approach, but

I hope this opposing viewpoint will carry forward the discussion of the proper role and mission of environmental ethics into the larger community of scholars.

It should also be said that Callicott has, on occasion, recognized the insidious associations of monistic inherentism when associated with a modernist conception of scientific objectivity with Cartesianism and he has even suggested that achievement of a truly adequate postmodern conceptualization of nature and value would require a different formulation. If I understand this postmodern version of Callicott's philosophy, it still includes a central role for inherent value. See *In Defense*, especially pp. 165–174, for a tentative trying-out of this alternative position. The arguments of this paper, however, have not been directed at the postmodern, non-Cartesian version of inherent value, but at the version defended in the vast majority of Callicott's published writings.

# 4

## Convergence Corroborated

### A Comment on Arne Naess on Wolf Policies

I have argued elsewhere that Deep Ecology, understood as a social movement seeking to redress currently unacceptable environmental policies by embracing the belief that nature has intrinsic value, has failed to articulate policies that are both plausible and significantly different from the policies implied by a broad and long-sighted anthropocentric viewpoint.[1] The convergence hypothesis, which I have offered as an alternative to the traditionally divisive characterization of environmentalists as split between "shallow", anthropocentric, resource managers and "deep", nonanthropocentric, environmental radicals, states that, *provided anthropocentrists consider the full breadth of human values as they unfold into the indefinite future, and provided nonanthropocentrists endorse a consistent and coherent version of the view that nature has intrinsic value, all sides may be able to endorse a common policy direction.*[2]

The convergence hypothesis is a general empirical hypothesis *about policy* – it claims that policies designed to protect the biological bequest to future generations will overlap significantly with policies that would follow from a clearly specified and coherent belief that nonhuman nature has intrinsic value. The convergence hypothesis is a contingent truth, a very general empirical hypothesis which shapes solutions sought by adaptive managers in particular situations. It is supported by facts both directly and indirectly; it could be falsified, but so far it has not been. The purpose of this "Comment" is to test that hypothesis through an examination of Arne Naess' thoughtful and interesting discussions of policies appropriate in "mixed communities" – communities that include both humans and large predators.[3]

If we are to understand the general importance of Deep Ecology, and especially its possible and likely contributions to policy, it is important to realize

From Nina Witoszelt and Andrew Brennan, eds., *Philosophical Dialogues: Arne Naess and the progress of philosophy* (Lanham, MD: Rouman and Littlefield, 1999). Reprinted with permission.

that the Deep Ecology philosophy of environmentalism, and the movement it has spawned, only make sense against a backdrop of empirical beliefs about environmentalists and their actions. I will call these background beliefs the "Divergence Theory of Environmentalism," for reasons that will become clear presently. The Divergence Theory is composed of two related elements. First, it includes a philosophical *definition*, elevated to the role of a metaphysical principle, marking the difference between anthropocentric and nonanthropocentric value systems. Second, the Divergence Theory uses this philosophical definition as an *explanation* of the behavior and policy advocacy of environmentalists, positing that there are two competing movements with discernibly different policy programs clustered around these different and exclusively defined systems of ultimate values. The philosophical theory is attributed importance in policy discussions because it *explains* what is taken to be an observable phenomenon – the clustering of environmentalists around two exclusively definable paths of policy advocacy. Environmentalists gravitate toward ultimately incompatible policy programs, according to the Divergence Theory, because they are split between those who accept and those who reject anthropocentrism.

Note that this is a distinctive application of the theory, originally due to the historian Lynn White, Jr., that Western societies have been so rough on the environment because our basic philosophical and religious assumptions are "anthropocentric."[4] Although White himself elaborated this point very little, philosophers, including Naess, subsequently developed the theoretical distinction between anthropocentric and biocentric systems of value, and biocentric egalitarianism became a formative principle in Deep Ecology. According to White's argument, *the whole of Western society* has been anthropocentric (recognizing only a few outliers in history such as St. Francis), and the philosophical distinction between anthropocentric and nonanthropocentric value systems was introduced to explain a society-wide phenomenon – that Western *societies* were failing to protect their environments. But the Divergence Theory's application of the anthropocentric–nonanthropocentric distinction does not explain the aggregated policy of the whole society, but rather explains a difference in policy programs among factions within a relatively small subgroup of society – the environmentally active.[5] I am not saying that this is a totally implausible application of White's hypothesis. It is arguable that Shallow Ecologists, though they call themselves environmentalists, advocate policies that are actually closer to those of advocates of unlimited economic growth than they are to the real advocates of environmental responsibility, the Deep Ecologists.[6] Indeed, there is historical precedent for this interpretation in that early in the environmental movement in the

United States there emerged two somewhat distinct groups – later labeled as "preservationist" versus "conservationist" – which advocated quite different policy programs.[7] So the Divergence Hypothesis remains arguable, though I reject it for reasons summarized elsewhere.[8]

But my point here is that it is an assumption about the *policy situation* – an empirical belief that environmentalists today can reasonably be sorted into two categories exhibiting predictably different policy *behaviors* that must ultimately give interest to the Deep–Shallow distinction in the policy arena. If there is no clustering of the policy proposals of environmentalists into two separable approaches advocated by reasonably stable groups of advocates who serve opposed ultimate values, then there is nothing for the Deep–Shallow distinction to explain, and the entire philosophical idea of separable and competing "Deep" and "Shallow" movements simply collapses for lack of interest. The question is not simply: Can we define, philosophically, a distinction between anthropocentric and nonanthropocentric value systems? The question is whether this distinction has *a useful application to the environmental movement as it exists today*. And if one believes, as I tried to show in *Toward Unity Among Environmentalists*, that it is false that American environmentalists today sort cleanly into two policy camps, the Deep–Shallow distinction collapses into a philosophical distinction looking for a policy difference to explain.

The Divergence Theory is not just an indirect expression of a false empirical hypothesis, however. The philosophical definition associated with it also implies a destructive heuristic; it urges us to address environmental problems by looking for *differences* between anthropocentrists and nonanthropocentrists. According to this heuristic, we enter policy discussions assuming that policies which are good for people will be damaging to the rest of nature, rather than assuming that there are policies which would be better for both. And it encourages us to attempt to implement policies by attacking the ultimate values of opponents, rather than by seeking policies that support the whole range and plurality of environmental values.

By arguing against the Divergence Theory, I do not, of course, deny that environmentalists disagree on many specific policies in many situations. What I deny is that these disagreements are meaningfully or consistently explained by attributing opposed ultimate values to the shifting groups and coalitions that weigh in on various policy discussions. Further, I believe that, underlying the cacophonous voices of environmentalists in disarray, there is emerging a common policy voice, including common explanations and justifications, that can be shared by anthropocentrists and nonanthropocentrists, despite their differences about ultimate values. If either of these empirical beliefs is correct, it is simply not true – as advocates of the Deep–Shallow distinction must believe if their distinction is to be more than a philosophical curiosity – that the positions of anthropocentric environmentalists are closer to those of development-oriented economists than they are to those of Deep Ecologists and other nonanthropocentrists.

Given my very different understanding of the situation among activist environmentalists, I prefer the "Convergence Theory," which states that as anthropocentrists articulate more broad-based and long-term policies, these policies will converge with the policies of Deep Ecologists and other nonanthropocentrists, provided that the latter adopt qualifications and clarifications necessary to achieve plausibility of their policy claims.[9] Admittedly, this theory has a somewhat curious status. It is *both* an empirically falsifiable predictive hypothesis about what will happen as environmentalists continue dialogue and advocacy and, on a deeper level, an article of faith that humans, being part of nature, share more common than divergent interests with the other elements of nature, and that the search for better policies will in fact benefit both humans and wild nature.[10] As an advocate of the Convergence Theory, I reject the Divergence Theorists' *factual belief* about environmentalists' behavior in the policy arena, so I find their philosophical definition/principle *uninteresting* as applied in the policy arena, because the empirical situation it is supposed to explain does not exist. It has the same validity as theories of witchcraft. Both are theories to explain the existence of groups of people who act according to behaviorally specifiable patterns; in both cases, however, attempts to establish empirically the existence of such behaviorally specified groups have failed.

Notice that it would be remarkable if it turns out that there is *no* convergence between what is good for humans and for other species. Is it at all surprising to claim that human activities that threaten other species are likely to pose threats to human beings? The evolutionary kinship of all the species has been a theme in biology since Darwin. Leopold, Carson, and many other environmentalists all operate on the expectation that caring for human well-being broadly understood usually converges with caring for the well-being of other species.[11] We share our evolutionary history with other species, especially the ones that are closely related to us, and it is hardly surprising that we share many of their vulnerabilities.

The convergence hypothesis implies, as noted above, important and falsifiable predictions. For example, amphibians around the world are disappearing at an alarming rate. Suppose that we learn the cause(s) of this demise of amphibians with some confidence. The convergence hypothesis predicts that, once that complex of causal processes is understood, those processes that threaten amphibians are more likely than random to eventually have negative impacts on humans. If in ten years we understand that a complex of processes is causing amphibians to disappear, the convergence hypothesis predicts that more of these processes will be harmful to humans than will be benign.

Let us then proceed to test the two opposing theories/hypotheses by examining Naess's own recommendations for policies toward bears and wolves in mixed communities. If the divergence hypothesis is correct, it should be possible to describe distinct policies, one of which is supported by consistent application of Naess' philosophy/argument form, and another set of policies that are supported by appeal to anthropocentrists, and these policies should be clearly distinct and not overlapping. If, on the other hand, one finds a continuum of policy positions and a variety of philosophical viewpoints and justifications, but a tendency to find compromise positions that may be acceptable to both anthropocentrists and nonanthropocentrists, then the example of Naess' wolf and bear policies tends to confirm the convergence hypothesis.

The policies advocated by Naess and Mysterud include a) continued attempts to reduce mortality of livestock by carnivorous wild predators, including, where possible, subsidized cooperatives to watch sheep more closely than is economically feasible otherwise; b) continued attempts to encourage local communities to find ways to gain economic advantage from the presence of wolf populations (because experience shows that local communities will tolerate otherwise hated animals if they have economic value); and c) expanded programs to compensate particular farmers who suffer livestock losses, funded by the nationally based advocates of wolf protection. When all else fails, Naess and Mysterud recognize that it may be necessary to designate some areas as agricultural areas and others as non-agricultural, allowing wolves to be removed from agricultural areas. They would clearly regret it if this process resulted in the elimination of all wolf areas from Norway, but they nevertheless discuss whether this outcome is consistent with Norway living up to its international agreements: "Would norms of international solidarity to some degree be satisfied if Norway contributed heavily to the maintenance of protected European (e.g., Swedish or Finnish) wolf areas that have a significantly less dense human population inside wolf ranges? Our tentative answer is 'yes'" (p. 32).

It seems, to me (as an anthropocentric advocate of wolf protection and reintroduction), that these policies are about the best one can articulate, given the many conflicting values involved in these situations. One area where I would perhaps differ with Naess and Mysterud has to do with their rather lenient interpretation of international law. When an international treaty designates a species for protection, it, in essence, declares special protection for that species, protection that should not be so easily balanced against other values, such as economic values or people's sensibilities toward sheep.[12] When international law declares a species for special protection, that special designation implies that further losses to that species are not "compensable" across borders or across generations. I believe that the loss of wolves and bears from Norway would be a terrible loss *to future Norwegians*. Members of future generations of Norwegians, if they feel profoundly the loss of wilderness experiences, might judge harshly those who have sacrificed their birthright of wildness for a few sheep.

So the question of what to do to protect bear and wolves turns out to be far more complicated than deciding what has intrinsic value. Worse, determining that the suffering of both sheep and wolves is intrinsically bad does nothing to tell us what to do when human activities have made it inevitable that either sheep or wolves must suffer. This quibble aside, I believe Naess and Naess and Mysterud arrive at policies that could easily be supported by anthropocentrists – they are compromise policies which attempt to protect wolves while being as little disruptive of established human communities as possible. It will not matter, if I am correct in this interpretation, whether wolf advocates value wolves intrinsically or if they value them for future generations, arguing that we ought not to destroy forever the possibility of humans having the experience of hearing a wild wolf howl.

Nevertheless, as Naess and Mysterud – probably correctly – conclude, we must give great weight to the psychological unfairness of preventing well-meaning and sensitive shepherds from protecting their highly valued sheep, sheep that are valued economically but also as the main support for a distinctive, and valuable, human culture that has survived for centuries. At the same time, we must recognize that wildlife belongs not only to that particular "time slice" of a community that currently occupies a place, but to a multi-generational community including persons not yet born. Here I think we can find a new common ground for the environmentalism of the future: the strongest element of Naess' work on wolf policies is his emphasis on human communities and the special status we must accord the sensitivities of traditional communities who have lived for generations – in this case, centuries – in a geographic region. This emphasis directs us toward the important issue of multi-generational communities and cultural institutions that create sustainable societies. As Naess realizes, there is an inherent tension between the attitude of more cosmopolitan populations, which value wolves while not having to live with the day-to-day consequences of sheep attacks – and who exert their environmental ideals through central governments – and

local people. These locally developed sensibilities and deeply felt responsibilities to protect sheep must be treated with the utmost respect, even by
those who consider themselves as more progressive, more forward-looking,
and more cosmopolitan in their values and attitude. Values and environmental problems understood in this way apparently reflect values that exist at
different spatio-temporal scales of the society.[13]

Resolving these disagreements and tensions will not be easy – but my point
is that these problems are unlikely to be resolved by reference to universal
principles such as biospecies egalitarianism.[14] Despite my general respect for
local values – shared with Naess and Mysterud – I would argue that in this
case, the local people should – again, as Naess and Mysterud suggest – be
pushed to change somewhat in the direction of wolf protection. My reasoning
would not be based on equality of species, but on the importance of holding
opportunities and options open for the future, and on the necessity in some
cases for environmentalists to be advocates for people not yet born, when
the interests of the future clash sharply with current attitudes of some communities. Too often, local communities have acted on the basis of short-term
interests, only to learn that they have irretrievably deprived their children of
something of great value. So, as in the case of the temporarily depressed
young adult who seeks to end his or her life, I would suspend the usual right
of self-determination of local communities in the hope that a period of regulation and pressure from the centralized government – pressure to avoid
irreversible outcomes like wolf extinction – will lead the local community to
see the value of cohabiting with wolves. If the local community recognizes
the value of wild wolves, they may resolve to protect the full complement
of their indigenous wildlife as treasures that should be passed on to future
generations of the culture. During this period of regulation, agencies of the
central government and national environmental groups can attempt, through
public education and perhaps economic incentives, to establish local support
for predators. This anthropocentrically based but long-sighted policy program
therefore converges with that of Naess and Mysterud.

So I conclude that Naess' position on wolf policy, despite the fact that the
Deep – Shallow distinction suggests a strong divergence in policies of advocates of differing ultimate values, in fact converges with that of a broadly
understood and long-sighted anthropocentrism. Naess' thoughtful policy proposals therefore corroborate my view that anthropocentrists concerned with
the entire range of human values, over many generations, will converge with
the viewpoints of nonanthropocentrists. The underlying source of this shared
policy direction rests not in a revolutionary theory of intrinsic value but in
a concern for sustainable communities. If we believe in the wisdom of local

cultures, and of their commitment to perpetuate their culture across time, then improving wolf policies – practically speaking, as Naess and Mysterud recognize – must be understood as a dialogue across generations, not between humans and nonhumans. The way to protect wolves in mixed communities is not to convince local sheep owners that the wolves have equal value with them or their sheep, but rather to convince them that the protection of wolves – and all the possibilities and options for experience and use of wolves that would be protected with them – will, in the true "Deep, Long-Range Ecological Movement," be in the interest of their offspring and their evolving culture.

NOTES

1. *Toward Unity among Environmentalists* (New York: Oxford University Press, 1991). While I realize that the interpretation I offer and criticize is much more simplistic than the entire body of Arne Naess' writings, I believe it is not an unreasonable interpretation of Naess' original idea, outlined as a list of seven statements of principle in the classic paper, and introduced as follows: "A shallow, but presently rather powerful movement, and a deep, but less influential movement, compete for our attention." Naess proceeds to characterize the two competing movements with lists of principles. The second principle of Deep Ecology is "biospherical egalitarianism – in principle" (quotation is from "The Shallow and the Deep, Long-Range Ecological Movement," *Inquiry*, 16 [Spring, 1973], pp. 95–100). I take this principle to be, in an important sense, the defining one; if all species are morally equal, and if humans have inherent value, natural objects have inherent value, as well. It is this expanded sense of value that changes one's value orientation, gives depth to the other principles, and inclines one to join the Deep Ecology movement rather than the Shallow Ecology movement.

    When I speak of Deep Ecology, I therefore refer to a movement that would never have existed had its founder not advanced the idea that an *important dichotomy* exists among environmentalists – some of whom are "shallow" and some of whom are "deep" – and which implicitly criticizes environmentalists who employ a broadly anthropocentric analysis to environmental problems for their failure to attribute nonanthropocentric intrinsic value to nature. If Deep Ecologists mean only to say that there are many ways to question our current abominable treatment of natural systems, and that we should strive to question existing assumptions as deeply as possible, then there is of course no controversy. But this version of Deep Ecology involves neither dichotomy nor exclusion. If Deep and Shallow Ecologies are not competing but complementary, most environmentalists would agree that we should think as deeply as necessary; so I guess it follows that all environmentalists are Deep Ecologists. I wish only that Deep Ecologists would – if that is all they mean to say – renounce (not just change labels of opponents from "shallow" to "reform") their exclusionary and insulting framework of analysis.

2. Bryan G. Norton, "Conservation and Preservation: A Conceptual Rehabilitation," *Environmental Ethics* 8 (1986), pp. 195–220; in *Toward Unity Among Environmentalists* (New York: Oxford University Press, 1991), especially pp. 237–243.

3. I will refer specifically to Arne Naess, "Self-Realization in Mixed Communities of Humans, Bears, Sheep, and Wolves," *Inquiry* 22 (1979), pp. 231–241; and Arne Naess and Ivar Mysterud, "Philosophy of Wolf Policies I: General Principles and Preliminary Exploration of Selected Norms," *Conservation Biology* 1 (1987), pp. 22–34. As far as I am aware, "Philosophy of Wolf Policies II" was never published.

4. Lynn White, Jr., "The Historic Roots of the Ecologic Crisis," *Science* 155 (10 March 1967), pp. 1203–1207.

5. See Lester Milbrath, *Environmentalists: Vanguard for a New Society* (Albany: State University of New York Press, 1984), for an argument that radical environmentalists and Deep Ecologists, who differ from the old-line wildlife and conservation organizations by having adopted nonutilitarian and nonanthropocentric values, from the "vanguard" of environmentalism and of a new society.

6. As is argued, for example, by Bill Devall and George Sessions, *Deep Ecology: Living as If Nature Mattered* (Salt Lake City: Peregrine Smith Books, 1985), esp. pp. 56–61, where modern "reform environmentalism" (Devall and Session's substitute term for "Shallow Ecology") is associated with the tradition of Gifford Pinchot, who advocated development of resources for human use, despite the fact that mainstream environmental groups have uniformly rejected Pinchot's developmental program as a failed set of policies.

7. I have argued, however, that this split into two groups did *not*, even historically, correspond to the distinction, as suggested by the Deep Ecologists, between anthropocentric and nonanthropocentric value commitments. See "Conservation and Preservation: A Conceptual Rehabilitation," *Environmental Ethics* 8 (1986), pp. 195–220. In general, the failure of Deep Ecologists and other nonanthropocentric environmental ethicists to recognize the immense and important differences between a broad and long-term anthropocentrism and narrow economism and consumerist anthropocentrism has perpetuated a misleading conception of the environmental movement.

8. See *Toward Unity*, especially Chs. 4 and 12.

9. Naess himself makes the point that the principle of biospecies egalitarianism must be interpreted as an in principle value only, and it should not be applied so as to rule out some exploitation. The "in principle" clause is inserted because any "realistic praxis necessitates some killing, exploitation, and suppression" (quotation from "The Shallow and the Deep, Long-Range Ecological Movement"). In "Self-Realization in Mixed Communities, "he says that strict application of this principle "is of course Utopian in the worst sense." So I take it that Naess accepts the necessary restrictions on biospecies egalitarianism. See *Toward Unity* for a more detailed discussion of the constraints placed on biospecies egalitarianism if it is to support practically employable policies.

10. See my *Toward Unity* and "Convergence and Contextualism: A Clarification and a Reply to Steverson," *Environmental Ethics*, 19 (1997), pp. 87–100, for more detailed discussion.

11. Norton, *Toward Unity*, p. 83.

12. See Edith Brown Weiss, *In Fairness to Future Generations* (Tokyo and New York: The United Nations University and Transnational Publishers, 1988), for an interpretation of international environmental law as a trust for future generations.

13. See B. G. Norton and B. Hannon, "Environmental Values: A Place-Based Theory," *Environmental Ethics* (forthcoming).
14. A point that is made quite directly by Naess and Mysterud in "Philosophy of Wolf Policies I," p. 33: "Without the slightest doubt, we recommend wolves as members of the nordic life community. But this clear theoretical acceptance of wolves on the basis of our philosophy of nature does not imply any definite practical wolf policy."

# 5

# Pragmatism, Adaptive Management,
# and Sustainability

INTRODUCTION

> The important thing is to not stop questioning.
> Curiosity has its own reason for existence. (Albert Einstein)

I would like to propose for discussion a claim that may seem quite surprising: that Charles Sanders Peirce's definition of truth provides a useful analogy, or template, for defining 'sustainable' and 'sustainable living'. This claim could never be fully justified in a single paper, of course, so I can only sketch a few elements of the complex case that would have to be made to fully justify it here. My purpose, then, is more to explore some new directions for environmental philosophy, and to provoke discussion of a set of hitherto ignored problems that are relevant to the search for a definition of sustainable living, than to offer definitive answers to the problems posed.

Representative versions of Peirce's definition are: 'Truth is that concordance of an abstract statement with the ideal limit towards which endless investigation would tend to bring scientific belief' (*Collected Papers*, 5.565) and 'Truth is the last result to which the following out of the (experimental) method would ultimately carry us' (5.553). In general, this definition presents a pleasing analogy to searchers for a definition of sustainability because of its 'forward-looking' temporal horizon. Exploring this analogy might uncover clues as to how to give a sustainability definition the kind of forward, normative thrust it needs. Further, Peirce understood his notion of the search for truth as the defining pursuit of a community of inquirers who start with diverse viewpoints, but who are carried forward toward the truth 'by a force outside of themselves to one and the same conclusion' (5.407). Surely any acceptable

From *Environmental Values* 8 (1999): 451–466. Reprinted with permission of The White Horse Press, Cambridge, U.K.

definition of sustainability must embody the idea of a forward-looking community that is normatively respectful of the pursuit, and also the perpetuation, of knowledge, so the analogy is suggestive in other ways as well. Thus Peirce's idea of truth-seeking suggests fertile ground for analogies and other forms of guidance in the perplexing task of defining 'sustainability'. I find the idea particularly attractive because it may provide a way through or around the fact–value gulf by establishing a normatively scientific notion of sustainability.

In Part 1, I introduce the idea of truth as temporal with quotations from Henry David Thoreau, who anticipated two important aspects of the pragmatic approach to truth. One might still ask, however, what pragmatists can possibly do for environmental philosophy, so in Part 2, it will be necessary briefly to consider the goals of environmental philosophy, and, in particular, the *practical* goals of environmental philosophy: what can philosophers contribute to activism, to the actual goal of protecting nature? Having outlined a possible contribution of philosophers to the real-world search for a sustainable future in Part 2, I will highlight the benefits of the pragmatic approach to understanding sustainable communities in Part 3.

## 1. 'CONFORM' VERSUS 'TRANSFORM' THEORIES OF TRUTH

Near the end of *Walden*. Thoreau (1854) says. 'No face which we can give to a matter will stead us so well at last as the truth. This alone wears well.' If one takes Thoreau the philosopher seriously – and I do – then this description of truth could be taken as a clear, if homespun, anticipation of Peirce's definition of truth. What Thoreau and Peirce share, in particular, is a tendency to address the philosophical problems of truth and objectivity, not in the usual terms of a time-bound relationship between thought and a chunk of the contemporaneous 'external' world, but rather in terms of an intertemporal relationship between present beliefs and the outcome of a complex process that occurs through time. Near the end of the explanatory chapter of *Walden*, 'Where I Lived, and What I Lived For', Thoreau says:

> Let us settle ourselves, and work and wedge our feet downward through the mud and slush of opinion, and prejudice, and tradition and delusion, and appearance, that alluvion which covers the globe, . . . , through poetry and philosophy and religion, till we come to a hard bottom and rocks in place, which we can call *reality*. . . . (p. 70)

This passage emphasises Thoreau's commitment to the existence of truth, and not just opinion, but it also links this idea to a process that takes time, a lifetime pursuit for a person, a pursuit that, when successful, also evokes more eternal connections:

> If you stand right fronting and face to face to a fact, you will see the sun glimmer on both its surfaces, as if it were a cimeter, and feel its sweet edge dividing you through the heart and marrow, and so you will happily conclude your mortal career. Be it life or death, we crave only reality. (pp. 70–71)

Thoreau starts the next chapter by once again extolling the eternal nature of achievements to truth:

> With a little more deliberation in the choice of their pursuits, all men would perhaps become essentially students and observers. . . . In accumulating property for ourselves or our posterity, in founding a family or a state, or acquiring fame even, we are mortal; but in dealing with truth we are immortal and need fear no change nor accident.

I quote Thoreau here because he anticipated two important aspects of Peirce's approach to truth and objectivity. First, as noted, he anticipated the temporal, forward-looking notion that Peirce later developed. But Thoreau also anticipated a more general feature of Peirce's thought, the idea that inquirers can struggle toward truth and objectivity, and that the struggle takes place entirely within human experience. Both Thoreau and Peirce clearly recognised that the problem of truth and reality cannot be addressed as a matter of correspondence with a reality external to experience. Thoreau said, 'I am not interested in mere phenomena, though it were the explosion of a planet, only as it may have lain in the experience of a human being' (Thoreau 1984, Vol. VI: 206).[1]

It has been characteristic of Western philosophy since Aristotle, and accentuated since Descartes, to seek truth and objectivity in a correspondence between thought and reality behind or beyond experience. Thoreau struggled to reconcile his neo-idealist view that nature must provide, within experience, adequate assurance of truth with his runaway individualism. Thoreau sought to combine his above-mentioned commitment to truth below the slush of opinion with a subjectivist viewpoint. In his journal for May 6, 1854, Thoreau wrote:

> There is no such thing as pure *objective* observation. Your observation, to be interesting, i.e., to be significant, must be *subjective*. The sum of what the writer of whatever class has to report is simply some human experience, whether he be poet or philosopher or man of science. The man of most science is the man most alive, whose life is the greatest event. Senses that take cognizance of outward things merely are of no avail. (Thoreau 1984, Vol. 6: 236–237)

90

His bold reconciliation was to see truth as a manifestation of the completeness of a process, an intertemporal relationship: to have a truth is to 'connect' with eternity. It is to have a belief that would be shared with an idealised individual who has had all possible experience. Operationally, Thoreau's conception suggests that assertions of truth are best seen as predictions: To say that a statement, P, is true is to predict that P will eventually be accepted by inquirers who have vastly more accumulated experience than we do.

Thoreau, I believe, placed undue faith in what he calls individual 'genius', and apparently thought the process of observation could result in truth by virtue of a spontaneous, conscious act of apprehension. As far as I can see, Thoreau never explained how and why such a look inside could be credited with such epistemological weight, and I have no intention to defend this aspect of Thoreau's philosophy. I nevertheless think Thoreau had a glimmer of an alternative approach to the problem of truth, and that his approach is worthy of mention in anticipation of Peirce.

It was left to Peirce to reconstrue the temporal relation more concretely as a *community* process, a process pursued by a very special community of scientific inquirers – the lovers of truth. This community has implicit norms and explicit methods for approximating the truth, and a study of the 'logic' of their enterprise, Peirce thought, would yield a method that will eventually zero in on truth. Peirce identified the point toward which this community would tend as a predestined point, which he identified with reality. He was able, therefore, to maintain that his brand of pragmatism was 'realist' and touted a 'correspondence' theory of truth. But he continued to view the search for truth as a temporally developmental process, a task undertaken by a succession of generations in the community of truth-lovers.

Peirce was a systematic thinker and I have been warned of the dangers of taking Peircean ideas out of their systemic context. Indeed, there are a number of aspects of Peirce's system that are clearly inhospitable to the practical task of defining sustainability criteria. I refer specifically to Peirce's sharp separation of science from practice and to the extreme abstraction of Peirce's logic, both of which result from his emphasis on deriving the rules of logic from thought itself. Peirce's virtual obsession to avoid 'psychologism' apparently explains his unyielding positions on these points. Given these specific problems and especially Peirce's penchant for metaphysics, environmentalists should perhaps focus more on later pragmatists, especially Dewey.[2] So my goal is not to link Peirce and his philosophy directly to modern environmentalism, but rather to focus attention on a set of productive philosophical problems that were raised by Peirce and then explored by pragmatists and their critics over the subsequent century.

For example, consider the distinction between 'conform' and 'transform' theories of truth, introduced in the 1940s as an attempt to clarify issues regarding truth (Ushenko, 1946: 2f).[3] According to this definition, there is a key philosophical distinction between those who believe that truth is a relation between a statement and an antecedent reality, and those who believe that truth emerges from a situation of uncertainly through the transformation of an unsettled into a settled situation. While some might identify transform theories as a characteristic of pragmatism in general, it is important to realise that the matter was debated heatedly, with Peirce steadfastly defending a conform theory throughout his career (Smith 1978: 52–53). Peirce identified the search for truth with a 'pre-destined' outcome, and asserted a correspondence between today's truths and that pre-destined outcome, which allowed him to retain the rhetoric, and perhaps the heuristic value, of a temporally emerging truth, even as he denounced his transformist critics as defenders of 'psychologism' and as 'nominalists'. Dewey, on the other hand, provides an excellent example of a 'transform' theory, and – not surprisingly – Peirce was critical of Dewey on exactly these points. What I want to emphasise here is that Peirce's definition of truth, and arguments regarding objectivity as an intertemporal relation, led to a lively debate about alternative approaches to truth and objectivity (Smith 1978). Attention to the debates provoked by Peirce's forward-looking, normative, community-based drive toward the truth might therefore provide interesting parallels and suggestions for sustainability theorists. These parallels will be the subject of Part 3. But first it is necessary to examine the role of philosophy in the larger, activist environmental movement.

## 2. OBJECTIVITY, ENVIRONMENTAL ETHICS AND ENVIRONMENTAL POLICY

What can environmental philosophers do for environmental activists? One way to answer this question, which I think allows us to state an important area of consensus, is to say that environmental philosophers can provide an essential ingredient in objectively supportable environmental policies.[4] But it is very important to be clear about what is meant by an 'objectively supportable' policy goal. If one thinks of truth and objectivity in terms of a 'conform' theory, then environmentalists are guided toward metaphysical solutions to the problem of justifying their goals. Indeed, many environmental ethicists, such as Holmes Rolston III, believe that, if environmental goals are to be justified, it must be through a representation, a correspondence to moral values that exist independent of humans and human cognition.[5] Down this road lie many insoluble metaphysical and epistemological problems, as I have argued elsewhere (Norton 1992); and it remains to be seen whether intrinsic value theory can be supported on less difficult-to-defend foundations (paper 3, this volume). One advantage of seeking for truth and objectivity within the transform tradition is that justifications can be sought within human cognition, and claims to truth can be understood as predictions of what beliefs will emerge from the rough-and-tumble of scientific and social debate. The pragmatists thus put their faith in countless new observations coupled with improvements in our ability to learn, counting on a self-reflective method to be capable of approximating truth from within experience.

Critics of pragmatism will be quick to say that this viewpoint is fraught with ambiguity, and it cannot be denied that very diverse positions are called 'pragmatist' today, ranging from Richard Rorty's almost-relativist version to present-day throwbacks to Peirce's bolder realism. While deep theoretical differences separate pragmatists – even as they did in the days of Peirce and Dewey – they retain the tradition's problems, and its way of formulating and addressing those problems – a shared emphasis on praxis – as well as a set of tendencies of thought that are often disputed or ignored by contemporary philosophers who do not share the pragmatic bent. My goal here is neither to police the boundaries of the use of the label nor to resolve all ambiguities, but only to show how pragmatists' problems remain relevant if discussed and disputed with a particular emphasis on the praxis of sustainable living in technologically advanced and socially fragile societies of today.[6]

The pragmatists' struggle toward a normative and intertemporal 'logic' of inquiry provides an alternative to the metaphysical approaches popular in contemporary environmental ethics, and to the interminable debates about who and what has 'intrinsic value'. Down the pragmatists' road toward truth there may be an alternative way to ground environmentalists' goals. If we can envision the search for sustainable living as a community-based struggle to learn, and to perpetuate a process of learning, then objective truth is a question of justifying goals and policies within a community of inquirers – of understanding and projecting a kind of transformation of subjective consciousness – not a matter of correspondence with an external reality.

Turning down this road shifts the main focus of environmental philosophy *away* from moral theory and *toward* epistemological issues of justification, and toward methods of inquiry, and how to improve them, more generally. If Peircean truth is understood as a prediction that the community of truth-seekers will also embrace our beliefs and goals, and endorse our policies, then the problem is to provide an epistemologically supportable scientific justification for environmentalists' goals. Once one relaxes the correspondence demand on inquiry and, accordingly, recognises that there are avenues to truth within subjective experience, attention can be focused on the development of methods to seek the truth and to speed the process of truth-seeking. The development of such methods contributes to the development and perpetuation of communities of truth-seekers organised into sustainable, decentralised units. Once this process of learning and of learning how to learn is located within human experience – the collective experience of many communities of truth-lovers – pragmatists can avoid the fact–value dichotomy and the debilitating anti-naturalism usually associated with it. Within the praxis-oriented tradition of pragmatism, one never separates 'fact' from 'value'. Facts gain their meaning within an action-oriented context. By positing a unity of the method of experience based on sociocultural learning normatively based in the love of truth, the pragmatists can argue for a unified experimental method that can be applied to values and purposes as much as to scientific, causal hypotheses. It is experience guided by logic and reflective methodology that encourages learning. Pursuit of proposed community goals is no less an experiment, and no more subjective, than the testing of a hypothesis in a laboratory. For example, when participants in an ecosystem management process articulate tentative goals, and revisit these in subsequent discussions, the goals are open to revision in the face of what has been learned, and what has been experienced, in the meantime (paper 27, this volume).

What is interesting is that Peirce and Dewey, despite a published dialogue on the subject, never resolved their differences over transform and conform theories, despite the availability of a fairly obvious 'pragmatist'/contextual reconciliation. As has been pointed out by Smith (1978: 117–118), Dewey and Peirce explicitly stated that they saw different purposes for 'inquiry'. Peirce often invoked the Scholastic distinction between '*logica docens*' and '*logica utens*' (Hookway, 1985: 43), with the former representing the precise logical analyses of logicians, while the latter represents the unformulated logic – the unspoken standards of reasoning used by individual agents, including individual scientists. Dewey drew a somewhat different distinction: between what he called 'common sense' and 'scientific inquiry' (Dewey, 1938, Ch. IV: 114–119).[7]

For Peirce, who most respected *logica docens*, inquiry is 'primarily a form of logical self-control which focuses on the manner in which beliefs are formed or, rather, should be formed'. Dewey, on the other hand, emphasised 'the motive to control the situation which evokes it and ultimately to reshape the environing conditions of human life' (Smith, 1978, p. 118). Because Peirce was so concerned to avoid psychologism and naturalism in his philosophical system, he often ignored and sometimes disparaged *logica utens*. But a more pluralistic approach to pragmatism might embrace the search for *both* an improved *logica docens and* an improved *logica utens*, treating these as separable tasks with separable goals, and applying different cognitive tools and different 'standards of proof' in different contexts in which different goals are dominant. The trick is to design an approach to environmental management that is 'adaptive' by playing *logica utens* off against *logica docens*. In environmental management, this would mean letting *logica utens* be dominant, in that the demands of particular situations may require actions before the scientific hypotheses on which they are based have been adequately verified. Also, in *logica utens*, demands of action may determine which experiments should be undertaken at a given time, and could legitimately affect criteria for funding ecological and biological research, for example. Managers, however, must also submit their findings from management-driven decisions to the more stringent rules of *logica docens*; expedient, policy-driven science must eventually pass muster within the more demanding strictures of the disinterested and timeless community of truth-seekers. The resolution is contextual; it depends on the extent to which action is forced in a given context, not on the inherent superiority of the academic's goal of enforcing stringent criteria of scientific verification over the practical goal of acting, on the best available evidence at a given time, to protect social values.

On this compromise it would be possible to pursue the two types of 'logic' of science simultaneously, with each complementing the other, and each having its appropriate domain. If, following Dewey, we have faith in the ability of science and method to address real problems, then *logica utens* is adopted as the logic of environmental management and *logica docens* remains appropriate for the 'academic' study of science – for the study of science in a context, that is, where action is not forced. To illustrate how this compromise position would function in practice, it is possible to cite at least two important differences between the operations of *logica docens*, the logic of truth-seeking science, and *logica utens*, the logic of problem-solving and adaptational living.

1. Value neutrality, which remains an ideal in *logica docens*, is no longer claimed or required when applying *logica utens* within the policy arena. The application of *logica utens*, in fact, demands the expression of many value viewpoints in the search for policies that fulfil, to the extent possible, the many and competing interests of the community.

2. *Logica docens*, in its application, abhors positive assertions of truth that cannot be fully verified. *Logica utens*, on the other hand, must balance the

concerns of too quickly asserting a nontruth against the possibility that inaction based on 'academic' uncertainly may prove calamitous (Lee, 1993: 74–75).

It may also be possible to suggest a third difference. It might be possible to reconcile Peirce and Dewey by arguing that, whereas the static, conform theory is the proper ideal of truth within *logica docens*, the more dynamic, transform concept of Dewey, with its experimental, problem-solving attitude toward truth-seeking in practical situations, is applicable in the practical disciplines such as conservation biology and adaptive ecosystem management.

Adaptive management, a movement toward more iterative and experimental management practice, was formally articulated by C. S. Holling (1978) and others, and has been further developed by numerous authors since. The philosophy of adaptive management is, I would argue, a very good first approach to developing a *logica utens* for environmental problems and policy. This is not surprising because, in fact, many of the ideas of adaptive management were anticipated by Aldo Leopold's multiscalar management model as illustrated in 'Thinking Like a Mountain' (1949; Norton, 1990; papers 1 and 3, this volume)[8] and earlier by Dewey's general approach to social learning in democracies (Dewey, 1984; Lee, 1993). Adaptive managers understand the search for improved environmental policies as one of designing institutions and procedures that are capable of pursuing an experimental approach to policy and to science. And it is hoped that, in the process of building such institutions and procedures, a process of social learning will move the community toward better understanding of their environment through an iterative and ongoing task, a task that will require not just unlimited inquiry, but also the encouragement of variation in viewpoints and the continual revisiting of both scientific knowledge and articulated goals of the community.

The roots of adaptive management have been traced by Kai Lee (1993) back to Dewey's form of social and experimental activism. It would be beyond the scope of this paper to explore adaptive management and its philosophical underpinnings in detail,[9] but we can summarise the principles of adaptive management in a couple of sentences. Adaptive managers believe that a path to sustainability cannot be charted by choosing a fixed goal or set rules at the start. We must start where we are; but we do have the ability to engage in experiments to reduce uncertainty and to refine goals through iterative discussions among stakeholders. Environmental management must be a *process* in which managers choose actions that serve as experiments with the capacity to reduce uncertainty and to adjust future goals and choices. In this tradition, the manager tolerates a variety of viewpoints, hypotheses, and proposals for action; this variety of viewpoints, and ensuing experimentation and political discussion, are all important parts of the process of selection of more and more 'adaptive policies'.

So this variability poses no serious problem for adaptive managers, provided they are understood as seeking truth and objectivity within the transform tradition. Within that tradition, variability of beliefs does not imply relativism; variability is an inevitable precondition of cultural and scientific, as well as biological, evolution. Commitment to a process, and to the progressive refinement both of beliefs and of the truth-seeking methods we develop, deliver the adaptive manager from the spectre of relativism. The truth is that which will emerge from an indefinite and open process of observation, and from the ceaseless application and improvement of the scientific method over time.

It can be noted, in passing, that – just as Peirce and Dewey differed regarding the advantages of adding a conform theory to the transform model of truth-seeking – the decision of adaptive managers to apply the transform theory need not commit them either way regarding scientific realism or anti-realism. They can, like Peirce, adopt at least some form of scientific realism. For example, the views of adaptive managers sketched here are consistent with the position, somewhere between Peirce and Dewey, that is referred to by A. F. Chalmers (1994: 163) as 'unrepresentative realism'. This position is realist in the sense that it assumes that the world is the way it is independently of our descriptions of it, and in the sense that it assumes that physical laws apply universally, both in natural and in experimental situations; but it does not embrace a correspondence theory of truth in the sense that specific sentences, taken singly, 'correspond' to any prelinguistic aspect of nature. It is at this point that those pragmatists who hope to find a middle-ground pragmatic epistemology between Rorty's near-relativism and the dualistic correspondence approach to objectivity emphasise *method*. If the process of truth-seeking itself is self-corrective – if, that is, we can learn *how to learn* even as we learn – specification of procedures and methods that rule out untenable hypotheses about reality may yield an adequate sense of 'objective' support for emerging truths.

Interestingly, the scientists who have developed adaptive management since Leopold have not explicitly embraced one important aspect of the program of Dewey and the pragmatists: they have generally emphasised the ability of adaptive management to reduce uncertainty through scientific management experiments, but have so far said little about Dewey's dynamic approach to value change.[10] For Dewey and his brand of pragmatists, philosophy is most vital when it is used to clarify and formulate questions of practical import; and ethics is most alive when it is testing, in practice, goals that have

97

been advanced in pursuit of consensus and social solidarity. If the scientific advocates of adaptive management more fully embrace the pragmatic movement and explicitly reject the artificial distinction between facts and values, then they may come to join Peirce and Dewey in declaring the unity of all inquiry and in including values in the purview of their 'experimental management'. If we feel comfortable working within the tradition of pragmatism, and if we feel that theorists in the 'transform' tradition may be able to specify in more and more detail how we might progress toward sustainability, then there is an exciting future for this thing called 'Environmental Pragmatism'. And if pragmatists, champions of the belief in the normative nature of logic and inquiry, can being the power of experimental reasoning to bear upon goals and values as well as facts, then environmental ethics may someday be seen as an important subfield of adaptive management science, rather than as an abstract, and sometimes abstruse, subdivision of 'the humanities'. Time will tell.

### 3. PRAGMATISM AND SUSTAINABILITY THEORY

I have suggested that American philosophers, beginning with Thoreau, and including especially the pragmatists, established an alternative approach to the problem of objective knowledge and objectively supportable goal-seeking; this tradition, which owes much to Darwin and also to the American Naturalists, is complementary to environmental thought in ways that the Modernist, representational model of perception and objectivity never could be. I base these comparisons on the following four points of similarity.

1. Peirce's future orientation, in the context of an inquiring and self-consciously methodological community of inquirers, provides an excellent precedent for the type of community that must be built if humans are to live sustainably on the earth. Lee (1993), for example, speaks of the importance of the development of 'epistemological communities' to develop guidance, trust, and support for managers who patiently undertake to use the scientific method to better tune our conception of sustainable living and sustainable policies. Peirce's respect for a community devoted to the search for truth characterises the type of respect that we must develop toward the future and toward the knowledge and wisdom that will be required if we are to live sustainably. Peirce's philosophy of truth and objectivity, and the controversies it spawned, provide interesting guidance for a discussion of sustainability goals.

2. Formation and effectiveness of an epistemological community is an essential aspect of sustainability because, as advocates of adaptive management

agree, sustainable outcomes are not definable in advance, but must emerge from a program of active social experimentation and learning. Both definitions are best understood as characterising evolving processes, rather than ideal outcomes. Thus the idea of adaptive management is connected – by virtue of the search for an emergent, temporally sensitive, transformative notion of truth – back, through Peirce, to Thoreau.

3. The pragmatic approach also has the advantage that it links contemporary environmental ethics historically to the Darwinian evolutionary idea so formative in Leopold's land ethic. In 1923, Leopold referred to the minor pragmatist, Arthur Twining Hadley, and embraced a definition that says, 'Truth is that which prevails in the long run'. This passage led directly into a concise but penetrating discussion of an ethic of sustainability based on broad anthropocentrism (Leopold, 1979: 141; Norton, 1988). Later, he began 'The Land Ethic' (1949) with the statement that a new ethic is 'an evolutionary possibility and an ecological necessity'. Leopold, as did the pragmatists, clearly sought both truth and right in adaptive behaviour, and clearly understood both of these in an adaptive, evolutionary sense. This interpretation establishes a connection, through Darwin and the pragmatists, to Leopold, and to the adaptive managers; this link through Leopold is especially strong if one emphasises Leopold's policy viewpoints, and de-emphasises his metaphysical and poetic speculations (Norton, 1999). Further, this interpretation favours a broadly Darwinian epistemology, an epistemology, that is enhanced with a commitment to the efficacy of methodology to improve the truth-seeking process. A Darwinian environmental ethic may provide a more unified basis for judgments supporting some choices as 'adaptive' and 'sustainable'. If so, the pragmatist interpretation of the land ethic avoids the deep tensions that are introduced into the Land Ethic by interpreters such as Callicott (1989: 166), who attribute to Leopold a Darwinian ethic and a 'modernist' epistemology.

4. The pragmatists' conception of logic and the study of inquiry as a self-sustaining and *normative* process provides a model for normative-descriptive sciences such as medicine, conservation biology, and sustainability studies and points the way around the fact–value dichotomy. That dichotomy, perhaps useful in academic science, is inapplicable to the practical problems of management science. Here, we need a *logica utens*, and goals as well as hypotheses must be understood and tested as hypotheses. To continue the quest, to ensure the continuation of the community and its truth-seeking ideals, the community must survive. Peirce's normative approach to logic thus points toward a more unified treatment of environmental knowledge, uncertainty, and goals for action. This aspect cries out for a more pragmatic approach to problems of value theory and applications of logic over multiple scales of time.

This fourth point of similarity and complementarity tempts me to go beyond speaking of analogies and templates, and to make a stronger statement to the effect that a pragmatist approach to knowledge and of obligations to the future may go a long way toward *justifying* some important goals of environmentalists. For example, many environmentalists, as well as a few philosophers, have argued that sources of new experience should not be summarily destroyed, providing an obligation not to destroy unique life forms and cause other irreversible simplifications of nature (Leopold, 1949: 108–112; Russow, 1981; Regan, 1986). This sentiment is often implied by advocates of biodiversity protection when they claim that every extinction of a species or every destruction of unique ecosystems is like burning libraries. Both actions irreversibly destroy unique opportunities to learn, thereby narrowing the possible occasions for us to observe and refine our current, incomplete belief systems.[11] We must, if we love the truth, accept a *prima facie* obligation to protect as many as possible of the particularities of the biological world for future study; we must also act so as to perpetuate the community of inquirers – so as to live sustainably, that is – or the search for truth will be prematurely interrupted. It would take egoism or 'generationism' of an extreme sort – not just anthropocentrism – to care not at all that the future will be prevented from studying and coming to know rain forests, millions of species, or natural ecosystems. Peirce also envisaged the progress toward truth as a self-driven process; the ceaseless search for truth, a commitment to contribute to a cosmic ideal of knowledge. I foresee a similar, or at least parallel, commitment to completion of a shared long-term enterprise of physical and cultural survival.[12] This would be the environmental application of the Einsteinian epigram at the beginning of this paper.

If a Darwinian, pragmatist epistemology is accepted as a unifying force in a broadly anthropocentric ethic, then it may be possible to embrace a more pluralistic theory of environmental values. *Broad anthropocentrism* would argue, based in pragmatism's analysis of value, action, and science, that a Darwinian epistemology complements pluralism with regard to ethics. If our most basic commitment is to survival – of our culture as well as our genes – we might emphasise a new, non-instrumental ethic as having survival value. While I stop short of embracing a pluralism so inclusive as to comprehend value independent of human cognition and motives (Norton, 1992), the Darwinian/Deweyan/Leopoldian approach encourages a variety of value hypotheses, and enthusiastically embraces a selection process based on results in the pursuit of improved environmental policies. The tradition of pragmatism, in other words, articulates a set of questions sufficiently comprehensive to encompass *both* the epistemological *and* the value questions that are essential for charting a course toward sustainable living, and for justifying environmentalists' goals to the broader population.

NOTES

I am indebted to my colleague, Jeffrey DiLeo, for helpful tutoring in Peirce's complex philosophy, for insightful discussions of my analogy, and for helpful comments on an earlier draft. I also thank two anonymous reviewers for *Environmental Values*, whose trenchant criticisms, I hope, led to improvements in the substance and the clarity of the paper. Some of the research for this paper was supported by a grant from the National Science Foundation, (NSF # SBR-9729229).

1.  See Richardson (1986: 309–310) for further discussion of these passages.
2.  Larry Hickman (1996) has begun the task of applying the thought of Dewey to environmental philosophy.
3.  The following discussion owes much to Smith (1978: 52f).
4.  J. Baird Callicott (1989: 163) states the case clearly when he says that his goal is to rescue environmental policy decisions 'from reduction to cost-benefit analyses in which valued natural aesthetic, religious, and epistemic experience are shadow priced and weighted against the usually overwhelming material and economic benefits of development and exploitation'. Callicott suggests that what is needed are objectively supportable moral values on which to base environmental action....
5.  See Rolston (1986: 96), where he says: '[W]e...believe that through [scientific judgments] we are accurately corresponding with the natural world. When we pass to judgments of value, we do not need to consider them radically different in kind....'
6.  Anonymous reviewers of earlier versions of this paper were concerned that I do not state unambiguously whether the brand of pragmatism I am defending is to be 'relativistic' or 'objectivistic'. I avoid stating a specific position on this issue here because my point is that there are interesting *parallels* between the philosophical dilemmas that have troubled pragmatists and today's problems of 'defining sustainability'. I fear that taking a dogmatic stand on the *solution* to the problems of pragmatism will detract from the parallels I want to highlight.

    Having said this, I also do not wish to be viewed as side-stepping such a crucial issue as the degree or type of 'objectivity' the pragmatists can ultimately deliver. I, and I expect a number of other pragmatists, would neither defend Peirce's strong version of 'constructivist realism' nor follow Rorty's 'philosophy-as-lifestyle' approach. These pragmatists seek a middle ground in a naturalistically based, unified approach to inquiry, and have faith that attempts to improve methods of inquiry can lead to an adequate and for a reasonable notion of 'objectivity'. To formulate the problem of objectivity as *either* there is objective knowledge (in the sense in which Descartes sought it) *or* we must embrace relativism, seems to me to beg the question against these 'middle-ground' pragmatists. Their main programme is to question and undermine the very dichotomies and methods that have driven the rationalist-empricist debate throughout the Modern period of philosophy, and that have led to all-or-nothing formulations of what they consider to be the mis-stated problem of 'objectivity'. For pragmatists, for whom truth is sought within experience, objectivity (or whatever non-relativism would be correctly called) emerges from a process of inquiry, and objectivity is not an all-or-nothing matter. Peirce, after all, maintained *both* that the truth emerges from a process over time (a transform understanding) *and* that this process has a unique outcome, which encourages him to assert that he has achieved, also, a 'conform' understanding of 'reality'. In this sense, Peirce *combined* a constructivist approach to specifying the truth with a more conformist understanding of the (practically, impossible) Final Result. Whatever one thinks of Peirce's heroic efforts to reconstruct a rationalistic epistemology within experience, it is better to think of pragmatists as arrayed across a continuum, with Peirce near one end and Rorty near the other. Dewey, I believe, sought, as I do, a middle ground between Rorty's near-relativism and Peirce's 'conformism'. Again, it is beyond the scope of this paper to *resolve* these problems; I only hope to raise them in a new context – the context of environmental policy debate – and to show that the problems spawned by a pragmatist examination of the problems of environmental policy and management may lead to better answers as well as better questions (Weston, 1992).

7. One interesting area for further research would be to explore these differences between Peirce and Dewey, and their implications for the logic of policy inquiry.

8. See Norton (1990) for an explanation of how Leopold's famous metaphor anticipates hierarchy theory, which has been endorsed as a major structural element of the conceptual system of the adaptive managers.

9. See (papers 3 and 27, this volume) for more detailed discussions.

10. An important exception is Lee (1993), but I do not find convincing Lee's brief discussion of value formation, expression and revision. For example, Lee introduces the ends–means distinction on the way to an explanation of social learning in Dewey (1993: 105–108), thus creating a muddle, since rejection of the ends-means distinction is a keystone of Dewey's philosophy.

11. The point cannot, of course, be expressed in simple quantitative terms. The claim cannot be that, if we extinguish a species, for example, we have reduced the number of possible experiences future persons can have. Since possible experiences are limited by one's time on earth and by the limits of information processing in the experiencer's brain, not by the (multiply infinite) possible objects of experience, no future persons should claim that we left them without enough possibilities of experience. It nevertheless seems intuitively true that, if the future were to be deprived of whole categories of experience – such as would occur if all naturally functioning ecological systems were converted to intense economic use – the future's ability to study and learn about natural systems and their self-organising behaviours would be impaired. In this case, I would argue, earlier generations would have harmed later ones.

12. By cultural survival, I do not mean survival of the dominant, expanding, Western European culture, but as many as possible indigenous lifestyles developed by non-Western people (Quinn, 1992). Again, once one shifts from a conform to a transform conception of truth, diversity of beliefs and values represent no threat to objectivity. In a Darwinian, adaptive worldview, variety and diversity is a necessary precursor to increasing objectivity.

## REFERENCES

Callicott, J. Baird 1989. *In Defense of the Land Ethic*. Albany, NY: SUNY Press.

Chalmers, A. F. 1994. *What Is This Thing Called Science?* Indianapolis, IN: Hackett Publishing Company.

Dewey, John 1938. *Logic: The Theory of Inquiry*. New York: Holt, Rinehart, and Winston.

_____1984 (originally, 1927). *The Public and Its Problems, In John Dewey: The Later Works: 1925–1927*. Volume 2, ed. J. A. Boydston, Carbondale, IL: Southern Illinois University Press.

Hickman, Larry 1996. 'Nature as culture: John Dewey's pragmatic naturalism'. In *Environmental Pragmatism.*, ed. A. Light and E. Katz. London: Routledge Publishers.

Holling, C. S. 1978. *Adaptive Environmental Assessment and Management*. New York: John Wiley & Sons.

Hookway, Christopher 1985. *Peirce*. London: Routledge and Kegan Paul.

Lee, Kai 1993. *Compass and Gyroscope*. Covelo, CA: Island Press.

Leopold, Aldo 1949. *A Sand County Almanac*. Oxford, U.K.: Oxford University Press.

_____1979. 'Some fundamentals of conservation in the Southwest', *Environmental Ethics* 1: 131–141.

Norton, Bryan 1990. 'Context and hierarchy in Aldo Leopold's theory of environmental management', *Ecological Economics* 2: 119–127.

_____1992. 'Epistemology and environmental values', *Monist* 75: 208–226.

_____1999. 'Leopold as practical moralist and pragmatic policy analyst'. In *The Essential Aldo Leopold*, ed. R. Knight and C. Meine. Madison, WI, University of Wisconsin Press.

Peirce, C. S. 1960. *Collected Papers of Charles Sanders Peirce*. Edited by C. Hartshorne and P. Weiss. Cambridge, MA: Harvard University Press.

Quinn, Daniel 1992. *Ishmael: An Adventure of the Mind and Spirit*. New York: Bantam Books.

Regan, Donald H. 1986. 'Duties to preservation'. In *The Preservation of Species*, ed. B. G. Norton, Princeton, NJ: Princeton University Press.

Richardson, Robert D. 1986. *Henry Thoreau: A Life of the Mind*. Berkeley: University of California Press.

Rolston, Holmes, III. 1986. 'Are values in nature subjective or objective?' In *Philosophy Gone Wild*. Buffalo, NY: Prometheus Books.

Russow, L-M. 1981. 'Why do species matter?' *Environmental Ethics* 3: 101–112.

Smith, John E. 1978. *Purpose and Thought: The Meaning of Pragmatism*. Chicago, IL: University of Chicago Press.

Thoreau, Henry D. 1854. *Walden*. New York: Penguin Books.

_____1984. *The Journal of Henry David Thoreau.*, Volume VI. Salt Lake City: Peregrine Smith Books.

Ushenko, A. P. 1946. *Power and Events*. Princeton, NJ: Princeton University Press.

Weston, Anthony 1992. *Toward Better Problems: New Perspectives on Abortion, Animal Rights, the Environment and Justice*. Philadelphia: Temple University Press.

# II

# Science, Policy, and Policy Science

My location in a school of public policy has the advantage that I, my colleagues, and my students are constantly dealing in practical case studies, in which scientists from multiple fields provide data and projections, and in which a problematic situation must be addressed, given imperfect science and in the face of serious disagreements among interest groups. I have learned that one of the most important areas in which philosophers can make a real difference is in understanding the complex role of science in policy processes. My involvement with the EPA illustrated how the anachronistic insistence on a sharp separation of science and value can skew the entire policy process. Insisting that "risk analysis" could be a pure science and that valuation, judgment, and decision making could be compartmentalized as "risk management," practitioners of risk analysis and EPA managers engaged in a charade that served their shared purposes. They were able to claim that agency decisions could be based on quantified, value-neutral data, provided that the politicians and social values are removed from the process, and provided that the scientists are well funded to do quantified analyses of "risk factors." In this way, the commitment to value-neutral science provided something for both scientists and bureaucrats. This charade, however comforting to scientists – who could eschew politics – and to decision makers – who could claim that their decisions were implied by quantified scientific models – rendered the EPA ineffectual. Analyses such as these led me to believe that there was indeed work for environmental philosophers who concentrate on the interactions of science and policy in environmental and regulatory agencies.

This charade represents only one example; working with conservation biologists and ecologists quickly revealed that while philosophers consider positivism and the idea of value-neutral science to be an exploded myth, positivism is alive and well in the biological sciences, including biological sciences that, like conservation biology and conservation ecology, have adopted a conservation mission. In a brief editorial, "What Is a Conservation Biologist?" and in a number of other papers, I tried to undermine the idea that any science – especially a mission-oriented science that has a goal of conservation – can be value-neutral, and to promote the alternative idea that, like medicine, conservation biology is a normative science. This brief editorial sets the stage for two additional papers that explore, in one way or another, how environmental science can inform policy without compromising the standards of science. In "Biological Resources and Endangered Species: History, Values, and Policy," I showed that the formulation of policy problems – for example, as problems of protecting endangered species – has remained oddly independent of readily available scientific knowledge that protection of species requires protection of their habitat. Similarly, whether we value species or ecosystems may not matter if science tells us that species depend upon ecosystems for their survival. This paper, then, recognizes that values affect science and science affects values; things really get interesting when one views these relationships from an action-oriented policy perspective. I continue exploring the relationship of science with value-directed policy efforts in a paper on the ways Leopold balanced the two. Interestingly, Leopold, in general a great integrator, seems to have held his speculations about ethics separate from his policy recommendations, basing political recommendation on the grounds of social benefit while treating the search for a new ethic as interesting and important speculation.

Finally, in "Improving Ecological Communication," I take ecologists to the woodshed regarding their reluctance to become involved in policy evaluation, which in my view has weakened environmental policy – especially communication about ecological factors affecting environmental policy – in important areas such as wetland protection and mitigation. Ecologists cannot, in my view, consistently argue that ecological processes and features are ecologically as well as economically important and then leave the task of evaluation of ecological change to economists. I am disappointed to report that ecologists have largely ignored my lecture; nor have they challenged its arguments or changed their ways. Apparently, I do not administer trips to the woodshed as effectively as my father did.

# 6

# What Is a Conservation Biologist?

What is a conservation biologist? According to one view, the conservation biologist acts as a mechanic. Our society demands "healthy ecosystems"; conservation biologists "adjust" the mechanism of nature to achieve, as efficiently as possible, this social goal. On this positivist view of science, facts and values can be separated, and science can operate in a world of pure description.

This mythical world is also attractive to social scientists – especially economists. Many economists claim that their science is value free, that they merely record social preferences as they are represented in markets, actual or hypothetical. According to this positivist view of science and society, Ph.D.'s granted to budding young conservation biologists serve as union cards that permit them to work as biological mechanics in the service of social preferences.

The danger in this whole view of science becomes apparent when economists and conservation biologists accept, without protest, the positivist picture. They will come to believe that good environmental management can somehow be cranked out of a decision procedure that merely records whatever preferences are registered in markets or in social opinion polls. The result is a bureaucratic brand of science (such as that which takes place in some national parks under the watchful eye of park management). This is not only bad science; in the end it does more harm than good.

The belief in value-free science is not just dangerous; it is also unrealistic. Popular among philosophers in the 1930s, 1940s, and 1950s, the positivist view is now seen by them as seriously oversimplified. Philosophers still follow the positivists in recognizing the importance of falsifiability in science, but they now emphasize that "falsification" always takes place against a backdrop of theoretical assumptions. Theoretical assumptions are often affected by the

From *Conservation Biology* 2 (1988): 237–238. Reprinted with permission.

values we hold – so scientists work, even when testing hypotheses, in a world permeated by values.

The positivist view of science gives an especially distorted view of conservation biology, because conservation biology is a *prescriptive science*. A physician thus provides a better analogy than a mechanic; just as physicians have a responsibility to participate in the social debate concerning the nature of human health, conservation biologists have a social obligation to participate in the public debate about the nature of ecosystem health.

The task is formidable. Stephen Kellert's work on social attitudes toward wildlife has demonstrated that most Americans relate to animals individually and that they are most often only willing to support conservation of "higher" animals. A national effort to protect biodiversity based on these social goals would ignore ecosystem functioning and would fail because species can only survive if their habitats are saved.

It is important that conservation biologists do not take a patronizing attitude toward the public, because conservation biologists have much to be humble about themselves. The discipline is hardly ready to announce with confidence and unanimity what the goals of conservation biology should be. The trained (though not infallible) members of the discipline must, nevertheless, contribute to an open and rational public discussion of conservation goals.

The first point of attack in this discussion should be on the positivist view of science itself – on its model of nature as a mechanism composed of interchangeable parts. Conservation biologists must reject the role of biological mechanics and embrace organicism. They must insist that whole systems of nature can be judged "healthy" or "ill."

Some, of course, still question the literal truth of organicism. Science, however, advances as much through shifts in operative analogies as through shifts in what is taken to be literally true. . . . As a model for discussing environmental management, the organic analogy is clearly superior to the mechanical one. Organicism justifies the analogy of conservation biologists to physicians and implies an obligation to engage the public in a search for an enlightened definition of ecological health. As Aldo Leopold saw when he contrasted the mechanist attitude with the organicist one in the section of *The Land Ethic* entitled "Land Health and the A-B Cleavage," the organicist analogy also encourages a belief that "A thing is right when it tends to preserve the integrity, stability, and beauty of the biotic community."

In practice, then, the organicist view requires a holistic approach to environmental management. Consensus may break down when more specific implementation is proposed but, whatever else holism implies, it implies that segments of nature cannot be managed without paying attention to the larger systems in which they are embedded, any more than one would treat a human organ with no attention to the overall person of which it is a part.

This implication is, in and of itself, sufficient to undermine the view that the health of ecosystems can be defined in the purely economic terms of market-revealed social preferences. If the search for economic efficiency, the application of production-maximizing principles to land use, threatens ecological systems, then conservation biologists must assert the precedence of the organic conception of ecosystem health.

The final result of the organicist analogy, in other words, is the implication that there are pre-emptive constraints, limitations placed on the search for economic welfare. These constraints protect the dynamic functioning of the larger systems in which productive units, whether farms, mines, or factories, are embedded. The range of economic choices open to our society must therefore be circumscribed by the limits of whole ecosystem health.

This brings us back to the awesome responsibility of conservation biologists. While often lacking crucial data and adequate theoretical models, conservation biologists must participate with the public in a debate regarding the very nature of ecological health, even while trying to protect it. This responsibility must be accepted squarely; our fledgling discipline must not hide behind a false facade of value-free science.

# 7

# Biological Resources and Endangered Species

## History, Values, and Policy

This chapter examines the role of the U.S. Endangered Species Act (ESA) of 1973 in the broader context of protecting biological resources by considering its historical, valuational, and policy aspects. Three historical "phases" of policy thinking on the environment are identified: single-species protection (1800–1980), biodiversity (1980–1988), and sustainability of ecosystem health (1988–present). The ESA is representative of the thought behind the first phase and, in this sense, is anachronistic; but the protection of biological resources is supported by many important human values. So the question remains whether the ESA is the best approach to protecting biological resources, given that current thinking goes beyond individual species. It is argued that, given our state of knowledge and managerial abilities, the ESA remains the most viable policy tool for protecting biological resources, but it should be applied so as to improve managerial knowledge and encourage more eco-systematic approaches.

The protection of biological diversity has become a central goal of environmental protectionism. On one level, this goal is noncontroversial, a motherhood-and-apple-pie issue – there are no advocates for species destruction or for accelerating species loss. Nevertheless, there remain important differences regarding how much we should do, what we should do, and even what is of ultimate value. Because the protection of biodiversity – no matter how strongly supported publicly – will certainly proceed within a context of uncertainty, this paper advocates an adaptive management approach to protecting biological resources.

From L. Guruswamy and J. McNeely, eds., *Protecting Global Biodiversity: The Converging Perspectives of Science, Politics, Economics, Philosophy, and Law* (Durham, NC: Duke University Press, 1997). Reprinted with permission.

The argument for a more adaptive, reactive approach is well summarized in an analogy offered by William Ruckelshaus, who contends that the search for a sustainable society in the future ought not to be likened to a crossroads – an all-or-nothing choice between two alternatives – but rather, to a canoeist shooting the rapids. Survival and sustainability will depend more on a willingness and ability to react to new information than on a single and forever-binding choice (Ruckelshaus 1989). This adaptive approach emphasizes the importance of choosing policies that fulfill two criteria: chosen policies should, given the best available science at the time of their implementation, protect both species and the ecological processes associated with them; and chosen policies should be designed to increase our information base to support further policy actions (see Walters 1986; Lee 1993). Philosophically, these two policy criteria are supported as minimal requirements for fairness to the future, although it will also be noted that there remain considerable disagreements regarding how to formulate such foundational values in environmental ethics.

This chapter attempts to put these disagreements into perspective, given the history, competing values, and immediately pressing policy issues faced by conservation biologists and environmental managers. Part 1 explains how the conceptualization of the problem of protecting biological resources in the United States has evolved in three stages since early efforts to protect particular game species from overhunting and overfishing. Against this historical backdrop, part 2 discusses whether the ESA, as currently written and interpreted, represents a reasonable response to threats to biological resources in the United States. Finally, part 3 will speculate on the extent to which lessons learned in the United States can provide guidance in addressing the more complex problems of protecting biodiversity worldwide.

## THREE PHASES IN THE PROTECTION OF BIOLOGICAL RESOURCES

### Phase 1: Single-Species Protection (1800–1980)

As early as the eighteenth century, local governments in New England began protecting particular food, fur, and game species against the depredations of overharvesting (Cronon 1983). This species-by-species approach led to legislation at the local, state, and federal levels to protect biological resources, and eventually to protect habitats for protected species (Fox 1981). These early attempts to regulate hunting and fishing, and later attempts to protect habitats in the main flyways for migratory waterfowl, remained "atomistic" in the sense that particular species were singled out as valuable and protection was conceived in terms of separate management plans for each protected species. This

111

approach to protection reached its culmination in 1973 with the ESA, which provides for a process of identification, listing, and recovery of species, both game and nongame, that are endangered or threatened. While the ESA retains the atomistic conception of species, it nevertheless represents a landmark in the development of biological resource policy. The act protects, at least to some degree, *all* species (rather than just a few valued game species), and it also pays attention to habitats, subspecies, and sometimes even populations. It is, then, important for its comprehensiveness.

Two intellectual-scientific aspects of this approach require comment. First, from a philosophical viewpoint, single-species management emphasizes protecting the elements of nature, and is therefore in keeping with long philosophical and historical traditions in Western civilization. It is often noted that Western philosophical traditions have stressed "being" – what exists now and how to understand it – over "becoming" – the processes of change and development that bring about and support those entities that exist at any particular time – (see Prigogine and Stengers 1984). Historically, this represented the triumph of the concerns of Plato and Aristotle for "substance" over the ideas of Heraclitus, who, in 500 B.C., declared that "All is in flux." This triumph was embodied in modern science, which emphasized explanation at all levels of nature that can be "reduced" to the motion of elementary particles. This concept of reduction to elementary particles as the essential form of explanation is ultimately inseparable from the mechanistic view of nature that has held sway throughout the modern period of Western civilization (Prigogine and Stengers 1984). For this reason, the "atomistic" idea underlying a focus on elements is so deeply ingrained in Western thinking that it was difficult even to imagine alternative conceptualizations of nature until relatively recently.

With the publication of Charles Darwin's evolutionary theory, however, the importance of systemic change and irreversible developments of complex, dynamic processes has reasserted itself (Dewey 1910). At first, this revolution was limited to the biological sciences and, even there, the full implications have still not been worked out. Now, the revolution has extended to physics, and physicists are leaders in an interdisciplinary effort to develop a more dynamic worldview. But again, the full implications of a dynamic world, one that creates diversity and complexity, apparently at the edge of chaos, are only just now being felt. It may be decades before these concepts are well understood, but creative work in nonequilibrium systems dynamics is already leading to new insights in ecology (Pimm 1991; Lewin 1992), and this direction holds promise for applications to environmental policy (Norton 1992). This much we know for sure: the full absorption of new thoughts on evolving systems into environmental management will have far-reaching impacts on the policies we advocate, and will almost certainly require more attention to interspecific relationships and system-level characteristics.

Second, the single-species approach to environmental protection is based in *autecology*: the study of individual species within their habitats. Autecology has been a powerful force in applied biology, since it was dominant in forestry and applied entomology. It represents a research program attempting to understand populations within their environments; this approach often sought to understand populations and their characteristics in terms of the physiology of its members. In contrast, *synecology* concentrates on the relationships between, and systems of, species. Because of the dominant place of autecology in management agencies and practical sciences, concern for the protection of biological resources naturally turned into a concern for species and, by extension, other taxonomic groups (McIntosh 1985). Synecology has been stronger in academic ecology, and today has given rise to "systems" ecology. Its continued study has, in turn, affected environmental management by encouraging more holistic and ecosystematic managerial experiments, giving rise to the current tension with species-oriented approaches.

Despite the growing popularity of ecosystem management, the single-species approach has several advantages: species are relatively easy to identify, they have a basis in biological fact, and they can be counted. Emphasis on species, therefore, provides fairly clear-cut management goals and at least the hope of achieving measurable success. Critics of autecology and single-species management, however, have mounted strong scientific and managerial arguments against this approach. Scientifically, autecology tells only one side of the story – certainly, the survival of species partly depends on the characteristics of individuals, but it also depends on a complex and changing set of relationships called their *habitat*. Managerially, emphasis on particular species in the short run can set in motion changes at the system level, resulting in strongly counterproductive management in the middle and long run.

This point was first made by Aldo Leopold in his trenchant criticisms of forest service management in the southwestern territories of the United States. In his brilliant metaphor of "thinking like a mountain," Leopold regretted his successful efforts to exterminate wolves in order to increase deer populations in the wilderness areas of the Southwest and argued for managing nature on multiple scales, at the ecological system scale of time and space (the mountain) as well as at the economic, short-term scale (Leopold 1949). In short, Leopold contended that attempts to manage biological resources from the perspective of autecology would fail, especially in fragile ecosystems, because autecology does not pay enough attention to relationships among species at the community level, and second, because it falsely assumes that nature maintains a static balance. Leopold, therefore, criticized traditional, single-species management as being too atomistic and not sufficiently dynamic in its understanding of nature (Norton 1991). Leopold's work has been corroborated and developed by C. S. Holling and his associates, who have shown that large-scale systems that are managed for regularized production of one or a few resources become more "brittle," susceptible to collapse or gradual degradation to a new, less-desirable stable state (Holling 1992; Norton 1991). Based on these broad intellectual and scientific trends, our strategies to protect biological diversity must pay more attention to system-level characteristics and the dynamic processes that they represent.

## Phase 2: Biodiversity (1980–1988)

By the early 1980s, it had become clear to policymakers that species diversity was only one aspect of life's variety. Species, genetic variation within species, and populations of species that are adapting to varied habitats were all recognized as components of *biological diversity*. This broader approach culminated in the Smithsonian Symposium on Biodiversity, which was followed by an important book and a traveling exhibition (Wilson 1988). The biodiversity phase represented a distinct advance in conceptualization because the introduction of multiple layers of diversity, and the emphasis on varied dynamics and habitats as well as species, significantly expanded our understanding of how complicated the process of protecting biological resources really is.

But this phase was conceptually unstable because the third element of diversity – the multiple dynamic processes within which different populations adapt and evolve – begins to shift attention away from the elements of nature, toward the processes that create and sustain those elements through time. Practically, this development is appealing; it promises to turn managerial attention away from a few desperately small populations of endangered species toward safeguarding processes on the theory that, if ecological processes remain intact, populations and species can take care of themselves. It is surely easier to keep species off the endangered list than to save them once they are in trouble. But the practical promise of a more dynamic approach has not led to decisive action or even very specific management proposals.

The problem is that the brilliant theoretical insights of Leopold have proven frightfully difficult to operationalize (Norton 1991). To manage an ecological system, one must know what processes are essential and which elements depend on each other. Unfortunately, dependency relations represent one of the most difficult and least understood areas of ecological study. Worse, the dynamic, nonatomistic approach is very information intensive. Unlike physics, which has achieved remarkable generality, biology and ecology are studies of particularity (Ehrenfeld 1993). The information demands necessary to model impacts on ecological systems are not only heavy, but the relevancy of information is highly dependent on local conditions. Most commentators hold out little hope that there will be a grand breakthrough in ecology, such as the discovery of some simple and general laws that will make understanding ecosystems and their complicated interactions causally transparent (McIntosh 1985; Sagoff 1988). Consequently, while ecosystem management will require a great deal of detailed and locally relative information about species and their interrelationships, ecosystem ecology offers neither quick nor easy models. Ecology applies to ecosystem management, but the applications must be at the local and particular level. The strong trend toward locally motivated attempts to articulate "ecosystem management plans" for areas and regions may represent also a trend toward local responsibility in understanding and managing ecological systems.

Arguably, the ESA, even as glossed with the multilayered emphasis of the biodiversity concept, remains element oriented. The act attempts to identify, list, and recover species, subspecies, and so on. Its friendly critics may justifiably maintain that the act is not sufficiently protective of processes, and that we need an "Endangered Ecosystems Act" or an "Endangered Processes Act." This line of reasoning has spawned the current phase of environmental policy debate regarding how to protect biological resources.

*Phase 3: Sustainability of Ecosystem Health (1988–present)*

The biodiversity phase – given that it steps on an apparently slippery slope leading toward more emphasis on processes and less on elements – may therefore prove to be only a transitional phase, as policy and management will endeavor to protect communities of species by formulating strategies for protecting essential processes. Species remain important in this phase, but with the recognition that species in varied habitats anticipate quite diverse ecological and evolutionary trajectories. The importance of these trajectories apparently implies that a system representing a significant habitat is an important management unit. This approach, which remains highly speculative, argues that policies to protect biological diversity must monitor and protect larger ecological units, such as ecological systems. This management program requires introducing descriptive/normative concepts, such as *ecosystem health* and *ecosystem integrity*, which apply at the ecosystem level and that emphasize processes rather than elements.

It is not yet clear whether this new emphasis will require efforts in addition to or instead of the ESA and the efforts it currently mandates. Indeed, it can be argued, as we shall see below, that doing a better job of achieving the goals and objectives of the current act may be the best we can do for the foreseeable future. There is continued support for protecting species; there is growing support for ecosystem management. The question remains: Can the key concepts of dynamic, ecosystemlevel management – ecosystem health and integrity – be defined with sufficient clarity to guide biodiversity policy?

## WHY WE NEED THE (ANACHRONISTIC) ENDANGERED SPECIES ACT

We are embarking on a new adventure in environmental management by integrating traditional element- and species-oriented management into a multitemporal, multiscalar, ecosystematic approach. While the ideas that spark the adventure remain somewhat speculative, the goal must be to create a new paradigm of environmental management. Given the lack of reliable data and scientific models, the only reasonable path to pursue is *experimental* or *adaptive* management (Lee 1993) focused at local levels (Chapter 19, this volume). Adaptive management is designed to function in an uncertain world; it devises managerial and other plans to make them learning experiences. We must make management proposals, try them out in carefully controlled situations, and then design pilot projects to determine the results of various methods.

There are two issues that are important to address in the uncertain context described above. First, we should examine the human values that drive the search for a better strategy for protecting species. Then we should discuss the practical, policy situation in the uncertain context that currently exists. At each reauthorization of the ESA there are attempts to rewrite the act. How should unflattering assessments of the act as a vestige of earlier thinking on the protection of biological resources affect concrete questions such as whether, and in what form, the ESA should be continued? What is the value of endangered species and biodiversity, and what kind of protection is appropriate given these values? Any discussion of the value of biological diversity should start with the recognition of the breadth of consensus favoring the protection of biological resources among serious scholars from every relevant discipline, and embracing virtually the entire intellectual landscape. If the present generation fails to protect biological resources into the middle and distant future, we will have committed a serious wrong.

116

The solidity of this consensus is sometimes obscured because the various academic disciplines and policy players often disagree on how to characterize that wrong, as well as how to weigh it against other obligations in comprehensive environmental decisionmaking. There are at least four quite distinct analyses of the values involved:

1. THE ECONOMIC/UTILITARIAN APPROACH. According to what may be the dominant view, the values involved in protecting biodiversity are fully represented in an accounting of the welfare of humans in the present and the future. Protection of species, genetic, and habitat diversity is unquestionably important as a source of future resources. Protection of biological diversity can be justified because of the many ways in which species and ecosystems provide services that we would otherwise have to supply. It is important to realize that human values derived from nature are broader than economic values; many people value nature for its positive impact on the human spirit, and studies show a significant willingness on the part of consumers to pay for the protection of species and ecosystems, simply because they want to know they exist (Norton 1987).

2. THE INTERGENERATIONAL EQUITY/STEWARDSHIP/SUSTAINABILITY APPROACH. The economic approach is often supplemented by an emphasis on intergenerational equity and an insistence that each generation should act sustainably. What exactly this emphasis adds to a longsighted application of economic or utilitarian reasoning is a matter of dispute, however. At issue is whether the obligation to the future is specific – there are particular resources that are essential and these should be protected – or general – we only owe the future a *just savings rate*, a rate of investment sufficient to ensure that future generations will have the opportunity to be as well off as we are. Economists and preference-based utilitarians tend to favor an *unstructured bequest package*, which aggregates all forms of capital – natural and human-made – into a single accounting system. They sum individual welfare within a generation and then compare it across generations. The future has no reason to blame us, so goes this line of reasoning, if we leave a world capable of maintaining a nondeclining stock of undifferentiated capital. Under this view, sustainability becomes a mere afterthought to economic growth theory (see Solow 1974, 1993).

Most advocates of sustainability argue that present actions are constrained not only by an obligation to maintain a fair savings rate, but also by more specific obligations to protect essential resources, such as tropical forests and the oceans' fisheries. These advocates of a more structured bequest package as a guide to intergenerational fairness – often called *ecological economists* – contend that we should keep separate accounts for natural and human-made capital, and that we have obligations to protect natural capital in an intact state. Intactness of ecosystems is generally referred to as *ecosystem health* or *ecosystem integrity*, so the ecological economics movement complements the call for a stronger and more specific criterion of intergenerational equity, usually associated with "strong sustainability" (Daly and Cobb 1989). This approach has as its defining intellectual task the problem of determining what resources are essential elements of natural capital.

3. THE BIOCENTRIC APPROACH. According to another line of moral reasoning, all living organisms are of equal value intrinsically, and humans are obliged to share resources equally with all other species (Taylor 1986). This approach, which is individualistic, focuses on the organism level; it values ecosystems and species, but only because they are made up of individuals. Denying human moral superiority, this approach argues that all living individuals, at least individuals who are morally considerable, have equal rights to the world's resources. This approach cannot, then, be applied to policy unless there are clear rules for defining which members of which species may exploit others in the struggle to survive. Various attempts to specify such rules have failed to gain broad consent. Until these problems are resolved, this approach is of little help in policy matters.

4. THE ECOCENTRIC APPROACH. Ecocentrists believe that ecological communities have inherent value and should be protected for their own sake. This approach to environmental valuation avoids many of the problems of biocentrism. It has a general answer to the question of which species are to be given priority: a species is valuable insofar as it contributes to the well-being of the ecosystem of which it is a part. This approach has been criticized for its "fascist" tendencies, in that it apparently sacrifices the interests of individuals to those of larger systems; it also appears, given the damage modern, technologically powerful humans now inflict upon ecosystems, to imply misanthropism. But most precise formulations of the principle are careful to limit the dominance of systems over individuals and to provide some special protections for human individuals (see, for example, Callicott 1989).

These various formulations are referenced not so much to emphasize the intellectual differences among them, although they are considerable, but rather to acknowledge and then consider them in policy contexts. While there is much discussion of nonanthropocentric approaches to environmental values (including 3 and 4, above), what is often not recognized is the extent to which policy prescriptions converge between ecocentric approaches and the less-controversial commitment to protect biological resources for future generations (Norton 1991). If one adopts formulation 3 or 4, in addition to 1 and/or 2, most of the goals of the nonanthropocentric approaches will be at least approximated by efforts to fairly discharge duties implied by requirements of intertemporal fairness. For example, whatever content is given biocentrism, it apparently requires protection of habitat for living organisms in the future. But that is also what is required by the obligation to protect ecological processes for the benefit of future generations. One could not, that is, fulfill the goals of biocentrists without, in the process, fulfilling obligations to maintain essential ecosystem processes (Norton 1991). While there may be some differences regarding how far policies must go to protect biological systems and species, advocates of all four approaches, with few exceptions (such as Kahn 1982; Simon 1981; Simon and Kahn 1984), believe that we should be doing much more than we are currently.

The important point at a public policy juncture, such as the periodic reauthorization of the ESA, is that we agree on what is valued and what our goals should be. In this context, it is not significant that various scholars and activists use different languages to characterize the values involved in protecting biological diversity – the important thing is that there exists a strong consensus of high value and urgent priority for the protection of biological resources, and that this policy consensus is based on the broad agreement of experts plus considerable public opinion. In order to focus more sharply on the policy questions, three related but crucially different questions arise:

1. Do we need an *Endangered* "Biological Resources" Act?
2. Do we need an Endangered *Species* Act?
3. What kind of protection should be mandated in an Endangered Biological Resources Act?

The answer to question 1 is implicit in the analysis just completed: We do need legislation to protect biological resources. This conclusion follows from the wide intellectual and scientific consensus that we have obligations to the future. Confusion regarding how to characterize values derived from ecological systems does not undermine the overwhelming consensus regarding the need/value of protecting biological resources at some level and scale.

Question 2, however, is much more difficult to answer, especially if one seeks *a scientific* response, because the ultimate importance of species in the overall picture is a matter of scientific disagreement and uncertainty. For example, it could be argued that, because of the significance of cross-population genetic variation, we need an "Endangered Populations Act"; or it could be argued, on the grounds that processes are really more important than species, that we need an "Endangered Ecosystem Processes Act." These difficult problems are really problems of scale. Once we have agreed that we need to save biological resources, there is still the question about the scale at which "protection" should be directed. These scalar problems have both a scientific and policy aspect (paper 16, this volume). Scientifically, it is certainly important to know how species extinctions and extirpations on large scales affect ecological processes; but there may be no general solution to this question. It appears that some extinctions are extremely significant ecologically, while others have very little ecosystemic impact (because their populations are rare, for example).

We cannot, however, afford the luxury of waiting for a full resolution of ongoing scientific differences. When the ESA comes up for periodic reauthorization, for instance, we face a *policy* question in the sense that a decision is forced, and must dominate the desire for further information regarding this decision. From a policy perspective, question 2 is actually easier to answer, once one accepts that the decision will be made with less-than-ideal amounts of scientific information. For example, the suggestion that we have an Endangered Populations Act can be ruled out because, while we might sometimes protect a population, it is too expensive to protect every population. Worse, saving every population would require "deep-freezing" nature, halting its constant dynamism.

Choosing between the ESA and an Endangered Ecosystem Processes Act is more difficult, however. As noted above, environmental management is entering a period of great flux, as the concept of "whole ecosystem management" is touted more and more in environmental protection and planning. The shift of focus to ecosystems involves the introduction of system-level characteristics – characteristics emergent on the ecosystem level – as measures of how well we are doing at protecting biological resources. We have noted that popular candidates include protecting the "health" and/or the "integrity" of ecological systems. One might ask: should we perhaps give up the attempt to save all species, and put more emphasis on ecosystems and their protection? While this ecosystems-sensitive approach has broad appeal, it might only be because nobody knows for sure what, exactly and operationally, would be necessary to protect the health or integrity of an ecological system. Yet there are interesting attempts in this direction (Costanza, Norton, and Haskell 1992). Advocates of economic development, for example, have shown interest in ecosystem management as a way of providing more "flexibility" to our efforts to protect nature. Protectors of nature, on the other hand, are beginning to realize that it is often better to concentrate on protecting ecosystems – which can harbor many species, including those that are rare and endangered – rather than making last-ditch efforts only after a species reaches the brink of extinction. There is no doubt, however, that any concrete measures to legislatively mandate the protection of ecosystem health and integrity, or descriptively/normatively characterize the changing state of ecosystems, would either be vague or controversial. While both Manuel Luhan, former Secretary of the Interior, and the deep ecologists have expressed an interest in more ecosystem management, it does not follow that they would agree on specific guidelines and actions to protect ecosystem-level characteristics.

For the present, there is simply not a detailed scientific-policy consensus regarding how to "define" ecosystem health and integrity, and it would be folly to dismantle the comprehensive endangered species protection system until there is a scientifically acceptable and politically viable definition of these ecosystem-level characteristics. The policy should be to protect species even as we continue to search for a more precise measure of exactly what we need to emphasize in the effort to protect biological resources over the long term.

This conclusion – that saving species may eventually play a less-central role in biodiversity policy and that we adopt the species-protecting policy as a temporary expedient – is not nearly as damaging an admission as it might seem. While the health/integrity movement emphasizes ecological processes, its advocates never question that species are necessary agents of those processes. An ecosystem that is rapidly losing species is most likely unhealthy; if steps are undertaken to reduce species losses, they will also contribute toward the goal of restoring ecological health to the system.

Indeed, the congruence of ecosystem and species-level objectives has a scientific basis. For example, some species are considered to be "keystone," since their loss would threaten other species and begin a cascading effect. Second, the plummeting population of one species can serve as an "indicator" for a whole ecosystem, as has been argued regarding the spotted owl and marbled mirulet in the Northwest. In this case, again, good ecosystem management will be identical to good species-level management. Third, virtually every definition of ecosystem-level, descriptive normative characteristics includes as an important condition that a "healthy" system, one maintaining its integrity through time, must be capable of sustaining biological diversity.

Most important, the ESA functions as an extremely useful "working hypothesis" within an adaptive management strategy to protect biological diversity. As currently amended, the ESA's goal is to protect all species; however, it contains an "exceptions" clause, whereby a high-level governmental committee – the "God Committee" – can judge that steps to save a given species are too costly if they conflict with overriding regional and national goals. Given this content, the act is well suited to encourage many experiments in species protection, while also allowing an escape hatch if the social costs of saving some species prove to be too great. The system has not worked nearly as well as it should (Tobin 1990), either in gaining information or in protecting species. Nevertheless, it has the potential to encourage a steep learning curve regarding how to protect elements, when it is important to protect elements, and also regarding how processes and elements interact in many specific situations. This is the heuristic value of having an act that focuses attention at a policy level that is manageable. Given current information, we usually have at least some idea of actions that can improve the survival chances of a species. While it is manageable, however, it does stretch current information-gathering techniques and capabilities, creating a scientific and managerial environment that encourages relevant learning.

Despite some scientific uncertainty and the disagreements about direction, there is no question that we should continue to protect species for the foreseeable future, even as we try to become more sensitive to ecosystem-level management. This adaptive approach to management would be furthered if the ESA were made more flexible – providing more latitude to introduce "experimental populations" – or if it encouraged protecting "suites" of species that tend to survive or fail together: for example, HB 2043, introduced by Congressperson Gerry Studds in 1994, provided a listing category of species that would make it less likely that other species "dependent on the same ecosystem" would require listing. This modest improvement would encourage more attention to and research about species interrelationships.

Ecological science – both established principles and accepted uncertainties – supports this sort of incremental experimentation with new ecosystem-level management. We know what does not work – it does not work to allow development, at great cost to biological systems, and it does not work to try to freeze ecological systems to protect every population in a community. Unfortunately, we do not know what does work, and so it is often best to protect species. As Leopold (1949) says, "The first rule of intelligent tinkering is to save all the pieces." Once we admit we're tinkering, the policy choice should clearly be to avoid irreversible losses, if at all possible. For this reason, it may be concluded – at least for now – that there is only one reasonable policy choice: to protect species while continuing to explore ways to be more sensitive to, and even regulate to protect, ecosystem-level processes and characteristics.

The reasoning behind this chapter strongly conflicts with the proposal to amend the ESA to include a "no-surprise" clause, a proposal put forth by critics of the current law. A no-surprise clause would allow negotiated settlements between landowners/developers, on one hand, and regulatory agencies on the other. The settlements would be binding for a long period of time, such as 50 to 100 years. Once in place, the owner/developer would be immune from further regulatory burdens or liabilities for the period of the agreement. Critics of the current law do validly argue that some private landowners in some situations may be required to accept an unjust burden to protect biological diversity. It would, however, be better to address this problem with a system of compensation for the real losses of harmed landowners than to lock regulatory agencies into inflexible protection and recovery plans. The increasing emphasis on dynamic processes in management, as well as the growing importance of flexibility and adaptive management in the face of uncertainty, imply that a no-surprise clause would be a big step in exactly the wrong direction – a step toward inflexible management that is unable to respond and adjust in the face of new information.

We can now turn to question 3, formulating it more precisely, in keeping with the foregoing argument: What kind of protection should be mandated in the ESA? Some environmentalists, such as David Ehrenfeld (1978), and deep ecologists have advocated a very strong and uncompromising line – that every species has a right to exist, and that we are bound to do anything necessary to prevent every possible extinction. Others would argue that we cannot save every species, but that we should try to save as many as possible, recognizing that species protection is not the only worthy environmental and social goal. Still others would consider the costs and benefits of protecting each species – calculating each case independently and, in essence, sorting through species, then saving only those that can pull their own economic weight.

It is difficult to defend either of the extremes. The hard line, that we should never give up on a species, is difficult to defend as a universal rule because there will, in some cases, be other competing environmental goods that override species protection. For example, it was reported in *Science* that an impasse was reached between efforts to restore water regimes to the Everglades ecosystem and attempts to save the Everglades kite, an endangered species (Alper 1992). Whether or not this case is an authentic clash of policies to protect ecosystems versus species, it does suggest that there may arise serious conflicts between those who emphasize species and those who emphasize ecological systems as the units of management. It is also possible to imagine rare instances in which some competing social good would override our obligation to protect a species. These uncertainties, far from undermining a commitment to adaptive management, support that approach precisely because it will bring these issues to light, encouraging the targeted study of particular systems and the feasibility of various conservation goals.

So any middle-ground position regarding the ESA moves significantly toward some balancing of the advantages and disadvantages of protection versus development in particular cases. But even the advocates of economic/utilitarian analysis have strongly advocated a presumption in favor of protection, because they recognize how difficult it is to identify and place dollar values on the benefits of species protection. Therefore, they argue, if the cost-benefit analysis is even close, policy should favor protection because of the greater difficulty in specifying the benefits of protection over those of development (Fisher 1981). The middle ground would seem to be the following position: there is a presumption of species protection, provided that the cost of protection is bearable; species will normally be saved, but interest groups have a right to challenge this presumption if important interests are at stake. This approach, which represents roughly the status quo in the ESA as amended to include the God Committee, can be theoretically formulated as an application of the "Safe Minimum Standard of Conservation" (SMS) due to Ciriacy-Wantrup (1952) and elaborated by Bishop (1978; also see Norton 1987, 1991). In situations where a resource may be irreversibly lost, always save the resource if the cost is bearable.

The matter of bearable costs, of course, is highly negotiable; one advantage – some may consider it a disadvantage – of this middle-of-the-road approach is that it makes justifiable costs a matter of degree. This means that most cases must be considered on their individual merits and that the question of protection is usually a matter for political compromise. Perhaps the scientific recognition that some species are much more valuable "ecologically" than others may enter into calculations regarding which species should be saved. Some species may deserve high priority because of their ability to support complexes of species; others may have little system-level significance. It may, therefore, make ecological sense to give up on some species and concentrate on those that are truly important for human or ecological reasons. But it is important that we not confuse ecological uncertainty with economic interest. Prior to amendments added in 1984, the ESA made a declaration of critical habitat for a listed species (which was required to complete the listing process) dependent on an analysis of the economic impacts of the listing. This requirement was misused early in the Reagan administration to virtually halt the listing process, requiring an amendment to separate economic from scientific considerations in determining whether a species should be listed. There have been several attempts to reestablish this linkage, but experience shows that this would be a mistake.

APPLYING THE U.S. EXPERIENCE TO GLOBAL BIODIVERSITY

While this chapter has mainly focused on policies to protect biological resources in the United States, it may nevertheless be useful to briefly discuss
the implications of the evolution of U.S. policy for attempts to protect species
worldwide. Indeed, the lesson of adaptive management is generalizable to
most situations. In many cases, efforts to protect biological resources are
initially undertaken in a context of relative scientific ignorance, where little
is also known about the social values associated with either the protection
or continued degradation of biological systems. Policies should be instituted
not only to deal with immediate, consensually accepted problems (since it is
possible to muster social resources to address them), but also to increase the
flow of information between scientists, policymakers, and the public. This
lesson surely applies worldwide.

It may be futile, however, to attempt to export the species-by-species approach embodied in the U.S. ESA to many developing countries, especially
tropical countries with rapidly increasing human populations, high degrees
of biological diversity and endemism, and rapid deforestation rates. Under
these conditions, so few species have been studied, and so many are threatened by rapid development, that policies directed at species would direct
scientific study and management efforts at an unrealistically small scale. It
makes more sense to focus immediate attention on "biodiversity hotspots"
(Forey, Humphries, and Vane-Wright 1994) in order to protect areas with
demonstrably high diversity and high rates of endemism. While this strategy
implies a shift to a larger scale in the system, it is still possible to design
policy responses to deal with immediate and recognizable problems *and* to
design those policies to increase our knowledge base. This recognition may
lead, eventually, to complementary scientific approaches – we maximize our
understanding of systems in general by studying many of them, and by studying them at multiple levels of organization, encouraging the supplementation
of autecology with a more system-level approach.

It is important to be realistic about what developing countries can afford to
undertake on their own. Many industrial nations have created their wealth by
destroying crucial elements of their own resource base. This wealth can then
be used – provided there is public support and a political will to do so – to
protect some areas as pristine representatives of historical systems that have
evolved with few impacts from human activities. In countries where the development process has begun more recently, and especially where development
has been retarded by international exploitation and colonialism, there may be
little public support for the goal of "total" protection of areas in their pristine

state. If environmentalists from the developed world hope to export this goal, they should recognize that they will be accepting responsibility to provide financial and scientific support as well. Assuming there will be limited funds to support such international efforts, it follows that most local and national strategies to protect biological resources will emphasize wise and sustainable use of resources, rather than protection of resources from any human use. Sustainable development may not lead in these cases to large, pristine reserves in developing nations; but it is becoming more and more widely recognized that some uses, such as sustainable eco-tourism, can protect natural areas from the utter destruction of clear-cutting and other highly destructive practices.

Other authors in this volume discuss various approaches to sustainable development, and a detailed discussion would be beyond the scope of this chapter, but it is important to ackowledge the importance of local context in developing management plans to protect diversity. The local context includes features of ecological communities – such as diversity and endemism – but it also must take into account local social and cultural conditions (see Norton 1994).

There is another apparent implication of the trends discussed above for international efforts: there will have to be experimentation with many new international institutions and perhaps the development of new concepts of property in biological resources, since public resources are often lacking in developing nations (see Vogel 1994). Again, the applicable lesson is local sensitivity to ecological and social conditions. One of the most important problems in developing an effective international policy is to create incentives to protect biological diversity on the part of private corporations. But here the international problem clearly outstrips the lessons that can be learned by examining the fledgling attempts of the United States to protect endangered species and other aspects of biological resources. These pressing issues must, then, be left to other authors in this collection, as well as those who are actively engaged in global biodiversity protection.

CONCLUSION

The protection of biological resources/diversity must proceed amidst considerable uncertainty. It has been argued that – despite uncertainties regarding the comparative role of elements as opposed to processes, confusion over which evaluative conceptualization to use in characterizing these resources, and significant recent changes in these conceptualizations – there remains considerable consensus in favor of protecting biological resources. It is, therefore, possible to chart a reasonable policy course by preserving and improving current practices as mandated by the ESA, even while striving to develop more comprehensive methods for protecting ecological processes as well as elements. First, it is important to maintain the ESA in something like its present form; second, it is wise to experiment with more ecosystem-related management initiatives, testing the new ideas of ecosystem management. These actions rest on the overwhelming support for protecting biological resources from many philosophical and disciplinary perspectives. While it may be necessary to modify and extend current efforts to protect species, this should be accomplished through incremental change, experimentation, and pilot projects, while maintaining the protection currently mandated by the ESA. Given current knowledge, it is almost always better to protect species because this goal contributes strongly, for a variety of reasons, to both the traditional and more speculative goals of ecological management. Whether or not the experiences in the United States – at state and federal government levels – can provide a useful template for developing a response to the biodiversity crisis, one thing is certain: as long as there is uncertainty in our knowledge base for managing biodiversity a high premium should be placed on an experimental approach that is both adaptive to local conditions and designed to increase knowledge even as it deals with pressing problems of biodiversity loss.

## REFERENCES

Alper, J. 1992. Everglades rebound from Andrew. *Science* 257:1852–1854.

Bishop, R. D. 1978. Endangered species and uncertainty: The economics of the safe minimum standard. *American Journal of Agricultural Economics* 60 (1):10–18.

Callicott, J. B. 1989. *In Defense of the Land Ethic.* Albany: State University of New York Press.

Ciriacy-Wantrup, S. V. 1952. *Resource Conservation.* Berkeley: University of California Press.

Costanza, R., B. Norton, and B. Haskell. 1992. *Ecosystem Health: New Goals for Environmental Management.* Covelo, Calif.: Island Press.

Cronon, W. 1983. *Changes in the Land: Indians, Colonists, and the Ecology of New England.* New York: Hill and Wang.

Daly, H., and J. Cobb. 1989. *For the Common Good.* Boston: Beacon Press.

Dewey, J. 1910. The influence of Darwinism on philosophy. In *The Influence of Darwin on Philosophy and Other Essays.* New York: Henry Holt.

Ehrenfeld, D. 1978. *The Arrogance of Humanism.* New York: Oxford University Press.

―――. 1993. *Beginning Again: People and Nature in the New Millennium.* New York: Oxford University Press.

Fisher, A. C. 1981. *Economic analysis and the extinction of species.* report no. ERG–WP–81–4. Berkeley: Energy and Resources Group, University of California Press.

Forey, P. L., C. J. Humphries, and R. I. Vane-Wright. 1994. *Systematics and Conservation Evaluation.* Oxford, U.K.: Clarendon Press.

Fox, S. 1981. *John Muir and His Legacy: The American Conservation Movement.* Boston: Little, Brown.

Holling, C. S. 1992. Cross-scale morphology, geometry, and dynamics of ecosystems. *Ecological Monographs* 62 (4): 447–502.

Kahn, H. 1982. *The Coming Boom: Economic, Political, and Social.* New York: Simon and Schuster.

Lee, K. N. 1993. *Compass and Gyroscope: Integrating Science and Politics for the Environment.* Covelo, Calif.: Island Press.

Leopold, A. 1949. *A Sand County Almanac and Sketches Here and There.* London: Oxford University Press.

Lewin, R. 1992. *Complexity: Life at the Edge of Chaos.* New York: Macmillan.

McIntosh, R. P. 1985. *The Background of Ecology: Concept and Theory.* Cambridge: Cambridge University Press.

Norton, B. G. 1987. *Why Preserve Natural Variety?* Princeton, N.J.: Princeton University Press.

―――. 1991. *Toward Unity among Environmentalists.* New York: Oxford University Press.

―――. 1992. A new paradigm for environmental management. In *Ecosystem Health: New Goals for Environmental Management*, edited by R. Costanza, B. Norton, and B. Haskell. Covelo, Calif.: Island Press.

―――. 1994. On what we should save: The role of culture in determining conservation targets. In *Systematics and Conservation Evaluation*, edited by P. L. Forey, C. J. Humphries, and R. I. Vane-Wright. Oxford: Clarendon.

Pimm, S. I. 1991. *The Balance of Nature? Ecological Issues in the Conservation of Species and Communities.* Chicago: University of Chicago Press.

Prigogine, I., and I. Stengers. 1984. *Order Out of Chaos: Man's New Dialogue with Nature.* Toronto: Bantam Books.

Ruckelshaus, W. D. 1989. Toward a sustainable world. *Scientific American* (September), pp. 166–174.

Sagoff, M. 1988. Ethics, ecology, and the environment: Integrating science and law. *Tennessee Law Review* 56:77–229.

Simon, J. 1981. *The Ultimate Resource.* Princeton, N.J.: Princeton University Press.

Simon, J. L., and H. Khan, eds. 1984. *The Resourceful Earth: A Response to Global 2000.* Oxford: Basil Blackwell.

Solow, R. M. 1974. The economics of resources or the resources of economics. *American Economic Review Proceedings.* 64:1–14.

―――. 1993. Sustainability: An economist's perspective. In *Economics of the Environment: Selected Readings*, edited by R. Dorfman and N. Dorfman. New York: W. W. Norton.

Taylor, P. W. 1986. *Respect for Nature*. Princeton, N.J.: Princeton University Press.

Tobin, R. J. 1990. *The Expendable Future: U.S. Politics and the Protection of Biological Diversity*. Durham, N.C.: Duke University Press.

Vogel, J. 1994. *Genes for Sale*. New York: Oxford University Press.

Walters, C. J. 1986. *Adaptive Management of Renewable Resources*. New York: Macmillan.

Wilson, E. O. 1988. *Biodiversity*. Washington, D.C.: National Academy Press.

# 8

# Leopold as Practical Moralist and Pragmatic Policy Analyst

Aldo Leopold's ideas and pronouncements on environmental policy, read fifty years after his death, establish how far Leopold was ahead of his – and our own – time. . . . Although Leopold was not a philosopher, he developed a remarkably complex and subtle "philosophy" of environmental management. He loved to speculate on "big" – or as he often said "general" – ideas, but he was much more than a prophet of a future environmental consciousness. The ideas he lived by were the ideas that were forced upon him by years of thoughtful and painful experience. His discussions of policy often read like briefings he might like to have given to his first boss, the eminently practical Gifford Pinchot. In these discussions Leopold generally eschewed "intangible" ideas, accepting common philosophical and religious commitments as constraints on his speculations; yet he gave – or struggled valiantly to give – carefully articulated reasons and justifications for all of his management precepts.

It may be helpful to list some of the ideas, articulated between 1920 and his death in 1948, that establish Leopold's claim to prescience in the area of management theory and process. First, he insisted – contrary to his contemporaries and in opposition to most of today's congressional representatives – that ethics, not economics, ultimately validate environmental policies. Second, in anticipation of the current trend toward public and stakeholder participation in policy process, Leopold expressed his progressivist-populist faith that it must be farmers, sportsmen, and other citizens themselves who accomplish conservation. This second belief led to a third idea, one that is anathema to many environmental managers today, as it was to Leopold's own contemporaries in government resource agencies. He believed that public-servant environmentalists should be just that; that the highest calling of resource

From Curt Meine and Richard L. Knight, *The Essential Aldo Leopold.* © 1999. Reprinted by permission of The University of Wisconsin Press.

managers was education and public involvement, rather than what he de-risively called the "ciphers" of management economics.[1] Fourth, Leopold recognized before others that management cannot simply be scientific in the sense of *applying* fixed principles of science; rather, and more important, we should be *managing* scientifically in the dynamic sense. Leopold thus insisted on policies designed to get results *and* reduce our ignorance through experiments with real controls.

Similarly, today's still-nascent but increasingly important ideas of ecosystem management were given shape by Leopold in his relentless attacks on atomistic management, which separated management of the land into "many separate field forces."[2] He advocated instead an integrated approach to the management of resources. To these innovations we could add mention of Leopold's extraordinary concerns for our resource legacy to future Americans and a coherent and reflective concept of "sustainable development."

Each of these ideas – and there are others – would have qualified Leopold as an important innovator. But the totality of them, and the way in which Leopold used his unparalleled powers of observation to illustrate, sharpen, and weave these points together, mark him as the premier genius in the field. We would do well to listen to him very carefully when we choose actions to alter or "improve" on nature and natural functioning of ecosystems.

I regard Leopold the policy analyst, the policy-maker, and the practical moralist as the originator and spiritual father of the flourishing tradition of "adaptive ecosystem management," so ably espoused today by C. S. Holling, Carl Walters, Kai Lee, and others.[3] Scientifically, Leopold anticipated the idea of ecological resilience, so prominent in the writings of Holling and the adaptive managers, when he described semiarid countries as "set in a hair-trigger equilibrium."[4] He clearly recognized that shortsighted management could render ecological systems and processes vulnerable to collapse. Leopold also anticipated a unifying theoretical idea, which later came to be called "general systems theory," or (in theoretical ecology) "hierarchy theory." Leopold's brilliant insight – that managers and agriculturalists must, to be successful, "think like a mountain" – was not (or at least was not only) a mystical vision.[5] It was hard-won wisdom, that (a) the manager, who observes and manipulates, is a part of the system, and not only views it but changes it from within; and (b) we can understand observed nature more coherently if we see it as a nested hierarchy of subsystems, with larger, slower-moving systems forming the "environment" for the smaller systems that compose it.

This scalar analysis, which would be incorporated wholesale into adaptive management, was embodied in Leopold's brilliant simile. In "Thinking Like a Mountain" Leopold provided a case study. The policy of killing wolves to increase deer herds seemed, from a short-term, human perspective to be a good idea to Leopold and others. He learned, however, that good policy must pass not only the short-term test of human economic reasoning but also the test of the mountain, which requires an ecological and evolutionary perspective. In addition to anticipating the adaptive managers' ideas of multiscalar analysis, Leopold also foresaw their emphasis on the need to include citizens and stakeholders in an iterative process that, at its best, involves social learning and the development of locally effective and cooperative institutions.

Leopold was a "scientific" manager who eventually, and sometimes through painful experience, came to appreciate both the strengths and the weaknesses of science in the practice of management. Because he had the professional mentality of a manager before becoming a professor, Leopold developed a nose for relevant science. Management, for him, asked interesting questions of ecology, and ecology provided useful tools for potential application. Accordingly, Leopold was scornful of pure theory when unrelated to practice, complaining that we argue over our abstract and conflicting ideas as to what needs to be done; instead, he suggested, we should "go out and try them."[6] He enjoyed speculation, but was disdainful of it when it was cut loose from experience and from pressing environmental problems.

Much has been written about Leopold's science and about his moral beliefs. Few authors and commentators, however, have acknowledged that Leopold expressed a quite sophisticated philosophy of science and epistemology, which was for him intertwined with his management philosophy. Because he functioned as a manager in both the political and scientific worlds simultaneously, he constantly faced both controversy and uncertainty. Consequently, he judged facts by their usefulness. In the 1923 manuscript "Some Fundamentals of Conservation in the Southwest" Leopold expressed a sophisticated version of Darwinian epistemology, using it to cut through uncertainty about broad theoretical principles. Ultimately, he reasoned, our scientific and managerial behavior must be adaptive or we will not survive as a culture; we will "be judged in 'the derisive silence of eternity.' "[7] However, his healthy respect for the uncertainties of management left him wary of general pronouncements that lacked tight connection to actual experimental test. His concluding discussion of conservation morality in "Some Fundamentals" includes no less than five cautionary statements about the limits of philosophical and speculative language.

Leopold implicitly practiced (even as it was being articulated by Herbert Simon and others) the decision method known as "satisficing."[8] In "Game and Wild Life Conservation" (1932) he concluded that the only way to protect any wildness is "to set up within the economic Juggernaut certain new cogs and wheels whereby the residual love of nature" may be fractionally protected.[9] For Leopold satisficing meant triangulating through waters made choppy by mindless devotion to "progress" and boosterism, while relying on nothing more than the experimental spirit of science, democratic involvement, and faith in the traditions of good sportsmanship and husbandry expressed as respect for self and land. Satisficing – even moralistic satisficing in the style of Leopold – recognizes that policy cannot be guided by grandiose prior plans or principles, but must seek, with an eye toward best alternatives and pilot projects, to gradually improve policies and keep track of what works.

Although Leopold was quite successful in articulating a practical *and integrated* philosophy of environmental policy, confusion may be avoided if one thinks of Leopold's theories of management as quite distinct from his environmental ethics. My point is not that Leopold's views on value and management were unconnected, but rather that he related them differently than most professional environmental ethicists do today. Leopold worked from observation toward theory, whereas most of today's environmental ethicists attempt to establish universal principles and then "apply" them to particular cases. Leopold's powers of observation were legendary. As illustrated in his first attempt to survey conservation ethics in "Some Fundamentals of Conservation," his discussions of morality were usually preceded by a careful empirical analysis of trends and problems in resource use. He spent the first two-thirds of that landmark paper explaining what he saw from horseback in the vast lands under his management as director of operations for the Forest Service in the Southwest. He concluded that simple observation revealed damage from human use and that the damage was having economic impact. Ethics, then, were layered onto Leopold's practical, day-to-day approach of experimentalism and his daily recognition of the importance of economic motives. Leopold waxes philosophic, then, in order to explain and make sense of the whole field of environmental management; but he takes this to include economic and other established values, as well as new and more speculative ideas that might guide us to a more complete understanding of values and policy.

In "Some Fundamentals of Conservation" Leopold reviews ethical and metaphysical ideas – including anthropocentrism, organicism (the view that the earth itself is a living organism), nonanthropocentrism, and the possibility that "God himself likes to hear birds sing and see flowers grow" – as possible moral bases for conservation.[10] But he treats these ideas not as a philosopher would, trying to ascertain their truth based on a priori reasoning. Instead, he "screens" ethical beliefs for their policy usefulness, weighing their interest, plausibility, and verifiability, as well as their political appeal. So Leopold, writing from the perspective of a policy-maker in 1923, was willing to speculate on many ideas (indeed, he did so with obvious delight), but he began and ended his speculation with an anthropocentric framework. In the end he emphasized that the nobility we humans claim requires, independently of any moral demands placed on us by nature itself, that we mend our ways and protect ecological communities as we develop our lands. Along the way he considered and specifically dismissed organicism *as a guide to management*, because most managers believe "this reason is too intangible to either accept or reject as a guide to human conduct." He also adopted an agnostic position on the anthropocentrism–nonanthropocentrism debate, resolving that he would "not dispute the point."[11]

Now we are in a better position to understand the relationship Leopold saw between ethics and policy. Most of the decisions he faced could, he thought, be guided by economic criteria, provided he took a long enough view of economics. He also knew, however, that there were other decisions that, if made on strictly individual, economic grounds, would irreversibly damage the things he loved most about the region – the trout streams, the wild vistas, and so forth. Leopold, most basically, believed in the "convergence hypothesis."[12] He believed that human interests and the "interests" of the natural world converge, and that if we were to protect humans – recognizing the full range of human values as projected into the indefinite future – one would also protect the natural world as an ongoing, dynamic biotic community. This leaves the hypothesis of intrinsic value in nature open, allowing Leopold to act in the long-term interests of humans as at least an approximation of what would be "good for nature." It also allows him, in policy contexts, to appeal to either human-oriented or nature-oriented explanations and justifications for the management goals he espoused, broadening the political base for conservation. He often speculated about values beyond the usual human-oriented values that dominated management in his day, but he carefully avoided resting controversial management proposals on these ideas. Leopold tried, whenever possible, to base real decisions on careful observation and experiment rather than speculation. This dualism, and this partial disengagement of Leopold's ethical thought from his management philosophy, explains how Leopold can both enjoy speculation and also wax disdainful of abstract thought.

Leopold saw, at least by 1933, that general philosophical solutions and "isms" – "Socialism, Communism, Fascism" and "Technocracy" – would fail, because they would not, or could not, adjust "men and machines to land."[13] Leopold knew long before the "isms" failed that "husbandry of somebody else's land is a contradiction in terms."[14] But he also knew – contra conservative privatizers of today – that many public interests must be protected on public lands. He appreciated that communities of users must, with the help of sympathetic agency managers acting as teachers, maintain control of those public lands, using their government to realize community-based values and goals that go far beyond economic ones.

NOTES

1. Leopold, "The River of the Mother of God" (1924), RMG, 126–27.
2. Leopold, "Weatherproofing Conservation," *American Forests* 39:1 (January 1933), 10–11, 48.
3. See Bryan G. Norton, "Integration or Reduction: Two Approaches to Environmental Values," in Andrew Light and Eric Katz, eds., *Environmental Pragmatism* (London: Routledge Publishers, 1996), 105–38.
4. Leopold, "Pioneers and Gullies" (1924), RMG, 112, *Aldo Leopold's Southwest* (ALS), 173. See C. S. Holling, "What Barriers? What Bridges?" in L. H. Gunderson, C. S. Holling, and S. S. Light, eds., *Barriers and Bridges to the Renewal of Ecosystems and Institutions* (New York: Columbia University Press, 1995), 3–34.
5. Leopold, "Thinking Like a Mountain" (1949), A Sand County Almanac (ASCA), 132. See Bryan G. Norton, "Context and Hierarchy in Aldo Leopold's Theory of Environmental Management," *Ecological Economics* 2 (1990), 119–27.
6. Leopold, "The American Game Policy in a Nutshell," *Transactions of the Seventeenth American Game Conference* (December 1–2, 1930), 281–83.
7. Leopold, "Some Fundamentals of Conservation in the Southwest" (1923), RMG, 97.
8. Herbert Simon, *Administrative Behavior* (New York: Free Press, 1945).
9. Leopold, "Game and Wildlife Conservation" (1932), RMG, 166.
10. Leopold, "Some Fundamentals of Conservation in the Southwest" (1923), RMG, 96.
11. Leopold, ibid., RMG, 95, 96.
12. Bryan G. Norton, *Toward Unity among Environmentalists* (New York: Oxford University Press, 1991).
13. Leopold, "The Conservation Ethic" (1933), RMG, 188.
14. Leopold, "Land-Use and Democracy" (1942), RMG, 298.

# 9

## Improving Ecological Communication

### The Role of Ecologists in Environmental Policy Formation

INTRODUCTION: DO WE NEED NEW TERMINOLOGY
FOR ECOLOGICAL COMMUNICATION?

I recently [1995] attended an excellent workshop on "Communicating with the Public on Ecological Issues"; the subject of the workshop was to discuss how well scientific ecologists, agencies, and the public communicate regarding ecological matters and how to improve these communication processes. Generally, participants were critical of the current levels of communication, but many positive and constructive ideas were contributed by breakout groups. One group was charged to discuss "Characteristics of Ecological Issues Affecting the Communication Process" and reported back with a list of caveats and recommendations. The first of these (as delivered by group leader Lawrence Slobodkin) was that we do not need any new scientific terms, such as "ecological health" or "ecological integrity." "Everything that needs to be said to the public," Slobodkin argued, "can be said in the existing terminology of ecological science." Since I know Larry enjoys a good philosophical argument, and I hope he will agree that the subject is of general interest, I here set out to provide a careful criticism of that recommendation. In the process, I hope also to provide, more positively, a brief glimpse of a somewhat different way of looking at the relationship between description and evaluation, and of a different way of relating social and natural science in policy process.

In order to understand the context of Slobodkin's recommendation, it is necessary to note the conceptual disorder that currently impedes communication among scientists and policy makers, a disorder that might be described as typical of extra-paradigmatic disagreement (Kuhn 1970; paper 12, this volume). A paradigm is a constellation of assumptions, methods,

From *Ecological Applications* 8 (1998): 350–364. Reprinted with permission.

and normative judgments, shared by the practitioners of a scientific discipline, that constitute and give shape to that discipline. Extra-paradigmatic communication can then be understood as proceeding without such a set of constitutive features. Sometimes extra-paradigmatic communication is "pre-paradigmatic," in which case one would conclude that the subject matter had not yet been made the subject of a proper scientific discipline. But this is not quite the situation in environmental science and policy, where there are competing natural and social scientific disciplines, each with its own paradigm; here, the problem occurs on the integrative level (paper 12, this volume). What is lacking is a paradigm, and an appropriate vocabulary to express it, at the integrative level of environmental management. What is to be the language/paradigm of this integrative, activist, management science? Is it to be the language of ecologists? economists? policy wonks? astrologers?

Slobodkin's prohibition on new terms might be taken in several ways, and it will be useful to list and explain their implications separately.

(1)   We do not need any new terminology, period. The existing descriptive and evaluative vocabularies are ideal. This view has the advantage of clarity and simplicity. But it raises difficult questions. Is it denied, contrary to the assumption justifying the Workshop, that there is a problem in ecological communication? If not, then this simple denial is hardly credible in the absence of an alternative explanation for the communication failure.

(2)   We do not need, the Slobodkin group might intend, any new descriptive terms to express the insights of ecologists. This may be true, and ecologists are no doubt better judges of its truth than I, but I believe the Workshop organizers intended to address a broader question: Is there a failure of communication in the more inclusive, normative-as-well-as-scientific discourse of environmental planning and management? If Slobodkin's group intended to confine its remarks to the communication needs of ecologists to other ecologists, I think its ban on new terminology is simply irrelevant to the problems of communication – between ecologists, policy makers, and the public – that was the topic of the Workshop. By focusing on this broader question, I hope to convince ecologists that they do have a responsibility to participate in the creation of means of communicating knowledge that is important for the public and policy makers to know. If ecologists try to affect policy by simply sliding research papers out from under their laboratory door, they are are leaving the interpretation of their data and generalizations to policy analysts who are much more likely to conceptualize the decision process as one affecting short-term human economics. Failure to participate in this interpretive process virtually ensures that long-term, ecologically based values will not be protected. Accordingly, I address these broader issues in the remainder of the paper.

(3)   Or perhaps the members of Slobodkin's group were mainly expressing their distaste for the specific terms, "integrity," and "health," as applied
to ecological systems (Costanza, Norton, and Haskell 1992; paper 17, this
volume). I do not wish to debate the choice of particular terms or phrases here,
because it is more important to address the broader question of the causes of
communication failure. If the terms "integrity" and "health" are tainted beyond usefulness in communicating with the public, then by all means, let us
jettison them and suggest alternative terms to link ecological data with social
values. If the criticism of the Slobodkin group was narrowly directed at these
particular terms, then again, I invite them, and other ecologists, to ponder
the more general question of whether we need some new terminology that
more explicitly links the discourse of ecology with the normative discourse
of evaluation.

For reasons discussed throughout this paper, I view descriptive-normative
terms as important elements in the discourse of policy integration, where the
problem is not simply to understand, but to make better decisions regarding
human use of resources, including scientific resources. This is not to say that
there is no place for disciplinary science; nor is it to say that we should not
use the best, and most "objective," science available in assessing policies; it
is rather to say that in public policy discourse there is a need for "indicator"
terms that have respectable scientific content and at the same time embody an
evaluative aspect. An indicator is "something that provides a clue to a matter
of larger significance or makes perceptible a trend or phenomenon that is
not immediately detectable" (Hammond, 1995). Note that this definition of
an indicator, when combined with arguments that "ecological significance"
cannot be determined without reference to social values (Harwell, Gentile,
Norton, and Cooper, 1994; Norton, 1994), entails that ecological indicators
embody normative judgments regarding social significance. On the approach
suggested here, ecological indicator terms would express a measurable quality
of ecological systems that exemplify socially desirable states of those systems.
One way of posing the general communication problem, then, is to ask whether
policy discourse would be improved by the introduction and general use of one
or more indicator terms that represent measurable characteristics of ecological
systems and also reflect impacts on social values.

The goal of this paper is to begin to analyze this communication problem
and to engage ecologists in a search for solutions to it. Accordingly, Part 1
undertakes a preliminary survey of the communication problem, discussing
the role of terminology in successful communication. In Part 2, two examples, considered representative of current efforts to link ecological science
and evaluative frameworks, are examined; it is argued that current attempts

are unlikely to succeed because they aim to create evaluative responses independent of, and subsequent to, ecological descriptions. Part 3 and 4 examine in more detail, using case studies, failures of communication between ecologists and the public. Part 3 provides evidence of the communication failure by examining the case of wetlands policy, establishing that the EPA Workshop posed a real and important problem at least in one important policy area. Part 4 examines another area of failure of communication – the choice of scale at which problems are characterized and studied – and shows that the problem is not just one of failure of ecologists to inform the public; ecologists are not sufficiently sensitive to social problems and values in their choice of research topics, which suggests that there is a failure of ecologists to accept signals from broader social and policy discussions. Finally, in Part 5, some positive characteristics of a more successful vehicle of communication are proposed for further discussion and criticism.

PART 1. ECOLOGY'S COMMUNICATION PROBLEM

Before addressing the question of whether new terminology is needed to express ecological insights in policy discussions, it will be useful to discuss the exact nature of the "communication problem" that occasioned the above-mentioned, EPA-sponsored meetings. Failures in ecological communication might, of course, be explained in a number of ways. Surely, the complexity of the message that ecologists often have to convey is an important factor. Surely, also, ecological illiteracy of the public and of many policy makers is a factor. But it is also possible that the communication problem is significantly due to the language and format in which ecological information is presented to the public; or it may be that most ecologists study aspects and processes of nature that have little interest or application to the public's concerns or to environmental management. All of these factors may have some effect, but it makes sense for ecologists to concentrate their attention on aspects of the problem over which they have some control. With regard to the first aspect, ecologists cannot, perhaps, reduce the complexity of their findings; but it may be possible for them to develop new techniques for presenting these findings less complexly, so that it is possible for nonspecialists to grasp essential ecological information and see its application to important public decisions. As for ecological illiteracy in the society, it may be unfair to blame this condition on ecologists; nevertheless, it seems that ecologists should, if anybody should, accept responsibility for increasing ecological understanding among the public and policy makers. These concerns point directly to questions about ecologists' methods, the linguistic patterns they use to present their findings in larger, policy-relevant contexts, and their own ability to understand and respond to policy discourse. These factors will be the main concern of this paper.

A number of authors have argued that there are features of ecology that greatly limit its usefulness as a guide to environmental policy (McIntosh 1985; Sagoff 1988; Peters 1991; Shrader-Frechette and McCoy 1993). For example, some authors emphasize that, unlike physics, which is a science with very general applicability, ecology is apparently a science of the particular. Its truths apply locally and it is often difficult to generalize across cases. This conclusion suggests, as a corollary, that the aspect of ecology that is likely to be useful in environmental protection is specific and localized knowledge of particular species and ecosystems, rather than general and abstract ecological theory (see especially Sagoff 1988). It is also argued that ecologists are divided into factions, members of which do not share even basic assumptions about the nature of biological communities (see, for example, O'Neill, DeAngelis, Waide, and Allen 1986). All of these and other arguments suggest that features of ecological science itself have resulted in the marginalization of ecology in policy processes. Despite the inherent difficulties of applying a complex and multi-faceted science such as ecology, however, the critics of ecology's current role are not arguing that ecologists should play a lesser role. They have argued, rather, that ecologists should abandon pronouncements based on grand theory and increase their involvement in the study and management of specific conservation sites. Adopting the role of working effectively as researchers with public agencies and the public in conservation projects, however, simply accentuates the need to improve communication among ecologists, communities, and policy makers.

As a preliminary step, I introduce a useful simplifying technique based on findings of logicians and philosophers of language (Carnap 1950; Quine 1964). Rather than considering the question of whether the communication problem is exacerbated by lack of "concepts" separate from the question of whether adequate "terminology" exists, it is possible to avoid reference to concepts at all and limit the question to whether there are adequate "terms," "vocabulary," or "language" to express what needs to be expressed. This ploy need not be considered as a refusal to consider concepts important. The existence of "concepts" is subsumed within judgments that a term is adequate – a term will be considered adequate to its task only if it is inter-defined with other terms within a theoretical framework, and only if it is accompanied by rules for its application in discourse, and so forth. Terms that are adequate in this sense get their meaning from their semantic relations to other terms within a "paradigm" which, as noted above, is a constellation of assumptions, theories, methods, and norms that unite a discipline of study. Further, a set of terms used to convey scientific information will, in turn, be considered "adequate" only if the paradigm which gives them meaning is capable of expressing that which must be communicated. This holistic approach to environmental science has emerged from developments in epistemology (Quine 1951; 1964) and also in the history of science (Kuhn, 1970), the latter of which has given rise to the current popularity of the idea of a paradigm.

To say that a term, or set of terms, is adequate for a given task, then, requires a) the existence of definitions that link the term to the theoretical core of a disciplinary paradigm relevant to the task at hand; b) that the disciplinary paradigm is sufficiently rich to express semantic content relevant to accomplishing the task; and c) that some elements of the paradigm must be sufficiently well connected to the natural world by operationalizations to render its key sentences subject to some level of verification or falsification. By directly inquiring whether these conditions are fulfilled, we can avoid confusing shifts between evaluating terms and evaluating concepts because, under this interpretation of a "concept," the existence of such will be entailed by the determination that the terms in question are "adequate" in the sense defined. It is unquestionably true that, despite this simplification, there remain many difficult issues to resolve regarding how to characterize the "task" at hand, the adequacy of the language to it, and other semantic matters. My point is simply that formulating the problem as one of adequate language is both simpler and more likely to lead to fruitful discussions, such as whether the current linguistic resources of ecologists are adequate to support communication in policy contexts.

Given that the task faced here is that of communication with the public and with policy makers, the goal should not be to choose the language or paradigm of any particular science, but rather to create an integrative language in a more comprehensive and more action-oriented science than either ecology or economics can be (paper 12, this volume). It is practically tautological to say it, but a disciplinary paradigm for the discussion of policy issues should be adequate to express all of the scientific information relevant to making good decisions in the specified policy arena, to integrate that information in a larger discourse that expresses social evaluations, and to employ that information to develop science that is responsive to real questions regarding how to protect social values. This activist, decision-making approach to seeking information differs fundamentally from the scientific mode because, in the world of environmental management, Type II errors – failures to draw a conclusion – can have costs equal to those of drawing a mistaken conclusion (Lee 1993). So the context – and the standards of scientific certainty and proof – can shift as one places science within a larger enterprise of management. The goal of the discourse of management is not just pure understanding; it is to learn how to protect and manage ecological systems we live within. So, unlike the special sciences, policy discourse – a discourse of action – must contain terms that are frankly value-laden. Their function in this broader action discourse is to express evaluations; the paradigm relevant to policy discourse is therefore not the discourse of descriptive science which, to the extent possible, is purged of explicit valuations. The problem of ecological communication, therefore, must be addressed within the activist discourse of the science of environmental management, not within the discourse of the special sciences. So if Slobodkin's group meant only to say that we need no more purely descriptive vocabulary, then there may be little grounds for disagreement on that point. But true ecological communication requires that descriptive results can be integrated into the comprehensive and action-oriented paradigm of policy and management. And the question remains: what should the ecologists' role be in shaping public policy discourse? A possible role would be the contribution of terms that have descriptive content and, through potent analogies or by connection with widely held human values, provide the public with a clear sense of why ecological science is useful in choosing better environmental policies.

To see how choice of terminology and the paradigm of a scientific discipline can have a huge impact on the larger arena where policy issues are formulated and decided, consider the impact on policy discussions of the terms "Gross Domestic Product" (GDP) and "growth," which have been given operationalizable and measurable meaning in the discourse of welfare economics. While GDP is in many ways a flawed measure, it exerts a powerful force because, given economists' identification of growth, improvement of welfare, and upward trends in the GDP, the person on the street does not have to ask whether a severe drop in the GDP is good or bad news. It is a term that summarizes an immense amount of empirical data and, at the same time, it wears its impact on widely shared social values on its sleeve. Because the public understands the negative impact of a weak economy on their lives and lifestyles, the public can be very hard on incumbent candidates who preside over disappointing economic growth. Again, my point is not to praise the GDP measure as an ideal, or even to suggest that ecologists should seek a single, over-all measure such as this, but rather to demonstrate the power well chosen terminology can have, especially when that terminology successfully connects trends in scientific data with an unquestioned public good – a strong economy in the case just mentioned.

By contrast, discussions by ecologists of long-term impacts of current activities are often expressed in language that suggests no clear connection between scientific assessments and identifiable social values. For example, Arrow et al. (1995), writing in a *Policy Forum* article, argue convincingly that achieving economic efficiency does not in many important cases guarantee adequate protection of resources. This same group, members of which were writing as distinguished representatives of economics, ecology, and related disciplines, then argue that sustainable use of resources requires not only economic efficiency, but also that human activities protect the "resilience" of ecological systems. Does the person on the street, or the average public servant, know the impact of having, or of losing, resilience of large-scale ecological systems? To their credit, Arrow et al. set out to explain its importance, but it is questionable that their explanations will be clear or convincing to non-ecologists (Norton 1995b). Again, the contrast with economic discourse is striking: whereas trends in the GDP are directly associated with trends people care about, resilience has no such connection, unless it is laboriously made in each and every context, and explained by appeal to anecdote and scientific generalization. Failure, then, to employ language that helps non-ecologists make connections between ecological trends and social values has a great cost: the public and policy makers know whether trends in data are good or bad only if they are willing to learn a body of scientific information and its application to "sectors" of public interest.

So, this is my central contention – that policy discourse currently suffers because, whereas economic data is easily associated with the well-being of citizens in our democracy, ecological data has no such resonance. Whether new descriptive, ecological terms are needed or not, I contend that existing language – expecially if policy discussants are prohibited from using "health" and "integrity" in this context – is inadequate to establish a meaningful, multi-directional dialogue among scientific ecologists, policy analysts, policy makers, and the public.

## PART 2. SERIAL THINKING VERSUS INTEGRATED THINKING IN ECOLOGICAL RESEARCH

In this part, I lay the groundwork for my analysis of the ecological communication problem by noting how ecologists and social scientists have conceived the task of integrating ecological science with analysis of social values in published work. To this end, I discuss two otherwise laudable research efforts in the area of ecology, economics, and policy and show

how this research exemplifies what I call "serial" thinking about science and policy.

First, the serial tendency is well illustrated by the organizational structure and content of an otherwise excellent essay by Batie and Shugart (1989). This essay provides a method of spatio-temporal scaling as an important part of a system for evaluating changes that may result from global climate change. The first half of the essay (which one assumes expresses mainly the contribution of the ecologist Shugart) provides general guidelines for describing the scalar aspects of physical systems and provides examples of useful models for interpreting the multiscalar activities of complex natural systems. There is no discussion in the paper of how information about social values might affect the descriptive models developed. Once the descriptive aspect of the problem is treated, the economist Batie brings the evaluative framework of economics to bear on the changing states of the physical system in the second half of the article. My point, here, is that their argument is developed in two distinct stages with distinct languages – ecological description followed by evaluative discourse. While this serial tendency may not be inconsistent with an iterative approach to information gathering, goal-setting, and managerial decision making – it may be possible to revisit the science and then discuss policy options once again – the serial approach does not facilitate the flow of information from evaluative discourse to scientific discourse. Batie and Shugart attempt to build an evaluation method on top of scientific description – but they do not close the loop and create a shared language that allows multi-directional dialogue about social goals and their impacts on scientific models. There is one language that describes the world and another that evaluates it, and these languages are deployed in serial fashion. In a truly effective and democratic process of environmental policy formation, the searchlight of science must also be sensitive to social values. One problem with serial thinking is that it does not facilitate the flow of information from policy discourse to ecologists – a topic to be discussed in Part 4 below.

As a second example, consider the remarkably complex and interesting effort at modeling the Patuxent watershed, reported in Bockstael et al. (1995) by researchers at the University of Maryland. This team of ecological and economic modelers has followed the strategy of building separate ecological and economic models of a river basin, one modeling the ecological impacts of land conversions and the other modeling the economic impacts of land conversions. The goal of associated interdisciplinary work, then, is to find or establish linkage points between the models. This effort, however sophisticated and successful in generating predictions and correlations, would not score well on the criteria I am using here, because the participants employ two separate descriptive languages – those of economics and ecology – to build two independent models. Even if this serial approach were adequate for providing information to professional managers, I would still question whether the use of fragmented descriptive languages will generate better public understanding. One cannot simply merge two descriptive vocabularies and create an evaluation. Good decision making is integrative, and these scientific languages are hardly suitable as tools of integration and broad social communication, because disciplinary languages are designed to describe one dynamic in separation from others, to abstract from other processes.

While the two efforts just discussed are important ones, and they no doubt expand our knowledge and methods in the area of environmental science and policy, they do not in my opinion establish an iterative and multi-directional discourse that could progressively inform a process of adaptive environmental management. The failure of serial approaches to integrate scientific information is no accident; they fail because they are based on a false image and an associated myth that is perhaps the greatest barrier to an improved understanding of the twin enigmas we call "ecosystem risk" and "ecosystem management." The image is that of an ideal environmental decision maker, one who has gathered all the descriptive information regarding the functioning of an ecological system, who has determined the likely outcomes of further impact from human activities, who has then polled the population to determine the values, goals and preferences in good democratic fashion, and who – armed with all the facts – decides what policy to pursue to maximize total welfare. This false image underlies much of our thinking about environmental management today.

We all know that the ideal of a fully knowledgeable manager is impossible and that the idea of a "complete" science is an oxymoron – science is an ongoing process of revision and improvement, not an existing body of fact. But the myth of a complete science nevertheless has a powerful influence on our view of policy process, because we simply substitute "the best available science" for "complete science" and sustain the damaging part of the myth – the assumption that there is a one-way flow of information from scientists toward policy makers and the public.

We also should question the serial approach because managing in open systems with high degrees of uncertainty and ignorance is very different from managing with a completed set of knowledge, because of the shift, mentioned above, in norms regarding Type I and Type II errors. As has been pointed out by advocates of adaptive management, further discussed below, once we realize that uncertainty and ignorance are an inevitable part of management, surprises are to be expected, and policies should be designed to survive surprises and to reduce ignorance in specified and useful ways in the process (Holling 1978; Walters 1986; Lee 1993; Gunderson, Holling, and Light 1995a). If we are to improve our management decisions, we must also use management efforts to direct future science, to reduce uncertainty, and to correct our policy course with new and relevant information. And here we see the inadequacy of the serial model, which appears as a mere after-image of a mythologically complete science. Science affects our management decisions; but on the serial model it is impossible for management decisions and the social values that drive them to direct the future path of science.

This shadow, this false image of a completed science, has been so hard to obliterate because it reinforces, and is reinforced by, another myth, the myth of a "value-neutral science" (Proctor, 1991). Value-neutral science supposes that the scientific enterprise can be purged of the corrupting influence of attitudes and social values, and posits a value-free space in which scientists can describe, objectively, the characteristics of a world beyond human value. Again, if pressed, we all admit that the path of scientific research is deeply affected by valuations – of the scientist, of funding agencies, of foundations and firms, and of other scientists. And yet we want our science, especially the science that affects policy, to be above all bias and reproach, so we pretend that the scientific description comes first, and that it is as complete as it will ever get before we begin policy analysis, goal-setting, and practical decisions.

PART 3. THE IMPACT OF ECOLOGISTS ON POLICY:
THE CASE OF WETLANDS

Serial, rather than multi-directional, communication patterns have failed to lead to integration of scientific information into policy processes; they have led, rather, to the development of separate ecological and economic models expressed in incommensurable languages. In this part, I explore the extent to which ecologists, dedicated to this serial approach, have been successful in bringing their science to bear upon policy by examining the case of wetlands policy.

It is unquestionable that ecologists have had some impact on public policy regarding wetlands; it was ecologists who created a flow of information about the functions and publicly enjoyed values of wetlands that eventually – in the 1970s and 1980s – reversed the longstanding governmental policy of allowing and even encouraging unlimited swampbusting on private lands. While this flow of information succeeded in changing the rhetoric of official policy, and led to a political commitment to a policy of "no net loss of wetlands," ecological science has not succeeded in creating a body of ecologically sensitive law and policy in the area of wetlands regulation. Mitsch and Gosselink (1993), in the second edition of their well-received study of wetlands, say (p. 542): "Wetlands are now the focus of legal efforts to protect them but, as such, they are beginning to be defined by legal fiat rather than by the application of ecological principles." Worse, there is no unified, specific national wetland law, and wetlands are not managed according to a coherent and comprehensive regulatory system, but rather according to a hodgepodge of water quality laws and terrestrially-based land use law (Mitsch and Gosselink 1993, p. 565)

The politically motivated formula, "No net loss of wetlands," has created a virtual market in "acres" of wetlands, with created or restored wetlands being bartered against permitted losses (Mitsch and Gosselink 1993; King 1997b; 1997d). But the entire idea of wetlands banking depends upon being able to assign comparative values to restored or created wetlands and wetlands lost to development. So far, ecological principles have had little effect on these calculations. Zedler (1996, p. 36) says, "Rigorous research on wetland restoration construction is inadequate to provide simple formulas for constructing one wetland to compensate for functions lost in destroying another." Simenstad and Thom (1996, p. 38) similarly state: "The ability of wetland managers to assess compensatory-mitigation success over short-term (e.g. Regulatory) timeframes depends upon the selection of attributes that can predict long-term trends in the development of the restored/created system. However, we are hampered by a basic lack of long-term data sets describing the patterns, trends, and variablity in natural wetland responses to disturbance, as well as natural variablity in natural wetland response to disturbance, as well as natural variablity in wetland attributes in presumably mature wetland communities."

The lack of relevant science is remarkable. According to Mitsch and Gosselink, despite thousands of 404 dredge and fill permits issued to allow trade-offs between constructed and mature wetlands, ecologists had (by 1993) published no relevant studies comparing undisturbed and constructed or restored wetlands. Preliminary data has since been published (Kentula et al. 1993). Preliminary studies apparently suggest that constructed wetlands are considerably more vulnerable to invasion by pests and less resilient in the face of bad weather, but nowhere near enough work has been done – or is in progress – to draw reliable generalizations and judgments about these important comparisons that are necessary in the current regulatory process.

Lack of good and relevant science in this area has opened the door to political solutions such as those proposed by Dan Quayle's Competitiveness Council, which simply count acres and allow wetland banking, creating a situation where "no net loss" means "no net loss of acreage." This approach is currently allowing monumental, but so far immeasurable, losses in social values that might have been derived from wetlands under a more ecologically informed policy. I will concentrate on two aspects of this failure here, emphasizing the failure of current models to encompass contextual features of wetlands and on the failure of academic ecologists to develop models useful in public decisions.

Failure to pay attention to contextual values is illustrated – and virtually ensured – by the wetland indicators used by the U.S. Fish and Wildlife Service – the Wetlands Evaluation Techniques (WET) and the Habitat Evaluation Procedure (HEP). These procedures rate wetlands according to a broad range of functionally defined attributes. Unfortunately these evaluation techniques and most of the information they embody are site specific and provide no information regarding the ecological context of wetlands under study (Mitsch and Gosselink 1993, p. 533; King 1997d). These techniques therefore cannot capture values that emerge at the level of ecosystems and at the landscape level. According to Mitsch and Gosselink, in only one area – evaluation for wildlife habitat – does there exist even one evaluative model, due to O'Neil et al. (1991), that takes into account large-scale ecological processes that affect wetland habitat quality. This model, however, applies only to bottomland hardwood forest as wildlife habitat (Mitsch and Gosselink 1993, p. 533).

Since contextual information is often crucial in making public decisions regarding what wetlands will be sacrificed to development and how to compare various sites, one might think that academic ecologists would be eager to work on these problems. King, however, divides relevant research into two categories and says: "With a few exceptions the analytical tools that hold the most promise for implementing Habitat Equivalency Assessment (HEA) and for evaluating scaling issues under OPA [Oil Pollution Act of 1990] are from the more practical ecosystem assessment literature, rather from the more rigorous scientific and landscape modeling literature" (King 1997b, p. 5). He states that most studies in the latter category, designed to find out how ecosystems work, is too theoretical and the cost of calibrating the landscape linkages necessary to apply the models are often prohibitive. He concludes: "At this time the methods of analysis described in the scientific literature dealing with ecosystems do not provide a practical basis for demonstrating that two ecosystems are equivalent or comparable in terms of services or values (the focus of scaling and and HEA under OPA)" (King 1997b, p. 4).

In practice, the strategy of limiting evaluation to capacity and function has been disastrous for policy discourse. By limiting their wetland classification methods to measurements of descriptive ecological values based mainly on site-specific judgments of capacity and function, ecologists have undercut efforts to include system-level ecological features as a part of the decision process (King 1997d). Thus, while ecologists argue explicitly that the methods of economists ignore "ecosystem and global-level values relating to clean air and water and other 'life-support' functions" (Mitsch and Gosselink 1993, p. 535, paraphrasing Odum 1979), they have nevertheless failed to expand the concept of "ecological value" of wetlands to include off-site features and the way these features affect public values. King (1997a, p. 6) insists upon a distinction between ecological assessments and ecosystem valuations, noting that the assessments normally undertaken by ecologists "provide only the front-end part of the analysis required to compare ecosystems on the basis of the services and values they provide." Mitsch and Gosselink (1993, p. 508) survey the techniques available for quantifying wetland values after acknowledging that ecologists are usually not interested in "value to society," but mainly in the ecological functions of "abstruse" wetlands processes. They proceed to provide an overview of ecosystem-derived services such as recreation and other socially valued commodities, recognizing that many of these represent public goods that are often not captured by owners. These public values, then, can be identified and measured only by use of contingent valuation (CV). But CV studies, however illustrative of particular public values, are too expensive to provide a comprehensive accounting of the vast number of public values affected by wetlands policy (King 1997a, p. 6).

King (1997c; 1997d) reports that ecologists have been slow to participate in attempts to establish a middle ground between economic, market valuation techniques on the one hand and morphological analysis on the other. Although long-term research continues on the analytical problems involved, it is unlikely to be useful in comparing wetlands in policy contexts because, as King argues (1997a, p. 7), "in most legal and regulatory contexts where decision-makers are asked to compare ecosystems on the basis of their values, they are asked to compare their expected future services and values on the basis of information about their current biophysical features and landscape contexts." But in order to be useful in these situations, "there must be some way for the assessment of observable ecosystem features to be linked forward to expected ecosystem functions and services, and for the valuation of ecosystem services to be linked backwards to observable or measurable ecosystem features. Most of the ongoing research to improve the scientific basis of ecosystem assessment methods and the credibility of ecosystem valuation methods will not fill this gap." King (1997a; 1997d) hypothesizes that ecologists avoid evaluative classifications that truly rank wetlands according to social value in order to avoid becoming entangled in value judgments and to avoid having their work used to justify lesser protection for wetlands they describe as having lesser value.

The strategy of limiting ecological study to assessment of ecosystem capacity has left key ecological values – values that can only be perceived by relating wetlands to their surrounding ecological matrix – in a linguistic no man's land, dangling between the assessments of capacity reported as measures of "ecological value" and the reductionistic, dollar-value approach of welfare economists (King, 1997a; 1997b; 1997d). For ecologists to push this problem of identifying valued features over to economists is folly. Standard willingness-to-pay economic models have simply not been developed to encompass system-level and landscape-relative values, the very values that might qualify as "ecological." Indeed, even supporters of economic evaluation have admitted that their techniques are unable to measure system-level values like biodiversity protection or ecosystem states and functions (Freeman 1993, p. 485). As a result of ecologists' evaluation strategy, wetlands advocates have no way to inject quantitative results regarding the value of ecosystematic features into the policy discussion. This is equivalent to showing up at a gunfight with a willow switch. Given the poverty of the evaluative language to express the social value of intact and functioning wetlands, it is unimaginable that a few vague references to "public goods" such as water purification, fish and shellfish incubation, etc., could overcome the effects of commercial and economic pressures that can point to unquestioned benefits of wetland conversion.

Furthermore, neither the capacity/function approaches of ecologists nor the dollar-value approach of economists address equity issues that emerge on the ecosystem and landscape scale (King 1997d). For example, if wetlands are evaluated only with regard to their capacity and function within a wetlands mitigation banking system of management, it is predictable that developers will request to dredge and fill wetlands in urban areas with high property values first, discharging their mitigation obligation by creating or restoring wetlands in areas of lower density and less value per acre. So the current approach to banking and mitigation is likely to be systematically transferring wetland function from urban and highly populated areas to lower-populated rural areas (King 1997d). Even if a greater than one-to-one ratio is required for replacement wetlands vis-à-vis original wetlands that are filled, this policy still represents a systematic transfer of social values – wetland services such as pollution reduction and amenities such as viewable wildlife – from lower-income urban areas to low density, relatively wealthy rural communities in many cases. Again, failure of ecologists to join a dialogue to bridge their descriptive information to important social values – in this case the value of environmental justice and equity – has left policy makers with no ecologically informed and quantifiable criteria for deciding when to accept and when to reject mitigation proposals offered by developers.

My analysis of the current status of wetlands policy suggests that additional evaluative terminology, such as indicators that reflect not just the capacity and function of wetlands, but also the landscape and society-wide impacts of specific wetlands conversion, would be of considerable value in policy discourse regarding wetlands policy. King (1997a; 1997b; 1997d) has provided a sketch of a more contextual and inclusive evaluation procedure (COPE), based on considerations of Capacity, Opportunity, Payoff, and Equity. This broader approach to valuation would help to bridge the gap between functional analyses of wetlands by ecologists and economic analyses by economists, but most ecologists, as noted above, have been slow to accept this approach because of the fear of becoming involved in value judgments (King, 1997c).

I conclude that there is a considerable case to be made that new evaluative terminology is needed to balance policy discourse, increasing the likelihood that the public will associate scientific data and empirical trends with social values other than short-term economic ones. Further, it seems to me that it is essential that ecologists contribute to the operationalization of such indicators. This, then, stands as an answer to the Slobodkin group's recommendation on terminology necessary for communicating ecological science to policy makers. I would, of course, welcome their response in turn.

But a further, more troubling aspect of this communication question emerges from our analysis. If I am correct that ecologists have been slow to provide comprehensive ecologically sensitive models for ranking wetlands systems, that the models and "evaluation" techniques are almost entirely site specific and therefore cannot capture truly ecologically-based values, and that ecologists have done only a small number of comparative studies of the thousands of conversions that have been allowed under mitigation banking, it begins to appear that there is also an "ecological communication" problem in the other direction – ecologists have not been sensitive to the needs of policy analysts and policy makers. The failure of ecologists to contribute effectively to the development of an ecologically-sound, positive wetlands policy certainly involves, to some degree, a failure of ecologists to communicate clearly information that they have gathered, but I hereby hypothesize that the larger problem is a failure of ecologists to pick up clues regarding what is really important from political discourse, and a failure to develop techniques and studies that can usefully inform policy discourse.

151

PART 4. SCALING AND VALUING IN SCIENCE AND POLICY

When we address Slobodkin's question of whether ecologists need new terminology, we must keep in mind that communication is a two-way street. The function of new terminology may not be solely to formulate information that will flow from ecological science into the policy process: its function might be to facilitate a flow of information from policy process to science, shaping the choices and approaches of ecological researchers. In this part, I explore the possibility that ecologists are marginalized in policy process because they are failing to pick up important signals that are coming back to them from policy discourse, signals that might guide them in setting their research agenda.

The belief that nature is, or can be, measured and described before one decides what is important – the serial view – is a dangerous illusion. It is an illusion because it assumes falsely that science can be without perspective and scale, and that it can be complete or finished prior to action; it is dangerous because the illusion of prior omniscience predisposes managers and scientists to under-value learning, social and technical, as a reason to pursue policies. Simply studying what they find "ecologically interesting," without engaging in dialogue regarding what is important in a larger management and public context, has left academic ecologists on the sidelines, providing almost no guidance as to what information and what studies are essential to improve wetlands policy.

The reason the serial approach is so damaging is now clearer. Even if scientific data is gathered and indicators are developed and applied iteratively, allowing successive studies and successive revisitations of a managerial issue, the iterations may not have the desirable effect of integrating scientific study into management unless there is some feedback ("learning") that occurs among scientists regarding which types of measures and which kinds of indicators are useful to managers in making difficult decisions. Only when ecologists and other wetlands scientists are able to learn from successes and failures of their methods in practice will they learn to tailor those methods to represent values that are important in management.

The problem in applying science in the policy process is not that correct description of natural systems is impossible. It is rather that there are too many correct descriptions of nature (Levin, 1992). Given the irreducible complexity of natural systems – which are composed of many dynamics on many interlocking scales – there are many, indeed, an infinite number of descriptively correct characterizations/models of what is going on in nature. Neutral science cannot tell us, however, which of these descriptive models is of interest to human beings and cultures. The result of ecologists studying nature at whatever scale they happen to choose is a Babel of too many voices and too many models. Ecologists have provided little help to the public and policy makers who must decide which of these many models have application to public policy. What is lacking, if we are to create science that is truly relevant to policy, is some way to identify those particular dynamics (out of the myriad possible ones) that should be monitored, encouraged, and in some cases altered in the interest of social values. Viewed in this way, a careful understanding of values affected by wetland alterations, and participation in public forums where social values are articulated, can act as a "filter," focusing attention of academic ecologists on dynamics that really matter to public decision making.

To illustrate this point, consider again the "scale" at which most ecologists study nature. Stuart Pimm (1991) has noted the discrepancy between the temporal scales of ecological studies and the ecological scales on which threats to species and communities occur, noting that in the ecology journals and in the eyes of the National Science Foundation, a ten-year study is considered "long-term." "So where is the ecology of tens to hundreds of species over decades to centuries, across hectares to thousands of square kilometers?," Pimm asks. He illustrates his point with a "caricature" in the form of a three-dimensional graph showing the small spatio-temporal scale of most ecological studies published in the literature. See Figure 9.1. Pimm notes that "The most pressing ecological problems involved many species and their fate across decades to centuries, over large geographical areas.... Managers of natural areas must protect their fragmented communities against a spectrum of possible threats. They may have many species in their charge, manage ten to thousands of square kilometers, and want to pass on their communities relatively intact to their successors a human generation in the future. In order to manage we must understand how populations and communities will change over the very organizational, temporal, and spatial scales about which we know least" (Pimm 1991, pp. 1–2).

Pimm's point illustrates one of the severe failings of ecologists as scientists, and in my opinion, it explains a huge proportion of the failure of their science to impact policy formation in positive ways; ecologists have not sufficiently understood the importance of scale in their own discipline and, even to the extent to which they have recognized the importance of scalar phenomena, they have not taken policy-relevant scale considerations as compelling or even useful information in developing their topics of study and in choosing their research methodologies and the scale at which their studies are directed. Admittedly, landscape-scale ecology is very difficult, and one must build such studies from analytic understanding of subsystems. But these facts do

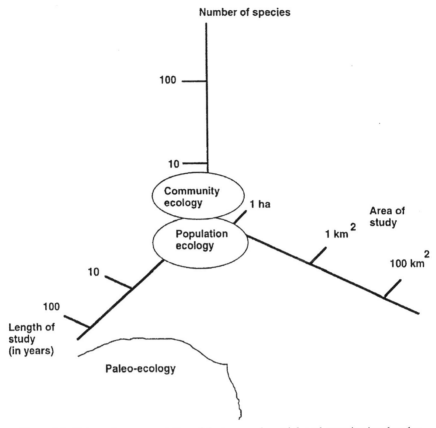

Figure 9.1. Schematic representation of the temporal, spatial, and organizational scales of ecological research. Most ecological research is done on short time scales and over small spatial scales, and even community ecology typically involves few species. The minimum time resolution of paleo-ecology is very much greater than that of the longest-term studies in the rest of ecology, and although biogeography deals with large areas, it rarely deals with how ranges change over time. (Reprinted from Pimm [1991:2], Fig. 1.1, with permission of The University of Chicago Press.)

not excuse the failure to study nature at scales that are relevant to social values.

Scale choices often express values because, given the myriad possible models that one could make of nature, we must choose to model some small subset of the actual dynamics describable there. Those choices express evaluations of ecological and social significance. And yet even the advocates of the most promising attempts to develop scale-sensitive models, such as hierarchy theorists, insist that they design/choose the scalar dimensions of their models only with an eye to improving "scientific understanding" (see, for example, Allen and Starr, 1982; Allen and Hoekstra 1992). This disclaimer represents, if it means anything at all, a thinly disguised appeal to the insulation of science from social issues and a claim of exemption from the value-laden nature of all inquiry, as if there is a conveniently value-neutral intellectual place where ecologists can stand as they develop their "descriptively useful" models in a social vacuum.

An example of an important failure to let social values guide scale selection in determining what to study has already been introduced and documented in connection with wetland evaluation. Expanding on her conclusion that rigorous research for evaluation of wetland restorations does not exist, Zedler (1996, as quoted in Simenstad and Thom, 1996, p. 51) states: "Most monitoring time periods for mitigation and restoration projects are not sufficient to encompass natural perturbations that would test persistence and resilience of the developing community, nor are monitoring attributes sufficient to predict resistance to natural or human-made disturbances or recovery rates thereafter" (also see Zedler 1987). King (1997b, pp. 18–19) reports that a survey of documents and interviews with professionals reveals that methods used to evaluate wetlands "have value judgments embedded in what are presented as objective technical criteria," and that "criteria for developing habitat quality indices for different species may favor waterfowl over fish habitat, resident waterfowl over migratory waterfowl, ... " and they do so with little or no public discussion. He applies the point directly to scaling decisions, noting that "habitats can be defined at different scales," depending on the individual species that are selected to be representative. "Discussions with practitioners suggest that these can be selected so as to favor certain restoration location or certain size restoration projects or can be manipulated in other ways to affect the outcome of ecosystem comparison." As an antidote to these clandestine value judgments, ecologists need to enter the public debate, show when decisions by policy makers and bureaucrats are biased, and create a body of knowledge and methods that allows decision makers to make more intelligent choices in situations of unavoidable uncertainty.

When ecologists fail to develop and improve wetlands indicators at useful scales and on the basis of contextual, landscape-matrix features, they doom their society to act in ignorance in wetlands management. Worse, this ignorance will remain most intense regarding system-level effects, the very effects that are usually cited as especially relevant to more holistic, ecologically sensitive management. Future generations will pay the price for this obstinate failure to develop new methods in response to a real social need. To say that ecologists are limited to studying the scales for which their methodologies are adequate simply begs the question – good science formulates a problem of interest and develops a methodology to gather data bearing on it. If ecologists put their mind to it, they would develop methods to fill in more spaces on Pimm's graph (Figure 9.1). It is a question of making a carefully considered choice as to what scale will be studied and then dealing with the methodological problems as they arise.

Fortunately, in the absence of long-term ecological studies by ecologists, a whole new field – that of ecological history – has emerged and illuminated long-scale, multi-generational change as it results from different management regimes (Cronon 1983; 1995; Crosby 1986) Borrowing methods from ecology and from history, and inventing a few methods of their own, ecological historians have succeeded in illuminating the long-term interactions of cultures and their ecological contexts, examining changing landscapes on a scale of decades and centuries, and linking large-scale ecological change to changing social institutions. One result of these new studies is greater recognition that very few natural systems are pristine, unaltered by anthropogenic impacts (Cronon, 1995). Hopefully, ecologists will accept this outcome and shift their focus from studying systems as if they are unimpacted by human actions, making possible more detailed study of multi-generational interactions of human cultures and their environments.

## PART 5. SOME CHARACTERSTICS OF AN INTEGRATED LANGUAGE OF MANAGEMENT

I turn now, more positively, to a discussion of six general characteristics that a more successful integrative language and discourse would exemplify. My goal in this part is to be provocative and to encourage an open and experimental approach to developing improved concepts for ecological communication, and I do so by offering some provocative suggestions. Based on the above discussion, I proceed to mention several characteristics of the language we seek. Readers seeking a more detailed explanation and justification of these ideas are referred to Norton (1996; papers 12 and 17, this volume) and to Norton and Ulanowicz (paper 16, this volume). These six features will be briefly discussed in turn.

A more acceptable language and associated paradigm for publicly discussing environmental goals and means would be: (1) adaptive, (2) perspectival, (3) multiscaled, (4) operationalizable, (5) normative in content, and (6) communication-enhancing.

156

## *1. Adaptive Management*

Fortunately, there is an alternative to the illusionary serial model, and many elements of this alternative are provided by writers in the tradition of adaptive management (Holling 1978; Walters 1986; Lee 1993; Gunderson, Holling, and Light 1995a). Adaptive management is experimental management; adaptive managers recognize that learning on an ecosystem scale can only be achieved as an interactive process involving scientists (physical and social), the community (including stakeholder groups), and environmental agencies and managers (Lee 1993). Advocates of adaptive management, adopting a multiscalar view of human populations and their environment, have for twenty years urged a monitoring and management effort that seeks to adapt human management to local conditions through creative use of management experiments to reduce uncertainty and adjust goals.

Unfortunately, adaptive management has itself been put forward without a well-articulated value theory. Despite this failing, adaptive management has developed a useful model for understanding human activities within ecosystems, and the model also points to the crucial role of human values in a comprehensive approach to environmental management. For example, Gunderson, Holling, and Light (1995b) have described the modeling problem as that of viewing and measuring human impacts from the inside of a multi-layered, open system in which information flows all directions in what they call a "panarchy." Individuals, who exist in some place, as a part of both natural and social dynamics, experience their surroundings as a mixture of opportunites for adaptation (survival) or as constraints on possible adaptive actions. All human actions which impact ecological systems can then be viewed as experimental manipulations at the interface of a human population and its environment. Once the problem is stated as one of adaptation at multiple scales, it seems reasonable to suggest that adaptive management needs a multiscalar approach to values and valuation to match its multiscalar models of natural systems (Norton 1996; Chapter 14, this volume). Much of the time, human individuals act to achieve short-term goals within an economic system. Often, these decisions will have only short-term impacts, in which case they can be modeled within an equilibrium system which assumes that individual decisions are made against a backdrop of assumptions about material and social systems. But cumulative impacts of the activities of whole populations, especially when there are pervasive trends in individual behaviors (such as sodbusting or clear-cutting), can spill over and affect larger-scaled, and normally slower-changing, dynamics. When management of resource-producing systems is pervasive over large areas, these management practices can make the system less resilient and more subject to rapid change (Holling 1992). Often such changes result in less desired states of the system, as in the progression of grasses to scrub on over-grazed semi-arid pasture-land. It follows that many decisions in adaptive management require that we recognize values that emerge on multiple scales and attempt to reconcile them.

The adaptive management model seeks information about impacts of human activities, as discovered by descriptive scientists, and treats this information as supporting or refuting hypotheses about human behavior. The point is that scientific information is important not just for itself, but also because it helps us to test hypotheses regarding which behaviors and policies are adaptive in the longer, multigenerational scale on which an economy and culture interacts with its larger environment. This outcome of adaptive management sets the stage for characteristic 5, below, according to which our approach must be normative as well as descriptive.

## 2. Perspective and Place

The integrative language chosen will have to be perspectival in the sense that it describes and values the systems of nature from some point inside a dynamic, self-organizing system, as the language of a community of actors who describe and evaluate from a "home" place. Unlike Newtonian science, which assumed an objective viewpoint outside nature for the describer of independent and objective reality, adaptive management sees the world from some place, and concentrates on the local and particular first. Perspective, then, is associated with place-preference, and I believe the language of management science must be rich enough to express many locally originating goals and ideals, and not to cancel out local interests in some grand national aggregation of everybody's national welfare (paper 19, this volume). How far one wishes to go toward emphasis on local values is currently, of course, a matter of much debate. But my point here is abstract: we at least need a language rich enough to encompass intercultural and regional differences, because environmental problems will increasingly become ones of integrating cumulative but large-scale impacts of various human activities on various levels of natural organization (Norton 1991). This can be a healthy debate provided that we can create a publicly understandable language of management that expresses the tensions between local concerns and regional and national attempts to address large-scale environmental problems. Such a language, I believe, will have to be scale-sensitive, both temporally and spatially.

### 3. Multiscalar Modeling

Models to describe environmental problems should therefore be "place-based" and perspectival; but a preference for home place must also be multiscalar. It must be balanced with a sense of the larger space around the place (Tuan 1974; 1977). A complete sense of place, one might say, inevitably includes a sense of space, a conception of how the home place relates outward, and also a sense of history, both cultural and natural, of the home place. So the integrative language we choose must also be capable of expressing spatio-temporal relations in a way that is clear and perspicuous. In this sense our language must be multiscalar (Norton and Hannon, 1997). Here, hierarchy theory, which assumes both a perspectival viewpoint and a principle for organizing multiscalar, spatio-temporal relationships, may provide important structure and methods for better understanding ecosystem management. But re-conceiving hierarchy theory within a frankly value-laden management context will undoubtedly prove controversial.

### 4. Operationalizable Terms

Another requirement of any integrative language is that it must contain operationalizable terms. It is necessary to specify certain measurable indicators that track the variables and dynamics essential to support and sustain social values. Adaptive management will succeed, if it succeeds at all, by designing management plans that allow scientific and social learning about human impacts on ecological and physical systems including perhaps changes in preferences and values (Chapter 14, this volume). Thus reducing uncertainty and allowing mid-course corrections. This style of management therefore requires forming hypotheses and testing various means to achieve stated management goals. Adaptive management, therefore, can only work if one adopts a scientific attitude toward management methods, and operational measures are essential to begin this process.

### 5. Normative Content

The language we use must express normative, as well as descriptive, content. In order to describe environmental goals and progress toward them, our policy-integrative language must possess sufficient semantic richness to express prescriptions as well as descriptions. How can a scientific description embody values as well as facts? Because we put the values there – in an activist, managerial science – the language must be designed to describe features of

159

the world that matter to us, to our culture, to our species and perhaps to others. And thus our values must in some sense be implicit in the descriptive concepts we develop. I am not, of course, advocating the arbitrary inclusion of particular values of individuals or interest groups in scientific terminology. Rather, I believe that these inclusions should be proposed explicitly and debated in public forums such as ecosystem valuation processes. Discussion of the goals and aspirations of a community can help us to choose descriptive concepts that focus our attention on those ecological processes that really matter socially.

One way of thinking of such terms is suggested by the philosopher B. A. O. Williams (1985), who argues that the fact–value dichotomy does not describe most of ordinary discourse; it has rather been "imposed" on the language of science and evaluation through an application of the failed theory of logical positivism. In ordinary discourse, Williams notes, there are many "morally thick" terms, terms that embody both descriptive and normative, moral content. Terms such as "depraved" are morally thick, combining descriptive content and moral opprobrium, while "stalwart" expresses both descriptive content and a positive evaluation. Applying Williams' idea (Nelson, 1995), we can ask whether management and policy discourse needs "normatively thick" terms, indicators that embody empirical content and also signal associations with important social values, values that express agreed-upon social goals. "Health" and "integrity" emerged, as noted above, as indicator terms that could be used to track large-scale and systematic characteristics of large-scale ecosystems. They are descriptive normative terms that summarize a great deal of factual information and also imply that important values are at stake. Perhaps the Slobodkin group, recognizing, as I do, that these terms also can be misunderstood, would prefer to offer alternative normatively thick terms. But I hope I have made a case that some new terminology – normatively thick indicator terms that will encourage more multi-directional discussion – is needed to encourage communication among the public, policy makers, and ecological scientists.

## 6. Communication Enhancement

Finally, the integrative language and discourse we develop must be communication-enhancing. Since it would be an unjustifiable imposition on biological researchers to require that they operate using only terms readily understandable by the general public, it follows that we need bridge concepts to create indices that are, and should be, of interest to concerned citizens.

Lee (1993) has correctly emphasized the importance of the interaction of scientists and the public in developing an "epistemological community" of

160

inquirers so that scientists serve the public interest in developing management plans, testing the results, and helping the public to choose and interpret
relevant data, so that the public gains a deeper understanding and a truer perception of the real problems and of possible solutions. All of these conditions
must be fostered to allow development of trust and true communication. Lee
discusses, for example, the surprising success in some aspects of management in the Mediterranean region and attributes this to the co-development
of an epistemological community of scientists who gained the trust of an informed public. In this case, an iterative public dialogue resulted in the building of trust and the articulation of shared goals, and the best science available
was brought to bear upon several important environmental problems in the
region.

### CONCLUSION: DO WE NEED NEW TERMINOLOGY?

I have summarized the reasons to believe that ecological science, as a discipline, has had an unfortunately limited impact on environmental policy, and I
have offered an analysis of the cause of this failure, here traced to ecologists'
implicit adoption of a serially organized mental model for relating facts and
values in the policy process. To explain the failure of ecologists to affect wetlands policy, I have hypothesized that the serial view of science and policy –
the view that we can and should describe nature prior to evaluating changes
in it – has encouraged ecologists to build separate languages for description
and evaluation. The result of this reticence has been the creation of a disci-
plinarily "closed" language, which bears few easily recognizable connections
to evaluative and policy discourse. This reticent strategy has been a failure,
both because it makes it more difficult for activists and policy analysts to find
the right and relevant data, and to interpret it properly in the policy decision
process, but also because the insulation of descriptive and evaluative language
has discouraged multi-directional communication, making it more difficult
for ecological scientists to recognize socially useful areas ripe for study and
applications of emerging science to real decisions that affect social values. If
the stronger version of the Slobodkin group's recommendation – that we stay
with the current terminological resources, period – is correct, or if there are
other ways that ecologists can correct the current intolerable situation while
maintaining the serial model, then my hypothesis can easily be disproven.
Ecologists need only demonstrate that they can in fact be effective in shaping
environmental policy in areas such as wetlands policy, and in doing so without
additional, descriptive-normative linguistic resources.

ACKNOWLEDGMENTS. Gary Cecchini assisted in the research for this paper. The author appreciates helpful discussions of issues and/or of an earlier draft with Dennis King, Rick Haueber, Alan Ringgold, Gary Cecchini, and David Policansky, as well as challenging comments from several anonymous reviewers for this journal. The project on Ecological Communication was organized by Michael Dover of Clark University and funded by the United States Environmental Protection Agency. Meetings were held in Washington, D.C., May 23–25, 1995. It should be noted that the final report of the Slobodkin group was more hospitable to concepts such as health and integrity than was Slobodkin's oral report, and only cautions that these terms be given careful definition and that they should be used with care.

LITERATURE CITED

Allen, T. F. H. and T. W. Hoekstra. 1992. *Toward a Unified Ecology*. Columbia University Press, New York.

Allen, T. F. H. and T. B. Starr. 1982. *Hierarchy: Perspectives for Ecological Complexity*. University of Chicago Press, Chicago, Illinois.

Arrow, K., B. Bolin, R. Costanza, P. Dasgupta, C. Folke, C. S. Holling, B. Jansson, S. Levin, K. Maler, C. Perrings, and D. Pimentel. 1995. Economic growth, carrying capacity, and the environment. *Science* 268: 520–521.

Batie, S. S. and H. H. Shugart 1989. "The biological consequences of climate changes: an ecological and economic assessment," pp. 121–131 in Rosenberg, N. J., Easterling, W. E., III, Crosson, P. R., and Darmstadter, J., editors, *Greenhouse Warming: Abatement and Adaptation*. Resources for the Future, Washington, D. C.

Bockstael, N., R. Costanza, I. Strand, W. Boynton, K. Bel, and L. Wainger. 1995. Ecological economic modeling and valuation of ecosystems. *Ecological Economics* 14: 143–159.

Carnap, R. 1950. Empiricism, semantics, and ontology. *Revue Internationale de Philosophie* 4: 20–40.

Costanza, R., B. Norton, and B. Haskell. 1992. *Ecosystem Health: New Goals for Environmental Management*. Island Press, Covelo, CA.

Cronon, W. 1983. *Changes in the Land: Indians, Colonists, and the Ecology of New England*. Hill and Wang, New York.

Cronon, W., editor. 1995. *Uncommon ground*. W. W. Norton and Co., New York.

Crosby, A. W. 1986. *Ecological Imperialism: The Biological Expansion of Europe, 900–1900*. Cambridge University Press, Cambridge, U.K.

Freeman, A. M. 1993. *The Measurement of Environmental and Resource Values: Theory and Methods*. Resources for the Future, Washington, DC.

Gunderson, L. H., C. S. Holling, and S. S. Light, editors, 1995a. *Barriers and Bridges to the Renewal of Ecosystems and Institutions*. Columbia University Press, New York.

———1995b. Barriers broken and bridges built: a synthesis, pp. 489–532 in L. H. Gunderson, C. S. Holling, and S. S. Lignt, editors, *Barriers and Bridges to the Renewal of Ecosystems and Institutions*. Columbia University Press, New York.

Hammond, A. 1995. *Environmental Indicators: A Systematic Approach to Measuring and Reporting on Environmental Policy Performance in the Context of Sustainable Development.* World Resources Institute, Washington, DC.

Harwell, M., J. Gentile, B. Norton, and W. Cooper. 1994. *Ecological Significance* in Ecological Risk Assessement Issue Papers. Risk Assessment Forum, U.S. Environmental Protection Agency, EPA/630/009, 2-1–2-49.

Holling, C. S., editor. 1978. *Adaptive Environmental Assessment and Management.* John Wiley & Sons, London.

―――1992. Cross-scale morphology, geometry, and dynamics of ecosystems. *Ecological Monographs* 62: 447–502.

Kentula, M. E., R. P. Brooks, S. E. Gwin, C. C. Holland, A. D. Sherman, and J. C. Sifneos. 1993. *Wetlands: An Approach to Improved Decision Making in Wetland Restoration and Creation.* Island Press, Covelo, CA.

King, D. M. 1997a. *Comparing Ecosystem Services and Values.* Report to the U.S. Department of Commerce, National Oceanic and Atmospheric Administration, Damage Assessment and Restoration Program, Silver Spring, MD, January 12.

―――1997b. *Using Ecosystem Assessment Methods in Natural Resource Damage Assessment.* Report to the U.S. Department of Commerce, National Oceanic and Atmospheric Administration, Damage Assessment and Restoration Program, Silver Spring, MD, January 31.

―――1997c. Personal communication, March.

―――1997d. Wetland values, wetland mitigation and sustainable watershed management. Unpublished manuscript.

Kuhn, T. 1970. *The Structure of Scientific Revolutions.* Second edition. University of Chicago Press, Chicago, Illinois.

Lee, K. 1993. *Compass and Gyroscope.* Island Press, Covelo, California.

Levin, S. 1992. The problem of pattern and scale in ecology. *Ecology* 73: 1943–1967.

McIntosh, R. P. 1985. *The Background of Ecology: Concept and Theory.* Cambridge University Press, Cambridge, U.K.

Mitsch, W. J. and J. G. Gosselink. 1993. *Wetlands.* Second Edition. Van Nostrand Reinhold, New York.

Nelson, J. 1995. Health and disease as "thick" concepts. *Environmental Values* 4: 311–322.

Norton, B. G. 1991. *Toward Unity Among Environmentalists.* Oxford University Press, New York.

―――1994. *Ascertaining Public Values Affecting Ecological Risk Assessment* in Ecological Risk Assessment Issue papers. Risk Assessment Forum, U.S. Environmental Protection Agency, EPA/630/009, 10-1–10-38.

―――1995a. Objectivity, intrinsicality, and sustainability. *Environmental Values* 4: 323–332.

―――1995b. Resilience and options. *Ecological Economics* 15: 133–136.

―――1996. A scalar approach to ecological constraints in P. C. Schulze, editor, Engineering within ecological constraints. National Academy Press for the National Academy of Engineering, Washington, DC.

Norton, B. G. and B. Hannon. 1997. Environmental values: a place-based theory, Environmental Ethics 19: 227–245.

Odum, E. P. 1979. The value of wetlands: a hierarchical approach, in P. E. Greeson, J. R. Clark, and J. E. Clark, editors, *Wetlands Functions and Values: The State of Our Understanding*, pp. 1–25. American Water Resources Association, Bethesda, MD.

O'Neil, L. J., T. M. Pullen, and R. L. Schroeder. 1991. *A Wildlife Community Habitat Evaluation Model for Bottomland Hardwood Forests in the Southeastern United States*. Biological Report 91(x). U.S. Department of Interior, Fish and Wildlife Service, Washington, DC.

O'Neill, R. V., D. L. DeAngelis, J. B. Waide, and T. F. H. Allen. 1986. *A Hierarchical Concept of Ecosystems*. Princeton University Press, Princeton, NJ.

Peters, R. H. 1991. *A Critique for Ecology*. Cambridge University Press, Cambridge, U.K.

Pimm, S. L. 1991. *The Balance of Nature? Ecological Issues in the Conservation of Species and Communities*. University of Chicago Press, Chicago, Illinois.

Proctor, R. N. 1991. *Value-Free Science?* Harvard University Press, Cambridge, MA.

Quine, W. V. O. 1951. Two dogmas of empiricism. *Philosophical Review* 60: 20–43.

———1964. *Word and Object*. MIT Press, Cambridge, MA.

Sagoff, M. 1988. Ethics, ecology, and the environment: integrating science and law. *Tennessee Law Review* 56: 78–229.

Shrader-Frechette, K. S. and E. D. McCoy. 1993. *Method in Ecology*. Cambridge University Press, Cambridge, U.K.

Simenstad, C. A. and R. M. Thom. 1996. Function equivalency trajectories of the restored Gog-Le-Hi Te estuarine wetland. *Ecological Applications* 6(1): 38–56.

Tuan, Y.-F. 1974. *Topophilia: A Study of Environmental Perception, Attitudes, and Values*. Prentice-Hall, Englewood Cliffs, NJ.

———1977. *Space and Place: The Perspective of Experience*. University of Minnesota Press, Minneapolis, Minnesota.

Walters, C. 1986. *Adaptive Management of Renewable Resources*. Macmillan Publishers, New York.

Williams, B. A. O. 1985. *Ethics and the Limits of Philosophy*. Harvard University Press, Cambridge, MA.

———1987. Why it is so difficult to replace lost wetland functions, pp. 121–123 in J. Zelazny and J. S. Feierabend, editors, *Increasing Our Wetland Resources*. Conference Proceedings, National Wildlife Federation Corporate Conservation Council, October 4–7. National Wildlife Federation, Washington, DC.

Zedler, J. B. 1996. Ecological issues in wetland mitigation: an introduction to the forum. *Ecological Applications* 6: 33–37.

# III

# Economics and Environmental Sustainability

My home field, environmental ethics, considers itself at war with economists – or at least with economists' ideas. Economics, you see, is unapologetically an anthropocentric scientific endeavor. If one observes the field of environmental economics from the viewpoint of a committed nonanthropocentrist – and the majority of environmental ethicists have committed themselves to some form of nonanthropocentrism – there is only one appropriate response to environmental economists and their models. Since all economic value is measured in units of human welfare, economists became the favorite targets of philosophers, who attacked them for dogmatically assuming an answer to the central question of environmental ethics. For someone like me, however, who doubts the importance of the anthropocentrism–nonanthropocentrism debate, it was possible to raise more nuanced questions about the philosophy of environmental economics and its contribution to understanding sustainability.

First, it seemed important to acknowledge that economics – which, after all, has the advantage of being able to quantify human values – is especially helpful in characterizing the ways in which people resolve important trade-offs. By representing environmental goods as consumer goods in competition with other opportunities to consume or invest, environmental economists provide one set of tools for examining the trade-offs between environmental protection and other social goods. In "Sustainability, Human Welfare, and Ecosystem Health," I first explored the possibility that different decision tools may be appropriate for different types of decisions, and that use of standard economic criteria such as cost-benefit analysis may provide a useful analysis of a wide range of private and public decisions, but that such decision processes are not appropriate for all types of decisions. In this paper, I introduced the idea of "risk decision squares," a method of representing the meta-level decision that is made when one decides whether to use cost-benefit analysis or some other decision rule, such as the precautionary principle, to evaluate particular

165

decisions. I argued that decisions affecting the health and integrity of eco-logical systems – certainly important aspects of sustainability – are almost certainly of the type that requires an ethical analysis in terms of intergenera-tional justice; decisions affecting the health and integrity of systems cannot be resolved simply in terms of costs and benefits.

The idea of risk decision squares, and an integrated analysis including economics and moral considerations, are refined and developed in two sub-sequent papers – "Evaluating Ecosystem States" and, with Michael Toman, "Sustainability: Ecological and Economic Perspectives." These papers es-tablished the difficulty of capturing ecological concerns as considerations in economic sustainability models and introduced the idea of a two-tier decision process. In a complex decision process, it is important to separate decisions made according to a given rule from decisions to use one or another rule. The two-tier process of recognizing multiple rules and adding a decision process for rationally choosing the appropriate rule, therefore, opens the door to a plu-ralistic but integrated approach to decision making. Multiple decision rules are available, and the use of risk decision squares provides a means by which the decision regarding which rule to apply can be rationalized and systematized according to empirically observable characteristics of the type of decision faced. In "Evaluating Ecosystem States," the idea of risk decision squares is elaborated and related to the idea of hierarchy theory, linking the temporal variable of ecological reversibility ("resilience") with spatial dimensions of impacts of decisions as a possible means to integrate economics and ecology. This theoretical connection sets the stage for a series of papers on scale in ecology and in public decision making, some of which are included in Part IV.

Because I did not take the anthropocentrism of economics to be a fatal flaw, and because I considered economic data to be one important kind of informa-tion relevant to policy decisions, I also turned my attention – in "Economists' Preferences and the Preferences of Economists" and, with Robert Costanza and Richard Bishop, in "The Evolution of Preferences" – to another constitu-tive assumption of contemporary welfare economics, what is usually called "consumer sovereignty." It is this methodological assumption – that individ-ual consumers should be regarded as the best judge of their own welfare – that makes "preferences" measurable (understood as units of willingness-to-pay) and allows economics to concentrate on aggregating individual preferences. These papers, however, show how this assumption – susceptible of many interpretations and varied justifications – places economists at odds with en-vironmentalists, who expect that achieving more sustainable environmental policies will require modern individuals to reconsider and in some cases re-form their preference sets.

To summarize, my work on philosophy and economics, which has for the most part left aside the issue of anthropocentrism and nonanthropocentrism, has instead focused on two related issues. First, I have challenged the completeness of economic analysis; I do not think that economics can provide a complete accounting of environmental values in terms of unquestioned individual preferences. Not willing to deny, however, the usefulness of information about people's actual preferences in some situations, I endorse a form of value pluralism, a recognition that our society values nature in many, conflicting ways and that we cannot today represent all values that appear in all situations in a single currency. In Part V, I return to this issue of pluralism and criticize the idea of "weak sustainability values." There I build on this negative conclusion and develop a more pluralistic – and more comprehensive – positive conception of sustainability. Second, as a means of seeking greater integration of economic and other considerations (as called for in "Sustainability: Ecological and Economic Perspectives), I explore the possibility that environmental policy can be best represented as having two "tiers" – I have lately come to favor the term "phases." Two-tier evaluation models embody two phases. In its simplest version, the two-tier system is conceived (ideally) as a public deliberative process having an "action phase" and a deliberative or "reflective" phase. This idea is further developed in the final paper in this volume, "Environmental Values and Adaptive Management," in which the two-phase model is given more substance by the provision of heuristics designed to help communities ask the right questions as they seek to identify and pursue sustainable policies.

# 10

# Sustainability, Human Welfare, and Ecosystem Health

The goal of 'sustainability' has emerged as a rallying cry for a broad spectrum of advocates of both environmentalism and rational development. The term sustainability was first popularized in the field of resource use, and it initially had a fairly precise application in phrases such as 'maximum sustainable yield', which represents the highest level of exploitation consistent with maintaining a steady flow of resources from a forest or fishery. Today, however, the term is used much more broadly to include, for example, levels of pollution and degradation of natural systems that are consistent with maintaining current levels of use and enjoyment of those systems. In the context of 'sustainable development', it must be used in the broader sense, and hence it is in this broader sense that the term has become a shibboleth of mainstream environmentalists.

It is no doubt useful, in policy discussions, to have a term like 'sustainability', which, like 'conservation' in days of old, can stand as a label for the many activities of environmentalists. The danger is that the term, like 'conservation' before it, will become a cliché.[1] Nobody opposes it because nobody knows exactly what it entails. To avoid this trap it will be necessary for environmentalists, with the help of scientists and philosphers, to develop, explain and justify a theory of environmental practice that gives form and specificity to the goal of sustainability. In particular, what is needed is a set of principles, derivable from a plausible core idea of sustainability, but sufficiently specific to provide significant guidance in day-to-day decisions and in policy choices affecting the environment.

As a first step in giving from to the definition, it is useful to note that the term implies sustainable *use*, so it would appear to exclude severely moralistic

From *Environmental* values 1 (1992): 92–111. Reprinted with permission of The White Horse Press, Cambridge, U.K.

approaches, such as positions of extreme deep ecologists who argue that the natural world ought not to be considered 'resources' for human use at all.[2] At the other extreme, advocates of unlimited economic growth, who argue that it is wrong to place any constraints on the ability of the free market to generate goods and services in response to consumer demands, would reject the implication that environmental concerns justify any constraints on the use of nature.[3] Between these extreme positions, however – and I think it is safe to say that these extreme positions have very few advocates – lie the vast majority of environmentalists,[4] who believe that use of the environment is morally acceptable, but that this use is constrained by obligations not to misuse the environment in unsustainable ways.

### PART I. SUSTAINABILITY AND HUMAN WELFARE

Today, the most often-cited definition of sustainability is that of the Brundtland Commission's report, Our *Common Future*: "Sustainable development is development that meets the needs of the present without compromising the ability of future generations to meet their own needs."[5] The Commission followed this definition with a formulation of the 'two key concepts', of their definition: "the concept of 'needs,' in particular the essential needs of the world's poor, to which overriding priority should be given", and "the idea of limitations imposed by the state of technology and social organization on the environment's ability to meet present and future needs".[6]

Since the exact meaning of sustainability will depend upon the specification of the 'limitations' mentioned in the second concept, it is notable that the Brundtland definition states these as determined essentially by "the state of technology and social organization". Sustainability is therefore defined as an intertemporal relationship between *human needs* and *human productive capacities*, as a relationship between human welfare at different stages of human development. While the environment is mentioned, it appears as a passive element in the equation – needs are human-determined, and limitations are seen as human limitations. The environment does not impose any non-negotiable limits on sustainable use, independent of limitations on the abilities of humans to control it. Any limitation on use of the environment may in principle be overcome by some new breakthrough in technology and social organization. Our obligation, on this view, is to balance present fulfilment of needs against the ability of future generations to fulfil their needs.

The Brundtland definition, then, can stand as characteristic of one broad approach to sustainability, which I will call the 'social scientific' approach, both because it is popular among social scientists, such as demographers and economists, and because it focuses most empirical attention on human demands and on characteristics of technical and social innovation.[7]

While the Brundtland definition was intended as a relatively 'neutral' definition, attractive to a broad range of environmentalists and developmentalists, we can now see that it may not be. The implication that there can be no insuperable shortages of resources precludes, by the very definition of sustainability, limitations imposed by characteristics of the environment itself: characteristics that might limit its ability to produce consumable goods or absorb human wastes. On the Brundtland approach, projections of economic and social growth can be calculated without accounting for the *scale* of human activities.

Intuitively, this implication that nature sets no natural limits on economic uses is implausible; it implies that no human activity will, in principle, be precluded by shortages of resources. This implication seems to contradict the obvious fact that the stocks of any given resource are finite, and that some of them, such as copper ore, are quite limited.[8] The denial of natural limits does not challenge this fact directly, however. It recognizes that stocks of non-renewable resources will decline and the price of raw resources will rise; the key to maintaining this position rests on a high degree of confidence in the intersubstitutability of resources. The finitude of copper does not cause a limit on economic growth because, as the price rises, a substitute resource will replace it. Similarly, as the cost of disposing of pollutants and wastes increases, entrepreneurs will be stimulated to develop alternative means of recycling and disposal. I am suggesting, then, that social scientific definitions of sustainability presuppose a very strong principle of intersubstitutability of resources, indeed, a Principle of Infinite Intersubstitutability (PII). This principle is inherent in the definition of sustainability as a simple balance of 'human welfare' across time. Environmentalists, I submit, will question PII. They should, therefore, be wary of attempts to define sustainability simply as a matter of human technology and welfare.[9]

It can be argued that the assumption of PII is intimately tied to the uni-dimensional value analysis of the mainstream economic paradigm. One will find PII plausible only if one assumes the interchangeability of labour, resources and capital, and that all value can be represented as prices in markets. Inter-changeability is essential to the central idea of mainstream economics: that all choices can be understood incrementally, as consumer choices at the margin.[10] If sustainability is to be a simple problem of balancing welfare across generations, then human welfare must be understood incrementally and interchangeably, as it is in mainstream economics. Provided we leave our descendants *richer* than we are, according to this analysis, we cannot have done wrong; the future can simply trade its wealth for amenities, substitutes for lost resources, or a pollution-free environment. In an incremental system of value in which all values are interchangeable and all resources have, with requisite capital, adequate substitutes, environmental constraints need be given no special pre-emptive status.

To recognize limits inherent in nature itself would be to introduce discontinuities into the analysis. If over-consumption of passenger pigeons were analysed in 1900, according to the mainstream economic paradigm, profits resulting from over-exploitation could have been deemed 'beneficial' to the future as capital capable of generating new sources of protein. If, however, one insisted that passenger pigeons represented an irreplaceable resource, one would have argued that continued consumption of squab, even as the stocks plummeted toward extinction, represented an unrecompensable harm perpetrated by one generation on subsequent ones.

It is tempting to set out to show that the economic paradigm, despite its unquestioned advantage of simplicity (in that it can represent all values on a single scale of welfare), is too simplistic to deal with questions of intergenerational equity. In particular, it could be argued that the incrementalist model of mainstream economics (which seems to be presupposed in the Brundtland definition) is ill-suited to deal with policy problems in which incremental choices can have irreversible effects that will have impacts over very long periods of time.[11] Space will not permit such an argument here. Instead, an alternative conception of intertemporal welfare will be proposed and explained. This conception, 'scientific contextualism', does not flatten out all decisions into interchangeable units of individual welfare, but instead retains a sensitivity toward different types and scales of impacts that the present can exert on the future.


PART II. A CLASSIFICATION OF RISKS


The flattening-out approach to judging intergenerational impacts, measuring intertemporal welfare according to a single scale of present valuations, usually dollars, ignores apparently important differences in the types of impacts the present can have on the future. A broad and inclusive conception of sustainability must gauge the ability of the future to deal with pollution and waste as well as with declining stocks of resources.[12] For the sake of a convenient terminology, and because it seems reasonable to treat some present activities as creating a 'risk' of future shortages of resources and sinks for waste

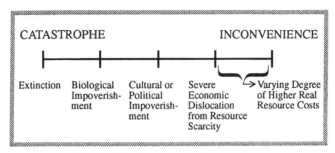

Figure 10.1. Typology of risk severities.

products, let us propose an intuitive scale for classifying types of risks that the present may impose on the future, as in Figure 10.1.

This scale recognizes the apparent difference between activities such as burning fossil fuel in great quantities, which may include considerable risk that the planet will become uninhabitable by future humans at one extreme, and less cataclysmic results, such as filling all available waste dumps, which might force future generations to give up disposable diapers and return to the old-fashioned practice of washing diapers. Because one of the apparent weaknesses of the incrementalist model is that it does not deal well with irreversibilities such as species extinctions, we can remedy this weakness by introducing a scale of comparative reversibility of present decisions. If some decisions we make today are easily reversible, then capital or know-how may be a reasonable substitute for some forms of environmental protection. Conversely, major cataclysms would be irreversible. Therefore, we can plot our intuitive scale of types of future risks against a scale of reversibility, creating a decision grid as represented in Figure 10.2.

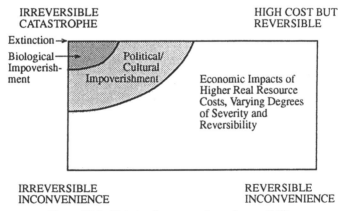

Figure 10.2. Risk typology: severity and reversibility.

172

The decision space here defined will include, in the far east, the southeast, and the far south portion of the space, decisions that we will consider simple trade-offs. If the negative impacts of our activities on the future result only in easily reversible changes, or if the impacts result only in minor inconveniences, we can figure that we have been 'fair' to the future, provided that we make available capital and techniques sufficient to reverse or counterbalance those effects. Decisions that have possibly cataclysmic impacts or irreversible consequences, on the other hand, will not be open to trade-offs. Decisions that fall in the northwest portion of the decision space will be governed by non-negotiable constraints. A slightly less constraining decision rule, the Safe Minimum Standard of conservation (SMS), might be applied in this 'red' area. This rule advises: Protect the resource (natural process, species, etc.) provided that the costs are bearable.[13]

I am hypothesizing, then, that our obligations will be least negotiable in the northwest corner of the decision space and that these obligations will decline along all vectors away from that corner. Figure 10.2 also represents the above-mentioned insight that the degree to which one believes in intersubstitutability of resources will determine the relative size of the decision space governed by non-negotiable constraints. In the limiting case of a belief in PII, the decision space governed by non-negotiable constraints will be null – all risks are recompensable with adequate capital and technological development. As one's faith in intersubstitutability decreases, the northwestern space will expand, representing more and more decisions as governed by non-negotiable constraints.

We can refer to approaches to sustainability that recognize some decisions with future impacts as governed by non-negotiable constraints and some decisions governed by trade-offs as 'hybrid theories' or as 'two-tier systems'.[14] They recognize at least two measures of value that cannot be aggregated together – one set of obligations may trump another. Two-tier approaches therefore differ from the single-tier systems of micro-economics and of other utilitarian approaches, which see only interchangeable units of welfare as the measure of sustainability, by recognizing some constraints that cannot be traded off. The two-tier approach eschews simple aggregation in favour of lexically ordered rules.

The moral status of these non-negotiable obligations in the northwest corner is, of course, open to much debate. Some of my colleagues in environmental ethics would insist that these obligations be formulated as protecting values 'intrinsic' or 'inherent' in nature, positing values independent of humans.[15] Others would posit basic rights of future persons, which would trump mere consumptive interests of present persons.[16] Another understanding of non-negotiable obligations, the one to be explained here, is morally anthropocentric but based in a holistic conception of the natural systems on which humans depend. If it turns out that nonhuman rights or rights of the not-yet-existent can be specified later, they could be added on, further strengthening the already strong obligations involved in contextualism. The point I wish to stress here is that the logic of two-tier, hybrid systems is similar, and this sets them apart from the single scale of values approach of many social scientists.

PART III. SCIENTIFIC CONTEXTUALISM

Consider an approach to sustainability that recognizes obligations of the present generation to future generations but views these obligations holistically – as not reducible to individual satisfactions or preferences.[17] These obligations are of the type that would be suggested by Edmund Burke's understanding of a society as "a partnership not only between those who are living, but between those who are living, those who are dead and those who are to be born",[18] and these obligations would be based on a belief that the continuance of the human species is a good thing.

Having posited such a value and cross-generational obligations not defeasible into individual satisfactions, let me now argue that, if we have such obligations – and I think we do – we can now posit an alternative approach to understanding sustainability. This approach recognizes that there are non-negotiable obligations regarding our use of resources (the NW corner of the decision space is not empty), and that those obligations can be understood as the obligations the present has to perpetuate the conditions necessary for the continuation of the human species and of its culture. The exact nature of these obligations must be determined scientifically, as we understand the impacts human activities have on their larger context. If, following Aldo Leopold's land ethic, we insist that this large context can only be understood as a complex ecological system, sustainable activities are activities that do not destabilize the large-scale dynamic, biotic and abiotic systems on which future generations will depend. Scientific contextualism applies a variety of moral rules, placing priority on different values in different situations. If plausible scientific models indicate that a realistic, if not necessarily probable, chain of processes could result in cataclysmic effects, we are in the 'red zone', and the SMS standard applies.

Admittedly, the information necessary to act sustainably in this sense would be very hard to obtain. But our concern here is mainly conceptual. Assume, for example, that models showing rapid and accelerating warming of the atmosphere in response to a build-up of greenhouse gases were strongly verified, and that these models showed increases, 50–100 years in the future, too rapid for civilization to survive. I believe that most people would say, once most scientific uncertainty was removed, that such a scenario would trigger non-negotiable constraints limiting current behaviour. Examples such as this are important, because they help to shift the burden of proof from those who would institute constraints on economic growth to those who would flirt with cataclysmic changes in the context of human adaptation. If there are clear examples in which non-negotiable constraints would exist, growth enthusiasts must show that their proposed activities violate no such constraints.

We know that undisturbed natural systems are able to maintain themselves across time, that they will keep their energy pathways open, and that they will maintain their productivity. Once a system is disturbed, effects cascade through the system; if those effects are of the sort that the system is used to and can assimilate, it does so. If, on the other hand, the disturbance is so pervasive or so new that the system has no means by which to damp out its effects, the system crosses a threshold; and humans, as well as members of other species, who are adapted to systems with a given set of characteristics, may be unable to adapt to accelerating and cascading changes in their habitat. For example, humans with light skin pigmentation, a trait which evolved in a time of relatively small exposures to ultraviolet light in temperate climates, may be unable to adapt to an earth with less upper-atmosphere ozone.

In a contextual analysis, individual behaviours are not the main focus of environmental ethics – it is trends in those behaviours that determine whether they will have intergenerational impacts. For example, if one farmer cuts and clears his woodlot to plant wheat, this is not wrong as long as his land is not on a highly erodible slope and provided that his action is not copied by all of his neighbours. The moral status of this activity depends not only on the content, but also on the context, of the action. If Farmer Jones plants wheat and Farmer Smith lets his wheat field go fallow, no trend is instituted, and there is little likelihood that the action will trigger non-negotiable constraints. If most farmers follow Jones, conditions ripe for a dust bowl or desertification may be created. Scale is crucial in determining when a red-zone decision is faced.

Expressed metaphorically, contextualism is organicism – the biota is a living system which has an internal, self-perpetuating organization – but organicism minus teleology. Contextualism need not posit a metaphysical value in the supraorganism, just as it need not posit independent value of wholes.[19] But contextualism does recognize the importance of protecting the processes sustaining self-organizing systems through time. For example, once Leopold fully understood the implications of a systems-oriented approach, he fell back upon the recommendation that we practice something akin to preventive medicine.[20]

Leopold's theory of sustainable management envisioned hierarchically organized systems, with human activities impacting them not individually, but in larger trends.[21] Technology and population growth have given human cultures the ability to alter larger and normally slowly-changing systems of ecology, climate and atmosphere, and to initiate oscillations and fluctuations in these systems. Since we have evolved to live within systems that change slowly, such activities play Russian roulette with the options of the future. Our generation could cause irreversible changes much too rapid for future generations to adapt to, either physiologically or culturally.

Scientific contextualism places a heavy burden on scientific models to help us determine which activities may have long-delayed, but potentially catastrophic, consequences. The contextualist paradigm of environmental management interprets the larger systems under impact from human activities in mainly ecological terms. An essential element of the contextualist approach to management will be a commitment (non-negotiable constraints) to protect the health and integrity of ecological systems. The contextualist paradigm is not, however, *simply* an ecological paradigm – as human impacts grow, biotic systems, and also atmospheric and climatological systems, are inexorably affected by the aggregated impacts of human economic and other activities.

While I agree with Leopold that these larger impacts should be understood through ecology, because humans are, after all, evolved animals who relate as living things both to the biota and to larger abiotic systems, contextualism also recognizes the role of non-biological sciences in defining the limits of our impacts on our larger context. For that reason, I call the approach to sustainability sketched here 'scientific' contextualism.

PART IV. HEALTH, INTEGRITY, AND SUSTAINABILITY

The idea that there is an obligation to protect the health and integrity of ecological systems rests firmly upon the premise that natural systems are self-organizing in an important sense. This is a difficult concept, and it must be carefully explained and qualified – a task that can only be begun here.[22] Recognizing that natural systems change constantly, intertemporal stability is conceived as a scalar relation between human activities and their larger environmental context. Contextualism assumes that the self-organization of large systems is essential to future generations and that the ability of those large, self-organizing systems to assimilate human impacts is large but not infinite. These systems provide the context within which we have evolved. Because they change more slowly than culture, these large systems set the 'stage' for human activities. They therefore give meaning to human culture. At the same time, the system is unquestionably dynamic. Stability only exists relative to differing scales of time; we might say that stability is a well-founded illusion.

Autonomous systems overcome entropy; autonomy is the characteristic of systems that allows self-organization. Given this operational definition of autonomy, we can define sustainability as follows. Sustainability is a relationship between dynamic human economic systems and larger, dynamic, but normally slower-changing ecological systems such that (a) human life can continue indefinitely; (b) human individuals can flourish; (c) human cultures can develop; but in which (d) effects of human activities remain within bounds so as not to destroy the health/integrity of the environmental context of human activities.

But how are we to define 'health' and 'integrity'? I doubt that one can understand a definition of sustainability without understanding the system of concepts and principles that surrounds it. Sustainability, when understood within the atomistic, incrementalist paradigm of welfare economics, reduces to a question of balancing interchangeable units of welfare across time. If, as we are hypothesizing here, there are non-negotiable constraints that mandate protection of large, autonomous systems of nature, those constraints must be expressed in a richer and more complex paradigm.[23]

So the goal of specifying practical guidelines will require a 'paradigm' of ecological/contextual management, a set of concepts and principles that can guide attempts to protect and restore ecological systems. Let me begin by citing and agreeing with the definition of Faber, Manstetten, and Proops, who note that 'ecology' combines, etymologically, the Greek ideas of 'house' with their idea of 'logos', which they translate as concept/structure, and define ecology as "the science of the principles of the self-organization of nature".[24] Thus defined, it is part of the specification of the field of ecological management that its subject matter is self-organizing. This definition also ensures that ecology, not economics, will be the new 'fusion point' of the sciences, because economic activities are understood as one type of ecological activity, one that takes place within the economic system of one (however dominant) species.[25]

Our current task is to build upon this approach to management by providing more elements of the ecological paradigm of environmental management. To that end, I suggest five Axioms of ecological management:

177

(1) The Axiom of Dynamism: Nature is more profoundly a set of processes than a collection of objects; all is in flux.

(2) The Axiom of Relatedness: All processes are related to all other processes.

(3) The Axiom of Hierarchy: Processes are not related equally but unfold in systems within systems, which differ mainly regarding the temporal and spatial scales on which they are organized.

(4) The Axiom of Creativity: The autonomous processes of nature are creative and represent the basis for all biologically based productivity. The vehicle of that creativity is energy flowing through systems which in turn find stable contexts in larger systems, which provide sufficient stability to allow self-organization within them, through repetition and duplication.

(5) The Axiom of Differential Fragility: Ecological systems, which form the context of all human activities, vary in the extent to which they can absorb and assimilate human-caused disruptions in their autonomous processes.

These Axioms function, in practice, in conjunction with a normative definition of ecosystem health/integrity. I begin by proposing a definition of integrity. An ecological system has maintained its integrity if it retains:

(a) the total diversity of the system, the sum total of the species and associations that have held sway historically[26] and

(b) the autonomous processes (systematic organization) that maintain that diversity, including, especially, the multiple layers of complexity through time.[27]

It is useful to have two related concepts to describe ecosystem well-being, as Leopold noted in his comparison of succession after the plough in Kentucky and in the American Southwest.[28] In both cases, the integrity of the system was compromised – including loss of total diversity and invasion by exotics. The difference, Leopold noted, was that bluegrass represented a new stable point, capable of maintaining itself across time. This system lacked the integrity of systems like Rio Gavilan, that maintained their 'aboriginal health' (including their historical mix of species), but maintained a 'healthy' equilibrium nonetheless.

I am suggesting that we use the term 'integrity' as the stronger term – though it certainly can admit of degrees – while we use 'health' to designate the somewhat weaker concept that describes the Kentucky bluegrass system. Integrity, in other words, emphasizes both clauses (a) and (b). System $S1$ maintains greater integrity than system $S2$ if $S1$ not only retains enough complexity to maintain autonomous functioning, but also maintains more of its original species, populations, micro-habitats, and processes of interaction. A system is healthy if it maintains its complexity and autonomy/self-organization.

Within this dynamic, contextualist paradigm we can understand the centrality of the goal of protecting biological complexity. Complexity is directly related to self-organization, and self-organization is an essential part of ecosystem health and integrity. And thus we understand the non-negotiable obligation to protect biodiversity: it is an obligation to future generations to protect the diversity and, even more important, the complexity of self-organizing systems. This obligation requires protection of complex processes of ecosystems.


CONCLUSION

I have sketched two broad approaches to understanding sustainability, recognizing that this keystone concept of modern environmentalism can only be given meaning as part of a constellation of concepts and methods – a paradigm, as some would say. The social scientific approach, which sees sustainability as a relationship between levels of welfare in the present and the future, defines sustainability within an incrementalist paradigm that interprets all values in a common measure, such as dollars or present satisfactions. The advantage of this paradigm, and its proposed approach to sustainability, is that this approach expresses the sustainability relation in interchangeable units; judgements regarding intergenerational fairness can therefore be understood as a balance between commensurable values across time. This incrementalist approach, however, has the attendant disadvantage that it does not deal very well with discontinuities and irreversibilities – and those who worry about global environmental problems such as the greenhouse effect and loss of species diversity emphasize concerns of precisely those kinds.

I have therefore sketched an alternative framework for understanding sustainability, based on a two-tier system of values, some of which are interchangeable and able to be traded off, and some of which are non-negotiable. Scientific contextualism relies on information and models from the natural sciences to determine which decisions carry significant risk of cataclysmic and irreversible results and, hence, when non-negotiable moral constraints trump interchangeable measures of individual welfare. This approach *balances* short-term economic and long-term ecological concerns but does not *reduce* them to a common metric. Environmental policy is constrained by *both* ecological and economic limits; economic concerns predominate when risks are not catastrophic or irreversible and when the areas affected are relatively small. Non-negotiable, intergenerational obligations predominate when decisions carry the risk of irreversible or catastrophic change in those large-scale systems on which the human species depends.

NOTES

1. See Caldwell, 1990, p. 177.
2. It seems to me questionable that anyone consistently holds this position in its most extreme form, even though certain passages in the writings of deep ecologists seem to consider all use of nature a violation of its intrinsic value. For a fuller discussion of the policy implications of deep ecology, see Norton, 1991, chapter 12.
3. See, for example, Kahn, 1982.
4. Data from the late 1970s showed only eight per cent of environmentalists advocating all of the tenets of deep ecology (Mitchell, 1980). I am unaware of more recent data on this subject. As for the other end of the spectrum, I have argued in *Toward Unity Among Environmentalists* that advocacy of some market constraints to protect environmental values provides a minimal defining characteristic of all environmentalists.
5. World Commission on Environment and Development, 1987, p. 43.
6. Ibid.
7. See Ehrlich and Ehrlich, 1986, pp. 8–10.
8. See Woodwell, 1985, for a useful, recent discussion of this difficult issue.
9. See Daly and Cobb, 1989, pp. 72–6, for a more positive characterization of the Brundtland definition. Daly and Cobb describe the Brundtland definition as "vague", but artfully so, and believe that it will give rise to more specifically biological criteria.
10. For a careful and detailed explanation of how the mainstream economic paradigm reduces environmental values to increments of willingness-to-pay, and how this paradigm consequently ignores problems of scale and magnitude of throughput, see Daly and Cobb, 1989.
11. This argument is made, for example, by Kneese and Schulze (1985), who can be considered, in general, proponents of the mainstream micro-economic paradigm.
12. Indeed, many environmental and resource analysts now believe that problems of waste disposal will prove far more intractable than will problems of resource availability. See Faber, Manstetten and Proops, 1990.
13. Ciriacy-Wantrup, 1952.
14. For a classic economic treatment, see Page, 1977. Also see Page, 1991, and Toman and Crosson, 1991.
15. See, for example, Callicott, 1989, and Rolston, 1988.
16. I have argued elsewhere that both of these approaches suffer from serious conceptual difficulties and will not repeat those arguments here. See Norton, 1982a and 1982b.
17. See Norton, 1989, for an explanation of a decision model that relies on general obligations to the future – obligations that are owed to no specifiable individuals.
18. Burke, *Reflections on the Revolution in France*, pp. 93–94.
19. See Ulanowicz, 1986, p. 25, for a concise explanation of how self-organization can be treated independently of teleology.

20. See Leopold, 1939, and Hargrove, 1989.
21. See Norton, 1990.
22. See Prigogine and Stengers, 1984; Gleick, 1987; and Ulanowicz, 1986, for a comprehensive examination of this and related concepts.
23. See Carpenter, 1990.
24. See Faber, Manstetten, and Proops, 1990.
25. See Leopold, 1939.
26. Total diversity (Gamma diversity) is defined as a function of within-habitat diversity (Alpha diversity) and cross-habitat diversity (Beta diversity). It is the diversity characteristic of a landscape composed of many habitats and micro-habitats. See Norton, 1987, pp. 32–33.
27. These axioms were introduced in Norton, 1991.
28. Leopold, 1949, p. 206.

### REFERENCES

Burke, E. *Reflections on the Revolution in France*. London, Dent, 1910 edition.

Caldwell, L. K. 1990 *Between Two Worlds*. Cambridge, Cambridge University Press.

Callicott, J. B. 1989 *In Defense of the Land Ethic*. Albany, State University of New York Press.

Carpenter, S. R. 1990 "Sustainability and Forms of Life", unpublished manuscript presented at the World Bank conference. The Ecological Economics of Sustainability, Washington, D.C., May 1990.

Ciriacy-Wantrup, S. V. 1952 *Resources Conservation*. Berkeley, University of California Press.

Daly, H. and Cobb, J. 1989 *For the Common Good*. Boston, Beacon Press.

Ehrlich, P. and Ehrlich, A. 1986 "Population and Development Misunderstood". *The Amicus Journal* 8: 8–10.

Faber, M.; Manstetten, R. and Proops, J. 1990 "Towards an Open Future: Ignorance, Novelty and Evolution", in *Ecosystem Health: New Goals for Environmental Management*, edited by R. Costanza, B. Norton and B. Haskell. Covelo, CA, Island Press.

Gleick, J. 1987 *Chaos*. New York, Penguin Books.

Hargrove, E. C. 1989 *Foundations of Environmental Ethics*. Englewood Cliffs, NJ, Prentice Hall.

Kahn, H. 1982 *The Coming Boom: Economic, Political, and Social*. New York, Simon and Schuster.

Kneese, A. V. and Schulze, W. D. 1985 "Ethics and Economics", in *Handbook of Natural Resources and Energy Economics*, edited by J. L. Sweeney. Amsterdam, North Holland.

Leopold, A. 1939 "A Biotic View of Land", *Journal of Forestry* 37: 727–730.

———. 1949 *A Sand County Almanac*. New York, Oxford University Press.

Mitchell, R. 1980 "How 'Deep,' 'Soft' or 'Left'? Present Constituencies in the Environmental Movement for Certain World Views", *Natural Resources Journal* 20: 352.

Norton, B. G. 1982a "Environmental Ethics and Nonhuman Rights", *Environmental Ethics* 4: 17–36.

_____. 1982b "Environmental Ethics and Rights of Future Generations", *Environmental Ethics* 4: 319–337.

_____. 1987 *Why Preserve Natural Variety?* Princeton, Princeton University Press.

_____. 1989 "Intergenerational Equity and Environmental Decisions: A Model Based on Rawls' Veil of Ignorance", *Ecological Economics* 1: 137–159.

_____. 1990 "Context and Hierarchy in Aldo Leopold's Theory of Environmental Management", *Ecological Economics* 2: 119–27.

_____. 1991 *Toward Unity among Environmentalists.* New York, Oxford University Press.

Page, T. 1977 *Conservation and Economic Efficiency.* Baltimore, The Johns Hopkins University Press.

_____. 1991 "Sustainability and the Problem of Valuation", in *Econological Economics*, edited by R. Costanza. New York, Columbia University Press.

Prigogine, I. and Stengers, I. 1984 *Order Out of Chaos.* New York, Bantam Books.

Rolston III, Holmes 1988 *Environmental Ethics.* Philadelphia, Temple University Press.

Toman, M. and Crosson, P. 1991 "Economics and Sustainability: Balancing Trade-offs and Imperatives", Resources for the Future Discussion Paper ENR 9105.

Ulanowicz, R. E. 1986 *Growth and Development: Ecosystem Phenomenology.* New York, Springer-Verlag.

Woodwell, G. M. 1985 "On the Limits of Nature", in *The Global Possible: Resources, Development, and the New Century*, edited by R. Repetto, pp. 47–66. New Haven, Yale University Press.

World Commission on Environment and Development 1987 *Our Common Future.* Oxford, Oxford University Press.

# 11

# Economists' Preferences and the Preferences
# of Economists

INTRODUCTION

One of the most pressing practical problems in environmental policy analysis
and formation is how to describe and measure environmental values. It would
be an ideal outcome if the various disciplines – economics, ecology, philosophy, environmental health, and environmental chemistry, to mention some
prominent ones – could speak about social values in a common evaluational
vernacular. My experience in many interdisciplinary discussions is that, at
present, no such common vernacular exists (Chapter 12, this volume).

One possibility, touted by economists and some other social scientists, is
to use the notion of individual preferences as a universal descriptive term to
characterise and eventually to measure the social values derived from protecting environmental quality, and the losses incurred when the environment is
not protected. The purpose of this paper is to examine preferences as they are
understood by economists and then to consider alternatives and/or additions
to this conceptualisation of environmental values.

Economists define an individual's preference set as constituted by 'all of
the hypothetical exchanges the individual would be willing to make at various
terms of trade' (Silberberg, 1978, p. 4). Preferences are then understood as
units of measure of the willingness of the individual to pay for a given outcome
or (reversing the property ownership aspect) as a measure of the compensation
an individual would require to give up some existing property right or privilege
(see Freeman, 1993). Preferences can therefore be associated with dollar
values (or any other units of exchange), permitting aggregation with other
commensurate units of exchange.[1]

From *Environmental Values* 3 (1994): 313–332. Reprinted with permission of the White Horse
Press, Cambridge, U.K.

## I. ECONOMICS AND THE STUDY OF PREFERENCES

The advantages of a unitary, preference-based system of analysis for environmental values are considerable. Five can be mentioned here. First, as just noted, this approach to values allows economists to associate economic behaviour with dollar values. The relationship is indirect, and requires a crucial assumption, but it is a relationship that is essential to the entire paradigm of neoclassical economics. Preferences are not directly observable because actual consumptive choices depend upon opportunities as well as preferences. Individuals' choices that constitute their market basket for a given budget period vary in response *either* to changes in preferences *or* to changes in income and opportunities. So, economists take preferences as givens in the analysis, and study changes in behaviour in response to changes in opportunities. On the basis of this assumption, neoclassical models claim to 'explain' changes in behaviour by reference to changes in opportunities. This set of assumptions – and the precise conception of preferences it employs – is therefore essential to all neoclassical modelling (Silberberg, 1978).

Second, if we assume that the values people really care about are ones that they are willing to act upon (if an appropriate situation arises) and if one accepts the triangulation-by-assumption implied in the last paragraph, economics establishes itself as a 'behavioural science'. The study of preferences can thus proceed in as value-neutral a manner as possible. Furthermore, aggregated accounts of the preferences actually held by all individuals in a society can stand as a general guide to 'democratic' decisions – policies that reflect the aggregated will of individuals in a free society.

Third, unlike voting behaviour or other yes-no indicators of values, preferences interpreted as choices to 'purchase' goods and services at a given price and given available endowments, can also register the intensity of feelings supporting those judgments (Page, 1992). It is often useful information to know not just *whether* individuals desire an outcome, but also *how strongly* they desire it. Thus, while a referendum on an issue reveals what a majority prefers, identification of the WTP of individuals tells us how strongly – measured in the willingness of respondents to forgo other opportunities – those individuals favour a given outcome.

Fourth, when embedded in a comprehensive theoretical framework of individual choice such as the valuation model of neo-classical economics, an accounting of preferences purports to be a theoretically completable study of individual and social values. Reasoning that, while individuals have many momentary and conflicting wishes and desires, *actual choices* represent, in

184

some sense, the true commitment of the individual as indicated by whatever other desires they are willing to give up in pursuit of a given value. The robustness of this approach to value is illustrated by economists' treatment of the view of some environmentalists that wild species have 'intrinsic value', meaning that they have value in their own right, not dependent on values any human individual derives from them (Mitchell and Carson, 1989; Freeman, 1993). While economists make no claim to capture the moral essence of this claim, they can nevertheless measure the intensity of these values by ascertaining the willingness of the person to pay to protect the object independent of any use that might be made of it. Economists thus 'solve' (dissolve?) the problem of incommensurate intrinsic goods by treating them as 'existence values', measuring the impact on behaviour of espoused values, in the absence of expected uses of the object in question. The point of this example is that the economists' system of evaluation has remarkable robustness, if one grants them the complex of assumptions outlined above.

Finally, if this theoretical completeness is accepted, a fifth, important methodological advantage is evident. If all values can be characterised in terms of tendencies to act in choice situations (interpreted as WTP or WTA), values derived in this manner can be *aggregated* into a single, monetarily quantified, system of analysis. Certainly one of the strongest considerations favouring the economists' methodology is the promise (seldom achieved in practice, but tantalisingly offered in theory) of a comprehensive scheme for counting and aggregating values actually held by individuals. Economists are usually quick to acknowledge that actual cost-benefit studies should be applied with judgment, citing the practical difficulties of achieving a complete analysis of costs and benefits given current theory and methods. They nevertheless defend comprehensive use of the methodology by noting that a computation of the benefits and costs of a policy – even if it is highly imperfect both methodologically and substantively – represents a 'hypothesis' regarding the social values involved in a decision (Randall, 1986). Thus, while many will disagree with specific value estimates, such an analysis constitutes a falsifiable hypothesis regarding what is of social value. This modest claim for cost-benefit analysis credits it with moving the discussion of social goals past ideology and toward empirical science.

Despite these weighty considerations in favour of a comprehensive accounting of environmental values as measurable preferences, I will state and explain some important reasons to question whether the economists' model provides a comprehensive approach to understanding environmental values. In doing so, I am careful to distinguish two quite different critical responses to economists' preferences and the models they use to measure them. I do *not* intend to argue that the economist model is not a useful one;[2] on the contrary, information about preferences can be one very important source of information regarding environmental values. My response, instead, is that preference models provide only one approach to the valuation problem and that the usefulness of preference explanations is actually enhanced if they are regarded as describing only one aspect of environmental valuation. When the study of preferences is supplemented with a broader, more comprehensive treatment of other aspects of environmental values, the overall picture of environmental valuation is clarified.

This argument implicitly distinguishes two somewhat different goals that might be pursued by those seeking a common, interdisciplinary vernacular for the study of environmental values. It was noted above that to some, the goal of expressing all environmental values in a single currency, a single, aggregable measure of all value, seems the ideal outcome of the interdisciplinary search for a theory of environmental values. An alternative goal would be to seek a common decision framework for the discussion and measurement of values, but one that uses different criteria in different situations (one criterion of which might be a cost-benefit criterion). This alternative goal would be pursued without insisting that data gathered to indicate conformity to one criterion of good management must be commensurate and aggregable with data relevant to other criteria. If we could provide a general theory that allows us to characterise fairly precisely the various situations in which people value and make choices, it may be possible to recognise different decision criteria that are appropriate in different situations. The point is that it is possible to have a common framework of analysis – one that is useful in interdisciplinary discussions – without having a single method of measuring all value. . . .

Single-currency systems such as the economists' model have obvious advantages, but they can serve as complete models only if they can capture the full range of environmental values. The question, then, is whether individual preferences, as defined and measured by economists, can provide a complete, or reasonably so, accounting of environmental values. My answer is that it cannot. One reason it cannot is that, the puzzling 'principle' of consumer sovereignty (CS), which takes human preferences as unquestioned and given for the purposes of the analysis, apparently disqualifies the economic decision framework from expressing an important aspect of the environmentalist viewpoint, the view that experience of wild places serves an important function in shaping a culture's self-image and sense of value (Sagoff, 1974; Norton, 1987, Chapter 10; Norton, 1991B). . . .

186

## II. CONSUMER SOVEREIGNTY

CS represents to a non-economist one of the most puzzling aspects of neoclassical models. Historically, modern welfare economics is of course an heir to the utilitarian tradition of ethics; but it has been clearly shown by Sagoff (1986) that modern welfare economics is the heir to one specific branch of that tradition, which traces to Jeremy Bentham (1948). Bentham argued that individuals are the best judge of their own well-being and that expressed preferences should be accepted at face value. The later utilitarian, John Stuart Mill, rejected this approach to welfare, arguing that there is an important difference between 'higher' and 'lower' satisfactions – that a life of the mind, for example, results in qualitatively better satisfactions than do the pleasures of the senses. Modern economics has unequivocally cast its lot with Bentham on this point by accepting CS, apparently without serious dissent among economic practitioners. CS states that 'what the individual wants is presumed to be good for that individual' (Randall, 1988).

This seems simple and straightforward enough on its face. But a closer look at this 'principle' raises many questions, questions that – surprisingly, given the importance of the principle in the foundations of economics – are hardly discussed by economists at all. One of the most puzzling aspects of this principle is its logical status. Is it an *empirical hypothesis?* If so, it seems curiously easy to refute, simply by noting the self-destructive choices of the drug addict or the virtually universal experience of wanting something very badly, only to find it led to personal disaster. Is CS, then, a tautology, true in some deeper sense not countered by these obvious counter-examples? Or is it more like a constitutive principle of the neo-classical paradigm in economics – in this case a sort of methodological choice? I have also heard CS expressed as a commitment to a democratic attitude and an antidote to totalitarianism. The principle, in its multiple guises, apparently entails quite different attitudes toward preferences, and especially toward the processes of preference formation and reformation. On Randall's formulation, just quoted, advocates of CS claim only that preferences are not to be criticised – that whatever a person expresses as a preference is *taken as given.* And Randall's version only *presumes* that they are in fact in the individual's self-interest. But George Stigler and Gary Becker (1977) defend a much stronger version of CS. They argue that preferences are both *given* and *fixed.* They seem to suggest not only that there is no point in questioning individuals' judgment that something improves their welfare, but also that very little could be learned by studying how preferences are formed and re-formed (Stigler and Becker, 1977, p. 89).

I will proceed by examining five distinct arguments for accepting the principle of CS, noting that different interpretations of CS seem to follow, depending on the argument that is taken as the rational, intellectual core of support for it.

## 1. The Competitive Advantage Argument

According to one argument, economists should accept CS because their methods are not appropriate for studying the development and change of preferences. The study of the formation and re-formation of preferences is a worthy intellectual task, according to this line of reasoning, but a task best left to disciplines such as psychology and sociology (see, for example, Silberberg, 1978; March, 1978). Economists, it can be argued, have a comparative advantage in studying the ways individual preferences should be aggregated into a measure of social welfare. This is an intellectual task well suited for their disciplinary training and given their expertise in mathematical modelling techniques. It is difficult, especially for a non-economist, to attack this argument – CS is in this guise put forward as a methodological choice of economists to engage in a particular type of analysis. If economists choose to define the intellectual boundaries and the explanatory scope of their field in a certain way, why should practitioners of other disciplines or policy makers care?

But it seems fair to note that this line of reasoning, though unexceptionable, also exhibits a certain circularity. It is not an *accident* that economists' economic models work best if they take consumer preferences as givens. Changes in opportunities track changes in behaviour *only on the assumption of stable preferences.* So the decision to accept preferences as givens is inseparable from a methodological commitment to accept data about how people spend their resources (or would do so under given conditions) as indicative of welfare. If one were to break this definitional link and forgo the conceptual simplification of holding preferences constant – the link that allows the empirical study of impacts on behaviour resulting from changes in opportunities – there would remain no connection, even theoretically, between preferences, welfare, and behaviour. CS is therefore a critical enabling assumption of the economists' explanatory paradigm.

The competitive advantage argument for CS, despite its apparent circularity, makes considerable practical sense. Surely it is an advantage to have a methodology capable of aggregating benefits and costs of a given policy into a 'bottom-line' analysis in which diverse considerations can be counted as commensurable and easily compared. Criticising a system of such computational power because it lacks the ability also to track the changes in preferences that provide the necessary data for the computations seems a bit like discarding a perfectly good hammer because it won't turn screws. But the circularity should also put us on notice about the limited implications of this form of the principle: CS, as defended by this line of reasoning, tells us nothing about the way the world is. This argument proves that, given assumptions essential to the system of microeconomic analysis, preferences must be treated as exogenous to that system. As is recognised by at least some who employ this argument (see, for example, Silberberg, 1978, p. 5), one cannot infer anything about preferences treated as characteristics of real individuals from this conclusion.[3]

Economists' preferences are 'theoretical' entities in the strongest sense. They are inferred not only in the sense that they are not observable given present tools (as, for example, were atoms before the invention of today's powerful microscopes), but also in the sense that it is *in principle impossible* to observe preferences directly. Preferences, as they exist in economists' models, are figments of theory, not directly linked to observable behaviour. This is emphatically not to say that they are worthless to science. Their worth to economic science will, *and logically must be*, determined by the success of economists in explaining and predicting real events. Preferences, as defined by economists in this logically unexceptionable argument, stand or fall with the paradigm of microeconomics.

## 2. The Direction-of-Analysis Argument

Another argument, closely related logically to the comparative advantage argument, treats CS as simply a decision to conceptualise the scientific problem as tracing causation *from* preferences *to* changes in behaviour,[4] which allows aggregation across individuals based on behavioural changes. The reaction to this argument should be identical to that of the comparative advantage argument in that, again, CS represents an assumption that constitutes the economists' paradigm of study. Here, CS is taken as a verbal representation of a particular map which cuts up the intellectual landscape into various disciplinary countries. It is a matter of drawing intellectual boundaries, and from it we should infer nothing about the motivations of real human beings. Again, the 'truth' of CS, given this defence, will be determined by the success of the economic paradigm in explaining actual human behaviour.

## 3. The Value Neutrality Argument

While it is seldom expressed explicitly as an argument, a strong motivation among economists for embracing CS derives from its association with the idea that economics can be a value-neutral science which describes preferences as held by subjects of study, but which invokes no value commitments beyond seeking the truth about these individuals' preferences (see McKenzie and Tullock, 1978, p. 7; Heyne and Johnson, 1976, p. 767). This argument might be articulated as follows. Any attempt, such as that of J. S. Mill, to identify some values/preferences as 'higher' or more worthy than others must introduce value judgments into the analysis of human preferences. Scientists, *qua* scientists, do not make value judgments. Therefore, acceptance of CS by economists is a necessary methodological commitment of any economist who hopes to be truly scientific.

A full analysis and evaluation of this argument would require more space than is available here, but I must make two points about it. First, it can be argued that choices of what to count within an analysis – even if they are justified by methodological reasons – can in fact embody values; so the search for purity of scientific analysis may be futile. Second, associating CS with value neutrality calls into question the status of the principle as suggested by arguments 1 and 2. Those arguments seem to leave open the possibility that there are important questions, worthy of study – such as how preferences are formed and whether some preferences are 'higher' than others – which are not included within the scope of economic explanation given the assumption of CS. The implication of this argument, however, is clearly that some of these questions are not 'scientific', and any disciplines that engage these questions are necessarily nonscientific in their methods. This argument therefore creates an ambiguity as to whether CS simply makes all study of the formation and re-formation of preferences exogenous to *economic analysis* or whether it makes this study exogenous to *all science*. It is therefore unclear whether CS is applied as a principle governing the methodology of economics – as is apparently implied by arguments 1 and 2 – or whether CS applies to all social sciences, economics included, as would follow from a general commitment to value-free science.

I have argued, elsewhere, that a science of sustainability, especially if it embodies a commitment to protect ecological integrity/health, like human and veterinary medicine, is necessarily a normative science (Norton, 1991C; paper 19, this volume). I think I detect an ambiguous, and in some cases hostile, attitude to the introduction of the idea of a normative science among economists. This ambiguity is at least partly due to confusion of the implications of the first two arguments with the implications of the value neutrality argument. Arguments 1 and 2 are consistent with recognising a normative science of sustainability; they simply imply that economics as presently constituted cannot fully contain such a science. If the argument applies to all science, it is an argument that normative analysis of values cannot be endogenous to any science. Under that stronger interpretation, the very project of developing a normative science of sustainability is considered misguided.[5] The questions of whether the study of sustainability can be a normative science, and in what sense it is one, have been conflated with the question of whether economics can accommodate such a science. This is not to say that there are no substantive disagreements regarding this issue, but only that addressing those substantive disagreements cannot occur until this conflation is recognised and avoided.

### 4. The Democracy Argument

When asked casually why they accept CS,[6] economists usually provide some variant of an argument considerably different from these first three, which we can call the 'democracy' argument. For example one economist friend responded to my request for reasons to support CS with arched eyebrows and a question: 'If individuals are not the judge of what is best for them, then who would you suggest *should* decide what is best for them?' The question was rhetorical and carried the clear implication that merely to question CS was to embrace elitism and to ally myself with Big Brother, mind control, and totalitarianism. But surely it is possible to distinguish the question of whether preferences should be *defined* within the economists' models as givens from the question of who decides what is best for individuals and what is right and wrong in a society. It is possible, for example, to say that the satisfaction of the preferences of the addict and the sexual predator are (a) not likely to result in increased welfare for them as individuals, and also (b) wrong, without deserting a democratic attitude or questioning the advantages of freedom over totalitarianism.

While the democracy argument is in one methodological sense similar to the first three arguments – its conclusion is, like theirs, registered as a methodological decision to take expressed preferences at face value – it is very different from the above arguments because the motivation for it has little to do with questions of effective explanation of behavioural phenomena. Justified in this manner, the commitment to CS apparently rests on a commitment to a democratic attitude favouring self-determination. If I understand this line of reasoning, it asserts what might be considered a moral commitment to respect the right of persons to the maximum right of self-determination that is consistent with like rights for others. This argument therefore appears to be incompatible with the value neutrality argument as developed above. The embrace of CS and this motivation for it are inconsistent with the decision to avoid value judgments. Economics cannot *both* be value neutral *and* based on a moral crusade for any basic value, however widely held in our society, even freedom.

## 5. The Positivist/Emotivist Argument

Another explicit argument for CS rests on an important debate about values that took place within philosophy during the 1930s–1950s. According to positivist ethicists, most of whom subscribed to some form of *emotivism*, values are arbitrary matters of individual taste, impervious to rational argument (see, for example, Stevenson, 1945; Ayer, 1936). Stigler and Becker (1977, p. 76), for example, emphatically assert (without argument) in their first paragraph that 'deplorable tastes, at least when held by an adult, are not capable of being changed by persuasion'. They proceed to argue that the domain of economics should never abandon any question to the realm of the non-rational, and conclude that the explanatory reach of economics could be expanded if economists simply assert that preferences are *fixed* and *similar across people*. The Stigler and Becker version of CS goes far beyond arguments about the most effective foundations for economic study, and therefore differs from arguments 1–4 in the scope of its conclusion. Stigler and Becker, apparently because of their positivist commitment to an extremely strong (and today mostly discredited) version of emotivism, essentially treat as trivial any study of values that does not employ the economists' model. This amounts to a sort of disciplinary imperialism by obliteration of all other competing forms of social scientific explanation, the equivalent of dropping a neutron bomb on the intellectual domains of other social scientists. They advocate seeking explanations of all choice behaviour – once they assert that preferences are fixed and unquestioned – in economic terms: 'The great advantage ... of relying only on changes in the arguments entering household production functions is that *all* changes in behaviour are explained by changes in prices and incomes, precisely the variables that organise and give power to economic analysis' (Stigler and Becker, 1977, p. 89).

So, in this particular form, CS implies a much more aggressive attitude toward disciplinary boundaries than is implied in the first three arguments. It also represents an important departure from traditional economic analysis which has endorsed the heterogeneity of preferences (Silberberg, 1978) and which has been undogmatic regarding the changeability of preferences, assuming them to be stable only for the period of analysis. The fact that economists can attribute such different characteristics to preferences (and never feel a need to submit their differences to empirical determination) is simply a symptom of the above-noted theoretical nature of preferences. Stigler and Becker are simply not making an assertion about psychological states; they are rather attempting to expand the predictive power of their paradigm through methodological assumption.

But Stigler and Becker's argument, based as it is on the failed philosophy of logical positivism, seems an anachronism to anyone who has read much philosophy of science or theoretical ethics since 1950. The positivist philosophy collapsed from within as epistemological arguments showed the conceptual impossibility of identifying any sentences that are wholly empirical in content, not dependent on choices of theory (Quine, 1953; 1960; Kuhn, 1962). Further, while ethical theorists still differ regarding the exact logical status of moral pronouncements, virtually all such theorists have now acknowledged that emotivism misses most of what is interesting in moral and persuasive discourse. If one refuses to accept the positivist assertion that value expression is entirely beyond the pale of rational discussion – the key premise in Stigler and Becker's argument – the argument cannot even be formulated, let alone persuade. I suggest that economists treat the Stigler/Becker version of CS as an interesting museum piece, analogous to an elaborate, but failed, time travel machine.

Given the variation in CS associated with different arguments, it is necessary to note two somewhat different departures that might be taken from CS by economists. One departure, which is apparently consistent with application of a methodological version of CS within economics, would be for economists to increase their emphasis on the importance of the scientific study of preference formation and re-formation, and encourage cross-disciplinary study of these intellectual questions with other social scientists. I expect that most economists would accept this departure and consider it simply an extension of collaborative work, already under way, between cognitive psychologists, survey researchers, and economists.

But I would not expect such ready acceptance of another departure, a departure that would undertake to *evaluate* preferences in a way that would mimic Mill's distinction between 'higher' and 'lower' preferences. I believe that some attempt to evaluate and alter preferences of consumers is an essential commitment of many, if not most, environmentalists (Norton, 1987; Norton, 1991 B). Economists have generally looked with suspicion on any attempt, such as Mill's, to qualitatively sort preferences into higher and lower satisfactions. For example, in the midst of an argument that sustainability can, essentially, be reduced to a question of wise investment because resources and capital of all forms are fungible, Robert Solow (1993, p. 182) says:

[S]ustainability is about distributional equity. It is about who gets what. It is about the sharing of well-being between present people and future people. I have also emphasised the need to keep in mind, in making plans, that we don't know what they will do, what they will like, what they will want. And, to be honest, it is none of our business.

Consider the final sentence of this quotation. Is it, indeed, none of our business what the future will do, what they will like, what they will want? Suppose, for example, I sincerely believe that violence on television is creating a generation of violent monsters. Have I no obligation to work to achieve changes in television programming to affect the values held in the future? Note that recognition of such an obligation carries no implication regarding what methods – censorship, persuasion, etc. – I would use to achieve these changes.

Suppose, to cite a case more relevant to environmental values, I sincerely believe that, if our generation converts all wilderness areas to mines and farmland or to other human uses, we will end the possibility of wilderness experiences for many centuries at least, probably forever. Suppose I also believe that people in the future would as a result of these acts be unable or unlikely to value wilderness. Instead, they would attend theme parks much like the 'rainforest exhibits' in today's zoological parks and botanical gardens. Is this a matter of indifference to us? If we have it in our power to stop the process of clearing, and simply prefer lower-priced timber and foodstuffs to future people's wilderness experiences, is it still a matter of indifference?

Since Solow is reluctant to make any judgments about what future people should be able to do or feel, he need only consider two outcomes: (1) the situation is as above, and many future people still love wilderness, and go to the theme parks in great sadness, pining for authentic wilderness experiences. Or (2) the people of the future do not value wilderness experiences and enjoy their theme park rainforests, feeling no loss. Solow reasons as follows: if they want wilderness experiences and can't have them, they are unhappy, but we are not responsible for their unhappiness because we have no way to know whether they will prefer wilderness or theme parks, and we cannot be expected to do what we cannot feasibly do (Solow, 1993, p.180). On the other hand, if they do not love and miss wilderness, then they will not blame us. Consequently, he reasons, we cannot be blamed because, if there is a harm to the future, we cannot be held responsible for it. He therefore concludes that we cannot harm the future, provided we make it possible for them to achieve the same level of welfare that we enjoy, and the wisest policy is to invest in economic development so that if they are harmed by our choices, future people will be compensated by being able to purchase alternatives. While it may seem self-serving to continue to destroy all wilderness in search of economic development, Solow reassures us that it is not really – it is the best way to maximise total welfare across time, given each generation's ignorance and indifference regarding the practices and values of subsequent generations.

But notice that this argument – which for many economists has stated the definitive view of sustainability – *assumes* rather than *proves* that we cannot and ought not to take steps to shape the values and preferences that future people will have. If we believe that we have some ability, and some responsibility, to affect what people of the future do and like, then we must consider another option: (3) we could act now to protect wilderness areas with the intention of providing future people with experiences of wilderness, experiences that we believe will ensure that they will both love wilderness and protect it for their own successors. On the assumption that our current activities *do* predictably affect the values held in the future, we must at least raise the question of whether we *should* protect wilderness and encourage future wilderness values. This turns out to be a possibility that is excluded by Solow's laissez faire approach to future values.

Solow's argument that we have no specific responsibilities to sustain specific resources therefore turns out to be just an intergenerational version of CS. And now the ambiguities uncovered above become crucial to the argument. If Solow believes, with Stigler and Becker, that values are just tastes and not worthy of study, evaluation, or re-formation, then he has a reason to dismiss these apparently important and interesting questions as meaningless or trivial. If he accepts Randall's version of CS, on the other hand, which merely presumes that individuals are the best judge of their welfare, the situation is more complex. If we interpret Randall's pronouncement as a methodological preference of economists (because it increases the power of economists' models), then it would only follow that future likes, activities, and preferences are none of the business of *economists*. They might, that is, be precisely the business of psychologists or sociologists or philosophers of environmental values or, more generally, voters who decide the fate of wilderness areas. I conclude that Solow's argument for a laissez faire attitude toward the options and values of the future is convincing only if we assume the highly suspect, positivist version of CS due to Stigler and Becker. . . .

Further, I believe we have now arrived at a clearly articulable difference between economists and most environmentalists and also between economists and members of other disciplines. Environmentalists are moralists, and one of the ways they show this is by having an active concern for both the options for experiences and the values of future people. They invest (or should invest, given their values) in environmental education, for example. Economists such as Solow believe no such obligations exist. This is a point worth further debate and discussion. It should not be resolved by methodological fiat.

The case that environmentalism is essentially moralistic is eloquently made by the environmental attorney Joseph Sax (1980), who argues that it is an essential part of the case for environmentalism that environmentalists believe we will be a better country, with stronger moral fibre, and with a more enduring moral identity if we protect natural areas and landmarks such as national parks (also see Sagoff, 1974; Norton, 1987). This may mean, for example, that environmentalists will attempt to change the preferences of owners of off-road vehicles and encourage people to prefer public transportation. These actions, and the motivations behind them, may make economists uneasy, and perhaps with good reason. There may be dangers of paternalistic excesses in the moralistic streak of environmentalism, but these are dangers shared by all moralists and persuaders. From the Gideon Society to zoos with conservation exhibits to the Exxon Corporation, there are countless individuals and groups out there attempting to change our values and preferences. Even if we follow Solow in arguing against moral education to affect the values and preferences of our successors, it seems this should at least be a matter for interdisciplinary debate, rather than the consequence of a methodological choice to take preferences as givens.

Indeed, the very issues that should be at the heart of a search for a sustainable society are the dynamic intergenerational questions that involve protecting and developing a connection to land. What is at stake is whether it makes sense to choose to shape the preferences, options, and opportunities available to future people. Faced with a decision, as we are today, of whether to manage national parks to protect biotic integrity or to develop them with more 'industrial tourism', it seems at least important to discuss the effect of our legacy on the attitudes, values, and preferences of the future. Protecting wilderness, and thereby protecting the future's possibility of experiencing wilderness, makes sense only if one believes that, if we protect the wilderness, future generations will come, also, to love it and they, too, will protect it for their successors.

... There are fascinating questions involved in the dynamic of intergenerational value formation: questions of how values – including conservation values – are passed from generation to generation and moral questions regarding whether it would ever be reasonable to use current resources to make various experiences and preconditions of value formation available to future people. Economists may choose not to address these questions, adopting CS as a methodological choice, but I urge them to consider whether they have arguments, beyond methodological constraints imposed by their conception of explanation, for dismissing these questions. It might make more sense for economists to loosen the assumptions of the neo-classical explanatory paradigm and remain open-minded about how to integrate preferences regarding future preferences and intertemporally emerging values into their conceptual and theoretical framework.

If one accepts my implication that these questions are important and that their discussion should be an important part of environmental policy discussions, we can see that interpreting CS is really about whether preference formation will be endogenous or exogenous to economic analysis. I have tried to leave that questions to economists themselves to decide; hence my somewhat impertinent title. If one follows environmentalists in believing we do have fairly specific obligations to protect landscape integrity and wilderness, then one way of expressing those obligations would be to insist that value formation and re-formation must be endogenous to the overall analysis of environmental policy. What we need most basically, it would follow, is a normative and activist science – much like medicine, including a public health service with a persuasive/educational function – for the study and treatment of environmental problems (Costanza, Norton, and Haskell, 1992). It may also follow that, insofar as they hope to contribute to this larger science of management, academic scientists, including economists, may have to step out of their narrower disciplinary paradigms and consider both the description and valuation of ecological systems (and changes in them) in a broader framework of analysis (paper 12, this volume)....

### III. CONCLUSION

We can now return with greater understanding to the point, made above, that preference-based economic analysis captures one aspect of environmental valuation, but that this analysis is nevertheless an abstraction that ignores other important aspects of valuation. Endorsement of some form of CS – the acceptance of preferences as givens – constitutes the neoclassical explanatory paradigm in such a way as to allow aggregation of values across individuals, abstracting from the complex processes that form and re-form preferences.... The decision to conceptualise preferences as sovereign is best thought of as a methodological decision. It is an enabling mechanism that permits the development of a powerful analytic tool, the computational paradigm of neo-classical economics. But I have argued that economists must be careful regarding the generalisations they might draw about actual human values from their elegant abstractions. If CS and reduction to present value are justified methodologically, as elements in a system that allows computation across persons and across time, these are characteristics of theoretical objects, objects which exist within the theoretical assumptions of neoclassical economics.

It may be useful to think of economists as emphasising a computational conception of rationality and choice. They believe we can learn a great deal about social values by measuring economic activity and aggregating across persons and times. They are therefore willing to make strong constitutive assumptions that will enable them to enhance their computational power and the scope of their explanations based on changes in opportunities. I have not challenged the interest or power of this approach. But, especially in the absence of complete data, many environmental decisions turn in reality not on *computation*, but on a *categorisation* – a categorisation of the problem at issue. Unless one can categorise an environmental problem/concern adequately, one does not know what is the context of the evaluation, and one does not know what criterion of good management is applicable. If, as argued here, shifting the context of a decision affects which decision criteria are appropriate, it will also affect what type of evidence is important. I believe that, given the complexity of environmental values, an adequate approach to environmental policy valuation must also be complex and multiscalar. However useful computations are in some situations, we must first decide what type of situation we face, and what we should be computing.

Can we say anything useful, in general terms, about the process of formulating an environmental problem? If there are values that unfold across multiple generations, does it not follow that somewhat different value criteria apply when they are threatened? All disciplines that hope to impact environmental policy should address these questions thoughtfully, and soon. These questions cannot be answered . . . by methodological fiat. These questions will require a discussion of the very foundations of environmental valuation and also a practical discussion of how environmental policy decisions can be understood in their full complexity, so that the multi-dimensional aspect of environmental values is represented in decision making.

NOTES

1. This system of evaluation assumes, also, the substitutability of goods for each other and of some dollar figure for a marginal unit of consumption for any good. See Freeman (1993).
2. As is argued by Mark Sagoff in a number of recent publications. See Sagoff (1988; 1993; 1994).

3.  See Fischoff (1991) for a discussion of the approaches and methodologies available to study real individual values and their elicitation.
4.  The importance of this line of reasoning was pointed out to me by Paul Wendt.
5.  These issues, which would require a deep-delving argument regarding the sharpness of the fact–value distinction itself, cannot be addressed here, though the reader is referred to Norton (paper 19, this volume) for the argument that a science of sustainability must treat moral evaluations as endogenous to its analysis.
6.  As I have concluded from a ridiculously informal and unscientific polling of some of my friends who are economists.

### REFERENCES

Ayer, A. J. 1936. *Language, Truth, and Logic.* London: Gollancz.

Bentham, Jeremy 1948 [1780]. *An Introduction to the Principles of Morals and Legislation.* New York: Macmilan.

Costanza, R., Norton, B. and Haskell, B. (eds.). 1992. *Ecosystem Health: New Goals for Environmental Management.* Covelo, CA: Island Press.

Fischoff, Baruch 1991. 'Value elicitation: is there anything in there?' *American Psychologist* 46(8): 835–47

Freeman, A. Myrick 1993. *The Measurement of Environmental and Resource Values: Theory and Methods.* Washington DC: Resources for the Future.

Heyne, Paul and Johnson, Thomas 1976. *Toward Economic Understanding.* Chicago: Science Research Associates.

Kuhn, Thomas 1962. *The Structure of Scientific Revolutions.* Chicago: University of Chicago Press.

McKenzie, Richard B. and Tullock, Gordon 1978. *The New World of Economics.* Homewood, IL: Richard D. Irwin.

Mill, John Stuart 1951[1863]. *Utilitarianism.* New York: Dutton.

Mitchell, Robert C. and Carson, Richard T. 1989. *Using Surveys to Value Public Goods: The Contingent Valuation Method.* Washington DC: Resources for the Future.

Norton, Bryan G. 1987. *Why Preserve Natural Variety?* Princeton: Princeton University Press.

———1991A. *Toward Unity among Environmentalists.* New York: Oxford University Press.

———1991B 'Thoreau's insect analogies: or, why environmentalists hate mainstream economists', *Environmental Ethics* 13: 235–51.

———1991C. 'Ecological health and sustainable resource management', in Robert Costanza (ed.), *Ecological Economics: The Science and Management of Sustainability.* New York: Columbia University Press.

———1995. 'Ecological risk assessment: towards a broader analytic framework', in R. Cotherne (ed.), *Handbook for Environmental Risk Decision Making: Values, Perceptions and Ethics.* Chelsea, MI: Lewis Publishers.

Page, Talbot 1992. 'Environmental existentialism', in R. Costanza, B. Norton, and B. Haskell (eds) *Ecosystem Health: New Goals for Environmental Management.* Covelo, CA: Island Press.

Quine, W. V. O. 1953. *From a Logical Point of View.* Cambridge, MA: Harvard University Press.

_____1960. *Word and Object*. Cambridge, MA: MIT Press.

Randall, Alan 1986. 'Human preferences, economics, and the preservation of species', in B. G. Norton (ed.) *The Preservation of Species*. New York: Princeton University Press.

Randall, Alan 1988. 'What mainstream economists have to say about the value of bio-diversity', in E. O. Wilson (ed.), *Biodiversity*. Washington, DC: National Academy Press.

Sagoff, Mark 1974, 'On preserving the natural environment', *Yale Law Journal* 84: 205–67.

_____1986. 'Values and preferences', *Ethics* 96(2): 301–16.

_____1988. *The Economy of The Earth*. New York: Cambridge University Press.

_____1993. 'Environmental economics: an epitaph', *Resources* No. 111: 2–7.

_____1994. 'Should preferences count? *Land Economics* 15(2): 127–44.

Sax, Joseph 1980. *Mountains without Handrails*. Ann Arbor: University of Michigan Press.

Silberberg, Eugene 1978. *The Structure of Economics: A Mathematical Analysis*. New York: McGraw-Hill.

Solow, Robert M. 1993. 'Sustainability: an economist's perspective', in R. Dorfman and N. Dorfman (eds) *Economics of the Environment: Selected Readings*. New York: W. W. Norton and Company.

Stevenson, Charles L. 1945. *Ethics and Language*. New Haven: Yale University Press.

Stigler, George J. and Becker, Gary S. 1977. 'De gustibus non est disputandum', *The American Economic Review* 67 (March): 76–90.

# 12

# Evaluating Ecosystem States

## Two Competing Paradigms

### 1. INTRODUCTION: ENVIRONMENTAL POLICY AND ANALYSIS: A SHARED CRISIS

Environmental management faces a crisis: legislation, high-level committees, and international agreements all urge that natural systems be managed to protect their health and systematic integrity,[1] but these widely employed analogies have not yet yielded a consensus in operational and specific management directives. Inaction may not be an option, however, because observable trends in the quality of the environment are forcing upon us the conclusion that traditional approaches to resource management are inadequate to the problems of today. Thus, while there is a growing consensus among ecologists, managers, and environmentalists that we must supplement traditional concerns for human health with more ecosystem-oriented management criteria, this consensus has not led to major changes in management or environmental decision-making.

Unfortunately, this crisis in management has an analogue in environmental policy analysis, the very place where one might first search for new directions for management. While there exists no shortage of arguments showing why we need more comprehensive, systematic management objectives, there is a foundational disagreement among policy analysts regarding how environmental objectives should be formulated, discussed, and analyzed. I believe these foundational disagreements can best be understood as "extraparadigmatic" disagreements. A "paradigm" is here defined as a constellation of concepts, values, and assumptions, as well as accepted practices, that give unity to a scientific discipline (Kuhn, 1970). The purpose of this paper is to compare

Reprinted from *Ecological Economics* 14 (1995): 113–127, Copyright 1995, with permission from Elsevier Science.

alternative approaches to valuing natural resources over time and to show that the disagreements that separate mainstream, neoclassical environmental economists from the growing insurgency of the "ecological economists" amounts to an extraparadigmatic disagreement. Once this is established, it will be possible to say something about the conduct and logic of extraparadigmatic disagreements, and take beginning steps toward characterizing models that would combine current economic reasoning with other informational sources within a more comprehensive decision framework.

The multifaceted problem of intertemporal comparisons of values turns on a cluster of conceptually related foundational questions (paper 11, this volume). These unresolved questions have blocked progress in developing a comprehensive and interdisciplinarily adequate account of environmental valuation. This interrelatedness can be observed by noting the highly connected nature of the opposed answers of mainstream and ecological economists to three related questions about time preference.

1. How should we *measure and compare values that are experienced at different times?*
2. How should we place a value on the *risk of irreversible loss* of a natural feature or productive ecological process?
3. How should we evaluate changes in the *scale* of an economy vis-a-vis its ecological and physical context? Do increases in throughput in the economy, which often result from economic growth, significantly impact the ability of future generations to achieve an equal or better life than is possible in the present? If so, how should we address the resulting questions of intergenerational equity in the use of resources?

I will argue that answers to questions in this cluster define an important "fault line" in the intellectual bedrock underlying environmental policy discussion and debate. While considerable disagreement and confusion exist on both sides of the fault line. I will argue that positions on these three questions are sufficiently inter-locked so that it is not inaccurate to argue that there are *two competing paradigms* for the discussion of intergenerational impacts. I will show how answers to these questions lead in turn to differing analyses of the idea of "sustainability." I introduce the device of "risk decision squares" as *maps* of the evaluative space in which environmental decisions are made. This neutral vocabulary exhibits the assumptions, conceptual commitments, and principles of economics and ecology. Ecological economics represents a new paradigm because it introduces a two-tier process of policy evaluation. Two-tier systems of analysis are characterized by their representation of some decisions as essentially "economic" in nature, but other decisions as requiring alternative methods of evaluation, such as consideration of equity among generations. These two-tier systems of analysis apply economic/utilitarian analysis within broad constraints set by long-term, physically described constraints.

## 2. DEFINITIONS OF SUSTAINABILITY LOCATE A PARADIGMATIC SPLIT IN POLICY MODELS

In a series of lectures and papers, Robert Solow (1974, 1986, 1992, 1993) has defended the view that "sustainability" can be fully defined, characterized, and measured within the neoclassical theory that shapes the mainstream economic tradition of resource analysis. Solow's basic idea is that the obligation to sustainability "is an obligation to conduct ourselves so that we leave to the future the option or the capacity to be as well off as we are." He doubts that "one can be more precise than that." A central implication of Solow's view is that, while to talk about sustainability is "not empty . . . there is no specific object that the goal of sustainability, the obligation of sustainability, requires us to leave untouched." Solow correctly acknowledges that the plausibility of this account of sustainability is essentially connected to the principle that no resource is irreplaceable, that every natural resource has an adequate substitute. In the jargon of economists, all resources are "fungible." Given this complex of essentially related ideas, Solow also, and validly, draws the conclusion that the problem of obligations to the future can be understood as simply the task of defining a rational and intergenerationally equitable *investment policy.* Sustainability, within this complex of principles and assumptions, is a matter of balancing consumption with adequate investment so that the future faces a nondeclining stock of total capital. Note that, on this view, monetary capital, labor, and natural resources are interchangeable elements of capital. Within this set of definitions. the future cannot fault us as long as we leave the next generation as able to fulfill their needs and desires as we have been in our generation. This nifty simplification of the intergenerational problem therefore succeeds, if it succeeds at all, on the heroic principle of intersubstitutability of resources because the intergenerational fungibility of resources follows from that foundational and constitutive principle. See Freeman (1986) for a brief explanation of the importance of fungibility in the economists' system of analysis.

Ecological economists have challenged the heart of this complex of assumptions and economic models. In particular, they have questioned Solow's principle that all forms of capital can be aggregated together and compared across generations. They argue that certain elements, relationships, or processes of nature represent irreplaceable resources, and that these resources constitute a scientifically separable and normatively significant category of capital – natural capital. This position directly contradicts Solow's central conclusion that sustainability is achieved, provided simply that the *total* stock of capital is not declining. We therefore face what is described by Daly and Cobb (1989) as the clash between "strong" and "weak" senses of sustainability. If there is to be an "accounting" and a judgment of the moral acceptability of one generation's bequest to all subsequent ones, on this view, there must be physically defined constraints on alteration of physical and ecological processes, which are apparently not associated in any measurable way with impacts on human welfare. The call of ecological economists for a structured bequest package cuts deeply against the grain of the mainstream paradigm because the acceptance of physically characterized constraints as a part of the definition of sustainability demands inputs of information that are not directly characterizable in the language of individual preferences. What is at stake, in particular, is what counts as relevant data for computing intergenerational balances – mainstream economists opt for a more highly aggregated data base with which to compute intergenerational fairness than would ecologists and ecological economists. The definition of sustainability is therefore practically important because it determines what type of information will be relevant in decisions affecting future generations. The conflict between paradigms also manifests itself in competition for research funding.

### 3. COMPETING PARADIGMS

The growth of the new field of ecological economics, the phenomenon of ecologists becoming more active in policy formation, and the expanding number of calls by political leaders for a more integrated and ecosystemic approach to environmental policy (see, for example, Gore, 1992) are all symptoms of an extraparadigmatic debate regarding environmental goals and objectives and, especially, of the types of information that are crucial for forming rational policy in the new millennium. In extraparadigmatic debates, there is always danger of the two sides talking past each other because, lacking a shared paradigm, communication is difficult. It is hoped that clear recognition of the extraparadigmatic nature of ecological economics, and tentative steps toward a more encompassing set of concepts, may increase communication and hasten the development of a new, more inclusive paradigm. The ideal outcome would be a set of integrative models of environmental problems in which information from multiple disciplines is integrated into a rational, long-term approach to environmental management.

I begin by sketching competing answers of mainstream and ecological economists to our three questions above, noting how the three answers are inter-related, and how crucial assumptions determine that positions on the three questions stand and fall together.

A. Mainstream economists have generally dealt with the first question, the question of intertemporal preference, by the device of discounting (see, for example, Lind et al., 1982). *Homo economicus* "discounts" values across time, preferring to enjoy benefits sooner rather than later and to delay losses. Economic analyses therefore discount values that will be experienced in the future; in business, costs or benefits expected in future years are discounted at a rate roughly equivalent to the rate of real interest. But at any significant positive discount rate, such as one a business might use to calculate the present value of future returns on investments, economic calculations imply that we care very little for even the next generation and not at all for distant ones. This viewpoint is consonant with Solow's simplification of intertemporal ethics. The value of all future outcomes and options depends on the willingness of present consumers to pay for those outcomes and to hold open those options. This result follows from the intertemporal fungibility assumption that all human values can be expressed within a single temporal scale, the present. Some economists, noting that significant public goods will be destroyed if all decisions are made on a "private" discount rate, have propounded the concept of a "social discount rate," a slower rate that applies to investments in public goods. But discounting, while undoubtedly a useful method in many shorter-term contexts, has been criticized, and at least one leader in the field has conluded that "searching for the 'correct' social rate of discount is searching for a will o' the wisp" (Page, 1988).

Ecological economists, by contrast, emphasize the importance of scalar issues. The hierarchical approach takes as its basic assumption that smaller-scale subsystems change at a more rapid pace than do the larger super-systems that form their ecological and physical context. According to one version of this approach, different valuational systems apply at different physical scales, with temporal policy horizons set by the social values impacted by the policies under consideration (Page, 1977; paper 16, this volume).

As ecological economists attempt to address these contextual and scalar issues that arise in addressing intertemporal changes in the ability of nature to produce a flow of goods and services, they have found Solow's simplification a barrier to modelling the full complexity of long-term environmental problems. Highly aggregated data is too coarse to retain information about impacts of human activities that affect the patchy, multi-scalar landscape on many different scales in space and time. Fungibility, which is essential if we are to aggregate these time scales together into the present, implies that information about the temporal and spatial scale of human impacts is lost. Solow's fungibility assumption is therefore rejected by ecological economists, and with it they reject the entire paradigm represented by Solow's definitions and simplifications of intergenerational obligations.

Just as intersubstitutability is a unifying principle in Solow's system, the ecological economists' position is unified by the belief that healthy ecological systems – defined as multiscalar patch works of inter-related processes – are essential elements of any "fair" bequest package. The future cannot be compensated for by technologies or financial/cultural capital as substitutes for creative and productive systems, according to ecological economists; they believe that the bequest we offer future generations must be structured, not unstructured. Further departing from Solow, the health of these large systems will be measured with physical descriptors rather than economic criteria. By abandoning Solow's project of reducing all intertemporal environmental values to present preferences, ecological economists have apparently exiled themselves from the discouse of mainstream environmental economics.

B. Mainstream environmental economists have recognized the importance of irreversibilities in decision making (Krutilla, 1967; Krutilla and Fisher, 1975; Smith, 1980; Fisher and Krutilla, 1985; Solow, 1993), arguing that significant public goods will be lost if questions of preservation are left to unguided markets. The highly regarded "Natural Areas Program" at Resources for the Future, for example, was based on the premise that natural resource scarcities will not hinder economic productivity, but that there is a danger of destroying "amenities" such as parks and natural wonders because individuals might have economic incentive to develop and irreversibly change outstanding natural features which should be protected as public goods. Note that this dual accounting system for private and public goods goes nicely with the dual conceptualization of private and social discount rates. If Page and others are correct that there exists no "correct" social discount rate, this approach lacks clear operationalization. Solow explicitly follows Krutilla in his description of our obligation to set aside natural wonders that are scarce. Solow incorporates this general approach into his program, arguing that there will be unique and irreplaceable natural features that we should set aside to protect "unspoiled nature as a component of well-being." He explains: "It is perfectly okay, it is perfectly logical and rational, to argue for the preservation of a particular species or the preservation of a particular landscape. But it has to be done on its own, for its own sake, because this landscape is intrinsically what we want . . . not under the heading of sustainability" (Solow, 1993). But this is not a surprising position: the conclusion that irreversible changes affect only amenities is conceptually equivalent to the Solow conclusion that future generations cannot complain as long as they are left with the capacity to be as well off as we are, in general.

C. Having seen that the mainstream economists' answer to questions 1 and 2, and the ecological economists' rejection of them, both turn on their oppositions regarding fungibility across resources and across time. It has now become evident that their positions similarly entail opposed positions on scalar issues. We have seen that Solow, following in the tradition of Krutilla and other writers on the valuation of natural areas, treats the protection of natural areas and natural features of the landscape as bounded decisions regarding how to use – develop or preserve – specific natural areas. Compare two examples. First, assume someone today proposes damming the Yosemite Valley to provide hydropower and irrigation water to jump-start the sagging California economy. Second, suppose someone suggests that, since corrective measures to "Save the (Chesapeake) Bay" are so expensive, the Chesapeake be "sacrificed" to development. Notice, first, that the scale of these two examples is importantly different. The Chesapeake consists of a bay stem plus a complex of tributaries, each of which is comparable in scale to the Merced River in Yosemite Valley. Because of the greater scale and complexity of the Chesapeake system, it is not implausible to refer to it as a "whole ecosystem," whereas the Yosemite Valley is more plausibly thought of as an element in a larger system.

Perhaps because the Yosemite Valley is encompassed within a National Park which has a long-established record of the willingness-to-pay of recreationists and others, it seems well within the capabilities of an economic analysis to produce plausible, and very high, use and non-use values that derive from protecting the valley. Given this established commercial and semicommercial productivity, and equally plausible measures of "existence" or "intrinsic value" enhanced by uniqueness or at least rarity, it is easy to see how the general outlines of an "economic" analysis of the decision to dam Yosemite might be constructed. But in the larger-scale case of the Chesapeake Bay, a collapse of the large-scale ecological system is more difficult to interpret in terms of losses to individual human welfare. This is not to say that there would be no impact on human welfare if the Bay is "written off," but rather to say that consumers and analysts would find it much harder to identify and classify as either "use values" or "existence values," the values they derive and expect to derive from the complex, beautiful, and productive system. Because the Chesapeake Bay is the keystone resource in the region, a failure of the

Bay system would have so many systemic effects on the broader ecology and economy of the region that calculation and aggregation of welfare measures – benefits and costs to individuals – are apparently beyond computation. Because the change is a change in a large-scale ecosystem process/function, one would have to calculate the outcomes of interacting series of changes in multiple layers of ecological processes over a whole landscape. That is, while the distinctive nature and well-documented history of use at Yosemite makes it a reasonable "unit" of economic analysis as a "rare" amenity, it seems much less likely that an overall collapse of the health of the Chesapeake Bay could be associated with measurable welfare impacts on individuals. But it is change at this ecosystem level that is the apparent concern of advocates of the ecosystem risk and ecosystem health paradigms. As more and more environmental policy is formulated in terms of ecosystem-level impacts, the question arises of whether the mainstream economic paradigm can forge any rational connection between the ecological processes that form the "production function" for ecological services and associate these with units of individual welfare.

Ecological economists have opposed the fungibility hypothesis on the grounds that, while small-scale alterations of particular resource- and service-bearing ecosystems may be overcome by improvements in technology and know-how, at some scale these alterations affect the health and integrity of the functioning system. Ecological economists believe that ecosystems are not equilibrium systems, but rather dynamic systems whose essence is to be self-organizing and self-creative. This emphasis on large-scale dynamics and ecosystem organization causes the ecologists to expect systems to behave in nonlinear ways and to consider the possibility that this behavior can be induced by gradual increases in the scale of changes to the landscape. Ecological economists cannot accept fungibility because they see the structure and processes of physical systems as vulnerable to incremental human choices, and they fear that, at some scale, the destruction of ecological systems and processes will result in direct impacts on human welfare. But they do not believe that this collapse can be correlated with present-value estimates of future utility because the analysis is carried out in a nonequilibrium system rather than an equilibrium system. It apparently follows that the differences between ecological economists and mainstream economists represent a serious conceptual disagreement, a disagreement closely related to the principle of irreversibility they respectively employ. Relying upon the principle of substitutability of resources, mainstream economists have concluded that resource scarcity has not and will not hamper future growth of the economy. Reduction of resource stocks and degradation of resource-producing systems are not expected to affect productivity. Therefore, the problem of irreversibility is reduced to the problem of protecting the *quality* of future life (Krutilla, 1967: Solow, 1993). This conclusion reduces the limits on economic growth to the "project" scale. It is not clear that many such "projects" of protection, designed to protect amenity and quality-of-life values of the future, add up to protection for whole ecosystems. On the contrary, it seems unlikely that large-scale ecological systems and habitats for far-ranging animals can be protected unless we manage the entire landscape, including privately used and exploited lands as well as nature preserves (Harris, 1984).

What is distinctive and important about recent calls for ecosystem management is a focus on larger-scale dynamics across whole landscapes and longer periods of time. The idea of ecosystem health/integrity, which attributes both descriptive and normative characteristics to varying ecosystem states, challenges the mainstream paradigm to develop an adequate method for associating changes in ecosystem states with changes in human welfare.

Indeed, it can be said that, so far, the mainstream economic paradigm has not been hospitable to values that emerge on the ecosystem level. Even when environmental managers such as officials at EPA wish to act to protect these whole-system characteristics they are stymied because there exist no accepted means by which to measure benefits derived from whole ecosystems, and they have little hope of justifying programs directed to this purpose in terms acceptable to auditors at the Office of Management and Budget. For example, in a National Academy of Sciences study, which attempted to weigh the relative costs of prevention versus accommodation to global climate change, zero value was assigned to the damages to natural systems. The panel justified this exclusion by stating that no adequate methods for ecosystem valuation exist (NAS, 1992). Again, this sort of exclusion – and protestations against it – are characteristic of extraparadigmatic debate.

A highly connected system of assumptions can often appear inflexible in the face of new problems and unexpected results. But it is also important not to conclude too quickly that a highly connected paradigm cannot account for new problems or data – the strength of a paradigm is measured by the ability of its practitioners to devise new theoretical and methodological means by which to encompass new information in the conceptual framework of the paradigm. It is not here claimed that the mainstream paradigm cannot ever develop a more comprehensive methodology for dealing with time preference and irreversibility. Indeed, it would be very interesting to see attempts to design contingent valuation studies to see just how much consumers and citizens are willing to pay to protect large-scale ecological systems and the landscapes they support. Or some new methodology for assessing values of ecosystem-level changes may emerge. In the meantime, I think it is fair to say on the basis of the arguments developed in this part that mainstream economists and their opponents among ecological economists exhibit disagreements that are extraparadigmatic.

Although it is perhaps too early to call the ecological economists' approach to intertemporal time preference a "new paradigm," I think it is no exaggeration to say that the ecological economists are uncomfortable with the tools offered by mainstream economics to estimate the values of protecting large, landscape-level ecological systems, and that their attempts to specify certain physical features of ecological systems as natural capital signal an intention to create an alternate paradigm that includes values and concerns that cannot be characterized in the mainstream paradigm.

#### 4. A NEUTRAL CONCEPTUAL GEOGRAPHY

What is lacking is a broader, but neutral, vocabulary and discourse in which these problems can be addressed. It is helpful to envision the intellectual terrain as represented in Fig. 12.1, which represents issues separating ecological and mainstream economists in neutral terms (paper 10, this volume). This "environmental risk decision square" plots the variables of reversibility against the magnitude of impacts, defining a decision space on which various policy decisions can be located, depending on types of possible risks.

The decision square of Fig. 12.1, as outlined here, represents the logical space in which environmental policy decisions are made and therefore provides a neutral space in which we can define possible positions/assumptions/ basic principles regarding the subject matter of the field we seek to understand: the overlap of economics and ecology, especially as they pertain to "policy."[2] The risk decision space can represent in neutral terms the three issues stated above that separate mainstream from ecological economists. It will therefore be an important metascientific tool for examining the conceptual commitments of various approaches to defining sustainability.

We can represent what we said above – that economists and ecologists employ a different "paradigm" – by recognizing that ecological economists would, at the expense of being unable to aggregate fully across the whole space, draw distinctions in *types of risks*, breaking the risk decision space into distinct regions where different considerations and criteria apply. They therefore differ from mainstream economists who assume that all types of risks can be expressed as dollar values which vary gradually and at the margins, representing their decision to treat all risks and benefits in "fungible" terms.

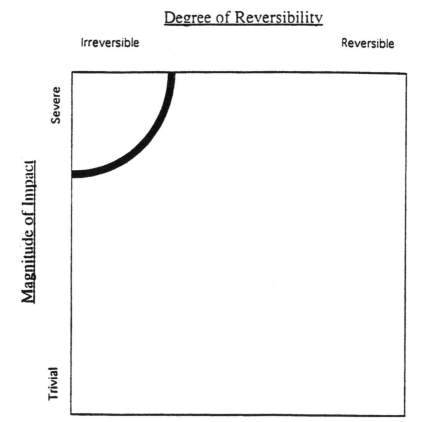

Figure 12.1. Neutral version.

For mainstream economists, then, options for preference satisfaction, and accordingly for future welfare, are not prejudiced by irreversible changes in ecological systems or other alterations of the physical world – the decision square therefore need not specify a time scale for reversibility. The horizontal axis can be understood simply as the degree of substitutability of a new resource for any damaged resource, with the goal being to ascertain the correct dollar figures a consumer would be willing to accept for the destruction of a resource that will require development of a suitable substitute. This simplified version of the risk decision square can be represented as Fig. 12.2 (Crosson and Toman, 1991).[3]

But this version of the decision space breaks conceptual neutrality because it treats reversibility as having no temporal dimension, and as a matter of increments at the margin of a system at equilibrium. It amends the traditional methods of mainstream economics only by recognizing that there may be

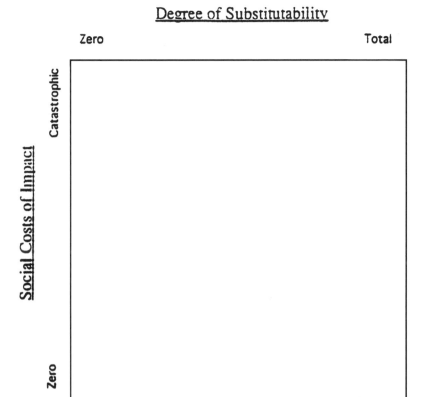

Figure 12.2. Economist's version.

decisions which, because of the potential scale of their impacts and because of the potentially large cost of "substituting" other means to attain the same level of welfare, may require special attention. With this amendment, the mainstream economic paradigm is made more sensitive to problems of scale and of irreversibility, but it maintains its commitment to a unitary decision space, positing no essential difference in the type of risks involved as one moves into the northwest corner of the space.

For mainstream economists, in other words, all risks remain of the fungible and compensable type. For the ecological economist, on the other hand, some risks can be understood as fungible and compensable, as aggregable with other costs and benefits, and others fall in the "red zone" and must be understood as involving a more fundamental (moral?) obligation to the future. The risk square, and the question of whether the space is continuous or not, dramatize the difference in the attitude of mainstream and ecological economists regarding intertemporal values. The technique of discounting is adequate to represent decisions made throughout a continuous decision space, whereas a multi-dimensional method of analysis is necessary if, as ecologists believe, information regarding temporal and spatial scales must be represented in decision processes. We have not, however, violated our policy of conceptual neutrality simply by adopting Fig. 12.1, because the doctrinaire economist who says that all decisions will be decided on costs and benefits as measured in markets can represent that position by asserting that the "red area" is empty. The economists' conceptualization thereby represents the limiting case in which there are zero cases of nonfungible resources.

Having introduced Fig. 12.2 as representing the economists' decision space by assuming reversibility can be treated as substitutability, we can elaborate Fig. 12.1 in a more ecological direction by incorporating the working principles of ecological economics, creating a more complex decision space which is represented here as Fig. 12.3. First, we can note that ecological economists,

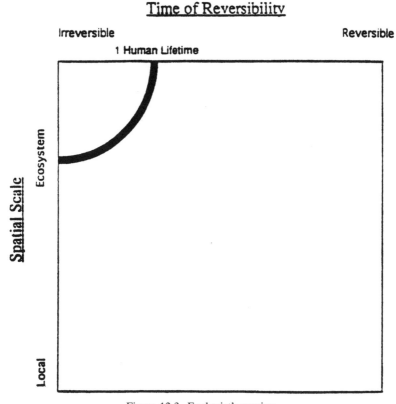

Figure 12.3. Ecologist's version.

purely by virtue of their emphasis on ecological systems and processes, will focus on the temporal aspects of change. For them, irreversibility is not just interpreted as an abstract concept of substitutability of one resource for another, as measured against units of welfare available to consumers – it will include the more concrete parameter of reversal of impacts in ecological time and space. The horizontal axis can therefore be calibrated as a measure of how long, given a particular impact or disturbance, it would take for natural processes to reverse that impact. The horizontal axis therefore locates decisions and policies which incur certain risks according to the restoration time necessary to repair damage if negative impacts occur as a result of those decisions or policies. The risk decision space therefore represents the range of positions separating economists, ecologists, and ecological economists, with version 1 representing a neutral space, and (versions) 2 and 3 representing the risk decision space as it would be interpreted, respectively, by economists confident of substitutability among resources and by ecologists who are not.

The ecological viewpoint here can be elaborated by introduction of "hierarchy theory (Allen and Starr, 1982; O'Neil et al., 1986; Norton, 1990, 1991), which has as its assumption that larger systems of nature change more slowly than do the smaller subsystems that compose them. We therefore assume that an ecologically adequate paradigm of management will be multiscalar, and that spatial and temporal scales of management systems will exhibit a positive correlation, with spatially limited subsystems changing at a faster pace and according to a different dynamic than do the larger systems that form their environment. Now, relying on hierarchy theory and its principle that larger, super-systems change more slowly than do smaller subsystems, we can correlate the vertical axis with spatial scale, creating a grid that will represent decisions as located on the space. Risks which threaten harm that is prevalent over a large geographical area and irreversible for a very long time will be clustered in the upper left of the square, suggesting that decisions in this region entail questions of intergenerational equity. We expect, given our model, that decisions in this area may be governed by moral constraints. Decisions that risk small-scale and quickly reversible impacts are not significant on the ecological scale and fall in the area governed by economic reasoning. Applying reasoning such as this, Norton and Ulanowicz (this volume), for example, have argued that since a commitment to sustain biodiversity is consensually understood to be a commitment to do so for many generations of humans (at least 150 years), we can conclude that the focus of biodiversity policy should be landscape-level ecosystems which normally change on a different temporal scale, an ecological scale that is slow relative to changes in economic behavior. If we can isolate the dynamics driving local economic opportunities from the dynamics supporting biodiversity, it may be possible to encourage both and avoid policy gridlock such as has occurred over the spotted owl in Northwest old-growth forests.

Human economics, which is paced to individual decision making, would on this perspective be understood as describing values that emerge in a subsystem of an encompassing physical system. Individual economic values are expressed on short scales and, provided they do not collectively add up to a trend that is significant on the larger ecosystem scale, these can be governed by individual free choice. In this sphere, economic analysis (perhaps corrected for considerations of interpersonal equity) accurately models individual decision making. When individual choices cumulatively impact large systems in an irreversible way, morally based considerations of equity across generations come into play. Good management, therefore, involves identifying and protecting processes crucial to the complex structure of the ecological system, which is to say that good management allows economic freedom, provided choices of individuals are damped out at a smaller scale and do not threaten to introduce ecosystem-level change that is irreversible. We can therefore define decisions that have impacts within two distinct spatio-temporal dynamics – individual choices affecting small subcomponents of a system that are reversible in one human lifetime, and decisions which threaten to create change that affects whole ecosystems. If these large-scale changes cannot be reversed within a single human lifetime, then they must be treated as decisions affecting intergenerational equity. The classification system corresponding to this ecological paradigm can be represented by the sorting diagram in Fig. 12.4. Sustainability of whole-system processes and the structure necessary to continue them is therefore one of those social decision areas that are a matter of intergenerational equity.

Given these contrasts we can, following Page (1977), begin to characterize a "two-tier" or "hybrid" approach to environmental decision making as an alternative to the unidimensional decision processes of mainstream economics. This position would see the decision space faced by environmental managers as split into regions, with the corner of the square that is characterized by major negative outcomes that are irreversible as representing an area of risk where values do not vary continuously with consumptive values. It could then be argued that moral strictures apply in these areas. This area of the decision space differs in kind from the regions where reversibility is high, the cost is low, or both. In these latter regions, we will be inclined to accept the usefulness of economic methods on the assumption that, in these decisions, future generations cannot fault us if we compensate them for destructive impacts

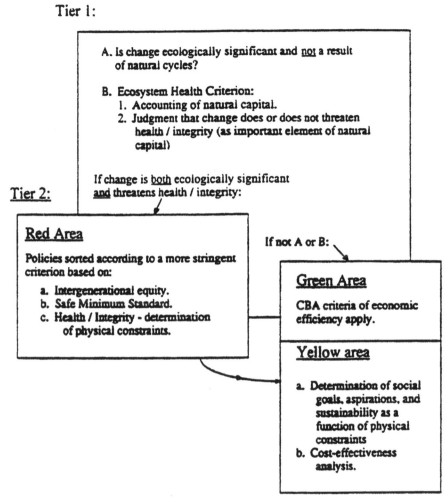

Figure 12.4. Sorting diagram based on risk decision square.

of our activities on natural systems with increased technological know-how, monetary capital, etc. In these regions, in other words, decisions are regarded as fungible.

The central issue facing the policy analytic community can now be formulated quite simply: Are there any environmental decisions that are located in the red area, and thereby governed by moral rather than economic criteria? If there are, the goal should be to determine how many decisions, and precisely which decisions, comprise these two broad categories, recognizing that introduction of constraints is always under a burden of proof to establish moral strictures because constraints on individual choice, invoked by placing some resource use decisions within the red area, are in principle limitations on the freedom of consumers to pursue their welfare as they see fit.

### 5. THE EXTRAPARADIGMATIC DEBATE ANALYSED

It would be comforting if the disagreements associated with these three issues were amenable to empirical confirmation or disconfirmation because then we could conclude that more scientific study would resolve the dispute between main-stream economists and ecologists. But these principles are, I think, too abstract to be directly supportable, or refutable, by empirical evidence. This is true because, as Kuhn (1970) and others before him have shown (Duhem, 1906; Quine, 1953, 1960), observation is theory-bound and because, as noted above, the central assumptions that serve to constitute a scientific discipline are accepted in clusters. The problem in extraparadigmatic disagreements is that there exist no shared conceptual basis, no shared assumptions, and no consensually accepted methodology according to which intellectual and policy agreements can be submitted to empirical resolution. It will be useful to discuss some of the factors that are important in understanding extraparadigmatic disagreements.

Speaking informally, the tendency of practitioners of different disciplines to talk past each other is typical of disputes that involve several scientific disciplines. These seemingly irreconcilable differences are exacerbated when two disciplines have both theoretical differences *and* differences regarding the role their respective disciplines should have in governance and in society. In this situation, intellectual debate often deteriorates into turf wars as competing disciplines jostle to gain more territory and more grant funding. More formally, the failures of communication involved can be characterized by distinguishing between scientific disagreements that take place within a paradigm, accepting its basic assumptions and principles, and those disagreements that arise among scientists, perhaps from different disciplines or adherents of a competing paradigm, when there are no accepted disciplinary principles and assumptions to which all disputants can appeal.

As originally stated in Kuhn (1962), the dichotomy between "normal" science (science undertaken within an accepted and largely unquestioned paradigm) and "revolutionary" science (science as it goes through a periodic and sometimes stormy transition to a new paradigm) was surely overdrawn. Even Kuhn himself has regretted some of the passages in which he overemphasized the nonrational elements that affect revolutionary science, and has since rejected interpretations of the revolutionary process as subjective (Kuhn, 1970). Nevertheless, I believe (and I think that most contemporary philosophers of science would agree) that Kuhn's argument at least established that there is a useful, if vague and messy-to-describe, difference in the "logic" of science in two different situations. In situations where important empirical results are being achieved within an accepted paradigm, and all practitioners can agree what results are important and why, then science is comparatively "normalized" – these conditions are conducive to further experimental work.[4] When these conditions do not hold, debate enters an extraparadigmatic ("revolutionary") phase.

According to convincing arguments showing that "crucial experiments" are impossible in science – which has led to important revisions in our understanding of hypothesis-testing and of the positivist approach to science – every experiment is based on "background assumptions." If an experiment does not result in the results "expected" given an experimental hypothesis, it is possible either to reject the hypothesis or to adjust the background assumptions (Quine, 1960). Practically, this implies that if these background assumptions are widely accepted within a field, research questions can be formulated precisely, tested, and results can lead to further research. The shared assumptions train our attention upon outcomes of experiments, and each result points toward new experiments. When a paradigm is under challenge, however, it may be impossible to find a sufficiently firm conceptual basis on which to define a hypothesis that will be both testable and yield results that all the disciplinary inquirers can accept as important. In this situation, the logic of scientific inquiry shifts. Many assumptions that could be built upon to form important hypotheses during normal science are now questioned. This is a time at which to loosen the intellectual hinges, to try many different models, and engage in disciplinary crossovers to increase communication across disciplines. Many hypotheses will ideally be floated and experimental work becomes more exploratory and hypothetical rather than precise and predictive. It is probably healthy, from the viewpoint of advancing science, that disciplines cycle through periods in this way because creativity of all types apparently results from a mixture of order and randomness.

One more cautionary note is necessary. It is important, in talking about a "paradigm," to be very clear as to what discipline, or level of resolution, one is pinpointing. Kuhn tended to use the term very loosely, applying it to disciplines, to schools of thought within disciplines, and sometimes to little more than groups of advocates for conceptual innovations of any magnitude (Toulmin, 1972). I therefore feel justified in using the term to refer two somewhat different, but also similar aspects of the total picture of science and management. First, I use the term narrowly to apply to accepted principles, etc. which govern professional activities within an established scientific discipline such as economics, ecology, or medicine. But I am also here extending the term to apply to the broader policy analysis and implementation community, treating this broader, multidisciplinary, and activist community as seeking a more inclusive paradigm that includes a scientific, a normative, and a political/economic component, hopefully in some reasonable balance. The search for a comprehensive approach to environmental management, then, represents a search on a broader level for an action-oriented paradigm that encompasses disciplinary paradigms in such a way as to integrate information from the natural and social sciences. Given these two uses, it will be possible to discuss, for example, the role of the disciplinary paradigms of economics and ecology, within the encompassing paradigm for environmental policies and management.

Having noted above that the logic of extraparadigmatic disagreements differs from intraparadigmatic inquiry, we can now note that the internal logical structure of paradigms may also differ significantly. The mainstream economic paradigm is reductionistic in the sense that most economists favor explanations and data that can be expressed in terms of a single measure – dollars as representing units of individual welfare. Their system of value measurement, because it is monistic in this sense, is "highly connected" – all of the statements of the discipline can be comprehended within a single measure and evaluated under a central principle.

In general, it is considered an advantage for a discipline to agree upon a small number of principles from which more particular consequences follow. The degree to which a discipline has reduced its concepts and belief structure to a few theoretical principles measures the "connectedness" of the system. Connectedness is valued because it represents theoretical elegance and reduces the likelihood of conflict among principles of action. Solow's proposed understanding of sustainability has the distinct advantage that it comprehends questions of sustainability within the highly connected and sparse theoretical assumptions of neoclassical economics. As a result of this high degree of connectedness, Solow's approach allows us to measure sustainability by measuring present preferences of consumers, providing a single metric (present dollars) as a measure of welfare across all generations. Using this approach to interpret and measure values has the tremendous advantage that, to the extent that price values represent patterns of real behavior reflecting an individual's willingness-to-pay, those values are concrete in the sense that they represent that individual's judgment of the value of a thing *in competition with other goods and services.* These features of economic analysis of public values combine to provide an extraordinarily powerful set of analytic tools for the analysis of public values.

While connectedness is a desirable characteristic when one is communicating within a paradigm, systems that are highly connected can appear inflexible in the face of new situations and problems. This characteristic of a discipline can be understood if we distinguish between (a) the "intuitive subject matter" of a discipline and (b) the "reach" of its methods. In clinical psychology, for example, antisocial behaviors, self-destructive behaviors, and perhaps also the moods that lead to them constitute events that require the attention of psychologists and form the intuitive subject matter of the discipline. It would be considered a criticism of a paradigm in this field if it could not classify, describe, and explain behaviors in these categories – its methodological principles, we would say, are inadequate to account for the intuitive subject matter of psychology. But advocates of a paradigm that is accused of insufficient reach may, rather than yield to criticism, answer by denying that some phenomena that are referred to in general conversation are "real," arguing that they are actually bogus entitities that are the ontological fallout, the theoretical dross, of failed paradigms. They believe, on this basis, that it is in fact an advantage of their paradigm that it makes no reference to these entities. Behavioral psychologists, for example, would argue that it is an advantage of their paradigm that they make no reference to unobservable mental states such as moods.

To return to the case at hand, advocates of the mainstream can either deny that there are important connections between states of ecological systems and the state of welfare of human populations in the present or distant future, or they can agree that there are impacts on welfare that predictably result from changes in ecosystem-level functioning and attempt to devise methods and means to describe and quantify this connection. In the first case, mainstream economists are denying that effects on ecosystems are real elements in the subject matter of environmental values; in the second case, they admit that ecosystem changes affect their measures of human welfare, and accept the burden of showing how those changes can be expressed as dollar figures.

Two points follow. First, extraparadigmatic disagreements often go in circles because there is agreement neither on basic principles nor on the scope of the true subject matter of the discipline. To put the point in philosophical terminology, methodological disputes become the manifestations of "ontological" disagreements – differences in the language and methods of measurement reflect differences in beliefs about what are really the constituents of the existing world. Second, sparseness of conceptual foundations has disadvantages as well as advantages. The burden of proof can shift against a highly connected paradigm if it repeatedly fails to encompass within its traditional, narrowly defined methods an element of its subject matter that is widely accepted as important.

## 6. CONCLUSION

Ecological economists, for their part, advocate a more structured bequest package containing an appropriate mix of natural and human capital. They do not believe that a set of "sustainable" policies can be specified without essential reference to the physical states of the large, functioning ecological systems that provide an ongoing stream of ecological services. The movement to introduce concepts such as ecosystem health and integrity into policy analysis and implementation is best understood as a movement to define a structured and appropriate bequest from the current to future generations, especially as that package affects the functional characteristics of whole ecological systems. This expanded paradigm implies a multi-tiered decision process, one that seeks protection of both economic and ecological resources. While little has been said here to resolve this dispute, it is hoped that the characterization of the debate as an extraparadigmatic one, and the development of a neutral vocabulary, will further the goal – central to the Ecosystem Valuation Forum – of increased communication and a more fruitful understanding of the current crisis in policy analysis and formation.

Let me end on a final conciliatory note, and state one plea. The idea that extraparadigmatic debates must be decided in nonrational forums unless the disagreements can be reduced to solving a conceptual or empirical "puzzle" within an unquestioningly adopted set of assumptions is based on a (useful) myth. Interdisciplinary cooperation can broaden debates in policy development if it is given a chance, and that chance is the creation of interdisciplinary panels that focus on problems and on finding solutions. These forums, in which experts from different disciplines pledge to work together, are capable of creating rational dialogue and developing a more interdisciplinary language. It is therefore very important to continue research into the theoretical questions isolated by the discussions in the first phase of the Ecosystem Valuation Forum, and to do them in conjunction with real case studies. To go to case studies only at this time would be to cut the theoretical head off the more organismic and dynamical models that could emerge from an integration of economics with ecology and the physical sciences. The point, after all, about all the discussion of "paradigms" is that if we are more aware of them and create dialogue across them, new, more inclusive conceptual frameworks will emerge. Most of the great scientific advances in recent years have come from disciplinary crossovers and interdisciplinary teams (the most publicly noted case being the discovery of the double helix structure of DNA molecules). There is no reason to think that a scientific approach to the environment, which is the study of complex, multilayered systems that provide the context of our lives, should be resolved within narrow disciplinary boundaries.

ACKNOWLEDGMENTS. Research for this paper was supported by the Army Environmental Policy Institute. The author thanks V. Kerry Smith. Michael Toman, William Anderson, and Claire Huppertz Miller for careful review and comments on an earlier version. R. Gordon Dailey, Jr., assisted in the research and graphics preparation.

## NOTES

1.  For example, the Clean Water Act of 1972 specifically calls for management of water bodies to protect their biological integrity. The Great Lakes Treaty is based firmly on a commitment to protect the integrity of the Great Lakes Ecosystems. In general, there have been many proposals to institute whole ecosystem management in diverse situations. The common element of all of these initiatives is emphasis on monitoring and protecting characteristics of whole ecosystems.

2.  It should be noted that the model as here presented is incomplete as a *decision* model because it does not incorporate a consideration of uncertainty/probability. Presumably, the obligation to act to avoid a risk to the future depends also on the degree of probability attached to the likelihood of the risk, given current knowledge, that a negative outcome will occur. That is, if the present faces two risks, the negative outcomes of which would be equally harmful to an equal number of people, we assume that there is a stronger obligation to act to avert a risk that has a 40 percent chance of a negative outcome than to avert a risk that has a 0.05 likelihood of occurring. It may be possible to model probability in a third dimension, but work in this area is too speculative to guide policy at this time. For simplicity we will use only versions that plot decisions in two dimensions. In effect, that means we are assuming that in comparing two risks, the risks carry equal levels of likelihood of negative outcomes. This means that the model must be used in conjunction with the judgment of managers, especially regarding the likelihood of various impacts.

3.  While this version of the decision square models the mainstream economic viewpoint, it should be noted that Toman (1992) goes on to develop an interesting and useful distinction between the context of ordinary economic thinking and the very different context of decision making that would be encountered in, for example, a constitutional convention.

4.  Note: This condition must be described as a *sociological condition* of a group of people, the disciplinary practitioners, rather than as a logical distinction between two types of discourse. Attempts to formally and logically specify a sharp logical difference between "empirical questions" and "linguistic/conceptual questions" have apparently failed and are likely impossible to achieve without arbitrary fiat

(Quine, 1953, 1960) Clear recognition that science is done by scientists who inevitably make judgments not reducible to relevant observations and deductions therefrom undermines any attempt to provide a logical characterization of a single rational response to any given body of data. Nevertheless, I think the distinction between paradigmatic and extraparadigmatic disagreements is a useful one in the sociology of science.

## REFERENCES

Allen, T. F. H. and Starr. T. B., 1982. *Hierarchy: Perspectives for Ecological Complexity.* Chicago: University of Chicago Press.

Crosson, P. and Toman, M., 1991. Economics and sustainability: balancing trade-offs and imperatives., Resources for the Future Discussion Paper ENR-9105, January.

Daly, H. and Cobb, J., 1989. *For the Common Good.* Boston: Beacon Press.

Duhem. P., 1906. La *Théorie Physique: Son Object et Sa Structure.* Paris: Chevalier and Rivière.

Fisher, A. C. and Krutilla. J. V., 1985. Economics of nature preservation. In: A. V. Kneese and J. L. Sweeney (Editors), *Handbook of Natural Resource and Energy Economics.* Vol. I. Amsterdam: North-Holland Press.

Freeman, III. A. M., 1986. The ethical basis of the economic view of the environment. In: Donald VanDeveer and Christine Pierce (Editors), *People, Penguins, and Plastic Trees.* Belmont. CA: Wadsworth Publishing Company.

Gore, A., 1992. *Earth in the Balance.* Boston: Houghton Mifflin Company.

Harris, L., 1984. *The Fragmented Forest.* Chicago: University of Chicago Press.

Krutilla, J. V., 1967. Conservation Reconsidered. *Am. Econ. Rev.*, 57: 777–786.

Krutilla, J. V. and Fisher, A. C., 1975. *The Economics of Natural Resources: Studies in the Valuation of Commodity and Amenity Resources.* Baltimore: Johns Hopkins University Press.

Kuhn, T. S., 1962, 1970. *The Structure of Scientific Revolutions.* Chicago: University of Chicago Press.

Lind. R. C. et al., 1982. *Discounting for Time and Risk in Energy Policy.* Washington, DC: Resources for the Future.

NAS (National Academy of Sciences), Panel on Policy Implications of Greenhouse Warming, 1992. *Policy Implications of Global Warming.* Washington, DC.

Norton, B. G., 1990. Context and hierarchy in Aldo Leopold's theory of environmental management. *Ecol. Econ.*, 2: 119–127.

_____1991. *Toward Unity among Environmentalists.* New York: Oxford University Press.

_____1994. Future generations, obligations to. In: *Encyclopedia of Bioethics.* 2nd edn. New York: Macmillan.

O'Neil, R. V., et al., 1986. *A Hierarchical Concept of Ecosvstems.* Princeton, NJ: Princeton University Press.

Page, T., 1977. *Conservation and Economic Efficiency.* Baltimore: Johns Hopkins University Press.

_____1988. *Intergenerational Equity and the Social Rate of Discount.* Washington, DC: Resources for the Future.

Quine, W. V. O., 1953. *From a Logical Point of View.* Cambridge, MA: Harvard University Press.

_____1960. *Word and Object.* Cambridge. MA: MIT University Press.

Smith, V. K., 1980. The evaluation of natural resource adequacy: elusive quest or frontier of economic analysis. *Land Econ.*, 56: 257–298.

Solow, R. M., 1974. The economics of resources or the resources of economics. *Am. Econ. Rev. Proc.*, 64: 1–14.

_____1986. On the intergenerational allocation of natural resources. *Scand. J. Econ.*, 88: 141–49.

_____1992. An Almost Practical Step toward Sustainability. Invited lecture on the occasion of the fortieth anniversary of Resources for the Future. Washington, DC, October 8.

_____1993. Sustainability: an economist's perspective. In: Robert and Nancy Dorfman (Editors), *Economics of the Environment: Selected Readings.* New York: W. W. Norton and Company.

Toman, M. A., 1992. The difficulty in defining sustainability. *Resources*, 61 (Winter): 3–6.

Toulmin, S., 1972. *Human Understanding*, Vol. I. Princeton, NJ: Princeton University Press.

# 13

## Sustainability

### Ecological and Economic Perspectives

with MICHAEL A. TOMAN

I. INTRODUCTION: ECOLOGISTS, ECONOMISTS, AND THE SEARCH
FOR SUSTAINABLE POLICIES

Decision makers are more and more often being told to "act sustainably" and
to pursue policy paths toward "sustainable development." These admonitions
and instructions appear to express a significant societal commitment to alter
current practices. And yet these widely supported admonitions provide little
guidance to policymakers and other actors, because the term "sustainable"
embodies deep conceptual ambiguities. These ambiguities cannot be easily
resolved because they rest, in turn, on serious theoretical disagreements that
transcend disciplinary boundaries. In particular, economists and ecologists
employ different conceptualizations for explaining the interactions of humans
with their environment. Nor are these differences easily ignored because they
pervasively affect the way we conceive and implement sustainable policies.

They also affect which data are gathered and considered relevant to policy
decisions, and they dictate quite different approaches toward aggregation of
data and the integration of information in the search for improved environmental policies. If we resolve to base environmental and economic policy on
the best scientific evidence available – and we think most scientists and policy
makers would agree, also, with this resolution – then it is essential that scientists co-operate across disciplines to encourage improved communication
among disciplines and also improved communication of information from the
various disciplines to decision makers. Data are just data; only "interpreted
data" will affect the decision process. Who, then, will interpret the data and

Norton, Bryan and Michael A. Toman. "Sustainability: Ecological and Economic Perspectives."
*Land Economics*, Vol. 73, No. 4. © 1997. Reprinted by permission of the University of Wisconsin
Press.

apply it in a decision process? Are the conceptualizations of one discipline more appropriate for integration of data? Or is this largely a post-scientific question, to be left to the environmental managers and decision makers?

The purpose of this paper is to explore the underlying theoretical difficulties that currently obstruct interdisciplinary communication and cooperation, and result in ambiguous admonitions to act sustainably. While we recognize the hazards of attributing a consensus to any academic discipline, we begin with an effort to characterize as fairly and clearly as possible the differing opinions that separate mainstream economists from advocates of a new paradigm of ecological economics.[1] Because economics is a social science devoted to understanding the values of the public and their behavioral consequences, it is essential that scientific questions regarding human economic and valuation behavior be formulated in a way that is, on the one hand, descriptive of values held and, on the other hand, expressed in concepts that can inform real decisions. It is apparent that the choice of methodologies and techniques for describing and recording social values – what we will refer to as the development of an accounting system for sustainability[2] – depends upon theoretical principles that are controversial in many interdisciplinary discussions of environmental policy.

Although there may be a number of issues separating economists and ecologists in evaluating environmental values, we will focus on two clusters of issues and intellectual problems that intersect at the edges of economic and ecological science as these disciplines bear upon environmental policy practice. For shorthand, we will refer to these as the problems of "reversibility and substitutability" and the "accounting problem." Our premise is that positions taken on these two issues, about which there is considerable cross-disciplinary disagreement even regarding their formulation as issues for debate, have important consequences for the way we model and act upon the problem of sustainability. We explore these two broad problem areas to highlight the central disagreements regarding the concept of sustainability, even as we caution that these two areas of cross-disciplinary disagreement cannot be resolved without making considerable progress in other areas of ecological and economic theory.

A central aspect of disagreements about the concepts of reversibility and substitutability among resources, among ecological systems, and among components of those systems centers on the different attitudes of economists and ecologists toward discontinuities and thresholds in understanding and managing the impacts of human culture on ecological systems. Economists favor marginal forms of analysis in practice and tend to pay less attention to the concept of the scale of an economy in relation to its resource base. Ecologists

226

believe that there are important thresholds of scale, and that human activities can, by stressing ecosystems in ill-advised ways, set in motion large-scale and irreversible losses in the functioning of ecological and physical systems.

Economists and ecologists also differ regarding how to measure and place value on changes in environmental quality, especially when these changes are long term in nature and may alter the intergenerational distribution of income. Economists tend to favor dollar-denominated present value calculations, perhaps with some concern for inter-generational impacts. Ecologists have often opposed both dollar valuation and present value calculations, especially as applied to decisions that may have long-term, highly negative outcomes that threaten the productivity of large-scale ecosystems. Their concern is stated in terms of the need for the present generation to protect essential ecosystem processes in order to protect important social values. However, these concerns arise from outside ecology, which does not have its own conceptual framework for addressing issues of human values.

Clarifying these issues is far more difficult than simply providing dictionary definitions because the semantic problems are in this case inseparable from theoretical issues dividing many economists from many ecologists. An open, ongoing multidisciplinary dialogue will improve our understanding of both conceptual and practical issues that arise in assessing environmental policy options. In the following sections we explain some of the theoretical and practical ramifications of the two crucial problems we have decided to focus on.

## II. REVERSIBILITY AND SUBSTITUTABILITY[3]

In this section we discuss contrasts between the views of ecologists and economists on the issues of resource substitutability and the reversibility of the consequences of ecological change. It should be noted at the outset that economists and ecologists often assign different meanings to these words in their analytical efforts and policy arguments. For economists, substitutability refers to the capacity to alter production and consumption activities in the event of increasing scarcity of some resource in order to maintain a desired overall flow of services. Production in this conception can refer to a human-engineered activity or the act of benefiting from values provided more directly by nature. Services and resources refer to anything valued by people, again both human-engineered and provided more directly by nature. Scarcity here is understood in the economists' sense of increasing relative cost, the amount of other valued goods and services that must be given up for the input in question.

Given these definitions, reversibility refers also to economic consequences broadly defined to include both market and nonmarket values, rather than to the physical states of ecosystems per se.

In most usage by ecologists, the terms substitutability and reversibility refer much more to physical properties of ecosystems themselves. The reversibility of a condition is related to the resiliency of the ecosystem, its capacity to return to a high level of function after being perturbed. Substitutability is a form of redundancy: if some system attribute is diminished, are there other sources of that attribute? The question of substitutability assumes particular prominence when the postulated natural degradation is somehow large in scale, an idea we attempt to make more precise below.

Economists are concerned with sustainability in the sense of maintaining acceptable levels of human well-being over time and thus are concerned with the capacity of the natural environment and other social assets to meet human wants and needs. These wants and needs can be conceived of very broadly, encompassing a variety of preservation and bequest motives in addition to direct interests in ecosystem use or resource consumption. Nevertheless, the conditions of ecosystems are but one avenue by which human well-being is affected. If economic substitution possibilities are high enough, natural disruption is not a special cause for concern in the economic model provided society's total savings rate is high enough to compensate for reduction of natural capital and thereby produce sustainable welfare paths. Even irreversible changes in the physical state of ecosystems are not that significant in this case, though the economic consequences of irreversible physical changes need to be accounted for.[4] However, the converse also is true: if substitution possibilities are limited, then satisfying both current consumption demands and intergenerational equity concerns can lead to a greater need for safeguarding natural capital. Here physical irreversibilities could raise concerns about irreversible economic costs that are a significant social concern.[5]

The substitution issue goes beyond substituting technological progress (human and knowledge capital) or investment (built capital) for depletion of mineral and energy resources, as important as this set of substitution questions is. Substitution also involves the ability to offset a diminished capacity of the natural environment to provide waste absorption, ecological system maintenance, and aesthetic services. Questions about substitution and technical progress versus thresholds and catastrophe risks are especially relevant when addressing large-scale damages to natural systems whose ecological functions remain poorly understood (Holling 1993; Arrow et al. 1995).

The literature on depletable resources and economic progress shows that a relatively substantial capacity to substitute other inputs for diminished natural capital services is needed to maintain consumption of final goods and services over time.[6] The difficulty with these conditions is that they seem to be inconsistent with physical laws.[7] Since the first law of thermodynamics requires conservation of mass and energy, the implication that, for example, the economy could run on a vanishingly small quantity of energy is problematic. It seems more plausible to assume minimum input requirements and bounded productivity of material and energy inputs, thus eventually limiting total output to a level consistent with the capacity of renewable resource inputs and waste absorption capacities. There remains the empirical question of how stringent these constraints might be over different temporal and spatial scales.[8]

In contrast to the economic perspective of ecosystems as "service factories," ecologists see these systems as complex, dynamic sets of processes that are organized on multiple scales (see Allen and Starr 1982; Holling 1986, 1992; O'Neill et al. 1986; Common and Perrings 1992; paper 16, this volume). Smaller-scale components or subsystems (e.g., a small forest patch) respond quickly to stimuli and can recover relatively quickly from shocks; moreover, there is much more redundancy at this scale. For these reasons, the substitution paradigm in economics can fit well with the function of lower-scale ecological systems as sources of services. In contrast, larger-scale systems (e.g., an entire forest system) respond much more slowly and with less redundancy. While even large-scale systems are resilient up to a point, it is possible to push them past that point and trigger rapid, discontinuous changes in function that may require large amounts of time for recovery (Holling 1996).

These concerns have led Holling (1992) to introduce the category of "keystone processes," analogous to the idea of keystone species but targeted at the level of important ecological processes. This concept incorporates the idea of species redundancy into the analysis, recognizing that in some cases an important ecological function is performed by a single species and, in cases of redundancy, by suites of species. The problem with these concepts in practice, however, is that determining the amount of redundancy in a system, and hence the likelihood of cascading effects from any change in an element of the system, is notoriously difficult.[9]

The introduction of scale sensitivity into the analysis creates a number of both quandaries and opportunities. Once one recognizes, for example, that some species are significant on the scale of ecological systems and others are not, study of the system – and evaluation of changes in it – must take this difference into account.[10] Ecological theory and economic analysis therefore require a treatment of interactions among scales in a system.[11] Moreover, given the importance of relationships across scales, precise judgments of the "importance" of any element to humans cannot proceed without knowledge of them.

Common and Perrings (1992) have shown that the concepts of resilience and stability as used by ecologists and economists, respectively, are largely disjoint. The concept of resilience taken as indicative of ecological sustainability describes the complex, nonequilibrium dynamics of ecosystems (Holling 1973, 1996; Ulanowicz 1986; Pimm 1984, 1991; Norton 1992; Waldrop 1992) while the notion of economic stability typically employed by economists focuses on steady-state behavior of economic organizations.[12] These two factors can change independently and at rates that differ by orders of magnitude in time (Holling 1992; paper 16, this volume). While the economic model typically abstracts from scale changes in the economy relative to the natural systems encompassing it, large-scale ecosystems subject to intensive human management can become brittle and more likely to shift into another level of functioning. These new "basins of attraction" in system functioning can be less supportive of human services and other human values.

The preceding discussion has several implications for ecological and economic contributions to concepts of environmental management. Recent work on hierarchical organization of systems attempts to track changes in dynamics at multiple levels, in which it is assumed that the smaller subsystems change more rapidly than do the larger systems that form their "environment." The larger system sets the constraints, patterns of stability, and opportunities which confront "actors" on smaller and faster levels. Changes in the system also must be described, understood, and valued from some point within the hierarchy. Unlike description in mechanical, Newtonian systems, which assumes a unified point of observation from outside nature, hierarchical description always specifies an observation point and a scalar context.

Hierarchical systems thus can be used to identify changes caused in larger systems by paying attention to behaviors at lower levels (cross-level spillover effects). Such models also provide one way to integrate individual preferences and ecological scale (Common and Perrings 1992; Holling 1994; paper 17, this Volume). Because larger-scale systems set the opportunities and constraints available at lower scales, concerns about the health and integrity of larger-scale systems may be different in kind, not just in degree. For example, safeguarding the functioning of larger-scale systems would figure prominently in satisfying concerns about intergenerational fairness. However, most individual choices are made at smaller scales. The hierarchical-model directs attention not just to the effects of these choices at smaller scales, but also to the cumulative effects of lower-scale decisions on larger-scale systems.

A further advantage of hierarchical analysis is that, if it is possible to associate important social values with various levels of the hierarchy, then it may be possible to devise policies that encourage the simultaneous attainment of human goals on more than one level by choosing policies that have positive impacts on multiple scales. Alternatively, one can seek opportunities to enhance one valued dynamic that at least has neutral impacts on other levels of the hierarchical system. Hierarchical assumptions therefore show some promise of integrating economic and ecological information by associating the information with different levels of organization of a dynamic system, since economies operate at smaller scales and faster rates than the ecological landscape within which they are embedded.

Leaving aside philosophical debates about the utilitarian underpinnings of economics (see the next section), one can represent limits on substitution in the economic production constraints that determine society's opportunity set. One could then attach economic values to different outcomes, even if there were limits on substitutability. This strategy doubtless would involve a departure from the marginal analysis generally favored in economic theory. The substitution paradigm that allows for incremental changes is less well suited to large-scale ecological impacts. However, this does not mean that the factory-of-services model is inherently wrong, though it does suggest that the rules governing the factory can be quite different from those often assumed in economic analysis. The basic logic of economic theory does not strictly depend on the capacity to make smooth marginal changes in all decisions and consequences (Scarf 1981a, 1981b).

It does depend, however, on an enumeration of the possibilities available for decisions and their consequences. In the present context, the approach just described would require a very sophisticated description of production possibilities that integrates built capital, natural capital, and knowledge at multiple scales and time frames. While some promising beginnings toward this end have been undertaken, it seems clear that in practice we still face gross uncertainties in applying standard economic practice to the problem at hand. Another key empirical issue in considering reversibility and substitutability is the capacity of technological progress to continue without bound and the capacity of humans to keep up with perpetually accelerating technical change.[13] To date, technical progress has been a powerful deterrent to absolute physical scarcity. However, the ecological resources that may be becoming scarce – such as waste heat absorption capacity – are not those for which we have a long record of experience in management, let alone innovation to alleviate constraints.[14]

The modeling of economic dynamics and ecological dynamics at different spatial and temporal scales gives rise to the question of whether there may be multiple measures of economic sustainability and ecological sustainability. Economists tend to argue in favor of dollar-based measures. This perspective can be rationalized by arguing that welfare-based economic models in principle have the capacity to capture all that is important in keeping sustainability accounts. Ecologists have argued that attention to the physical state of ecosystems as well as to economic information, is needed in forming judgments about substitution constraints and their implications for management, especially in light of the value judgments embedded in standard benefit-cost analysis and the limited empirical information available for valuing ecological impacts (see, e.g., Costanza 1991; United Nations 1993). However, proponents of using ecological information in policy analysis need to make clear both the scientific basis for such applications and the value judgments that are embodied. As an example of the latter point, Norton (1989) asserts that no generation should destabilize the ecosystem functions that underlie and provide the context for all human activity (see also Weiss 1989). Similar statements about science *and* value judgments need to be enunciated and debated to undergird ecological indicators at smaller scales.

The discussion in this section regarding reversibility and substitutability also can be linked to the different approaches mainstream economists and ecologists adopt for specifying a fair bequest package for future generations, given that one has accepted some ethical obligation to the future (see Section III). Solow (1993), for example, argues that, since resources are fungible, intergenerational obligations reduce to a concern for a fair investment policy. There are no particular things that we owe. Solow therefore concludes that while we owe the future opportunities equal to our own – which requires that we maintain a nondeclining stock of aggregate capital – the obligation to the future can be discharged simply by maintaining adequate investment in order to compensate the future for use of degradation of particular resources. This position can be understood as advocating an "unstructured bequest package" between generations – sustainability can be represented as an obligation to honor a fair savings rate across generations.

This approach is not accepted by those who question the fungibility of human investment and the natural endowment and believe that the latter – in particular, the preservation of a variety of ecosystem functions – is crucial to the well-being of future generations. Advocates of this view argue that there is an ethical obligation to protect these processes in order to provide adequate options to future generations (Page 1977; Norton 1989; Weiss 1989; Chapter 12, this volume). They thus advocate a more "highly structured bequest package" that keeps account of "natural" and "man-made" capital separately and pays special attention to the former (Daly and Cobb 1989; Costanza et al. 1993). Moreover, many ecologists doubt that the welfare benefits of protecting large-scale processes are amenable to economic measurement, and thus tend to favor protection of physical processes even if changes in these objects of protection cannot be calibrated in terms of individual welfare benefits.

This disagreement over the characterization of the bequest package is important because it affects what counts as data relevant to measuring progress toward sustainable resource use paths. If the bequest package is unstructured, highly aggregated data regarding national income accounts and the rate of savings provides crucial information to assess the sustainability of current policies. If the bequest package must be more structured to ensure fairness across generations, more disaggregated data regarding changes in particular resources and regarding effects on ecological processes will be necessary.

## III. VALUES, VALUATION, AND ACCOUNTING

One very practical area of disagreement between ecologists and economists is the whole area of valuation studies and modeling, which is sometimes referred to as the problem of "environmental accounting." How should we assign values, and what values, however measured, should be assigned to changes in states of the world that result from human actions that degrade or protect the environment? Advocates of more stringent environmental policies, and those who oppose them as sometimes going too far, often cannot agree on what really counts. And you can't *ac*count until you know what to count! If the disputants could agree on what to count, the problem of evaluation would be, if not easy, at least significantly advanced. In this section we survey some of the difficult moral and conceptual problems that bear upon the question of how to value environmental changes, paying special attention to changes that are large in physical scale and unfold slowly – over decades, generations, and millennia.

If sustainability means anything – and we believe there is at least agreement on this point from all quarters – it represents a concern for the future, especially including horizons beyond the length of a human generation. But there exists very little agreement across the humanistic or social science disciplines regarding how to formulate and evaluate our concern for the well-being of future generations or our obligations to them (Norton 1995). No discipline – not ecology, not economics, and not philosophy – provides a coherent and complete understanding of human values as they are applied across multiple generations.

Economists do not embrace a single approach to assessing intertemporal distributions of well-being. Generally economists argue that discounting is

233

essential to account for the time preferences exhibited in individual behavior and the potential for the future to be better of than the present as a consequence of economic growth. However, many economists also argue that this standard approach to discounting is inadequate for analyzing intergenerational welfare trade-offs, as distinct from intertemporal allocations across one person's life (see, e.g., Kneese and Schulz 1985; Page 1988; and Howarth and Norgaard 1990, 1993).

One way this concern can be addressed is to define a "social" discount rate for intergenerational allocations that is different than the rate of personal time preference. The social rate of discount reflects societal value judgments about the appropriate intergenerational distribution of well-being, including the potential for future generations to be better off than the present from economic progress or worse off due to environmental harm. It is thus different in concept as well as magnitude from the rate of individual time preference (Burton 1993). Given this social discount rate, positive and negative effects can be evaluated in terms of their consequences for utility streams.[15]

Philosophical ethics encompasses a variety of conceptualizations of moral obligations. At least three types of general theories are usually recognized: (1) teleology, including utilitarianism of various sorts, welfare economics, and various interest theories; (2) rights theory / deontology; and (3) contractarian theories. Each of these major types of theories suggests a somewhat different formulation of intertemporal obligations, and the profusion of formulations creates disagreements and confusion.

Utilitarians would emphasize all generations having a chance to maximize their welfare as individuals, and have proposed various criteria for intertemporal welfare comparisons. Advocates of the utilitarian line of reasoning oblige themselves to measure welfare in terms that are fungible across persons and across time, and address the intergenerational opportunities by specifying rules that will guarantee that future generations will have the opportunity to enjoy a standard of living (measured in units of individual welfare) comparable to that of preceding generations (WCED 1987). In practice, utilitarians often address intergenerational obligations by aggregating welfare within generations and then comparing it across generations, but neither this aggregation nor discounting is in principle required in a utilitarian approach (Broome 1992).

In a rights theory, the question is whether we should act as though future people have rights that we must respect. A deontologist might argue that, if we act as though future persons really have rights, these rights would trump any mere enjoyments that the present gains at the expense of future well-being (see, e.g., Howarth 1995). The deontological line of reasoning has usually been premised on the assumption that the time at which a harm occurs is irrelevant to its moral status (Parfit 1983; Page 1997), implying that if the decisions and actions of the present terribly damage the life prospects of future generations of persons in avoidable ways, the present will have been guilty of intertemporal despotism. Finally, contractarians have emphasized the importance of at least implicit "consent" which, applied intergenerationally, would imply an acceptable rate of exploitation and savings in every generation.

These deontological and contractarian approaches to intergenerational obligations encourage a formulation of policies that distinguish "vital" from "nonvital" needs, placing obligations on the present to respect those resources essential to the future's ability to supply its "basic needs." There is, however, considerable disagreement regarding what constitutes basic needs and what level of infringement obliges intervention. Moreover, future people are only potential persons. Specification of any rights or contracts spanning our generation and generations not yet born necessarily represents a thought experiment carried out by us in the present.

Notwithstanding these differences in perspectives, there appears to be one significant area of agreement among adherents of all three views. All seem to agree that our ability to affect the way the future will live imposes a responsibility that involves issues of justice, fairness, and equity, not just questions of economic efficiency (Rawls 1971; Page 1977; Weiss 1989; Solow 1993; Howarth 1995). There remains the question of how these multiple formulations interact with each other in decision making. Should we formulate policy in response to threats to the basic rights of future people or in terms of a fair investment policy within a welfare-based analysis? Can we pursue policies in response to threats to both rights and to basic welfare, or would attempts to follow these rules simultaneously point us toward contradictory goals or different priorities in protecting resources for the future?

Putting these policy questions in more philosophical terms, the question of how to identify what resources to save for the future raises the abstract question of "moral monism." Theories of value are "monistic" if they resolve all philosophical quandaries according to a single principle or criterion. Interestingly, most rights theorists and utilitarians share a commitment to monism, employing a welfare criterion or a rights-based criterion, respectively.

Pluralistic approaches to ethics recognize multiple principles and criteria. Stone (1987) has argued that environmental ethics is best approached pluralistically, that wise management of the environment will require the application of differing principles in various situations. Callicott (1990) has criticized Stone's pluralism, however, arguing that pluralistic principles would result in the arbitrary, or self-interested, application of principles in particular cases. In Part IV we explore a two-tier approach to environmental valuation – what might be called a form of "integrated pluralism."

## IV. A TWO-TIER DECISION MODEL

Since disciplines place high value on developing a unified theoretical frame-work for analyzing phenomena under their purview, it is not surprising that disciplinary practitioners tend to approach, at least initially, interdisciplinary policy discussions within their own vocabularies or paradigms. Given the depths of the cross-disciplinary conceptual differences surveyed in this paper, continued pursuit of this strategy by all parties will guarantee that policy discussion and debate will remain balkanized for the foreseeable future. Given this situation, we think it is time for a new strategy. We suggest that for a period of conceptual experimentation, we pursue more pluralistic systems of decision making.[16]

The "two-tier" approach we suggest operates by establishing a categorization of problems, which determines the kinds of decision rules that should be applied in different cases, as well as identifying the decision rules themselves. This implies that the decision rules applied depend on the context. This approach, while pluralistic, does not inherently lead to arbitrary application of decision criteria.

Two-tier approaches have been advocated by a number of analysts. Page (1977) argues that materials and environmental policies should be judged by two criteria, an *efficiency criterion* and a *conservation criterion.* He applies these rules according to a categorization of problems as being intragenerational or intergenerational in their impacts: when intergenerational issues figure more prominently, more attention is given to conservation. There have also been attempts to integrate multiple criteria for action according to temporal and spatial context (Norton 1991, Chapter 10, this volume; Toman 1994; Chapter 16, this volume). These approaches attempt to formulate the question of which criterion to apply in a given problem situation by characterizing situations in which a standard cost-benefit criterion is relevant, for example, and other situations where we should apply the Safe Minimum Standard of Conservation criterion in the weighing of benefits and costs. This criterion places a larger burden of proof on those who would destroy important ecosystem functions by asserting that the resource be saved "provided the social costs are bearable" (Ciriacy-Wantrup 1952; Bishop 1978, 1979; Randall 1986; Norton 1987; Toman 1994). This rule requires those who would undertake such a risk to a resource to show that the costs of protecting it are unacceptably high before undertaking the risk.[17] It is a concrete expression of a moral judgment that large-scale negative environmental effects may have unacceptable consequences for the intergenerational distribution of opportunities and well-being.

The two-tier approach, which allows multiple possible action rules, provides conceptual room to recognize obligations that go beyond striving to implement a wise aggregate investment policy. To apply this approach, it is necessary to categorize problems according to the type of risk involved and then to identify what criteria apply within different categories. While the two parts of the process cannot be carried out in isolation, we believe that making these questions explicit can improve the quality of current discussions of sustainability by separating issues of which criteria are applicable from questions of whether current policies are achieving results according to various criteria.

An advantage of a two-tier approach is that it can logically comprehend the cost-benefit approach of welfare economics as well as other options, and encourage public discussion of what criteria to apply in sustainability calculations and measures. It can also encourage a more adaptive, experimental process in which scientists, local communities, and policymakers participate in an ongoing discussion of both what to do in specific situations and what criteria are appropriate in various situations. This kind of discussion may lead to a process of value articulation, criticism, and experimentation with multiple schemes for valuing environmental goods.

The pluralistic approach to valuation also combines well with the ecologically based concept of "adaptive management" (Holling 1978; Walters 1986; Lee 1993; Gunderson, Holling, and Light 1995), which emphasizes that, in situations of high uncertainty, management plans should be formulated so as to improve knowledge and reduce uncertainty by approximation. Combining adaptive management with a pluralistic approach to evaluation will encourage discussion of the likely results of applying various valuation criteria, and thus of what criteria should be employed.

Pluralism in environmental values and valuation, as a heuristic hypothesis, therefore, can sharpen the learning curve regarding environmental valuation – both among members of the public and within scientific disciplines. However, the hypothesis could be proved false if alternative criteria all tend to yield comparable results. This unifying outcome would then be established by a self-correcting process of interdisciplinary inquiry and public discussion, rather than by initial assumption. Whatever the outcome of such a process, it is likely to improve interdisciplinary communication regarding the evaluation of environmental goods and policies.


V. SCALE-SENSITIVE EVALUATION OF ENVIRONMENTAL POLICIES

We have not yet addressed the practical applicability of a two-tier approach. How are demarcations of the tiers to be identified in real management situations? If monistic approaches are unacceptably narrow, how can this pluralistic approach be grounded in facts so that it is not arbitrary?

These are not easy questions, and we do not propose to answer them here. Our goal in this section is to identify features of ecological systems whose destruction may give rise to especially large and long-lived risks, suggesting that the affected natural features warrant greater protection. This information may be useful for determining the categorization of risks within a two-tier system. For example, the idea of an intergenerational trust or an organic, communitarian commitment could be combined with measures of ecological health or integrity to argue that we owe to the future protection of productive processes embodied in large-scale ecological systems, because these processes are inherently required in order to provide valued options to future generations to fulfill their wants, whatever their wants turn out to be.

Protection of such "ecological production functions" might be seen as analogous to what Rawls (1971) calls "primary goods," things that every rational person is assumed to want. Note, however, that the identification of such goods cannot be achieved on scientific grounds alone; value-laden views about the nature of the risks involved are an inherent part of the identification. The need to depart from a scientific commitment to value neutrality makes this effort controversial (Chapter 11, this Volume). This challenge arises on top of the basic scientific uncertainties that currently complicate the delineation of the decision tiers in our proposed approach. The presence of these uncertainties and value controversies further underscores the need for experimentation to better understand the performance of pluralistic approaches.

There have been many and varied attempts to define ecological integrity and health in an operational and measurable manner (Karr 1981, 1991; Haskell 1992; Rapport 1992; Schaeffer and Cox 1992). In one view (Callicott 1989; Westra 1994) references to integrity imply a commitment to the intrinsic or inherent value of ecological communities. According to this view, the interests of humans must be understood as limited by the legitimate, countervailing interests of other species and ecological systems.[18] A second suggestion, by Ehrenfeld (1992), is that these concepts should have limited theoretical, ecological content, but that they should embody the general goals and values of a community more than identifiable and measurable characteristics of physical systems. Above, we expressed a similar point by claiming that health and integrity are better understood as terms in public policy discourse rather than as scientific terms. These theoretical commitments have the purpose of making scale-sensitivity integral to the models. But once more, there is the need to identify values as well as ecosystem traits, and to explain why some traits of ecosystems should be protected, while others are not.

There have also been attempts to use these terms to link scientific assessment and valuation (Edwards and Regier 1990; Norton 1992; Ulanowicz 1992). In these approaches, "integrity" and "health" are used as terms that can be defined with some level of scientific precision, but which also indicate characteristics of systems that are thought to be socially valuable. For example, Norton and Ulanowicz (this volume) have experimented with embodying social values and perspectives in scalar representations of physical systems by associating social values (such as biodiversity protection) with particular levels of landscape functioning within a hierarchically ordered system. Adaptive management must, in this case, involve processes of inquiry that are *both* scientific *and* evaluative. *Both* the processes of value articulation, formulation, and reformulation *and* the process of scientific hypothesis testing must be, for this approach to be possible, endogenous to the management process.

These scientific/normative theories can be thought of as analogous to trust funds (Weiss 1989; Brown 1994), with each generation serving as the agent for future generations. Weiss argues that such an analogy has precedent in international law. According to this approach, the goal of long-term environmental policy should be to specify ecological processes, features, or characteristics that are of lasting social value, and to develop policies that will perpetuate those features by training the energies of science and management on identifying and protecting those physical dynamics that are associated with important social values.

Since this approach requires agents in any generation to preserve resources for the future as well as to use them in the present, it can be applied only if there is some criterion to separate intergenerationally significant changes to physical systems – changes that will unacceptably harm future generations – from benign use of resources that might enhance, or at least not diminish, their lives. Ecological economists and exponents of adaptive ecological management alike tend to see the marginalist, equilibrium models of mainstream economics as not adequate to describe and evaluate changes in the complex, non-equilibrium behaviors of complex ecological communities. However, there are also debates both among and between ecological economists, more conventional economists, and ecologists advocating adaptive management regarding which ecophysical characteristics should be identified as important "social capital," and how to measure these characteristics.

One notable interdisciplinary attempt to specify characteristics of ecosystems that should be protected is Arrow et al. (1995). The authors, a distinguished interdisciplinary group, argue first that conventionally applied economic criteria are not adequate to protect some important environmental goods. Among these goods they emphasize the importance of ecosystem resilience – the magnitude of stress a system can absorb before it flips into another local equilibrium. Resilience is important, the authors argue, not just because of its biological consequences, but also because of its effects on the options available to future generations and the uncertainties about how human activities affect the environment (which complicate the analysis of options).

Resilience can be connected to a two-tier system of management through measures that reflect not just changes in ecosystem states, but also the consequences of such changes for the options available to human beings. As noted previously, larger-scale systems determine the opportunities and constraints confronted in smaller-scale subsystems. Greater resilience of larger-scale systems therefore provides increased options to both the current generation and future generations. A measure of the resilience of larger-scale systems could function both as a descriptor and, given a set of community values associated with the resilience of the environment, as a normative measure of how well the society is doing in sustaining opportunity across generations. The challenge is in developing such an index in practice, with appropriate connections to economic as well as biophysical information.

## VI. CONCLUSION

The two-tier, pluralistic approach, which patches together ideas from several sources and disciplines, may be an alternative to the various monistic systems of value that have hitherto competed haphazardly for attention in the monitoring and evaluation of environmental trends. There are clearly many "ifs" associated with this approach, and tremendous uncertainties that would obtain if we were to attempt to employ it in decision making. In particular, this approach must somehow identify "essential" processes in terms that complement conventional measures of economic welfare. Any system chosen to account for intergenerational equity and values likewise rests on assumptions and involves tremendous uncertainty, however, especially in judging multi-generational impacts of proposed policies.

In this paper we have tried to assess realistically the prospects for early reconciliation between economic and ecological models as they bear upon sustainability across multiple generations. We have acknowledged that at present the conceptual chasms between the two disciplines remain deep in at least two areas – the concept of reversibility and substitutability and the development of adequate accounting systems. We believe it is futile to hope to resolve these deep theoretical disagreements quickly if the current strategy of asserting, defending, and applying opposed, monistic systems of value in exclusive disciplinary contexts is continued. As an alternative, we suggest a more experimental, open approach in which various measures are suggested, applied, refined, and compared with other evaluation methods. We believe that such a two-tier system of analysis, if applied within a publicly accountable, iterative process of policy formation, could lead to increased understanding of cross-disciplinary theoretical differences, and also broaden and deepen our understanding of the impacts of human actions on the environment.

NOTES

An earlier version of this paper was prepared for a Colloquium on Sustainability in the United States, sponsored by the Environmental Protection Agency and organized by the Environmental Law Institute. The authors are grateful to C. S. Holling, R. Howarth, R. Norgaard, and T. Tietenberg for very useful comments on different drafts. Research underlying this paper was partially supported by a grant to Resources for the Future from the Office of Exploratory Research, United States Environmental Protection Agency. Responsibility for its content is the authors' alone.

1. It should, of course, be recognized that there may be quite articulate minorities within disciplines, whose members articulate alternative conceptions. So our comments should be understood as directed at dominant conceptualizations within disciplines and not as implying perfect unanimity.

2. This use of the term "accounting" should be distinguished from the emphasis in many sustainability discussions on the much narrower question of revising national income accounts to incorporate environmental values.

3. The discussion in this section draws to a considerable degree on Toman, Pezzey, and Krautkraemer (1995).

4. In particular, permanent changes in physical states should be assigned an opportunity cost that reflects the loss of future as well as current services (see Krutilla and Fisher 1985).

5. In this connection, the terms "weak sustainability" and "strong sustainability" are often invoked, though these terms can be interpreted in various ways. An interpretation that refers directly to substitution among different production inputs is found in Pearce and Atkinson (1993) and Victor (1991). In this interpretation, if

natural and other capital are substitutable, then the weak sustainability criterion of preserving aggregate capital can be applied, but if there are limits on substitution, then the strong sustainability criterion of preserving natural capital may be relevant. Another interpretation of weak and strong sustainability, one with an intertemporal element, is found in Barbier, Markandya, and Pearce (1990). These authors treat strong sustainability as requiring that net damages to environmental capital be nonpositive along the whole time path of resource exploitation, while weak sustainability requires only that the present value of damages be nonpositive; both of these definitions allow for some substituability among various capital inputs.

6. See Dasgupta and Heal (1974), Solow (1974), and Stiglitz (1974).

7. For discussion of physical limits see Ayres and Kneese (1969), Kneese, Ayres, and d'Arge (1971), Perrings (1986), Anderson (1987), Gross and Veendorp (1990), and Holling (1994).

8. In addition, dissipation of production potential (implied by the second law of thermodynamics) may limit long-term production, though here arguments over the importance of the constraint are even more controversial (cf. Daly 1992a, 1992b; Young 1991, 1994).

9. The Risk Assessment Forum, for example, in its workshop on Ecological Risk Assessment, has provided a discussion of how to evaluate the "ecological significance" of a risk (see, e.g., Harwell et al. 1994 and Norton 1994). Two points were emphasized. First, there will always be difficulty in determining whether a change represents an ecological discontinuity – whether, that is, a human-induced change exceeds the range of natural variation – because natural systems fluctuate at many scales. Some very subtle changes can have huge impacts over multiple generations (Pimm 1991). Second, this report acknowledges that, in a management process, human values as well as scientific descriptions are an important part of any assessment of ecological significance (Harwell et al. 1994; Norton 1994).

10. See Allen and Starr (1982) and Allen and Hoekstra (1992) for arguments that scale relationships are an essential feature of how we *understand* complex physical systems. On this view, we can remain agnostic regarding the actual existence of irreducible scalar structure in nature. But Holling has recently introduced empirical evidence that there are unavoidable scalar structures as an element in the actual structure of ecological communities, if viewed from the perspective of animals with characteristic body sizes (Holling 1992).

11. This commitment of ecologists to detailed empirical observation of particular systems is illustrative of a deeper disciplinary difference with many economists. Economists often rely on mathematical models employing highly aggregated data (Waldrop 1992); ecologists usually build site-specific, observation-based models (Ehrenfeld 1993).

12. The consequences of these differences in economic models and ecological models warrant further exploration. In particular, how does one relate information derived in an equilibrium system to information embodied in dynamic systems with feedbacks and redundancies in which processes and change are modeled?

13. For discussion of these issues see Smith (1979), Ayres and Miller (1980), Baumol (1986), Daly (1992a, 1992b), Lozada (1991), Young (1991, 1994), and Pezzey (1992).

14. Efforts to perpetually accelerate technical progress ultimately also may tax our genetic ability to process information (Pezzey 1992).

15. One practical way to implement a social discount rate is to treat members of all generations equally by ignoring the component of the individual discount rate that reflects impatience for current consumption over future consumption (Pearce 1983; Cline 1992). An alternative perspective is that intergenerational discount rates should reflect the willingness of the current generation to undertake collective sacrifices for the future (Schelling 1995). Page (1988), on the other hand, argues that the search for a correct social discount rate is futile.

16. Norgaard (1989) argues that complex ecological and human systems can be understood only through multiple disciplinary perspectives, and that both economics and ecology have wrongly attempted to assert primacy in the search for such understanding.

17. Others have suggested a Precautionary Principle, which argues for erring on the side of caution when there are risks of highly adverse outcomes and uncertainty is high (e.g., O'Riordan and Jordan 1995). Stated in this way, the Precautionary Principle involves less scope for balancing benefits and costs than the Safe Minimum Standard and thus is a priori problematic for most economists.

18. These theories, prominent in the field of environmental ethics, assert that nature has "intrinsic value." These philosophical theories – sometimes associated with "deep ecology," although there are many advocates of intrinsic value in nature who would not so describe themselves – posit nonanthropocentric values, values that exist independently of human values and motives (Devall and Sessions 1985; Taylor 1986; Callicott 1989). We will not address intrinsic value theories in this article for two reasons. Deep ecologists have generally rejected a sustainability ethic, especially if it includes a commitment to development of resources for present and future human use, because these commitments are viewed as human-centered; also, no clear policy implications have emerged from the pronouncements of nonanthropocentrists.

## REFERENCES

Allen, T. F. H., and T. W. Hoekstra. 1992. *Toward A Unified Ecology.* New York: Columbia University Press.

Allen, T. F. H., and T. B. Starr. 1982. *Hierarchy: Perspectives for Ecological Complexity.* Chicago: University of Chicago Press.

Anderson, C. L. 1987. "The Production Process: Inputs and Wastes." *Journal of Environmental Economics and Management* 14 (Mar.):1–12.

Arrow, K., B. Bolin, R. Costanza, P. Dasgupta, C. Folke, C. S. Holling, B. Jansson, S. Levin, K. Maler, C. Perrings, and D. Pimentel. 1995. "Economic Growth, Carrying Capacity, and the Environment." *Science* 268 (Apr. 28):520–521.

Ayres, R. U., and A. V. Kneese. 1969. "Production, Consumption, and Externalities." *American Economic Review* 69 (June):282–297.

Ayres, R. U., and S. Miller. 1980. "The Role of Technological Change." *Journal of Environmental Economics and Management* 7 (Dec.):353–371.

Barbier, E. B., A. Markandya, and D. W. Pearce. 1990. "Environmental Sustainability and Cost-Benefit Analysis." *Environment and Planning* 22 (Sept.):1259–266.

Baumol, W. J. 1986. "On the Possibility of Continuing Expansion of Finite Resources." *Kyklos* 39:167–179.

Bishop, R. C. 1978, "Endangered Species and Uncertainty: The Economics of the Safe Minimum Standard." *American Journal of Agricultural Economics* 60 (Feb.): 10–18.

———. 1979. "Endangered Species, Irreversibility and Uncertainty: A Reply." *American Journal of Agricultural Economics* 61 (May):376–379.

Broome, J. 1992. *Counting the Cost of Global Warming*. Cambridge, UK: White Horse Press.

Brown, P. G. 1994. *Restoring the Public Trust I*. Boston: Beacon Press.

Burton, P. S. 1993. "Intertemporal Preferences and Intergenerational Considerations in Optimal Resource Harvesting." *Journal of Environmental Economics and Management* 24 (Mar.):119–32.

Callicott, J. B. 1989. *In Defense of the Land Ethic*. Albany: State University of New York Press.

———. 1990. "The Case Against Moral Pluralism." *Environmental Ethics* 12 (Summer):99–124.

Ciriacy-Wantrup, S. V. 1952. *Resource Conservation*. Berkeley: University of California Press.

Cline, W. R. 1992. *The Economics of Global Warming*. Washington, DC: Institute for International Economics.

Common, M., and C. Perrings. 1992. "Towards an Ecological Economics of Sustainability." *Ecological Economics* 6 (July):7–34.

Costanza, R., ed. 1991. *Ecological Economics: The Science and Management of Sustainability*. New York: Columbia University Press.

Costanza, R., C. Folke, M. Hammer, and A. M. Jansson. 1993. *Investing in Natural Capital: Why, What, and How?*" Solomons, MD: ISEE Press.

Costanza, R., B. G. Norton, and B. D. Haskell, eds. 1992. *Ecosystem Health: New Goals for Environmental Management*. Washington, DC: Island Press.

Daly, H. E. 1992a. "Is the Entropy Law Relevant to the Economics of Natural Resource Scarcity? – Yes, of Course It Is!" *Journal of Environmental Economics and Management* 23 (July):91–95.

———. 1992b. "Steady-State Economics: Concepts, Questions, Policies." *GAIA* 6:333–38.

Daly, H. E., and J. Cobb. 1989. *For the Common Good*. Boston: Beacon Press.

Dasgupta, P. S., and G. M. Heal. 1974. "The Optimal Depletion of Exhaustible Resources." In *Review of Economic Studies Symposium on the Economics of Exhaustible Resources*, vol. 41, pp. 3–28. Edinburgh: Longman Group Limited.

Devall, B., and G. Sessions. 1985. *Deep Ecology: Living as if Nature Mattered*. Salt Lake City: Peregrine Smith Books.

Edwards, C. J., and H. Regier, eds. 1990. *An Ecosystem Approach to the Integrity of the Great Lakes in Turbulent Times*. Ann Arbor, MI: Great Lakes Fisheries Commission Special Publication 90–94.

Ehrenfeld, D. 1992. "Ecological Health and Ecological Theories." In *Ecosystem Health: New Goals for Environmental Management*, eds. R. Costanza, B. G. Norton, and B. D. Haskell. Covelo, CA: Island Press.

———. 1993. *Beginning Again: People and Nature in the New Millennium*. New York: Oxford University Press.

Gross, L. S., and E. C. H. Veendorp. 1990. "Growth with Exhaustible Resources and a Materials-Balance Production Function." *Natural Resource Modeling* 4 (Winter):77–94.

Gunderson, L. H., C. S. Holling, and S. S. Light. 1995. *Barriers and Bridges to the Renewal of Ecosystems and Institutions.* New York: Columbia University Press.

Harwell, M., J. Gentile, B. Norton, and W. Cooper. 1994. "Ecological Significance." In *Ecological Risk Assessment Issue Papers*, Risk Assessment Forum, U.S. Environmental Protection Agency, Office of Research and Development, Report: EPA/630/R-94/009, Washington, DC.

Holling, C. S. 1973. "Resilience and Stability of Ecological Systems." *Annual Reviews of Ecology and Systematics* 4:1–23.

———, ed. 1978. *Adaptive Environment Assessment and Management.* London: Wiley.

———. 1986. "Resilience of Ecosystems; Local Surprise and Global Change." In *Sustainable Development of the Biosphere,* eds. W. C. Clark and R. E. Munn. Cambridge: Cambridge University Press.

———. 1992. "Cross-Scale Morphology, Geometry, and Dynamics of Ecosystems." *Ecological Monographs* 62:447–502.

———. 1993. "New Science and New Investments for a Sustainable Biosphere." In *Investing in Natural Capital: Why, What, and How?,*" eds. R. Costanza, C. Folke, M. Hammer, and A. M. Jansson. Solomons, MD: ISEE Press.

———. 1994. "An Ecologist View of the Malthusian Conflict." In *Population, Economic Development, and the Environment,* eds. K. Lindahl-Kiessling and H. Landsberg. New York: Oxford University Press.

———. 1996. "Engineering Resilience versus Ecological Resilience." In *Engineering within Ecological Constraints*, ed. P. C. Schulze. Washington, DC: National Academy Press.

Howarth, R. B. 1995. "Sustainability under Uncertainty: A Deontological Approach." *Land Economics* 71 (Nov.):417–427.

Howarth, R. B., and R. B. Norgaard. 1990. "Intergenerational Resource Rights, Efficiency and Social Optimality." *Land Economics* 66 (Feb.):1–11.

———. 1993. "Intergenerational Transfers and the Social Discount Rate." *Environmental and Resource Economics* 3 (Aug.):337–358.

Karr, J. R. 1981. "Assessment of Biotic Integrity Using Fish Communities." *Fisheries* 6:21–27.

———. 1991. "Biological Integrity: A Long-Neglected Aspect of Water Resource Management." *Ecological Applications* 1:66–84.

Kneese, A. V., R. Ayres, and R. d'Arge. 1971. *Economics and the Environment.* Baltimore: Johns Hopkins University Press for Resources for the Future.

Kneese, A. V., and W. D. Schulze. 1985. "Ethics and Environmental Economics." In *Handbook of Natural Resource and Energy Economics,* eds. A. V. Kneese and J. L. Sweeney. Amsterdam: North-Holland.

Krutilla, J. V., and A. C. Fisher. 1985. *The Economics of Natural Environments: Studies in the Valuation of Commodity and Amenity Resources.* 2d ed. Washington, DC. Resources for the Future.

Lee, K. N. 1993. *Compass and Gyroscope: Integrating Science and Politics for the Environment.* Covelo, CA: Island Press.

Lozada, G. A. 1991. "Why the Entropy Law Is Relevant to the Economics of Natural Resource Scarcity." Draft manuscript, Energy and Resources Group, University of California Berkeley (October).

Norgaard, R. B. 1989. "The Case for Methodological Pluralism." *Ecological Economics* 1 (Feb.):37–58.

Norton, B. G. 1987. *Why Preserve Natural Variety?* Princeton: Princeton University Press.

———. 1989. "Intergenerational Equity and Environmental Decisions: A Model Using Rawls' Veil of Ignorance." *Ecological Economics* 1 (May):137–159.

———. 1991. *Toward Unity among Environmentalists.* New York: Oxford University Press.

———. 1992. "A New Paradigm for Environmental Management." In *Ecosystem Health: New Goals for Environmental Management*, eds. R. Costanza, B. G. Norton, and B. C. Haskell. Covelo, CA: Island Press.

———. 1994. "Ascertaining Public Values Affecting Ecological Risk Assessment." In *Ecological Risk Assessment Issue Papers*, Risk Assessment Forum, U.S. Environmental Protection Agency, Office of Research and Development, Report: EPA/630/R-94/009, Washington, DC.

———. 1995. "Future Generations, Obligations to." In *Encyclopedia of Bioethics*, 2d ed. New York: Macmillan.

O'Neill, R. V., D. L. DeAngelis, J. B. Waide, and T. F. H. Allen. 1986. *A Hierarchical Concept of Ecosystems.* Princeton: Princeton University Press.

O'Riordan, T., and A. Jordan. 1995. "The Precautionary Principle in Contemporary Environmental Politics." *Environmental Values* 4:191–212.

Page, T. 1977. *Conservation and Economic Efficiency.* Baltimore: Johns Hopkins University Press for Resources for the Future.

———. 1988. "Intergenerational Equity and the Social Rate of Discount." In *Environmental Resource and Applied Welfare Economics,* ed. V. K. Smith. Washington, DC: Resources for the Future.

———. 1997. "On the Problem of Achieving Efficiency and Equity, Intergenerationally." *Land Economics* 73 (Nov.):580–596.

Parfit, D. 1983. "Energy Policy and the Further Future: The Social Discount Rate." *Energy and the Future*, eds. D. MacLean and P. G. Brown. Totowa, NJ: Rowman and Littlefield.

Pearce, D. W. 1983. "Ethics, Irreversibility, Future Generations and the Social Rate of Discount." *International Journal of Economic Studies* 21:67–86.

Pearce, D. W., and G. D. Atkinson. 1993. "Capital Theory and the Measurement of Sustainable Development: Some Empirical Evidence." *Ecological Economics* 8 (Oct.):103–108.

Perrings, C. 1986. "Conservation of Mass and Instability in a Dynamic Economy-Environment System." *Journal of Environmental Economics and Management* 13 (Sept.):199–211.

Pezzey, J. 1992. "Sustainability: An Interdisciplinary Guide." *Environmental Values* 1 (Mar.):321–362.

Pimm, S. L. 1984. "The Complexity and Stability of Ecosystems." *Nature* 307:321–26.

———. 1991. *Balance of Nature?* Chicago: University of Chicago Press.

Randall, A. 1986. "Human Preferences, Economics, and the Preservation of Species." In *The Preservation of Species*, ed. B. G. Norton. Princeton: Princeton University Press.

Rapport, D. J. 1992. "What Is Clinical Ecology?" In *Ecosystem Health: New Goals for Environmental Management*, eds. R. Costanza, B. Norton, and B. Haskell. Covelo, CA: Island Press.

Rawls, J. 1971. *A Theory of Justice*. Cambridge: Harvard University Press.

Scarf, H. E. 1981a. "Production Sets with Indivisibilities – Part I: Generalities." *Econometrica* 49 (Jan.):1–32.

―――. 1981b. "Production Sets with Indivisibility – Part II: The Case of Two Activities." *Econometrica* 49 (Mar.):395–423.

Schaeffer, D. J., and D. K. Cox, 1992. "Establishing Ecosystem Threshold Criteria." In *Ecosystem Health: New Goals for Environmental Management*, eds. R. Costanza, B. Norton, and B. Haskell. Covelo, CA: Island Press.

Schelling, T. C. 1995. "Intergenerational Discounting." *Energy Policy* 23 (Apr./May):395–401.

Smith, V. K. 1979. *Scarcity and Growth Reconsidered*. Baltimore: Johns Hopkins University Press for Resources for the Future.

Solow, R. M. 1974. "Intergenerational Equity and Exhaustible Resources." *Review of Economic Studies, Symposium on the Economics of Exhaustible Resources*, Vol. 41, pp. 29–45. Edinburgh: Longman Group Limited.

―――. 1993. "Sustainability: An Economist's Perspective." In *Selected Readings in Environmental Economics*, 3rd ed., eds. R. Dorfman and N. Dorfman. New York: Norton.

Stiglitz, J. 1974. "Growth with Exhaustible Natural Resources: Efficient and Optimal Growth Paths." *Review of Economic Studies Symposium on the Economics of Exhaustible Resources,* Vol. 41, pp. 123–137. Edinburgh: Longman Group Limited.

Stone, C. 1987. *Earth and Other Ethics*. New York: Harper and Row.

Taylor, P. 1986. *Respect for Nature*. Princeton: Princeton University Press.

Toman, M. A. 1994. "Economics and 'Sustainability': Balancing Trade-offs and Imperatives." *Land Economics* 70 (Nov.):399–413.

Toman, M. A., J. Pezzey, and J. Krautkraemer. 1995. "Neoclassical Economic Growth Theory and 'Sustainability.'" In *Handbook of Environmental Economics*, ed. D. W. Bromley, Oxford: Blackwell.

Ulanowicz, R. E. 1986. *Growth and Development: Ecosystems Phenomenology*. New York: Springer-Verlag.

―――. 1992. "Ecosystem Health and Trophic Flow Networks." In *Ecosystem Health: New Goals for Environmental Management*, eds. R. Costanza, B. Norton, and B. Haskell. Covelo, CA: Island Press.

United Nations. 1993. *Integrated Environmental and Economic Accounting*. United Nations publication ST/ESA/STAT/Ser.F/61. New York: United Nations.

Victor, P. A. 1991. "Indicators of Sustainable Development: Some Lessons from Capital Theory." *Ecological Economics* 4 (Dec.):191–214.

Waldrop, M. M. 1992. *Complexity: The Emerging Science at the Edge of Chaos*. New York: Simon and Schuster.

Walters, C. 1986. *Adaptive Management of Renewable Resources*. New York: Macmillan.

Weiss, E. B. 1989. *In Fairness of Future Generations.* Dobbs Ferry, NY: Transnational Publishers.

Westra, L. 1994. *An Environmental Proposal for Ethics: The Principle of Integrity.* Totowa, NJ: Rowman and Littlefield.

World Commission on Environment and Development (WCED). 1987. *Our Common Future.* New York: Oxford University Press.

Young, J. T. 1991. "Is the Entropy Law Relevant to the Economics of Natural Resource Scarcity?" *Journal of Environmental Economics and Management* 21 (Sept.):169–179.

————. 1994. "Entropy and Natural Resource Scarcity: A Reply to the Critics." *Journal of Environmental Economics and Management* 26 (Mar.):210–213.

# 14

## The Evolution of Preferences

### Why 'Sovereign' Preferences May Not Lead to Sustainable Policies and What to Do about It

with ROBERT COSTANZA and RICHARD C. BISHOP

### 1. INTRODUCTION

Conventional, neoclassical economics is based on assumptions that, while yielding useful models for understanding the problems of efficient allocation of resources in the short term, are misleading and potentially dangerous in dealing with the long-term consequences of economic choices. This problem arises more specifically in evaluating the impacts of various policies that are proposed to promote 'sustainable development', because the temporal horizon of sustainability analysis is multi-generational. Sustainable development involves three hierarchically inter-related problems. These are maintaining (1) a sustainable scale of the economy relative to its ecological life support system; (2) a fair distribution of resources and opportunities, not only among members of the current generation of humans, but also among present and future generations (and even in some formulations among humans and other species); and (3) an efficient allocation of resources over time that adequately accounts for natural capital (Daly, 1990).

Over generations, the economy is expected to grow significantly in its material scale as a result of population growth and probably also as a result of the increasing expectations of consumers. As the scale of economic activity increases there is also, in general, an increase in the impacts of economic activity on the environment.[1] This focuses attention on both population growth and the material consumption levels of advanced societies. In this paper the focus is on the material consumption vector, questioning whether there may be a social interest in influencing individual preferences toward less material consumption-oriented forms of satisfaction. Any comprehensive effort to

Reprinted from *Ecological Economics* 24 (1998): 193–211, Copyright 1998, with permission from Elsevier Science.

address problems of the scale of the economy and per capita consumption of resources must somehow address this fundamental problem of preference formation.

Neoclassical economists have largely ignored scalar problems and intertemporal distribution issues as being 'outside the domain' of economics, because they have assumed infinite substitution possibilities among resources and unlimited technological change. Economics is, on these assumptions, limited to solving the technical issues surrounding the efficient allocation problem. The problem is that optimal allocation does not guarantee sustainability. Mainstream economics has assumed that the goal is to manage environmental resources as efficiently as possible over time. It has largely ignored the theoretical principle that there are an infinite number of time paths for resource use and preservation that would satisfy Pareto efficiency criteria. It is a relatively easy step to show that not all efficient time paths are sustainable. Additional criteria beyond Pareto efficiency will be needed if the dual goals of efficiency and sustainability are to be recognized in environmental policies (Bishop, 1993).

If economics is defined more broadly as the 'management of the household', as the Greek root of the word implies, then it must address all the problems attendant on that management, including scale and distribution problems, even if those problems do not submit to the mathematical models and prescriptions that have been used to solve the allocation problem.

The management of a household has both internal and external relations. One way of addressing the relationship of ecology to economics is to consider economics as the analysis of decisions within the 'closed' internal system of the family/household, while ecology provides tools for analyzing the interrelations of the family to the outside world. Imagine the problems with a household that operates under the assumption that its members' preferences will be 'taken as given' and 'unlimited' for the purposes of setting each successive family budget. Limits imposed by the ecophysical system producing resources may impose limits on the scale of an economy, just as the income limits of a family inevitably force family members to reduce their demands for consumption.

In this paper, the conventional assumption that individual preferences can be accepted as given and stable for the purposes, and for the duration, of an economic study, and how the usefulness of this assumption changes as the time frame of the study increases, are addressed. The range of policy options which are deemed 'acceptable' is dependent on this assumption, so it has a large impact on long-term ecological and socioeconomic sustainability. The objectives of this paper are (1) to examine the assumption of consumer sovereignty and the role of this principle in the analysis of environmental values; (2) to enhance the understanding of how values change over time in a society; and (3) to suggest some new directions for analysis and modeling that may enhance the understanding of the social changes necessary to achieve a social commitment to sustainability in democratically organized societies.

## 2. FIXED TASTES, PREFERENCES, AND CONSUMER SOVEREIGNTY

Conventional economic methodology is based on consumer sovereignty, the assumption that tastes and preferences are givens and that the economic problem consists of optimally satisfying those preferences (Silberg, 1978). Tastes and preferences usually do not change rapidly, so, in the short term at least, this assumption makes sense. But tastes and preferences do change, especially in the longer term. Therefore, economists' models, which treat preferences as exogenous, cannot be expected to correctly characterize or guide decisions that have potential impacts over decades, centuries, and longer.

Insisting that preference formation must be endogenous to models of resource decision making presents a very disturbing prospect for economists because it blocks easy definition of what is 'optimal'. If tastes and preferences are fixed and given, one can adopt a stance of consumer sovereignty and just 'give the people what they want'. There is no need to know or care why consumers want what they want; their preferences just have to be satisfied as efficiently as possible. However, if preferences are expected to change over time and under the influence of education, advertising, changing cultural assumptions, etc., a different criterion for what is 'optimal' is needed; and how preferences change, how much they are a function of 'nature' and how much of 'nurture', how they relate to this new criterion, and how they can or should be actively influenced to satisfy the new criterion need to be figured out.

One alternative for this new criterion is sustainability itself, or, more completely, sustainable scale, efficient allocation of resources in the short term, and a reasonably fair distribution of nature's assets over multiple generations. This model implies a two-tiered decision process (Page, 1977; Chapters 10, 12, and 13, this volume) with interactive tiers (Fig. 14.1). The search for improved policies is assumed to be an iterative process, with the model cycling through both tiers many times. In one tier, the 'reflective' tier, the central objective is to characterize and categorize environmental 'problems' by building models of human communities and their ecological, contextual physical system. These focus attention on physical dynamics that are socially valued or are associated with important economic or cultural values (such as salmon

---

**Tier 1 (Reflective)**
Social consensus on broad goals and vision of the future, combined with scientific models of dynamic, non-equillibrium, long-term ecological economic interactions. *Here, environmental problems are classified according to the risks to social values they entail.*

**Tier 2 (Action)**
Resolution of conflicts mediated by markets, education, legal, and other institutions, combined with short-term equillibrium models of interactions and optimality. *Here, particular action criteria are applied, acted upon, and tested in particual situations.*

---

Figure 14.1. Two-tiered decision structure.

runs or forest regeneration). The goal of the first tier is to identify measurable physical processes that are relevant to locally defined goals for environmental management. One cannot discuss goals without discussing social values, so this reflective tier of the model must connect policies with impacts on social values. And while individual preferences of consumers may be a starting point for determining what ecophysical processes have social value, the model also contains a set of democratic processes for discussion and re-evaluation of preferences (see Slovic, 1995; Sagoff, 1998). The output of this reflective tier is a determination of the type of environmental problem that is faced, an analysis of the social values at risk, a set of goals in dealing with that problem, and a proposal that one or more of the available 'action criteria' be applied. The second tier, the 'action' tier, generates policy guidance by applying one or more of several action-level criteria, such as cost benefit analysis (CBA) or the safe minimum standard of conservation criterion (SMS) within a scale-sensitive model. In the action tier, appropriate action criteria, given the classification of the risk or problem to be addressed, are applied to specific management situations. This two-tier, iterative model integrates the processes of social valuation, environmental science, and environmental monitoring and management in a more comprehensive way.

Besides threatening the theoretical and computational elegance of economists' welfare models, questioning consumer sovereignty also raises legitimate concerns regarding the possible manipulation of preferences. If tastes and preferences can change, then who is going to decide how to change them? There is a fear that a 'totalitarian' government or narrow special interests might be employed to manipulate preferences to conform to the desires of a select elite rather than the society as a whole. Two points need to be kept in mind however: (1) preferences are already being manipulated every day; and (2) we can just as easily apply open democratic principles to the problem as hidden or totalitarian principles in deciding how to influence preferences. So the question becomes whether it is better for preferences to be determined behind the scenes, either by a dictatorial government, by big business acting through advertising, or in some other way. Or do we want to explore and shape them openly, based on social dialogue and consensus, with a higher goal in mind? Either way, this is an issue that can no longer be avoided, and one which is best handled using democratic principles and innovative thinking.

Another way of looking at this issue is to draw a distinction between 'positive' and 'normative' analysis. 'Positive' refers to the way things are, while 'normative' refers to the way we would like things to be. Economists have attempted to make this a clear-cut distinction, and have often tried to confine their work to 'positive' analysis. However, like the mind–body distinction, the positive–normative distinction is not clear-cut. Our vision of how we would like things to be influences how things are and vice versa. Therefore, purely 'positive' analysis is impossible. A two-tier process allows us to move back and forth between an examination of social values and likely impacts of acting on those values, which points toward appropriate criteria for judging policies with varying potential impacts (in the reflective tier), and applications of those criteria in actual situations (in the action tier). When actions guided by a chosen action criterion lead to unwelcome consequences in a particular situation, it is possible to return to the reflective tier and re-examine the choice of an action criterion on the basis of new evidence. This re-examination may result in a re-classification of the problem and choice of a new criterion and, at least in some cases, further reflection on the appropriateness of social values as stated. In this way, experimental models and management experiments undertaken in the action tier provide a sort of laboratory for testing both management techniques and goals. When viewed in this way, management experiments may provide information that will provoke public discussion of community norms and social values.

### 3. FOUR DEGREES OF CONSUMER SOVEREIGNTY

Due to the fact that the principle of consumer sovereignty is so intimately entwined, theoretically and methodologically, with other constitutive elements of the mainstream economic paradigm, economists have often accepted the principle without providing it the intellectual scrutiny it deserves, especially given its decisive impact on the scope of questions, addressed in 'economic' analyses (Norton, 1991; Chapter 11, this volume). In fact, consumer sovereignty should not be thought of as an all-or-nothing state of reality or of policy, but rather as a continuum of possible assumptions regarding the extent to which questions regarding preference formation are held to be endogenous to the system of analysis of environmental values. Indeed, it has been shown that the various arguments to defend consumer sovereignty as a basic assumption of economics would support quite different interpretations of the principle in theory and practice (Chapter 11, this volume). The survey of several possible interpretations, and the apparent policy implications, of four 'degrees' of the consumer sovereignty principle follow. While these four degrees of consumer sovereignty do not exhaust the plausible positions on this continuum, they are useful because they each require somewhat different assumptions about the boundaries between economics and other traditional academic disciplines.

### 3.1. Degree 1: Unchanging Preferences

Some economists (Stigler and Becker, 1977) argue that individual preferences are not susceptible to rational analysis and that they should be considered both (a) given and (b) fixed for the purposes of analysis. To say that preferences are given is to say that, for the purposes of any analysis, preferences of individuals will be accepted, at face value, as indicative of the individual's actual 'good' or welfare. This implication will be discussed in more detail in connection with degree 2. To say that preferences are fixed is to make a much stronger claim, that, at least from the viewpoint of economics, preferences of individuals do not change through time. According to this view, preferences are best thought of as tastes that, by the onset of adulthood, are locked in, at least in the sense that they are impossible to change through rational considerations (Stigler and Becker, 1977).

If it were to be assumed that this is a correct empirical generalization about human preference development (the point will be disputed below), it still would not follow that preferences are in fact fixed over time. They might, consistent with the hypothesis that there is no rational route to preference change, change in response to non-rational factors such as subliminal advertising or other forms of 'propaganda'. Also, it seems obvious that the preferences of individuals do, in fact, change over time in response to some variables, e.g. witness changing attitudes toward smoking and sexual freedom.

Whether such a process is rational is perhaps debatable, but the phenomenon of preference change is itself not in serious doubt.

Sympathetic interpretation of Stigler and Beckers's claim, then, requires that they are not taken to be making the apparently false empirical claim that individuals never change their preferences, but rather the claim that, since preference change is not susceptible to rational analysis, it makes methodological sense to decide to treat preferences as fixed because this brings all consumer behavior under the explanatory scope of conventional economics, treating all such behavior as a function of opportunities. As Stigler and Becker (1977) say: "the great advantage . . . of relying only on changes in the arguments entering household production functions is that all changes in behavior are explained by changes in prices and incomes, precisely the variables that organize and give power to economic analysis." However, this argument rests on the crucial assumption that there cannot be alternative ways to rationally understand or evaluate processes of preference change. And these are exactly the assumptions that are addressed in this paper. In fairness to Stigler and Becker, it should be said that they might have appealed more explicitly for justification of their premise that preferences are unresponsive to rational argument to the philosophical, ethical theory of emotivism. According to logical positivisms, a philosophical movement that had some currency between 1920 and 1950, evaluative utterances are simply expressions of emotion; their linguistic import and meaning are not cognitive but emotive. Despite the usefulness of this theory in reminding us, as Hume (1888) had demonstrated two centuries ago, that the logic of descriptive and prescriptive sentences differs importantly, emotivism as a theory of ethics has been thoroughly discredited on the grounds that it misses most of what is important in processes of moral justification (Feigl, 1952; Williams, 1985).

Since it is obviously true that preferences do in fact change, it must be remembered that this strongest form of sovereignty for consumers is purely a methodological commitment having as its justification the expansion of the explanatory strength of economic theory; it does not have any implications whatsoever for the nature of preferences as psychological states of individuals or as tendencies to actually behave in specific ways in specific situations (Silberberg, 1978). Preferences, interpreted as both given and fixed, are, however useful, highly theoretical entities that cannot be regarded as anything more than hypothetical constructs. It is asserted, then, that preferences interpreted in this manner, however useful they are in unifying disparate elements of theory in economics and in simplifying aggregations, are not useful at all in creating an economics that is truly explanatory of behavior in the long run.

## *3.2. Degree 2: Preferences as Given*

A majority of economists adopt a somewhat weaker version of consumer sovereignty according to which preferences are assumed to be given and fixed in the methodological sense, but not necessarily in the ontological sense. By this is meant that, other things being equal, individual choice is the best available measure of what is good for a person. Economists find it reasonable to assume that individual preferences are fixed for the duration of the analysis or experiment. Preferences are aggregated from 'snapshots', not as dynamic processes. The analysis abstracts from the question of changing preferences because of the advantages afforded in the ability to aggregate. If preferences are given and fixed for the duration of the analysis, then they are not influenced by changes in other people's behavior (and preferences) and can easily be aggregated. This represents an acknowledged trade-off of reality for mathematical elegance and explanatory power. As noted in Section 1, analysis of policies to promote sustainable development must necessarily involve very long time horizons. Strict adherence to consumer sovereignty thus makes economic analysis less relevant to the evaluation of such policies.

Many economists who are advocates of methodologically supported consumer sovereignty would state (and have stated) that they of course recognize that preferences do in fact change, and that (speaking non-economically) the question of preference change is an important one. Their point is only that, for disciplinary comparative advantage, given economists' expertise in modeling and aggregation of welfare, they are choosing to define their disciplinary boundaries to exclude questions of preference change (Silberberg, 1978). Their argument, essentially, is that preferences are exogenous to economics, but endogenous to social science in the broader sense. On this view, it is logical for economists to seek interdisciplinary contact, to undertake interdisciplinary studies, and to seek broader theories of value that might unify their data and theory with data and theory from psychology, anthropology, and sociology. Consumer sovereignty in this form is simply a methodological assumption in economics and, so understood, it decisively supports our goal of addressing individual preferences in ways that incorporate the insights of economists who aggregate preferences-as-given into a broader understanding of value.

## *3.3. Degree 3: Consumer Sovereignty as Commitment to Democracy*

If most mainstream economists accept the importance of preference formation and reformation, then what is the issue? Why not just plunge into interdisciplinary dialogue, and see what implications for social science theory emerge? But mainstream economists have shown reluctance to engage in this broader dialogue. We believe, however, that this reluctance rests mainly on a misunderstanding. A third degree of consumer sovereignty admits that preferences change, takes given-ness as an expression of a methodological decision, but nevertheless expresses skepticism regarding the evaluation of preferences and attempts to change preferences in an explicit or systematic manner. This position, which is not inconsistent with degree 2, insists that there are dangers involved in evaluation and criticism of individual preferences. Such economists express fear that, if we set out to evaluate preferences, we have taken a giant step down the road toward paternalism, expertism, and perhaps even totalitarianism (Randall, 1988). According to this version of consumer sovereignty, it is recognized that individual preferences in fact change, but changes in preferences are highly individual, and nobody, not politicians, not philosophers, not social scientists, and certainly not environmental activists, is justified in telling individuals what their preferences should be. A commitment to democracy, and a rejection of any role for philosopher kings, scientific experts, or, especially, totalitarian manipulators of opinion, demands that preference formation be a highly individual, non-coercive process, according to this view. In this sense the individual consumer is sovereign, even as his or her preferences change, because the process of preference change is directed by the individual, rather than by an outside agent (this, of course, flies in the face of the fact that preferences are being manipulated by outside agents every day).

Note that this position has both methodological and evaluative elements. The argument is that the aggregation of revealed and expressed preferences should be interpreted methodologically, as given, because this strengthens the role of the public, taken as individuals whose preferences are sometimes stable and sometimes changing, and hence reinforces democratic trends in policy formation and implementation. Consumer sovereignty in this sense is a methodological stipulation in service of a commitment to the moral ideal of democracy. We feel great sympathy with the intent of this position, but we also think that it is essential to examine several of its underlying assumptions, because we doubt that, properly understood, a commitment to democracy requires rejection of rational analysis of individual preference change, or that a careful examination of reasons individuals should change their preferences must lead to elitism or totalitarianism.

Consider an example: imagine that we live in a nation, the vast majority of the population of which are members of a particular religious sect, but that we are members of the minority. Surely, a commitment to democracy demands that we respect these individuals' right to their own belief as an element of their right to freedom of belief and of speech. But what if one of the majority's beliefs becomes the basis of an onerous policy? Suppose, to take an extreme case, that it is firmly believed by a majority (the believers) that the world, as we know it, will end in a final Armageddon in exactly 10 years, and that this holocaust can be averted only by the sacrifice of the first-born of every nonbelieving family in the nation. While in this case we might, along with economists, be uncomfortable with 'expertism' and 'totalitarianism' in dismissing the majority's religious beliefs, and even if we would respect their right to their beliefs, we would also be opposed to taking the 'obligation' felt by the majority for human sacrifice as a given and the obviously (on their empirical hypothesis) welfare-maximizing policy of mass human sacrifice as a fait accompli.

This (hopefully implausible) scenario helps us to make a distinction between this third degree of sovereignty, expressed by economists who wish to guard the individual right to choose one's own preferences, and a fourth degree, which recognizes a stronger role for public discussion, expert input, and leadership, but does not run rough-shod over individual preferences. If we were a member of the nonbelieving minority in the fictional scenario, we surely would not accept the policy as a fait accompli, nor would we accept the current preferences of the majority for mass human sacrifice as a preventive for Armageddon. We would surely try, by scientific, rational, and whatever other means necessary, to establish that there exists no plausible scientific evidence to support either the hypothesis of coming cataclysm or the suggestion that human sacrifice would avert the cataclysm. Respecting the right of persons to their own beliefs and preferences does not preclude judgments that these beliefs lack sufficient support to justify policy actions, or that the public would be better off if the preferences of many of its citizens were to change.

### 3.4. Degree 4: Democratic Preference Change

It is now possible to address the crux of the issue regarding preference formation and reformation. How is it possible to respect individual self-determination of preferences and at the same time to address the possibility that sincerely felt preferences of many individuals in a society, if pursued as a public policy, will nevertheless be extremely detrimental to the public interest or to the rights of a minority? In particular, since our concern is with considering what would be necessary and permissible to promote a sustainability ethic, an ethic that attempts to articulate and defend the interests of generations yet unborn (and consequently unable to reveal or express their own preferences), it is asked whether it is possible in a democratic society to bring scientific, rational, moral arguments to bear on the question of whether some preference sets are more defensible than other preference sets.

It is possible to retain a commitment to democracy and to discuss the appropriateness of values because the democratic commitment is mainly procedural, while assertions of appropriateness are put forward as empirically and morally supportable theses regarding what our obligations to the future are. Hoping we will never face a situation so dire as to live in a society solemnly and with due legislative process committed to human sacrifice (as in our hypothetical example above), one hopes that policy will be set in a situation of open debate, with experts weighing in, and with interactions between the public, experts, and political decision makers. If a democratic process, including safeguards for individual rights of present people, is in place, then surely it makes sense to inject into the debate moral concerns about the well-being of future generations, even if these arguments require questioning and criticizing individuals' sincerely felt preferences.

We suspect, then, that a combination of commitments, to mathematical and methodological simplicity, and to democracy as noted, has inclined mainstream economists to favor consumer sovereignty of degree 1, 2, or 3 to degree 4. Once questions regarding the decision process are addressed, and allowing that in particular cases there may be real reasons to doubt the likelihood of predictions of dire consequences of current activities, there is in principle no reason why democratically inclined policy analysts cannot conclude that it would be better in the long run if certain current preferences of individuals were to be reconsidered and amended. Advocacy of such criticism and education programs to change preferences need not be coercive; criticism of particular preference sets, based on the implications for those preference sets for the welfare of under-represented individuals such as future persons, may rather be in the form of rational suasion, of pointing out to people the consequences of their desires, and of showing them alternative paths to personal satisfaction that have less severe impacts on the future of society.

We can also think of the democratic process as a multilevel one, rather than as flat and monolithic. Evidence that current behavior has negative impacts on other individuals, other species, or the future may require re-consideration of that behavior and the preferences that generate it. We can come to a democratic consensus about our shared preferences for a sustainable society through a process of discussion and debate, and then use these principles as guides to encourage people to see the inappropriateness of some preferences, given the scientifically demonstrable impacts of acting on those preferences (Fig. 14.1).

In the remainder of this paper, how preferences of individuals change through time, and the means available to criticize, and perhaps encourage change in, patterns of current individual preference, will be examined. Only after such a discussion will it be possible to assess the prospects for advocates of sustainability to advocate and effect changes in attitudes and preferences of citizens in a democratic society.

### 4. HOW PREFERENCES CHANGE

To summarize the arguments so far: preferences change and the questions are, how? why? and does any subgroup within society have any business consciously participating in the process? In this section we concentrate on how and why preferences change (since these seem to be inextricably interlinked). In the following section the more difficult question of society's role in changing preferences is addressed. To address the how and why questions, we refer to research from three areas: (1) psychology and economics, in particular recent research on revealed preferences and preference reversals (Tversky et al., 1988, 1990; Fischoff, 1991; Irwin et al., 1993; Knetch, 1994), constructed preferences (Gregory et al., 1993; Slovic, 1995), and decision making under uncertainty (Heiner, 1983); (2) social psychology and sociology, in particular research on social traps (Platt, 1973; Cross and Guyer, 1980); and (3) anthropology, especially research on coevolutionary adaptation of cultures and ecosystems, or ecological anthropology (Harris, 1979).

From these sources we conclude that preferences are formed in humans (and many other animal species) by selection acting on traits that are transmitted both genetically and (in the case of humans) culturally in a coevolutionary way. Preferences are but one more in a long list of traits that are 'formed' in this way. Preferences, as expressed, for example, in the economist's indifference map, are determined by the combination of forces that are often summarized under the headings of 'nature' and 'nurture'. That is to say, each human being enters life with a certain genetic makeup. Encoded there are certain basic needs and drives, including food, shelter, clothing, the sex drive, and the need for interaction with other people. These needs and drives are fundamental to our make-up as human beings. They are not now and may never be fully understood. Furthermore, they play themselves out in our daily lives in ways that are profoundly influenced by the social and natural systems we live in. Our preferences for food, for example, are influenced in some ways by our genetic make-up, but are also determined by our social groups (e.g. Jews and Muslims have an aversion to pork which they learn through their religious education) and by the alternative foods that their social and natural systems have in combination made available to them. Thus, over time, preferences may be affected by human genetic evolution, education, technological change, the evolution of social systems, and the changing availability of environmental and other natural resources.

A large and growing literature in cognitive psychology and related field highlights the 'lability' of preferences, which refers to the variability of expressions of preference under varied contextual conditions. The findings of this research challenge any interpretation of actual preferences as fixed or invariant, and therefore call into question economists' conceptual model of preference formation. Summarizing evidence that preferences are often "constructed, not merely revealed, in the elicitation process", Slovic (1995) says: "psychologists' claims that people do not behave according to the dictates of utility theory are particularly troubling to economists, whose theories assume that people are rational in the sense of having preferences that are complete and transitive and in the sense that they choose what they most prefer". While Slovic and his colleagues have mostly concentrated on preference variance under different conditions of elicitation, their conclusions, by inference, suggest that preferences are very unlikely to remain constant over time, and that it is misleading even to suggest that preferences exist independent of the occasions of their elicitation and/or expression in a specific context. Although again the economist need not consider the news to be bad news, it turns out that psychologists have already learned a great deal about how preferences are constructed. If economists simply accept that consumer sovereignty is only a methodological consequence of their chosen models, a whole new field of study opens up at the edge of economics and psychology, e.g., the study of how preferences vary according to context and how they change across time and circumstances.

Further understanding of preference formation can be gained by examining the role of preferences in the larger human decision process. To again quote Slovic (1995), actual human decision making is not an exercise in the rational calculation of utilities, but is better understood as a form of 'mental gymnastics' (a highly contingent form of information processing) and is sensitive to all sorts of contextual pressures. Empirical evidence from the social sciences thus supports our introduction of a two-tier and iterative decision model. Meanwhile, data regarding preferences as described in fixed-preference models can still provide snapshots of opinion, and this information provides one important input into the larger process of measuring, examining, and re-examining preferences and social values. What is gained by economists is many interdisciplinary opportunities to study preferences, their formation, and their relationship to behavior, and to embed this information in a more realistic model of decisions involving public values associated with the environment.

## 5. A COEVOLUTIONARY EXPLANATION FOR PREFERENCE FORMATION

In modeling the dynamics of complex systems it is impossible to ignore the discontinuities and surprises that often characterize these systems and the fact that they operate far from equilibrium in a state of constant adaptation to changing conditions (Lines, 1989; Holland and Miller, 1991; Kay, 1991; Rosser, 1991, 1992). The paradigm of evolution has been broadly applied to both ecological and economic systems (Boulding, 1981; Arthur, 1988; Lindgren, 1991; Maxwell and Costanza, 1993; Norgaard, 1994) as a way of formalizing understanding of adaptation and learning behaviors in non-equilibrium, dynamic, complex systems. The general evolutionary paradigm posits a mechanism for adaptation and learning in complex systems at any scale using three basic interacting processes: (1) information storage and transmission; (2) generation of new alternatives; and (3) selection of superior alternatives according to some performance criteria.

The evolutionary paradigm is different from the conventional optimization paradigm popular in economics in at least four important respects (Arthur, 1988): (1) evolution is path dependent, meaning that the detailed history and dynamics of the system are important; (2) evolution can achieve multiple equilibria; (3) there is no guarantee that optimal efficiency or any other optimal performance will be achieved, due in part to path dependence and sensitivity to perturbations; and (4) 'lock-in' (survival of the first rather than survival of the fittest) is possible under conditions of increasing returns. While, as Arthur (1988) notes, "conventional economic theory is built largely on the assumption of diminishing returns on the margin (local negative feedbacks)", life itself can be characterized as a positive feedback, self-reinforcing, autocatalytic process (Kay, 1991; Günther and Folke, 1997) and increasing returns, lock-in, path dependence, multiple equilibria and sub-optimal efficiency are the rule rather than the exception in economic and ecological systems and should be expected.

## 6. CULTURAL VERSUS GENETIC EVOLUTION

In biological evolution, the information storage medium is the gene, the generation of new alternatives is by sexual recombination or genetic mutation, and selection is performed by nature according to a criterion of 'fitness' based on reproductive success. The same process of change occurs in ecological, economic, and cultural systems, but the elements on which the process works are different (Toulmin, 1972). For example, in cultural evolution (1) the storage medium is the culture, e.g., the oral tradition, books, film or other storage media for passing on behavioral norms, and belief systems; (2) the generation of new alternatives is through innovation by individual members or groups in the culture; and (3) selection is again based on the reproductive success of the alternatives generated, but reproduction is carried out by the spread and copying of the behavior or ideas through learning and imitation rather than biological reproduction. One may also talk of 'economic' evolution, a subset of cultural evolution dealing with the generation, storage, and selection of alternative ways of producing things and allocating that which is produced. The field of 'evolutionary economics' has grown up in the last few decades based on these ideas (cf. Day and Groves, 1975; Day, 1989). Evolutionary theories in economics have already been successfully applied to problems of technical change, to the development of new institutions, and to the evolution of means of payment.

For large, slow-growing animals like humans, genetic evolution has a built-in bias towards the long run. Changing the genetic structure of a species requires that characteristics (phenotypes) be selected and accumulated by differential reproductive success. Behaviors learned or acquired during the lifetime of an individual cannot be passed on genetically. Genetic evolution in humans is therefore a relatively slow process requiring many generations to significantly alter the species' physical and biological characteristics.

Cultural evolution is potentially much faster. Technical change is perhaps the most important and fastest-evolving cultural process. Learned behaviors that are successful, at least in the short term, can be almost immediately spread to other members of the culture and passed on in the oral, written, or video record. The increased speed of adaptation that this process allows has been largely responsible for the amazing success of *Homo sapiens* at appropriating the resources of the planet. Vitousek et al. (1986) estimate that humans now directly control 40% of the planet's terrestrial primary production, and this is beginning to have significant effects on the biosphere, including changes in global climate and on the planet's protective ozone shield.

Thus, while the benefits of this rapid cultural evolution are significant, the costs are also potentially significant. Like a car that has increased speed, humans are in more danger of running off the road or over a cliff. Cultural evolution lacks the built-in long-run bias of genetic evolution and is susceptible to being led by its hyper-efficient short-run adaptability over a cliff into the abyss.

Another major difference between cultural and genetic evolution may serve as a countervailing force, however. As Arrow (1962) has pointed out, cultural and economic evolution, unlike genetic evolution, can at least to some extent employ foresight. If society can see the cliff, perhaps it can be avoided.

While market forces drive adaptive mechanisms (Kaitala and Pohjola, 1988), the systems which evolve are not necessarily optimal, so the question remains as to what external influences are needed and when they should be applied in order to achieve an optimum economic system via evolutionary adaptation. The challenge is to first gain foresight, and then to respond to and manage the system feedbacks in such a way as to avoid any foreseen problems (Berkes and Folke, 1994). Devising policy instruments and identifying incentives that can translate this foresight into effective modifications of the short-run evolutionary dynamics is the challenge (Costanza, 1987).

One of the possible modifications is to the preferences of individuals, which drive short-term dynamics in economic systems. Individual preferences can have a huge impact on ecological resources. In one particularly dramatic example, the rapid spread of popularity of New Orleans Chef Paul Prudhomme's blackened redfish dish caused a rapid expansion in demand and threatened to destroy the redfish fishery in the Gulf of Mexico. Less dramatically, but perhaps more importantly, the preference of consumers for pre-packaging and small, individual containers has a very large impact on landfills throughout the United States; similarly, a taste for expanses of green lawns in the suburbs affects water quality and water availability for natural systems in arid areas. Looked at physically, these are problems of scale, e.g., human impacts on natural systems increase as a function of population and consumption, with the volume and type of consumption being a function of preferences. Market forces, supplemented with concerted attempts to internalize environmental costs, can, of course, have an effect on consumption. If, for example, the full costs of irrigated lawns, including damage to wildlife, stream-water quality, etc., were paid by customers, they might turn off the spigot and 'suffer' with a brown lawn. The difference between charging full costs and changing preferences is that in the former case, consumers end up feeling deprived and unhappy, whereas they may feel enlightened and happy after being educated about the joys of a xeriscaped lawn. Given that populations in many areas of the world will continue to increase for at least decades, any attempt to address the problem of the scale of an economy vis-a-vis its limiting ecological factors must be addressed through reducing the impacts of per capita material consumption.

Fortunately, once the possibility of encouraging more appropriate preferences is introduced, there need be no necessary link between impacts of consumption on the environment and the levels of welfare experienced in a society. The good news is that, in evolutionary modeling, it may be possible to make small social investments that will affect which types of consumption bring enjoyment to consumers, reducing the scale of human impacts without decreasing, perhaps even increasing, levels of welfare of consumers. To continue with the xeriscaping example, a desert city with a shrinking aquifer might rationally consider a program to increase the number of residents who enjoy xeriscaping and use of native plants. They might, for example, establish a center for xeriscaping with a botanical garden of native plants, encourage courses on xeriscaping as a part of the agricultural extension service, or help to establish a program in xeriscaping in the state university's landscape architecture department. These might be wise public investments that realize increasing returns as 'trends' are created toward lower-impact consumption patterns as a result of neighbors teaching neighbors the beauty and enjoyment, and lower maintenance costs, of native-plant landscaping. Once xeriscaping becomes an element of the community's identity, and citizens encourage a change in the tastes of their neighbors, a trend toward less water use and more native habitat might build on itself, providing increasing returns on a small investment. Investments such as this could pay increasing returns in lowering per capita demand for scarce resources and buffer the economy against shortages and rising prices.

And who is to say, to the gardener proud of his or her xeriscaping and its appropriateness to the surroundings, that a low-impact yard/garden yields less satisfaction than the currently prized lawns or tropical gardens with plants from rainforests and other ecologically inappropriate sources? But here, again, concerns about manipulation are encountered. The response, once again, is that there is no inconsistency with democracy if the goals chosen to guide preference reformation were arrived at through a democratic process such as a well-run ecosystem management plan, a community project on xeriscaping, or some other process that includes public input and free exchange of information.

## 7. PROSPECTIVE: VALUES AND THE FUTURE

In the last part, we emphasized that preferences, especially preferences that affect the community and long-term public goods such as environmental protection, are best not thought of as fixed, or even necessarily stable, over years, decades, or generations. Community processes that encourage articulation of values and associated management goals therefore may need to be iterative and political in nature (Fig. 14.1). Snapshot views of individual preferences understood as market behavior or shadow prices can yield important aggregations that are useful in analyses of policies with predictably limited long-term impacts. But we have advocated an analysis of policies, especially policies with long-term or irreversible impacts, that allows for a process of public articulation, discussion, and evaluation of public values. It has been shown above that there are interesting research and policy questions surrounding public processes of value articulation and management participation. The question of whether some of the currently felt preferences of individuals in industrial and post-industrial societies would be better changed is now addressed. In other words, is it possible to develop, within democratic institutions, a set of processes that would encourage the reconsideration and reformation of the preference sets of individuals? Again, a question that makes economists very concerned, and not without reason, as noted above, is faced. So far these concerns have been lumped under several types, e.g., concerns about 'manipulation' of preferences, about totalitarianism, etc., but what all of these concerns have in common is a suspicion that 'paternalism' regarding preferences is itself unacceptable. In this part, it is analyzed in more detail whether, and under what circumstances, a judgmental attitude regarding preferences is appropriate, and whether a certain form of paternalism might represent an important tool for developing and implementing sustainable institutions and policies.

As noted above, economists have generally favored self-determination of preferences because they see free preference formation as an important element in democracy. Solow (1993), for example, dismisses any attempt to affect the preferences of the future in one sentence (the preferences expressed by members of future generations are 'none of our business'). But we think that, if we leave theoretical assumptions and definitions of economics aside, most of us would admit that there are some situations in which attempts to shape future preferences both make sense and are defensible. For example, consider wilderness preservation, historic preservation, and other attempts to protect important features of our landscape. Solow acknowledges that our generation may want to set aside particular 'places' or scenes because they are intrinsically valuable to us because of their beauty, historical significance, etc. But decisions such as this are more complex than the analysis suggests because if we assume we know nothing, and care nothing, about what future generations will want, we would be unlikely to invest in preservation. For example, would our generation be willing to set aside wilderness, or to forgo consumption that threatens species such as whales, if we believed that the next generation will not value these things for themselves and would destroy them whenever it is profitable? No, the decision to invest in protecting a wilderness area carries with it a commitment, or at least a desire, to influence the future to continue to value wild places and naturally evolved species. While we may share Solow's distaste for some forms of paternalism, it also seems inevitable that huge investments to save natural or historical landmarks must be viewed as part of a cultural dynamic in which the choices of one generation affect both the choices available to individuals in the future (maintaining or expanding their options) and the set of preferences that will be expressed by individuals in the future. In a very important sense, the preferences of future generations cannot be independent of current preferences. We pass on preferences to our children and we must in some cases decide which preferences to pass on in a very literal, paternalistic sense.

Consider again the example of wilderness protection. Suppose we choose in the present to alter all wilderness areas and use them for commodity production and apply a multiple use policy everywhere. Would future people retain or develop a taste for wilderness hiking, would they experience 'existence values' for such areas (values attributed to feelings that, even if the wilderness area is not visited or used, express a preference that it continue to exist)? Or would they never miss these lost opportunities to value wild things and, perhaps, be happy with theme parks and virtual reality wildernesses? The laissez-faire attitude of Solow (1993) toward future preferences now seems disingenuous. Either future individuals will desire to experience real wilderness or they will not. If they do value it and wish to experience it, but cannot because we developed it, they would have reason to fault us for our choice. In this case it is at least important that we ascertain to the extent possible what preferences the future will in fact have so that we can take this into account in our computations of intergenerational fairness and in related decisions (Bromley, 1998). However, consider the case in which all wilderness is compromised irreversibly, and the more likely outcome that all future people have as a result lost the ability to value wilderness. In this case, the current decisions in fact have had a huge impact on the future and its range of options for enjoyment, whether future people recognize this impact or not. Solow's remark that the preferences expressed by members of future generations are 'none of our business' seems in either case to ignore issues of considerable importance and complexity, and to naively favor policies likely to reduce cultural continuity and social meaning over time.

We suspect that Solow has arrived at his attitude because of a professional commitment to 'positive' economics and a form of methodological individualism involving a commitment to analyze decisions as faced by individualistic 'homo economicus'. His viewpoint on the intergenerational dynamic of preference formation might change if he were to treat at least some decisions from the viewpoint of the community. While the search for cultural continuity may become more and more difficult in a highly mobile society, it may still be important to build a sense of shared community values, both for the fulfillment of the aspirations of present people and also for the options available to members of future generations. The geographer Tuan (1974, 1977) has developed a powerful case that all people exhibit 'topophilia', a commitment to a given geographical place, even if that place is seen as economically challenging and as a worse choice than other places to live by usual economic criteria.

Let us return to the case of attitudes toward, and preferences regarding, smoking. Since scientific evidence is now overwhelming that smoking increases the risk of morbidity and mortality, and that smoking-caused illnesses have large social as well as personal consequences, this is an area where the Surgeon General's office accepts a public duty to affect behavior. Leaving aside for the moment that smoking is addictive, it still seems rational for a society to take steps to discourage young people from acquiring a preference for smoking cigarettes. If one believed, analogously, that individual preferences for more consumable goods have high social costs, it would by analogy make sense to invest society's resources in discouraging such preferences. The economist, at this point, might invoke a distinction between presenting consumers information regarding risks and influencing the values or preferences of individuals, and then argue that presenting information is appropriate, but that neither the government nor anybody else should attempt to influence preferences. According to this version of consumer sovereignty, preferences are sacrosanct and, while the government can justifiably dispense information, it ought to leave the 'final decision' regarding what to purchase and use to the consumer. The problem is that such an insulation of information from influence on preferences is impossible in practice. The choice of information dispensed must be designed to be effective in changing behavior – it would be silly and counter-productive if the Surgeon General were required only to present information in a way that would never affect youngsters' preferences, but only give them information about what might happen if they act on those preferences. On the contrary, a successful campaign to reduce the number of teenagers who smoke must, ultimately, mark its success in the loss of, or failure to develop, a taste for smoking. Even if a public service announcement never mentions, in addition to the fact that smoking causes cancer, that 'smoking isn't cool, anyway', the factual content of the public service message is nevertheless chosen for presentation because of its likely direct and indirect impacts on the future preferences of prospective smokers. The idea of a public campaign to reduce smoking among teenagers is necessarily, even if implicitly, expressive of a commitment to alter consumption patterns in service of the social goal of having a healthy population.

By analogy, a public campaign to reduce lawn watering in desert areas, based on scientific information regarding the negative impacts of irrigated lawns on the aquifer and surface water quality, and praising xeriscaping as a satisfying option, cannot be represented as simply an 'information campaign'. The fact that it is undertaken as a 'campaign' to change behavior already belies value neutrality; the insistence on presenting facts only, and leaving the decision up to consumers, cannot alter the value-ladenness of the enterprise. Nor can such a campaign proceed intelligently without adopting the goal of affecting the preferences that are finally held by individuals in the society in the future. So we are led back to the empirical question, beyond the scope of this paper, of whether the preference for irrigated lawns in the desert is, like smoking, a preference that is sufficiently destructive of public goods to warrant 'information' campaigns to alter those preferences.

It is suspected that the case for a suggested campaign to reduce irrigated lawns seems weaker than the justification for an anti-teenage-smoking campaign because, in the latter case, health risks to the individual are involved, whereas in the case of lawns, the impacts are on more diffuse public goods such as clean and plentiful water. However, this difference is actually irrelevant in this case. To see this, suppose that it is scientifically established that second-hand smoke is actually worse for nonsmokers than first-hand smoke is for smokers. This change in the understanding of the facts would strengthen, not weaken, the case for a campaign to change the behavior and preferences of smokers. The point of these examples, then, is that there are possible cases in the area of public health, and in the area of protection of environmental public goods such as clean water, in which there could be a legitimate public interest in affecting individual behavior and the preferences associated with it. In such cases, a democratic process could lead to a legitimate public-spirited decision to alter, through information and rational suasion, individual preferences and behaviors in the service of a social good.

Having argued that such situations are possible, even plausible, we must hasten to add that, sharing economists' worries about the potential dangers of preference manipulation, we believe claims that the public interest demands 'campaigns' of this sort must be submitted to the most disciplined analysis and that they must be considered carefully on their merits. In many cases, we suspect that calls for such public campaigns to affect public values and change individual behaviors in the service of a social agenda may be thinly veiled attempts to manipulate opinion in the service of narrower-than-public interests. Nor is there a foolproof way of separating justified cases of public examination of values and unjustified ones. Having said this, however, it is nevertheless recognized that a major part of the sustainability platform rests on the existence of such cases and we therefore proceed to discuss how these cases might be addressed, provided it is established that preference change is truly in the public interest.

In what remains, we will continue to elaborate a general model for understanding situations such as these in an environmental policy context. This model makes value formation and reformation an endogenous element in the search for a rational policy for managing the impacts of human economic activities on the ecological and physical systems of nature (Fig. 14.1). The model will appear rather simple, abstract, and conceptual, but this generality signifies how much of the specification of a sustainability metric must be accomplished on a local, place-based level. The goal is merely to provide the general outlines of a sustainability policy that fulfills a minimal moral requirement of intergenerational equity. The details of such a policy must in all cases be worked out from a local perspective, with special attention to details that make local places distinctive, and with local issues that seem very important from a given perspective in space and time.

The model proposed is more comprehensive than either economic or ecological models of the relationship, but it should not be thought that it merely places economics and ecology side by side in an attempt to integrate normative elements and descriptive elements in an iterative process. The model is also more comprehensive in that it is an action-based model that includes economic models and ecological models in a larger, iterative system of monitoring, analysis, and action, followed by continued monitoring, etc. This larger system operates by assuming 'working hypotheses' relating social values with ecophysical processes, states goals, and engages in experiments and pilot projects in pursuit of those goals, monitors progress toward those goals scientifically, and then factors scientific results back into an ongoing public process. The importnat point is that preferences, values, and goals are open to revision, just as scientific hypotheses are. Economic models represent large subsystems that are embedded in larger-scale models of ecological and physical systems, and this structure is embodied in the two tiers of our model. Economic behavior is modeled in an equilibrium system, while ecological models, which encompass multiple equilibrium points, apply at larger scales (Common and Perrings, 1992; Holling, 1995). A two-tier system of analysis therefore sorts possible environmental problems and risks according to the likely temporal and spatial scale of their impacts in the first tier, and applies an appropriate action criterion, such as a cost-benefit criterion or a safe minimum standard criterion, given the scope and scale of possible risks of a policy in the second tier.

The two levels are interactive in social processes because we hypothesize that certain features of ecological communities support various social values, and we invite the public to specify goals for environmental management. Scientists, preferably local scientists, should also be involved to encourage the development of realistic, observable measures that can be hypothesized to track values that have been articulated by the community. Once goals, however tentative and difficult to measure, are set, the iterative process can be begun. It is expected to result in improved approximations of sustainability, because each experiment will be designed as a 'pilot' solution to a particular local problem, and also designed so as to increase our knowledge in some way that improves future management, and our ability to measure its successes and failures.

If we are correct that re-examination of individual preferences will be an important part of any model that represents a socio-physical dialectic capable of attaining sustainable institutions and sustainable policies, its role in a general model such as this would be represented as the interplay of scientific findings with the ongoing social dynamic as particular communities attempt to specify social goals, enlist scientists in an 'epistemological community', and set about a process of scientific inquiry and social learning (Lee, 1993). This process could submit policies to rigorous re-examination both with regard to progress toward stated goals and with regard to the 'appropriateness' of preferences under various models.

Unlike most models for evaluating environmental policies, our conceptual model embeds both economic and ecological models in a larger social process. The first step in that process, however, is political, not scientific. It is necessary for the various elements of a community, perhaps through representatives of stakeholder groups prominent in the community, to propose and discuss various visions, or scenarios, that they would set as positive outcomes of a process of economic development over generations. An important part of this will be the ranking of risks and attempts to set some priorities in addressing risk problems. But comparative risk processes are not as important as public discussions of their positive, long-term aspirations for their region. As Sagoff (1994) has noted, development of a deep sense of place has been interrupted by the tremendous mobility of populations in the United States. It is suggested that using ecosystem management plans and other public processes to build concern and responsibility for resources, and that one important role of public agencies and private environmental groups is to build a place-based sense of responsibility for sustainable management. Pilot projects and management experiments can play an important role in these public political processes, and the articulation and questioning of social values must be an important part of them.

One advantage of this approach is that, ideally, it may be possible through experimentation and scientific testing to find policies that have positive impacts on both the short-term economic dynamics and the longer-term social and ecological dynamics that affect longer-term goals. For example, tree-planting programs in deforested areas may contribute to local economic goals (by reducing fuel wood shortages as planted areas are pruned), while simultaneously reducing erosion and improving stream water quality, and even contributing to slowing global climate change. While such policies cannot cure all ills, the important message is that, with an experimental spirit and involvement of a committed public, it may be possible to encourage development that is consistent with longer-term, as well as shorter-term, criteria. If such a positive development were to begin, it may be possible to intertwine the processes of monitoring and measuring impacts of human economic and protective efforts on physical systems with the more social process of developing a nature-based community identity in various regions. Such a system will not optimize one variable on any particular scale, but it will seek policies that are robust and effective on many scales. It will seek them by building community support at multiple levels, and by joining in a cooperative venture with local physical and social scientists to describe and evaluate both the means to the goals and the goals themselves.

## 8. CONCLUSIONS

The search for new, sustainable policies must, in addition to finding ways to internalize market externalities in economic activity, also address the question of the over-all scale of economic activity. It is difficult to see how this question can be addressed within a value system, such as that exemplified in neoclassical economics, that (a) does not allow the rational questioning of individual's value sets and also (b) makes the assumption that, whatever people will prefer, their desires are unlimited. If individual preferences change (in response to education, advertising, peer pressure, etc.) then value cannot completely originate with preferences. We need to distinguish at least two kinds of value within this context: (1) short-term or current value based on current individual preferences and (2) long-term or sustainable value that emerges from a community process and encourages preferences that promote long-term sustainability (sustainable scale, fair distribution, and efficient allocation). Instead of being merely an expression of current individual preferences, sustainable value (at least in the mid- to longterm) becomes a system characteristic relating the item's evolutionary contribution to the survival of the linked ecological economic system. Current value is the expression of individual preferences in the short term and locally, while sustainable value is the expression of community preferences in the long-term and globally. Achievement of clearly articulated and integenerationally equitable goals that are in fact the expression of values experienced in local communities, their aspirations as well as their preferences, will require an iterative, democratic, and public process in which those communities develop goals and community values that are valid expressions of their ongoing culture. In this paper it has been argued that it is a legitimate activity of the policy community to encourage and participate in such an iterative process, which will require development of analytic tools that go beyond registering and aggregating currently expressed preferences. This change would bring policy analysis and discussion more into line with the constructivist approach to values and environmental valuation discussed above.

Preference change can be thought of in this context as an alternative to price change. Both influence behavior, and both are subject to imperfections. We may wish to (and need to) influence both prices and preferences in order to achieve our long-term social goals. To go back to our example of lawn watering in arid regions, both higher prices on water and public campaigns to encourage preferences for xeriscaping are potentially important ways to change behavior. An important distinction between these two policies is that reduced water consumption due to higher prices would lead to people feeling worse off, while reduced water consumption due to changed preferences would lead to people feeling better off as they experience pride at behaving in a more environmentally appropriate way. Thus, reduced consumption can lead to either a decrease (through price increases) or an increase (through preference change) in welfare, depending on the method by which we achieve the desired result.

Actively seeking to influence preferences is not inconsistent with a democratic society. Quite the contrary; in order to operationalize real democracy, a two-tiered decision structure must be used (Fig. 14.1). This is necessary in order to eliminate 'preference inconsistencies' between the short term and the long term and between local and global goals, a phenomenon described in the social psychology literature as a 'social trap' (Platt, 1973; Cross and Guyer, 1980). There must first be general, democratic consensus on the broad, long-term goals of society. At this level 'individual sovereignty' holds, in the sense that the rights and goals of all individuals in society must be taken into account, but in the context of a shared dialogue aimed at achieving broad consensus. Once these broad goals are democratically arrived at, they can be used to limit and direct preferences at lower levels. For example, once there is general consensus on the goal of sustainability, with agreement by all the major stakeholders in society, then society is justified in taking action to change local behaviors that are inconsistent with this goal. It may be justified, for example, to attempt to change either people's preferences for driving automobiles or the price of doing so (or both) in order to change behavior to be more consistent with the longer-term sustainability goals. In this way the foresight that we do possess in order to modify short-term cultural evolutionary forces toward achieving our shared long-term goals is utilized. If economics and other social sciences are to adequately address problems of sustainability, it will be necessary to develop evolutionary models that make preference formation and reformation an endogenous part of the analysis, and to develop mechanisms to modify short-term cultural evolutionary forces in the direction of long-term sustainability goals.

ACKNOWLEDGMENTS. This paper was originally presented at the SCASSS workshop on economics, ethics, and the environment, Upsalla, Sweden, 25–27 August 1995. We thank Carl Folke, Ann Bostrom, Michael Farmer, and two anonymous reviewers for helpful comments on earlier drafts, and the Beijer International Institute for Ecological Economics for support during the preparation of this manuscript.

NOTE

1.  Although there is evidence that some environmental impacts can decrease with increasing economic activity beyond a threshold, this effect is limited and does not outweigh the larger trend (Arrow et al., 1995).

REFERENCES

Arrow, K., 1962. The economic implications of learning by doing. *Rev. Econ. Stud.* 29, 155–173.

Arrow, K., Bolin, B., Costanza, R., Dasgupta, P., Folke, C., Holling, C. S., Jansson, B.-O., Levin, S., Maler, K.-G., Perrings, C., Pimentel, D., 1995. Economic growth, carrying capacity, and the environment. *Science* 268, 520–521.

Arthur, W. B., 1988. Self-reinforcing mechanisms in economics. In: Anderson, P. W., Arrow, K. J., Pines, D. (Eds.), *The Economy as an Evolving Complex System.* Addison Wesley, Redwood City, CA, pp. 9–31.

Berkes, F., Folke, C., 1994. Investing in cultural capital for sustainable use of natural capital. In: Jansson, A. M., Hammer, M., Folke, C., Costanza, R. (Eds.), *Investing in Natural Capital: The Ecological Economics Approach to Sustainability.* Island Press, Washington, DC, pp. 128–149.

Bishop, R. C., 1993. Economic efficiency, sustainability, and biodiversity. *Ambio* 22, 69–73.

Boulding, K. E., 1981. *Evolutionary Economics.* Sage, Beverly Hills, CA.

Bromley, D. 1998. Searching for sustainability: The poverty of spontaneous order *Ecol. Econ.* 24, 231–240.

Common, M., Perrings, C., 1992. Towards an ecological economics of sustainability. *Ecol. Econ.* 6, 7–34.

Costanza, R., 1987. Social traps and environmental policy. *BioScience* 37, 407–412.

Cross, J. G., Guyer, M. J., 1980. *Social Traps.* University of Michigan Press, Ann Arbor.

Daly, H. E., 1990. Toward some operational principles of sustainable development. *Ecol. Econ.* 2, 1–6.

Day, R. H., 1989. Dynamical Systems, Adaptation and Economic Evolution. MRG Working Paper No. M8908, University of Southern California, Los Angeles.

Day, R. H., Groves, T. (Eds.), 1975. Adaptive Economic Models. Academic Press, New York.

Feigl, H., 1952. Validation and vindication: an analysis of the nature of ethical arguments. In: Sellars, W., Hospers, J. (Eds.), *Readings in Ethical Theory.* Appleton-Century-Crofts, New York.

Fischoff, B., 1991. Value elicitation: is there anything in there? *Am. Psychol.* 46, 835–847.

Gregory, R., Lichtenstein, S., Slovic, P., 1993. Valuing environmental resources: a constructive approach. *J. Risk Uncertain.* 7, 177–197.

Günther, F., Folke, C., 1997. Characteristics of nested living systems. *J. Biol. Syst.* (in press).

Harris, M., 1979. *Cultural Materialism: The Struggle for a Science of Culture.* Random House, New York.

Heiner, R. A., 1983. The origin of predictable behavior. *Am. Econ. Rev.* 75, 565–601.

Holland, J. H., Miller, J. H., 1991. Artificial adaptive agents in economic theory. *Am. Econ. Rev.* 81, 365–370.

Holling, C. S., 1995. Engineering resilience versus ecological resilience. In: Schulze, P. C. (Ed.), *Engineering within Ecological Constraints.* National Academy Press, Washington, DC, pp. 31–43.

Hume, D., 1888 (1776). *A Treatise on Human Nature.* Selby-Bigge, L. A. (Ed.). Clarendon Press, Oxford.

Irwin, J. R., Slovic, P., Lichtenstein, S., McClelland, G. H., 1993. Preference reversals and the measurement of environmental values. *J. Risk Uncertain.* 6, 5–18.

Kaitala, V., Pohjola, M., 1988. Optimal recovery of a shared resource stock: a differential game model with efficient memory equilibria. *Nat. Resour. Model.* 3, 91–119.

Kay, J. J., 1991. A nonequilibrium thermodynamic framework for discussing ecosystem integrity. *Environ. Manag.* 15, 483–495.

Knetch, J. L., 1994. Environmental valuation: some problems of wrong questions and misleading answers. *Environ. Values* 3, 351–368.

Lee, K., 1993. *Compass and Gyroscope.* Island Press, Covelo, CA.

Lindgren, K., 1991. Evolutionary phenomena in simple dynamics. In: Langton, C. G., Taylor, C., Farmer, J. D., Rasmussen, S. (Eds.). *Artificial Life, SFI Studies in the Sciences of Complexity*, vol. X. Addison–Wesley, Reading, MA, pp. 295–312.

Lines, M., 1989. Environmental noise and nonlinear models: a simple macroeconomic example. *Econ. Notes* 19, 376–394.

Maxwell, T., Costanza, R., 1993. An approach to modelling the dynamics of evolutionary self-organization. *Ecol. Model.* 69, 149–161.

Norgaard, R. B., 1994. *Development Betrayed: The End of Progress and a Coevolutionary Revisioning of the Future.* Routledge, London.

Norton, B. G., 1991. Thoreau's insect analogies: or, why environmentalists hate mainstream economists. *Environ. Ethics* 13, 235–251.

Page, T., 1977. *Conservation and Economic Efficiency.* Johns Hopkins University Press, Baltimore, MD.

Platt, J., 1973. Social traps. *Am. Psychol.* 28, 642–651.

Randall, A., 1988. What mainstream economists have to say about the value of biodiversity. In: Wilson, E. O. (Ed.), *Biodiversity.* National Academy Press, Washington, DC.

Rosser, J. B., 1991. *From Catastrophe to Chaos: A General Theory of Economic Discontinuities.* Kluwer, Amsterdam.

_____1992. The dialogue between the economic and ecologic theories of evolution. *J. Econ. Behav. Organ.* 17, 195–215.

Sagoff, M., 1994. Settling America: the concept of place in environmental ethics. *J. Energy, Nat. Resour. Environ. Law* 12, 351–418.

_____1998. Aggregation and deliberation in valuing environmental public goods: a look beyond contingent pricing. *Ecol. Econ.* 24, 213–230.

Silberberg, E., 1978. *The Structure of Economics: A Mathematical Analysis.* McGraw-Hill, New York.

Slovic, P., 1995. The construction of preference. *Am. Psychol.* 50, 364–371.

Solow, R. M., 1993. Sustainability: an economist's perspective. In: Dorfman, R., Dorfman, N. (Eds.), *Economics of the Environment: Selected Readings.* W. W. Norton and Company, New York, pp. 179–187.

Stigler, G. J., Becker, G. S., 1977. De Gustibus Non Est Disputandum. *Am. Econ. Rev.* 67, 76–90.

Toulmin, S. J., 1972. *Human Understanding*, vol. I. Princeton University Press, Princeton, NJ.

Tuan, Y.-F., 1974. *Topophilia: A Study of Environmental Perception, Attitudes, and Values.* Prentice-Hall, Englewood Cliffs, NJ.

_____1977. *Space and Place: The Perspective of Experience.* University of Minnesota Press, Minneapolis.

Tversky, A., Sattath, S., Slovic, P., 1988. Contingent weighting in judgement and choice. *Psychol. Rev.* 85, 371–384.

Tversky, A., Slovic, P., Kahneman, D., 1990. The causes of preference reversal. *Am. Econ. Rev.* 80, 204–217.

Vitousek, P. M., Ehrlich, P. R., Ehrlich, A. H., Matson, P. A., 1986. Human appropriation of the products of photosynthesis. *BioScience* 36, 368–373.

Williams, B. A. O., 1985. *Ethics and the Limits of Philosophy.* Harvard University Press, Cambridge, MA.

# IV

## Scaling Sustainability

### Ecology as if Humans Mattered

While ecologists number many devoted conservation activists among their ranks – and many of these have contributed mightily to environmental causes – I have sensed a reluctance among ecological scientists to become too deeply involved in ethical issues or even to discuss, at least professionally, environmental values. Accordingly, I have tried to establish a dialogue with ecologists about how to make the evaluation of environmental goods, especially ecologically based goods, more ecologically sensitive. This effort is represented here by a series of papers on how to understand social values and associated political processes within a natural ecological context.

The related themes of scales and social values run throughout the papers – written for ecologists or dealing with ecological themes – that appear in this part. First, I argue that we need to rethink the boundaries of scientific and evaluative discourse because sciences such as conservation biology are, inevitably and unabashedly, mission oriented; they, like medicine, are normative sciences. Besides needing to abide by the methodological norms of their discipline, physicians – and conservation biologists – are obligated to seek relevant and important information. "Importance," here, means social importance, so a science limited to mere description cannot tell us what is important to study. So, science must be value-laden at least to the extent that it can be responsive to demands for data and information needed to manage ecological systems. A third, related theme is my assertion that scale issues, as understood in ecological science, may provide a starting point for developing a new approach to evaluating the impacts of environmental policies as they unfold at multiple levels and scales. Ecologists, it turns out, have developed an important theoretical tool – hierarchy theory – that shows promise for enlightening deliberation about social goals and objectives as well as improving our understanding of natural dynamics.

I first broached the subject of ecological scale and environmental policy goals in a paper exploring Leopold's interest in temporal scale, as evidenced by his brilliant simile, "thinking like a mountain." I argued that, minus the label, Leopold implicitly embodied elements of hierarchy theory – which was not named or formalized until the early 1970s – in his model of environmental management. Encouraged by this historical antecedent, and favored with a grant from the Ethics and Values in Science Program of the National Science Foundation, I engaged the thoughtful ecological modeler Robert Ulanowicz in a project ("Scale and Biodiversity: A Hierarchical Approach") to see how we might link reasoning about social goals to the structure of ecosystems by articulating value questions within a hierarchically shaped social decision space. In this space, environmental problems – and the threatened social goals that define these as problems – present themselves as associated with different dynamics that can in turn be associated with various time-sensitive social values. Once so presented, the goals of environmental management sciences – to protect biological diversity, for example – can help to identify management objectives by monitoring the natural system on the scale appropriate for the generation of the valued outcome – protecting biological diversity, in the case in point. This line of reasoning also links my work on ecological scale to work on the philosophy of economics, as discussed in the last part of this book; once one has a means to "scale" environmental policies in terms of their impacts at multiple levels of time and space, hierarchical modeling also invites us to operationalize limits on the application of present-based systems of values. By applying hierarchy theory at the stage where one decides which impacts to monitor and how serious to consider risks that such impacts negatively affect the future, it is possible to use hierarchical organization to begin to operationalize a multiscalar conception of human valuation. In "Ecological Integrity and Social Values: At What Scale?" I explore in more detail this connection between various scales in natural systems and the temporal and spatial horizons of values associated with protecting those systems.

The problem of scale is further explored in a paper on "Change, Constancy, and Creativity," where it is noted that recognition of the prominence of change in natural systems has led some ecologists to despair of setting any standards for good ecological management. In response, I point out that, indeed, the problem of understanding stability in a sea of change and, conversely, understanding how change takes place even in "stable" systems, has bedeviled philosophers and natural scientists since long before Plato. But hierarchy theory is helping to tame this problem as it recognizes multiple scales and dynamics in nature, providing us with a concept of "relative stability" – stability that can be measured against a backdrop of slower-changing environmental

variables. With the help of careful reasoning about social goals, we can identify those physical processes that are most useful to monitor and protect given
social goals. And, if it is possible to identify dynamics that are closely associated with processes necessary to maintain an important social value, it may
be possible to forge a connection between ecological processes and social
policy goals.

Returning, with the help of Bruce Hannon, to the theme of political and
economic aspects of scale-sensitivity I next explored the nature of sense
of place values in the paper "Democracy and Sense of Place Values in
Environmental Policy." Attempting to apply consistently the idea that many
environmental values, as experienced, are highly dependent on local contexts,
and that democracy apparently implies respect for the desires of those affected
by a decision, we examine whether values present in local contexts are likely
to be captured in economic measures of value, such as market prices of land.
We argue that market prices of land must systematically underestimate the
actual value of land because sale prices cannot include value accruing from its
sometimes being valued by a seller as "home." For the buyer, the land remains
a commodity because the buyer has not yet learned to value the land through
habitation. We use these arguments to conclude that we should place more
weight on input from local users in identifying environmental values and in
setting conservation goals. If I were writing this paper today, I would have
more to say about how to reconcile local values and goals with broader and
larger-scale environmental goals. Nevertheless, I stand by our claim that, in
some fundamental sense, the responsibility for local environmental problems
must rest on local commitments, even though it is clear that competing local
commitments and clashes with regional goals will demand coordination and
integration with larger-scale policies in many cases.

Taken as a whole, the papers in this part seek to relate the human scales on
which human values are experienced with the physical dynamics that might
be studied by ecologists. Further, the papers ask whether, having made such a
connection, ecology and value studies could begin to illuminate one another.
The ideas developed here range from the speculative – that hierarchical modeling could enhance our understanding of how we value natural systems – to
the didactic – as I urge ecologists to address certain policy-relevant questions.
All of the papers in this part, however, are designed to encourage a dialogue
with ecologists and to further the integration of ecological science into a new
approach to evaluating anthropogenic environmental changes.

# 15

## Context and Hierarchy in Aldo Leopold's Theory of Environmental Management

Perhaps the most pervasive trend in environmental management today is a movement toward holism. As one example. I quote from a U.S. Environmental Protection Agency document on the protection of Chesapeake Bay: "The Bay is, in many ways, like an incredibly complex living organism. Each of its parts is related to its other parts in a web of dependencies and support systems. For us to manage the Bay well, we must first understand how it functions" (USEPA. 1982).[1]

The problem, historically, has been to develop a system of scientific concepts adequate to express holistic concerns while avoiding the introduction of metaphysical and speculative ideas such as a belief in a supraorganismic being corresponding to the biosphere-as-a-whole. Aldo Leopold, working in the field of wildlife and range management, recognized the importance of complexity and wholeness but felt uneasy attributing literal truth to the organicist metaphor (paper 1, this volume).

The purpose of this paper is to show that Leopold developed a theory of *environmental management*, to be distinguished from *resource management*. Resource management, which operates on a generally utilitarian criterion, deals with productive subsystems of larger functioning systems; rapid trends in land and resource use may force the environmental manager to pay attention to the larger context of those subsystems. Leopold significantly anticipated important insights of hierarchy theory, a new approach to understanding complexity in ecological systems. More importantly, hierarchy theory may provide conceptual tools for formalizing and studying problems that limited the concrete applications of Leopold's elegant theory of environmental management.

Reprinted from *Ecological Economics* 2 (1990): 119–127, Copyright 1990, with permission from Elsevier Science.

Through the 1920s Leopold operated on a simple utilitarian management model, destroying predators to maximize hunting and domestic grazing opportunities (Chapter 1, this volume). During the 1930s, Leopold decisively rejected Gifford Pinchot's "economic biology," and incorporated community ecology into his management model. He argued that the land system in the Southwest was "ill" as a result of too many wild and domestic grazers; he also generalized his analysis to explain that the dust bowl was a predictable result of ignoring the limitations inherent in fragile land. Leopold equated health with "integrity," which is closely related to complexity. Assuming energy flows to be the essence of a system, Leopold modeled ecological systems as pyramids of species, considered functionally as producers and consumers, and organized to pump energy through all its levels. Leopold thought that the goal of management should be to protect the integrity of the system, understood as the complex set of relationships that keeps the energy pathways open (Leopold, 1939; 1949, pp. 214–218).

These considerations led Leopold to develop what might be called a "contextualist" approach to environmental management. According to this approach, which Leopold explained with organicist metaphors, resource management for maximization of production will tend to threaten the larger system in which it is embedded. The "cell," such as a field or farm, must be considered a part of a larger context, a community. In cases like the dust bowl, pervasive trends in the management of cells over a large area can simplify the system, reduce its multiple levels of complexity, and destroy the integrity of the pyramidal system. The community model supported Leopold's hypothesis by explaining two aspects of ecological breakdowns like the dust bowl. The pyramidal structure and emphasis on energy flows explain what ecological sickness and breakdown is: it is a simplification that inhibits the flow of energy through the complex system. The community model also explains why breakdowns often occur only after years of successful exploitation: there is redundancy on each level of the hierarchy; this redundancy is gradually reduced during normal years. When the system is stressed, as in cyclinal droughts, the weakened system may collapse (Leopold, 1979).

Despite these theoretical breakthroughs, Leopold was unable to apply his management model as a positive guide because of three weak, or at least undeveloped, aspects of the community ecology of his day. First, management to protect the dynamic integrity of a system would require an understanding of the ecological fragilities of the particular system in question. Ecologists could not then, and still cannot, model relationships in natural systems adequately to state confidently the degree of fragility of a system prior to human-induced disturbances.

Second, Leopold's management goals, which required that a plan be developed for a "whole ecosystem," raised a serious problem of "parts and wholes." (See, for example, Chase, 1986, pp. 319ff.) Community ecologists tended to treat "ecosystems" as abstract mathematical models of temporary methodological convenience and were reluctant to apply these models to large natural systems under management.

Third, community ecology had not developed, and still has not developed, a truly dynamic conception of systematic integrity, which was essential to applying Leopold's contextual model. The relationship between the cells managed for productivity and the larger, also changing, ecosystematic context can only be modelled dynamically.[2]

Recently, however, there has emerged within ecology a new and highly promising theoretical approach, "hierarchy theory," which bears a striking resemblance to Leopold's community model and contextualist approach to management. This new approach promises to make the concepts that plagued Leopold's theory of environmental management more precise. Hierarchy theory is based on general systems theory and focuses on the complexity and internal organization of systems – what Leopold called "integrity" (Leopold, 1949, p. 224; Allen and Starr, 1982, pp. xi–xvi; O'Neill et al., 1986, p. 6). According to hierarchy theorists, natural systems exhibit complexity because they embody processes that occur at different rates of speed: generally speaking, larger systems change more slowly than the microhabitats and individual organisms that compose them, just as the organism changes more slowly than the cells that compose it. Hierarchy theory's sliding scales of systems within systems, changing at varied rates of speed, provides a formal model corresponding to Leopold's idea that productive units – cells – must be understood as existing in a larger context – the community. Because the community survives after individuals die, the individuals and microhabitats themselves are unlikely to affect the larger system because the individual is likely to die, or the microhabitat to change, before the slow-changing system in which it is embedded will be significantly altered.

This is not, of course, to say that elements have no impact on systems that provide their context. The elements, often called "holons" in the context of the analysis, are "two-faced"; each holon "has [a] dual tendency to preserve and assert its individuality as a quasi-autonomous whole, and to function as an integrated part (of an existing or evolving) larger whole" (Koestler, 1967, p. 343: Allen and Starr, 1982, pp. 8–10). As a part, the holon affects the whole, but scale is very important here – the "choices" of one element will not significantly alter the whole – but if that part's activities represent a trend among its peers, then the larger, slower-changing system will reflect these changes on its larger and slower scale. One cell turning malignant will not affect an organism significantly unless a trend toward malignancy is thereby instituted. Individuals and subsystems, therefore, affect the larger whole mainly through participation in trends exhibited by cohorts on their same level.

This multi-scalar approach to time and space in ecology is reminiscent of Leopold's metaphorical discussion of differing scales of time and our perception of them in *Thinking Like a Mountain* (Leopold, 1949). Hierarchy theory provides a more precise conceptual model for what Leopold called "the land," which was for Leopold a slower-changing system composed of many faster-changing parts. He explicitly commented that our failure to see deterioration in the land community is due to our failure to recognize that ecological and evolutionary changes take place on a slower scale of time than the scale perceived by humans. Agriculturalists and game managers focus on the production of annual crops. The mountain, as Leopold explained metaphorically, must look at the value of wolves in a longer perspective (Leopold, 1949, pp. 133, 206).

Leopold's much discussed references to the "integrity" of ecosystems can now be understood as the set of complex relationships between parts and wholes as these relationships unfold through time. A system retains its integrity when the larger environing system maintains autogenic functioning – it is changing, in other words, according to the slower processes that reflect a time scale appropriate to its position in a hierarchical structure. While resource management maximizes productivity within productive cells, environmental management keeps a watchful eye on the parameters of normal, slower changes in the environing system.

When nature is viewed hierarchically, as a system of parts embedded in larger and larger wholes, holons higher on the hierarchy represent the environment of lower holons and constrain their activities (Allen and Starr, 1982, pp. 11–13). Allen and Starr argue that "The positive aspects of organization emanate from the freedom that comes with constraint. The constraint gives freedom from an unmanageable set of choices: regulation [from above] gives freedom within the law" (Allen and Starr, 1982, p. 15). According to hierarchy theory, then, natural systems are seen as organized units embedded in a hierarchy, with larger, slow-changing systems (the environment, or "context") determining the range of choice available in smaller systems. This abstract hierarchical model can be applied on various levels, and provides a sliding scale of concepts for analyzing relationships among the parts and wholes of living systems.[3]

In ecology, which emphasizes relationships among systematic elements, hierarchy theory provides a tool for analyzing the multi-layered complexity of natural systems, and shows promise to model the dynamic relationships among their parts. The hierarchical model may also point the way toward a managerially useful concept of dynamic stability and ecosystem health. This concept will relate functions of faster- and slower-changing systems. Accelerating changes in normally slow-changing systems may indicate deterioration – illness or destabilization – in the land community. Viewed from above, the part is constrained by its larger environmental context. Viewed from below, "normalcy" or "stability" is achieved as cells follow laws intrinsic to their behavioral capabilities – actions of one short-lived cell will normally cancel out the actions of others, achieving stability (understood as statistically probable regularities) on the larger systematic level (Pattee, 1973, pp. 105–107). The birth of an individual, for example, will not affect population stability if birth and death rates remain relatively constant.

Leopold recognized that modern human societies, which combine astounding technological capabilities and unprecedented individual mobility, can rapidly alter trends that affect the larger context. If farmer A clears his woodlot and plants wheat, this may be cancelled out as farmer B lets his wheat field go fallow. If, however, in response to trends in grain prices, most farmers imitate A and none imitate B, the larger system can undergo destabilizing change. Hierarchy theory can model these occurrences as exceeding parameters set by statistically understood patterns in the actions of parts. Radical, or systematic, change occurs when unbounded behavior of individuals within part of the system begins to cause accelerated change in the larger system.

Accordingly, Allen and Starr note that the dust bowl represents an example in which activities of individual farmers had little effect on the large-scale environment of the plains as long as they occurred on a limited scale. As more and more plainsland was converted to monocultural agriculture, however, a threshold was reached and the plains system, which had been changing slowly for millennia, underwent rapid change (Allen and Starr, 1982, p. 219; O'Neill et al., 1986, p. 211).

This explanation is isomorphic with Leopold's: the dust bowl resulted from intensive and pervasive application of monocultures and intense grazing in a bioregion ill suited to intense use because of its cyclical patterns of rainfall and aridity. Similarly, Leopold, as a game manager early in his career, saw deer/wolves/hunters as a cell that could be managed. Experience taught him that that cell must also be considered as a part of its larger environment. Removal of wolves caused an irruption of deer, which overgrazed their browse, exceeding the equilibrative capacity of the larger system. A parameter (viewed from above, a constraint) was exceeded and the larger system became unstable.

Hierarchy theory raises again the question of parts and wholes: how seriously should we take any particular designation of "parts" and "wholes," of "cells" and "contexts?" Hierarchy theorists avoid this problem by insisting that their model resides not in reality, but in our understanding of reality (it is "epistemological," not "metaphysical") (Allen and Starr, 1982, p. 10). Whether natural systems are themselves hierarchical is not the point, they say; viewing them as hierarchical makes their complexity intelligible to us. Any designation of parts and wholes is therefore a temporary expedient, a result of choosing a particular "observation set" – "a particular way of viewing the natural world" (O'Neill et al., 1986, pp. 7, 13: Allen and Starr, 1982, p. 6). This approach avoids commitment regarding which aspects are fundamental: "since [no viewpoint] encompasses all possible observations, neither can be considered to be more fundamental. When studying a specific problem, the scientist must always focus on a single observation set. However, when developing theory, many observation sets must be considered" (O'Neill et al., 1986, p. 7). I interpret this noncommittal position as roughly equivalent to Leopold's conclusion that organicism is metaphorically, not literally, true.

But O'Neill and his co-authors also recognize that "the task of choosing an appropriate system for investigating a particular phemomenon is inseparable from consideration of underlying organization and complexity" (O'Neill et al., 1986, p. 39). Applications of hierarchy theory to environmental management will, therefore, focus on the organizational complexity and the inherent fragilities of the particular system to be managed. Whereas hierarchy theorists are seeking an abstract model that applies at every level of biological organization, the scale of concern to the manager will be indicated by the problem at hand.

Again, Chesapeake Bay management provides an example. Governments in the region are firmly committed to holistic management of the Bay ecosystem. They now accept that the specific resource-producing systems surrounding the Bay must be managed with an eye to not exceeding parameters imposed by the health of the larger Bay system. Agriculture, fishing, recreation, tourism, etc., can be conceived as socially driven systems partly parasitic on the larger, normally slower-changing Bay system.

The most acute current problem in managing the Bay, the algal blooms resulting from too much run-off from yards, fields, and pastures, can then be interpreted as rapid change in the normally slow-changing Bay system, change that results from trends in residential development patterns, agricultural production, and consumer taste. Fishermen, farmers, and home builders choose according to short frames of time – a season, a year, a few years – and trends in their choices can cause changes in the Bay; perhaps the most threatening

of these is a significant reduction, over the last decade, of the highly productive submerged grass beds. These changes are apparently related to the
algal blooms mentioned earlier, as cloudy water and algae setting on leaves
reduce photosynthesis; nutrient loading, which results from trends in human
behavior, has overwhelmed the Bay's historical means to absorb nutrients.

The goal, as understood in the terms of hierarchy theorists, is to achieve a
relationship in which the Bay's subsystems function without disturbing, or disturbing minimally, the rate or trajectory of the normally slower changes in
the larger context.[4] Can the illness in the larger contextual system, on which
so many productive activities depend, be reversed? That will depend on three
factors: (a) careful monitoring of trends in the component subsystems to determine whether these trends are likely to have major spill-over effects; (b) developing a scientific ability to monitor the Bay system and model changes in
it so as to distinguish destabilizing trends from cyclical and other autogenic
changes in the larger system; and (c) achieving a grassroots political consensus
that restrictions on activities in the component systems are necessary and in
the interests of all concerned. Each of these three factors represents a major
challenge; each must be met if the health of the Chesapeake is to protected.

CONCLUSION

Environmental management attempts to protect the complex organization of
large natural systems. Hierarchy theory may provide conceptual tools for modeling management activities within the context of larger systems by modeling
relationships among levels. A policy-analytic approach consistent with a hierarchical conceptualization of management cells within a larger ecosystem
context would involve two hierarchically organized stages. In the first stage,
constraints based on an understanding of the fragility of the larger contextual
system will be determined. Efficiency criteria, such as the benefit-cost test,
would then pick the most productive management option that fulfills the biologically based constraints of the first stage. Benefit-cost criteria, in other
words, would apply *within* appropriate management cells, but would not attempt a synoptic representation of all benefits and costs. Primary attention
should first be placed on the goal of developing biological criteria of ecosystem health and integrity. These biological criteria would guide management
of the larger natural context in which these productive cells are organized
(Norton, 1987). This contextual feature of Leopold's management theory,
which can be modelled by positing biologically determined, hierarchically
imposed constraints, represents the operational meaning of the Land Ethic.

ACKNOWLEDGMENT. The research for this paper was undertaken while the author was Gilbert White Fellow at Resources for the Future.

## NOTES

1.  For another example of holistic management proposals, one applied to protecting old growth and biodiversity over the entire region of the Pacific Northwest, see Harris (1984).
2.  Contemporary ecologists still tend to model ecosystematic stability across time as *resilience*, the ability of a system to return quickly and completely to its prior state following a disturbance. Strong evidence shows that natural systems are not, in reality, stable in this sense. For a fuller discussion of this topic, see Norton (1987, chapter 4).
3.  It is important to recognize that there are many alternative hierarchical systems which might be posited to explain ecosystem structure. My use of familiar concepts in this overview is for illustrative purposes only. It would be a mistake to assume that the most effective hierarchy will correspond to intuitive conceptions or any particular scientific model. See O'Neill et al. (1986, chapter 4).
4.  See Allen and Starr (1982, chapters 3, 8, and 9) for a discussion of scales and "filters" (which represent conversion functions between systems operating on different scales).

## REFERENCES

Allen, T. F. H. and Starr. T. B., 1982. *Hierarchy Theory: Perspectives for Ecological Complexity*. University of Chicago Press, Chicago, IL.

Chase, A., 1986. *Playing God in Yellowstone: The Destruction of America's First National Park*. Atlantic Montly Press, Boston/New York.

Harris, L. D., 1984. *The Fragmented Forest: Island Biogeography Theory and the Preservation of Biotic Diversity*. University of Chicago Press, Chicago, IL.

Koestler, A., 1967. *The Ghost in the Machine*. Macmillan, New York.

Leopold, A., 1939. A biotic view of land. *J. For.*, 37: 727–730.

——. 1949. *A Sand County Almanac: With Other Essays on Conservation from Round River*. Oxford University Press, Oxford/New York.

——. 1979. Some fundamentals of conservation in the Southwest. *Environ. Ethics*, 1: 131–141.

Norton, B. G., 1987. *Why Preserve Natural Variety?* Princeton University Press, Princeton, NJ.

O'Neill, R. V. et al., 1986. *A Hierarchical Concept of Ecosystems*. Princeton University Press, Princeton, NJ.

Pattee, H. H., 1973. The physical basis and origin of hierarchical control. In: H. H. Pattee (Editor), *Hierarchy Theory*. Braziller, New York.

USEPA, 1982. *Chesapeake Bay: An Introduction to an Ecosystem*. Environmental Protection Agency, Washington, DC.

# 16

## Scale and Biodiversity Policy

### A Hierarchical Approach

with ROBERT E. ULANOWICZ

### INTRODUCTION

There exists a broad consensus supporting the protection of biological diversity; but the exact meaning of this consensus for policy is not clear. In the United States, for example, the *Endangered Species Act* emphasizes protection of species. But this emphasis has led to the question: Since approximately 99% of all species that have existed on earth are now extinct, how can it be so urgent that we reduce anthropogenic species extinctions? The standard answer to this question – that extinction itself is not bad, but rather that the accelerated *rate* and broadened *scale* of extinctions are unacceptable – likewise raises more questions than answers. One might ask, what would be an "acceptable" rate of extinctions? If species are not sacrosanct, what then *is* the proper target of protection? These questions are important because our inability to answer them indicates huge gaps in our understanding of environmental management and of biodiversity protection: It is not clear at what scale the problem of biodiversity loss should be addressed, Nor is it clear that measuring rates of species loss is the only or best criterion for measuring the success or failure of protection efforts.

In this paper we explore the policy implications of a hierarchical approach to protecting biological diversity. The hierarchical approach, which represents a specific application of general systems theory,[1,2] models natural complexity as a hierarchy of embedded systems represented on different *scales*. A major assumption of hierarchy theory is that smaller subsystems change according to a more rapid dynamic than do larger systems.[1,3–5] We believe that this correlation between system size (hierarchical level) and rate of change introduces some conceptual order into discussions of the proper scale on which

From *Ambio* 21 (1992): 244–249. Reprinted with permission.

to address environmental policy goals, and we illustrate our approach by applying it to biological diversity policy.

While much of our conceptual apparatus is adapted from theoretical ecology, we do not consider our work to be scientific in the narrow sense that it consists of value-free descriptions and explanatory hypotheses. We, on the contrary, believe that conservation biology is a normative science – like medicine it is guided most basically by a commitment to important social values. Just as medical research must fulfill both a criterion of methodological rigor *and* a criterion of relevance – usefulness in healing patients – conservation biologists are likewise obligated to characterize ecological systems in ways that are not only accurate, but useful in protection and recovery programs. The goal of conservation biology should therefore be to examine dynamics that affect environmentally important goals. Social values, and our attempts to understand how to protect them, direct conservation biology by pinpointing crucial natural dynamics that should be understood and protected. Thus, while species are of course important because species are essential participants in natural dynamics, we intend to shift the focus of biodiversity policy to protecting the *health* of socially important natural processes.

This approach eschews purely scientific delineation of goals for conservation biology and departs from the pure science paradigm. But this approach can be regarded as value-free in another and more realistic sense. Whether elements of nature are valued for themselves (intrinsically) or for future humans (instrumentally), we can provide a scientific argument that it is multi-generational, ecosystem-level dynamics that should be the target of protection policy. Because protecting ecological processes that unfold across multiple generations is the only way to sustain species diversity for future generations, and because we are committed to this policy goal, the question of whether species or future humans are ultimately valued is rendered moot.

A THEORY OF SCALE FOR BIODIVERSITY PROTECTION

Contemporary philosophy and physical theory have converged to show that there exist many consistent and coherent accounts of reality as we experience it.[6,7] One manifestation of this more general result directly affects scalar questions. Newtonian physics assumed that the world could be understood on a single, unified scale, there being no universal constants in the Newtonian system of description. "Scale therefore becomes all-important," in the words of Ilya Prigogine and Isabelle Stengers, "because the universe is no longer homogeneous, and the synoptic perspective is abandoned in favor of a hierarchically organized, multi-scalar and dynamic world."[6]

Choice of system boundaries and scale is therefore an essential part of describing a system that is to be managed for a given purpose, and thus the best description of a system is one that describes dynamic processes on a scale determinative of priority social goals. We are therefore not bothered by the recognition that ecologists use the concept of an ecosystem loosely and variably. We recognize choices to bound a given system in time and space as decisions based broadly on the usefulness of certain models in understanding targeted physical processes. Since there are many useful ways to understand a system, articulated social goals must direct choices as to how natural systems are described. Choice of the proper scale on which to address an environmental problem such as species loss is therefore an interactive process in which definitions of policy goals guide choices of system boundaries, even as scientific descriptions of processes, and human impacts on them, help us to refine our understanding of policy goals. Determining the correct scale and perspective from which to address environmental problems therefore involves a complex interaction of value definition, concept formation, and scientific description – an interaction in which the articulation of environmental goals drives science (Fig. 16.1).

We emphasize the development of a physical scale for conservation biology, and assume a high social value on protecting biological diversity. We proceed to combine this assumption with hierarchical principles and to explore the implications of this combination for biodiversity policy. We believe that the emerging concept of ecosystem health, understood in conjunction with hierarchy theory, should guide policy debate. The outcome of that debate, admittedly a political affair (as is any process of value articulation), should in turn guide biological diversity policy.

The difficult theoretical problem we have posed for ourselves is as follows: Given that the scale of ecosystem description is relative to choices regarding the concepts and values we operate with – and these, in turn, are relative to goals and value determinations – how can ecosystem scale and boundaries be constructed on a rational basis? Implicit in this question is the recognition that a choice can be *relative* to certain factors, including public values, without thereby becoming *subjective* (not amenable to rational analysis). Choices to employ certain concepts to describe an ecosystem, and choices to view it on a particular scale, involve tremendous latitude and depend on the goals of the researcher. Nevertheless, these decisions are in fact constrained by the goals of managers as well as those of the researcher. To understand a natural

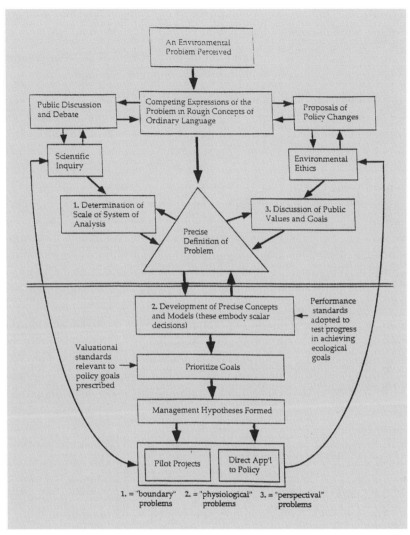

Figure 16.1. The environmental policy process. Environmental problems are not clearly formulated when they first emerge in public discourse. Determination of the proper scale at which a problem should be "modelled" requires an interactive public process in which public values guide scientific development of models. Once the problem is precisely defined and models are developed, the process of experimentation with solutions can begin.

dynamic in order to protect it requires that the dynamic be modelled at a scale relevant to social values.

## Scale and Biodiversity Policy

The goal of conserving biological diversity for the benefit of future generations determines the temporal horizon of biodiversity policy. Relying on hierarchy theory, we reason that since the policy horizon for this social value is many generations, we must concentrate on a large-scale dynamic such as the dynamic that determines total diversity over landscape level systems. This approach squares with what we know biologically: landscapes are essentially patchy. Many populations of plants and animals are ephemeral and we cannot save every population of every species, nor can we save every species. As the scale of human activities on the earth increases and human-dominated landscapes prevail more and more, it is inevitable that the rate of ecological change will accelerate. The goal of policy should be to maintain the health of the dynamics that support and retain diversity on a large geographical scale. The proposed approach therefore agrees with advocates of species protection in most cases, although for different reasons. Policies should usually protect species because an accelerating rate of species loss is the best available benchmark of illness in ecological systems. According to our approach, however, the value of species is mainly in their contribution to a larger dynamic, and we do not believe huge expenditures are always justified to save ecologically marginal species. The problem, of course, is to specify what is too large an expenditure and to define "ecologically marginal." On the approach developed here, these definitions must be built upon a theoretically adequate conception of system scale, one that is also useful in guiding protection and restoration efforts and in communicating with the public as it articulates goals and values.

The limits to any dimensional description of a system have been formalized in the discipline of *dimensional analysis*, for which the Buckingham–Pi Theorem provides the foundation.[8] This theorem states that there are a limited number of dimensionless groupings of the physical parameters of a system (expressed in terms of fundamental dimensions of the system) that are sufficient to control the dynamics of the system. A corollary of the Buckingham–Pi theorem, which states that only those dimensional groupings near order unity are important to the dynamic system description, is of special relevance here. That is, if any of the characteristic parameters becomes disproportionate with respect to the scale of the other system phenomena, then it becomes irrelevant to the system description. Either it is so slow as to appear constant or so fast as always to be in equilibrium with other limiting factors. Thus, in a real and quantitative way, the Buckingham–Pi theorem allows us to circumscribe the domain of applicability – the focal level in a hierarchy – for any given system feature.

Let us, for example, look at how the Buckingham–Pi theorem would apply to a specific management problem. German foresters of the 19th century emphasized production of timber and converted huge areas of the German forest to monocultural spruce. Initially, yields of high-quality timber increased. After three or four iterations, however, yields plummeted. Young trees could not penetrate the soil with their roots and a condition called "*soil sickness*" developed.[9] Analysis showed that soil composition had been altered because essential microorganisms had been lost. In this example, descriptions at a particular scale – the scale of economic forestry – had been assumed to provide a complete and unique description of reality. Failure to recognize that timber production is a process that exists as a part of a system that has evolved over centuries, and that that system is supported by processes existing on a longer scale than is registered in the language of production forestry, resulted in a serious management failure.

The Buckingham–Pi theorem provides a tool by which we can pinpoint the proper scale at which to formulate policy in cases such as this. Suppose at the outset that we are ignorant of the actual dynamics of litter decay and wish to determine which parameters of the system, at what scale, are pertinent. We could list the following parameters as candidates: $D_o$, the initial density of litter on one square meter of forest floor; F, the rate of litter fall per year; $B_o$, the initial density of bacteria among the litter; r, the instantaneous rate of litter decay; L, the characteristic length of the patch we are observing; and h, Planck's constant. There are three fundamental dimensions (mass, length, and time) among the six parameters. The Buckingham–Pi theorem says that there will be 3 (= 6–3) dimensionless groupings that characterize the system dynamics. Without going into the details about how they are determined, those three groupings (pi numbers) may be taken as $B_o/D_o$, $D_o r/F$ and $h/FL^4$.

Now we go into the field and laboratory and actually measure these parameters. Our (hypothetical) estimates are:

$$D_o \quad 1.2 \, \mathrm{Kg\, m^{-2}}$$
$$F \quad 0.5 \, \mathrm{Kg\, m^{-2} y^{-1}}$$
$$B \quad 0.00009 \, \mathrm{Kg\, m^{-2}}$$
$$r \quad 0.41 y^{-1}$$
$$L = \quad 1\mathrm{m}$$
$$h = \quad 2.09 \times 10^{-26} \, \mathrm{Kg\, m^{-2} y^{-1}}, \; \text{Planck's}$$

constant which means that:

$$B_o/D_o \qquad 7.5 \times 10^{-5}$$
$$D_o r/F^\circ \qquad 0.99$$
$$h/{}^\circ FL^4 \qquad 4.18 \times 10^{-26}$$

As one could have guessed, the grouping that contains Planck's constant, h, is much less than one, and the corollary to Buckingham–Pi says that this grouping will not influence the dynamics we are observing to any visible extent. That is, phenomena at the submolecular level occur so fast and over such a small space that we need not concern ourselves with those details. What is less obvious is that the first pi number, $B_o/D_o$, is also very small, so that one need not be concerned with following bacterial concentrations. What this result implies is that some unknown factor (e.g., nitrogen concentration, soil moisture, soil temperature, aeration rate, etc.) is limiting the breakdown of the litter. The bacteria themselves grow very quickly, and their densities will rise and fall in very short order to track the unknown limiting factor.

The dynamics of forest litter are best described in this case by the second grouping, $D_o r/F$, which characterizes the ratio of the decay rate to that of the supply. That the two are very comparable indicates that litter buildup should be a slow process. In fact, one can calculate a characteristic time for the process by dividing the difference between the supply and decay rates by the stock of litter present, i.e.

$$(F - D_o r)/F \quad 0.00667 y^{-1}$$

The reciprocal of this accretion rate is 150 years, which accords with our intuition that soil buildup is a very slow process. The slowness of that process explains how, by concentrating on forest production of timber in a single cycle of planting and harvest (8–40 years), the relevant scale for economics, the German foresters ignored a crucial factor in sustainable forest use and, in the process of simplifying the system, actually impoverished it on a longer scale of time. The scale relevant to a policy of intergenerational sustainability correlates very closely with the ecological parameter of soil build-up because this parameter is associated with maintaining both the diversity and the productivity of the system over multiple generations. According to this hypothetical example, a public concern for long-term sustainability of forest products and for maintenance of forest health implies an approximate horizon of 150 years. A policy horizon of this length is suggested by the recovery time from damage to soil composition and we arrive at a scale based on the production function for leaf litter, the crucial variable if our public concern is indefinite intergenerational sustainability of forest productivity. The characteristic dynamic of forest development is that of carbon retention, and the rate of soil build-up is probably the rate-limiting factor most relevant to maintaining and encouraging long-term forest productivity. In this way, an understanding of the production function affecting a public value of intergenerational sustainability, when coupled with a hierarchical understanding of ecosystem structure and functions, determines the proper scale for addressing an environmental problem.

## WHOLE ECOSYSTEM MANAGEMENT

The goal of biological diversity policy should be, given the long time horizon of the policy of intergenerational sustainability, to protect *total diversity at the landscape level of ecological organization.* While we do not intend to reify any given system description as complete and uniquely correct, we do believe that one can scientifically determine ecosystem boundaries and membranes *provided a priority social goal such as protecting biodiversity* is specified. Eco-system-level management is distinguished by its concern for characteristics of whole systems – characteristics that cannot be reduced to aggregated characteristics of its parts. The decision to emphasize whole ecosystem management is a decision to employ hierarchical, rather than aggregative models – it is to seek models that integrate policy goals on multiple levels or scales, rather than simply counting bottom-line costs and benefits.[5,10,11] To add a whole eco-system level to a management plan is to resolve to manage, in addition to managing components of the system for resource production, the system as a system. Successful ecosystem management will necessarily be management that has conceptualized the system in a way that focuses attention on the central features of the system – features that are important to supporting important public values. The intuitive idea of ecosystem health is valuable because it focuses attention on the larger systems in nature and away from the special interests of individuals and groups. Competing and special interests, and the goals they articulate, must be integrated into the larger-scale goal of protecting the health and integrity of the larger ecological system.[12] While decisions regarding particular elements of the landscape, especially those in the private sector, will be managed according to economic goals and criteria, biodiversity policy focuses on the larger scale. The regulative idea of a healthy ecological system organizes, tests, and integrates these special interests on a landscape level of organization. A priority goal of conservation biologists must be educating the public toward a better understanding of ecological management, and helping citizens to articulate their values and to express those values in management decisions.

But the analogy of ecosystem health/integrity is best understood as an intuitive guide, rather than as a specific determinant of policy choices[13] because, like all analogies, it eventually breaks down. The strength of the medical analogy is that it focuses attention on the overall organization of the system: just as a good physician would not treat a specific organ without paying attention to impacts on the health of the entire organism, whole ecosystem managers must constantly monitor impacts of human activities on the larger ecosystems that form the human environment. The medical analogy is important in emphasizing the importance of systems thinking, and of a recognition of multiple levels and scales on which systems change dynamically.[14] But the medical analogy has an important drawback[5]. Whereas human medicine and veterinary medicine focus on individual organisms and are guided by the unquestioned goal of protecting the health of patients, ecosystems are multiscalar and have no obvious identity. No prior overriding consideration like the Hippocratic Oath determines which level of the complex hierarchies of nature should be considered the "whole," organismic level of the system. Managers have considerable latitude in choosing the boundaries, and hence the scale, of the systems they monitor and manage. We believe that choices within this latitude will remain indeterminate until a viable consensus regarding management goals has been articulated. In cases where public goals have been clearly formulated, scientific description of the internal functioning of ecological systems will provide guidance regarding the location of boundaries, and regarding which internal compartments/membranes of the system to emphasize (Fig. 16.1). A whole ecosystem is a system whose boundaries include essential elements of a dynamic relevant to important social values for a region.

Ecosystem health/integrity therefore stands as the central policy concept to guide ecologically understood environmental management; and, we are arguing, public values – aesthetic, economic and moral – all depend on protecting the processes that support the health of larger-scale ecological systems. These systems create the context for those activities, and in this sense are crucial elements in their value.[5,11,15] The local, cultural goal of protecting the capacity of systems to react creatively and productively to disturbance, whether footprints of hikers or harvests by oystermen, therefore can sometimes take precedence over the short-term goals of individuals and economic interest groups.

## AUTOPOIETIC SYSTEMS

Whole ecosystem management must be understood as management of a self-organizing system – a system that creates and maintains itself by homeostatic and homeorhetic responses to changing conditions. We describe the creative feature of ecosystems as *autopoiesis* (from the Greek term meaning "self-making").[16–18] Emphasis on autopoiesis implies that the macroscalar boundaries separating the system from its surroundings as well as the smaller-scaled boundaries that separate the system into subsystems or "organs" are chosen to accentuate dynamics essential to sustaining biodiversity. It is not claimed that the features we emphasize are the only features of ecosystems that could be spotlighted; it is claimed only that the scalar choices (boundaries and membranes) represent conceptualizations (models) of the system that are *managerially relevant* and *naturally appropriate* given the goal of protecting healthy ecosystems and their elements over many generations.

While we agree with those, such as Botkin,[14] who emphasize the dynamism of natural systems, we recognize also that dynamically creative change requires a certain amount of stability in the form of larger, slower-changing systems that provide "stable" backgrounds for the processes of iteration and reiteration that allow evolutionary development. Evolutionary creativity on long scales requires creative solutions to environmental constraints that are essentially "fixed" on the scale of individual specimens.

A commitment to ecological sustainability, the resolution to protect complex and creative ecological systems for future generations, assumes the possibility of stability across multiple generations. Stability, here, is treated as a "well-founded illusion of scale." It is an illusion because the system on all levels is constantly dynamic. But this illusion, from a human perspective, is nevertheless well founded because large-scale ecosystems have historically changed sufficiently slowly that there existed continuity of landscape across human generations.

From the environmental standpoint a most important attribute of self-supporting units is their ability to adapt to new circumstances in *creative* ways. This creativity supports the ability of natural systems to rebound in response to heavy economic exploitation and also explains their ability to absorb human wastes. As human activities become ever more intrusive in the systems of nature, these creative adaptations will become even more crucial. Creativity has been perceived as relevant only to conscious, goal-forming agents. But as new developments in physical theory have made clear, the process of creation is ubiquitous in the universe and at times can even transpire in systems not containing living members.[6] Ulanowicz[19] has argued that the capacity for creativity constitutes the crux of what is normally referred to as *ecosystem health*. But the capacity for creativity is too often misperceived, which comes as no surprise, given the difficulties in describing it in semantic, much less quantitative terms. The emerging consensus[19–21] indicates that creative action is contingent upon two mutually exclusive properties of the performing system.

First, it is necessary that any system capable of solving a novel problem possess a requisite amount of ordered complexity. Order implies constraints – events impinging upon the system or subsystem must initiate a channeled sequence of reactions (which may be and probably must be reflexive to some extent) that culminate in the response of the system to that input, e.g., compensation, indifference, counteraction, co-option, etc. Without such coherence, creativity is impossible, and Atlan[20] demonstrates how thresholds in ordered complexity must be surpassed before a system is capable of creative action. This side of creativity is widely understood. It is unquestioned that an organism or system must possess enough "apparatus" before creativity is possible. But some of the most tightly ordered objects in the universe are machines – artifacts that are incapable of truly creative actions, primarily because they lack an adequate degree of inherent disorder.

It is not so universally acknowledged (or, in many cases, even suspected) that incoherence is also a prerequisite for creative action. Before creativity is possible, a system must possess a potential "reservoir" of stochastic, disconnected, inefficient features that constitute the raw building blocks of effective innovation. In the course of normal functioning such disutility appears as an "overhead" or an encumbrance. However, when faced with a perturbation or problem, it is this background of dysfunctional repertoires that is utilized to meet the exigency. Background species or marginally extant trophic pathways – system redundancies – can be activated in response to a disruption of the normally dominant means an ecosystem employs to process material and energy. If the disturbance is recurrent or persistent, the new response eventually will be incorporated into permanent coherent structure. This idea of freedom resonates in public values with the emphasis on wilderness protection and with the importance placed on protecting wildness wherever possible and appropriate.[5,22]

The concepts of order and incoherence may seem subjective to some, but Ulanowicz[23] has suggested that it is possible to employ results from information theory (quantitative epistemology) to estimate the relative amounts of each of these attributes possessed by a given system. To attach numbers to these system properties, it is necessary first to describe the system as a collection of subunits linked together by processes that can be quantified. For example, ecosystems are often described as a collection of species or other aggregations of organisms linked one to another by exchanges of material or energy. These exchanges can be assigned physical units and measured or otherwise estimated in the field or laboratory.

Once the ecosystem has been bounded and then characterized as a network of palpable flows, one can employ information theory to quantify the diversity of flows in this ensemble as if each flow were independent of all others. Of course, the exchanges do not occur in random, unconnected fashion. There is an order in the pattern of trophic connections and temporal sequences. Such order gives rise to a component of the overall diversity of flows as computed by a variable called the *average mutual information of the network topology*.[24] Ulanowicz[25] has given the name *ascendancy* to a scaled version of the mutual information. Systems with more clearly defined pathways of cause and effect will exhibit higher values of ascendancy. One can rigorously prove that the mutual information and linkages can never exceed the measure for the diversity of flows. This condition has led Ulanowicz[25,26] to call the latter term the system *capacity* for growth and development. System capacity obviously is tightly coupled with the biodiversity of the system. We are here hypothesizing that this idea may also serve as the link between the intuitively understood policy concept of ecosystem health and the more precise, quantitative disciplines of systems theory and information analysis.

The amount by which the capacity exceeds the mutual information has been called the system *overhead*. All those system features which contribute nothing to its order and coherence by definition add to its overhead. These include redundant and inefficient pathways, stochastic and illphased events, etc.

In terms of these three concepts – ascendancy, capacity, and overhead – one can enumerate the requirements for a system to act creatively in response to a novel circumstance: (i) The system must have a high capacity for growth and development, i.e., its biodiversity and complexity must remain high. (ii) Most of this capacity needs to be expressed as ordered and coherent ascendancy. (iii) Some capacity must remain as unstructured and incoherent overhead to afford the system the degrees of freedom necessary to respond to novel environmental stimuli.

A biotic system satisfying all three requirements can be termed "healthy."[19] Thus, we can suggest a definition of ecosystem health for public policy consideration: "An ecological system is healthy and free from 'distress syndrome' if it is stable and sustainable, i.e., if it is active and maintains its organization and autonomy over time."[27] The goal of sustaining ecosystem health so defined therefore involves maintaining a capacity for autopoietic activity on the scale relevant to many human generations.

## THE VALUE OF BIODIVERSITY

One advantage of the approach to scale and policy goals sketched here is that it bypasses intransigent value questions and focuses attention on concrete and achievable goals. It does so by reversing the usual valuational methods of utilitarians and economists, who place a price value on species and then aggregate toward a total value for ecosystems. On our approach, the policy-driving values are ecosystem-level processes; we save species *both* because we value them directly (at least in many cases) *and* because of their roles in ecosystem processes. But since the processes must in the long run protect the species, the question of ultimate value of species or ecosystems will arise only in those cases where large expenditures are required to save an ecologically marginal species.

If we are committed to saving species/biodiversity for future generations and wish to introduce dollar figures into policy debates, we should estimate the total value of the ecosystem dynamic that protects species to be equivalent to the costs that would be incurred to maintain individual species in alternative ways. If we do not protect species in the wild, they must then be protected in zoos or other artificially managed areas. The cost of artificial protection would be prohibitive for more than a few species. We therefore adopt the intermediate goal – which is instrumental to the goal of protecting ecosystem processes – of protecting as many species as possible. But this is not to say that one would never declare a particular species too expensive to save, given its ecological role. The obligation to protect species is therefore best understood in the terms of the *Safe Minimum Standard* as formulated by Ciriacy-Wantrup and developed by Richard Bishop.[28,29] Endangered species policy should be governed by the rule: protect all species as long as the costs are bearable.

This approach to valuation also suggests a new sort of partnership between biologists, economists, and the public. Emphasis on the self-perpetuating features of ecological systems and their role in achieving social goals such as species preservation implies highest priority for studies that promise to characterize the structures, functions, and processes that make an ecological system a habitat capable of perpetuating species for many generations. Economists also have important roles. By developing new methods of valuation for deciding policy priorities and by determining costs of various alternatives for maintaining functioning habitats, economists can make protection efforts more efficient. Especially, they must develop incentive systems that will encourage healthy economies that are compatible with protecting ecosystem health. Since ecosystem health is as much an evaluation as it is a descriptive concept, both economists and ecologists must work to inform the public about management options and work to develop scientific models that both express and, through an interactive process, improve values. It is therefore a high priority to develop new methods of valuation that are sufficiently interactive to contribute to the dynamic process of defining and protecting ecosystem health.

## CONCLUSION

Ecosystematic, hierarchical management recognizes that many choices we make, both individually and collectively, will introduce disturbances on a local scale, as when a field is plowed, a fire set, or a forest plot harvested. The recommendation that we manage for ecosystem health as well as for productivity in the various cells of the system implies that, when we disturb a wetland, for example, we will look also at the impacts of the disturbance on the larger level of the landscape. This approach would recognize that, in managing *particular* fields or wetlands, we usually seek to maximize productivity and economic efficiency. One might call management on this cellular level *resource management.* But the hierarchical approach also recognizes more inclusive levels of management, levels where we are concerned about the healthy functioning of the creative systems of nature and about the continued existence, across the landscape, of indigenous species and distinctive ecological communities – what is popularly called *biodiversity.* We have emphasized that a hierarchical, whole ecosystem approach to management recognizes multiple levels of system organization; the scale on which an environmental problem is addressed must depend on the public goals that are given prominence.

Because the public derives many values – economic, cultural, and aesthetic – from the landscape, no single ranking of environmental goals can be adequate to guide public policy. Hierarchical thinking helps us to avoid policy gridlock, however, if we recognize that successful policy will encourage a patchy landscape. On the level of field or farm, economic criteria will predominate, while on the ecosystem level we must manage for total diversity and complexity. Here, macroscalar criteria must guide the development of incentives that protect ecosystem health. This general approach seeks integration of levels; it places priority on finding new and various methods and procedures, and on arranging economic incentives to encourage economic development that has minimal negative impact on large ecological systems. This process will be political. A variety of economically efficient policies will be delineated; simultaneously, expectations will be set for maintenance

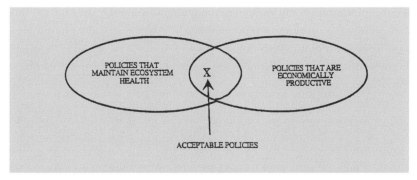

Figure 16.2. A hierachical system of analysis does not seek a single, synoptic "bottom-line" determination of the best policy from every perspective, but rather defines (through public debate and political processes) policies that will protect ecosystem health and policies that fulfil essential ecological *and* economic criteria. If no such policies have been proposed, there should be an intense effort to develop them.

of the health of the larger system that perpetuates complexity and total diversity. Good policies will be those that fulfill key criteria on both levels (Fig. 16.2).

The political process of developing a biodiversity policy for any given region should be guided by three central principles.

I.  Efforts at maintaining biodiversity should be directed at maintaining the total diversity of the landscape over multiple generations. Landscape-level goals must be defined more precisely by increased articulation of biodiversity values within a locale. Good management will require public dialogue as much as expert opinion because the definition of goals and development of scientific understanding is an interactive and experimental process.

II.  Diversity must be understood dynamically, in terms of healthy processes, rather than merely as maintenance of current elements of the system. The development of landscape-level models will involve choosing a scale and a perspective from which to both understand and manage large ecological systems. Dimensional analysis, combined with information theory and applied to hierarchical models, can provide techniques to help pinpoint dynamics associated with important public values and their support.

III.  Economic activities that complement and enhance, rather than oppose and degrade, ecological processes are to be preferred and encouraged. Recognizing that natural systems will react creatively to change, we should develop economic incentives to encourage economic development that mimics natural disturbances.[30]

REFERENCES AND NOTES

1. Allen, T. F. H. and Starr, T. B. 1982. *Hierarchy: Perspectives for Ecological Complexity.* University of Chicago Press, Chicago.

2. Hayden, F. G. 1989. *Survey of Methodologies for Valuing Externalities and Public Goods.* Prepared for the U.S. Environmental Protection Agency, Office of Environmental Planning. Contract number: 68-01-7363.

3. O'Neil, R. V. 1988. Hierarchy theory and global change. In: *Scales and Global Change.* Rosswall, T., Woodmansee, R. G. and Risser, P. G. (eds). John Wiley and Sons, New York.

4. Chapter 15, this volume.

5. Norton, B. G. 1991. *Toward Unity among Environmentalists.* Oxford University Press, New York.

6. Prigogine, I. and Stengers, I. 1984. *Order Out of Chaos: Man's New Dialogue with Nature.* Bantam Books, New York.

7. Quine, W. V. O. 1960. *Word and Object.* MIT Press, Cambridge, MA.

8. Long, R. R. 1963. Dimensional analysis. In: *Engineering Science Mechanics*, ch. 9. Prentice-Hall, Englewood Cliffs, NJ.

9. Meine, C. 1988. *Aldo Leopold: His Life and Work.* University of Wisconsin Press, Madison, WI.

10. Page, T. 1977. *Conservation and Economic Efficiency.* Johns Hopkins University Press, Baltimore.

11. Page, T. 1991. Sustainability and the problem of valuation. In: *The Ecological Economics of Sustainability.* Costanza, R. (ed.). Columbia University Press, New York.

12. Leopold, A. 1949. *Sand County Almanac.* Oxford University Press, London.

13. Ehrenfeld, D. 1992. Ecosystem health and ecological theories. In: *Ecosystem Health: New Goals in Environmental Management.* Constanza, R., Norton, B. and Haskell, B. (eds). Island Press, Covelo, CA.

14. Botkin, D. B. 1990. *Discordant Harmonies.* Oxford University Press, New York.

15. Page, T. 1992. Environmental existentialism. In: *Ecosystem Health: New Goals in Environmental Management.* Constanza, R., Norton, B. and Haskell, B. (eds). Island Press, Covelo, CA.

16. Maturna, H. R. and Varela, F. J. 1980. Autopoiesis: The organization of the living. In: *Autopoiesis and Cognition*, Maturna, H. R. and Varela, F. J. (eds). D. Reidel Publishing Co., Boston.

17. Rees, W. R. 1990. The ecology of sustainable development. *The Ecologist* 20, 18–23.

18. Callicott, J. B. 1989. *In Defense of the Land Ethic: Essays in Environmental Philosophy.* State University of New York Press, Albany.

19. Ulanowicz, R. E. 1986. A phenomenological perspective of ecological development. In: *Aquatic Toxicology and Environmental Fate: Ninth Volume.* Posten, T. M. and Purdy, R. (eds). American Society for Testing and Materials, Philadelphia, ASTM STP 921, pp. 73–81.

20. Atlan, H. 1974. On a formal definition of organization. *J. Theor. Biol.* 45, 295–304.

21. Wagensberg, J., Garcia, A. and Sole, R. V. 1990. Connectivity and information transfer in flow networks: Two magic numbers in ecology? *Bull. Math. Biol.* 52, 733–740.

22. McNamee, T. M. 1986. Putting nature first: A proposal for whole ecosystem management. *Orion Nature Quarterly* 5, 5–19.
23. Ulanowicz, R. E. 1986. *Growth and Development: Ecosystems Phenomenology.* Springer-Verlag. New York.
24. Rutledge, R. W., Basorre, B. L. and Mulholland, R. J. 1976. Ecological stability: An information theory viewpoint. *J. Theor. Biol.* 57, 355–371.
25. Ulanowicz, R. E. 1980. An hypothesis on the development of natural communities. *J. Theor. Biol.* 85, 233–245.
26. Ulanowicz, R. E. and Norden, J. S. 1990. Symmetrical overhead in flow networks. *Int. J. Sys. Sci.* 21, 429–437.
27. Haskell, B. D., Norton, B. G. and Costanza, R. (eds). 1992. *Ecosystem Health: New Goals for Environmental Management.* Island Press, Covelo, CA.
28. Ciriarcy-Wantrup, S. V. 1959. *Resource Conservation: Economics and Politics.* University of California Division of Agricultural Services, Berkeley and Los Angeles.
29. Bishop, R. 1978. Endangered species and uncertainty: the economics of safe minimum standards. *Am. J. Agric. Econ.* 60: 10–18.
30. This research was supported by the U.S. Environmental Protection Agency, Office of Policy, Planning and Evaluation, through a cooperative agreement with Chesapeake Biological Laboratory. Benjamin Haskell contributed to the research for this paper.

# 17

## Ecological Integrity and Social Values

### At What Scale?

INTRODUCTION: THE SEARCH FOR A TRANSDISCIPLINARY SCIENCE
OF SUSTAINABILITY

A number of recent writers and speakers have asserted, and I strongly agree
with them, that ecosystem health and ecosystem integrity require normative
definitions (Norton 1988; Norton 1991a; Callicott 1992; Rapport 1992). Since
it is not questioned that these terms must also have descriptive content,
it is fair to say that they are both normative and descriptive and that they
form the central terms of an emerging science of ecologically sensitive environmental management. It is, therefore, imperative that these definitions
be clear and operational; but it is also necessary that the terms reflect, in
some important way, human and social values. The purpose of this paper
is to explore what it really and practically means to propose a science capable of both describing and evaluating changes in the physical and biotic
systems that form the human environment. In the process, I illustrate how
social values might be integrated into a science of management by outlining
a scalar approach to valuing and describing changes in states of ecological
systems.

Providing definitions for the key terms, *ecological health* and *ecological
integrity*, is far more complicated than providing dictionary-type definitions
because these concepts must be embedded in a new, more comprehensive conceptual framework – what is sometimes called a new paradigm. (In this paper
I will discuss integrity mainly, but many points made here about the logic of
descriptive-normative concepts apply as well to ecosystem health. I believe
that the term ecosystem health performs an important function in public policy
because it emphasizes the evaluative, goal-directed aspect of environmental

From *Ecosystem Health* 1 (1995): 228–241. Reprinted with permission.

policy, by analogy to human medicine. But I think integrity shows more promise in building links between social valuation and ecological science, which is the goal here.) Defining these concepts will require no less than the creation of a new, transdisciplinary science capable of integrating information from many disciplines in search of scientifically founded and morally supportable environmental policy. With some trepidation I undertake to provide a language for the characterization and analysis of environmental problems, a language supported by a set of axioms, definitions, and operationalizations that have sufficient semantic richness to express both descriptive and evaluative judgments. This bare-bones language may provide a general direction for developing scientific, analytic models of environmental problems in which values, as well as factual information and hypotheses, are endogenous to the analysis. Progress in this direction may, in turn, contribute to the larger task of providing a comprehensive theory of environmental science, evaluation, and policy.

First, a note about disciplinary terminology. My understanding of scalar issues is guided by systems thinking and in particular by what is sometimes called hierarchy theory, an application of general systems theory to ecology. Further, I include economic values as part of the analysis, so it is tempting to call my approach ecological economics but that would also be inaccurate in two senses. The concerns are broader than ecological relationships – human systems must be integrated into physical systems such as into the atmosphere as well as into biotic communities; and the values to be explored apparently require a richer moral vocabulary than the preferred utilitarianism that dominates contemporary mainstream welfare economics (Norton 1991a; Chapter 11, this volume). Thus, while my ideas have much in common with those of ecological economists, such as Herman Daly and Robert Costanza, I see my approach as more directly complementary and contributory to the tradition of adaptive management as developed by C. S. Holling, Carl Walters, and colleagues (Holling 1978; Walters 1986; Lee 1993). It is an advantage that this is an active, management tradition and that the approach to environmental policy proposed by adaptive managers is openly linked to John Dewey (1987) and the pragmatist tradition of the use of science in solving public problems (Lee 1993; Holling, Gunderson, and Light, 1995). A management science must include elements of economic and ecological information. But management is necessarily normative and policy oriented. So far, the adaptive management tradition has taken values as givens in an otherwise dynamic model of human activity within ecosystems. I argue, in the tradition of Dewey, that valuation as well as information must be a part of the experimental adaptive process if we are to understand environmental values, policy, and management. This can be thought of as closing the action-environment-value feedback loop and will represent a further extention of recent advances in the conceptualization of adaptive environmental management to include a political dimension (Lee 1993). Both economics and ecology, as scientific disciplines, will be part of the broader approach, of course. But only a truly transdisciplinary language – one that integrates information, theory, and evaluation – can serve as the basic vocabulary for an activist management science.

Aside from wishing to avoid the inaccuracies of the phrase, ecological economics, I also doubt whether a truly ecological economics can be created simply by harnessing the traditional models of ecological science in tandem with descriptive economic models. Simply merging two descriptive sciences cannot create a normative one. Nor will rational and defensible policy result if ecology and economics feed information to policy analysts and policy makers, leaving to them the task of integrating bodies of unrelated science to propose and implement rational policies. A major problem of government agencies is that the information available is not the information they need to make intelligent decisions. For example, biologists and ecologists usually study dynamics of natural systems at smaller spatial scales and over shorter time frames than are relevant to long-term ecosystem planning (Pimm 1991). How is ecological and economic information to be incorporated into a more inclusive treatment of both description and prescription that is both scientifically respectable and normatively adequate to guide an environmentally sensitive public policy? It is, of course, essential that both physical and social-scientific information have an important input into the formation of policy. But, as long as the sciences remain insular and value-neutral, values and value formation will remain exogenous to the process of social learning through experimental action. Ecological and economic information comes to bear on environmental policy through action, reaction, reflection, and renewed action. Social learning must be achieved through active experimentation and public involvement in decision making. What is needed is a unified theory of environmental management – a plan of action, not a pure description of the stage on which humans and governments act. The particular scientific disciplines must find their role in this larger, normative and activist picture according to their contribution to practical environmental goals, including the achievement of sustainable use of the environment.

The new transdisciplinary science of management must accordingly fulfill at least two conditions: (a) it must be multiscalar/integrative (in the sense that it analyzes events and environmental problems, and applies proposed policies on more than one spatial scale) and (b) it must be normative (in the sense that it must go beyond the standard, modernist conception of science as a value-free description of natural processes). In the remainder of this paper I explore the reasons for, and some consequences of, these two conditions.

ADAPTIVE MODELING OF ENVIRONMENTAL PROBLEMS

Holling and his colleagues have provided much of the basic picture for a scale-sensitive treatment of environmental management. Holling refers to the multileveled system of nature as a "Panarchy," which refers to the Greek god, Pan, god of field, forests, and wild animals, and evokes, through the prefix "pan," meaning "all-embracing," the important idea that dynamism and information flow pervade the entire system (Holling, Gunderson, and Peterson, no date). Holling's system describes a Panarchic, self-organizing system viewed from the inside and requires a different paradigm than that of Newtonian, objective and detached science; it also requires a departure from mechanistic models in favor of dynamic, nonlinear models of reality. Holling's idea of a panarchy incorporates the formal rules of hierarchy to characterize, from the inside, the scaled environment as it is encountered by human perceiver/ actors.

In a panarchy, free choices by level-one individuals respond and adapt to patterns and processes at level two. These choices and activities are analogous to one's sense of individual freedom – the freedom to act independently of constraint, the possibility to choose between available options, to adapt in an effort to survive, prosper, and leave successors. In a dynamic, irreversible system, time is necessarily asymmetrical. Constraints flow down the hierarchy, but information flows upward in the system as well – the aggregated choices of level-one individuals are component processes in the larger, landscape-scale environment. Individual, free choices are made against a stable backdrop, or environment, which appears to individual choosers as a mix of opportunities and constraints. In essence, individuals appear both as individuals on one level and also as parts in a larger dynamic system. In a longer frame of time, the cumulative impacts of individual choices reverberate as changes in the environment – and alter the ratio of opportunities to constraints faced by future persons. The intergenerational impacts of a culture on the landscape can, therefore, be understood in terms of cross-generational exchanges in which the operative currency is freedom (options) expressed as a ratio of opportunities to constraints. These opportunities are stored in the structure of an ecological system. If growth in human impacts erodes natural capital stored in ecological systems on larger and slower time scales, future generations will face more constraints and fewer opportunities as cumulative impacts of many human individual actions skim off opportunities (such as capital stored in ancient forests) and leave constraints (poor and eroding soils exposed by clear-cuts). Alternatively, if resources are protected and if important and productive ecosystem processes are maintained, the people of the future will, as we have, face an abundance of opportunities.

Ecological systems are thus modeled as dynamic, open systems that are organized asymmetrically in space and time. Economic systems are understood as dynamics driven by choices of individuals who live, choose, and die on relatively rapid cycles by comparison to the rate of change in the surrounding environment. Regularities and predictable patterns on these larger levels provide the opportunity for biological and cultural evolution. Human individuals interact with their environment, usually through the mediation of an economic system. Individual perception, accordingly, is geared to short-term changes that occur in economic time. Large-scale ecological impacts of human activities must be understood both as results of cumulative, individual actions on level one, and also as having spill-over impacts on the larger scale of environmental systems – systems that would normally change so slowly that their dynamic is unnoticed by short-sighted humans. One requirement is to analyze our impacts on these slower scales because accelerated change at these scales may affect the context in which future members of our society will face choices and adapt to their environment.

Using his own spruce-budworm research and supplementing it with several other examples, Holling (1992) argues that human activities, when they simplify processes by concentrating productivity into a few crops or species, can make large-scale systems more ecologically brittle. Because this style of management reduces the redundancy of pathways fulfilling such essential functions as energy transfer, the system becomes more prone to shift into a new steady, functional state, one that is less likely to be supportive of the cultural and economic behaviors/adaptations that have emerged in response to the opportunities available in local environments. Accordingly, Common and Perrings (1992) and Holling (1995) proposed that we recognize two concepts of resilience. One concept, which operates at the economic/engineering level of the system, assumes a single, stable point to which the system, once disturbed, will return. This concept is appropriate for individuals and governments to apply in making day-to-day economic decisions. It assumes that system organization and behavior will be unaffected by an individual decision and develops equilibrium models to understand human behavior on the individual level. The other concept of resilience – ecological resilience – becomes relevant when the concern is to understand cumulative impacts of individual decisions on the larger, ecosystem scale, where the concern is with risk that the larger system will shift into a new stable state of functioning. Because such new stable states are almost always less desirable for humans, it is important that environmental management be formulated and evaluated on both the short-term, single-state model and on the longer-term model that monitors whether cumulative impacts are threatening to exceed a threshold and cause a flip into a new system state.

INTEGRATING ECOLOGY AND ECONOMICS: CURRENT APPROACHES

Any successful integration of ecology and economics must involve questioning the assumptions of both ecology and economics in the search for a new, transdisciplinary science (Norton 1992). The new science, being normative, departs from the traditional idea of pure or value-free science. That traditional disciplinary ideal encourages a view of science in which independent disciplines develop explanatory models which are subsequently combined within a policy-oriented discipline such as environmental management to evaluate and justify particular policy options. Description is completed before prescription is undertaken.

The serial approach is well illustrated by the organizational structure of an otherwise excellent essay by Sandra S. Batie and Herman H. Shugart (1989). "The Biological Consequences of Climate Changes: An Ecological and Economic Assessment." This essay provides a method of spatio-temporal scaling as an important part of a system for evaluating changes that may result from global climate change. The method employed, a form of hierarchy theory, is explored below; here, I want to contrast my approach to description and evaluation with that of Batie and Shugart. The first half of their essay (which one assumes expresses mainly the contribution of the ecologist Shugart) provides general guidelines for describing the scalar aspects of physical systems and provides examples of useful hierarchical models for interpreting the multiscalar activities of complex natural systems that change and evolve on many different temporal and spatial scales. Once the descriptive aspect of the problem is treated, the economist Batie brings the evaluative framework of economics to bear on the changing states of the physical system in the second half of the article. Similarly, efforts to create ecological economic models of the Patuxent Watershed, a case study designed to apply ideas developed in the Ecosystem Valuation Forum, take a similar approach by developing economic and ecological models independently and then finding points of contact between the independent models (Bockstael et al. 1995).

This approach is reassuring to scientists because it suggests that our valuational opinions are formed on the basis of a scientifically respectable objective description of the system being analyzed. But I doubt that building bridges to connect insular disciplines of science – social and physical – will be sufficient to yield a truly normative science of environmental management. Creative solutions to environmental problems will emerge only within a paradigm that is rich enough to integrate science and values within a single descriptive, normative, and action-oriented panarchy.

### ENVIRONMENTAL PROBLEMS AS PROBLEMS OF SCALE

Environmental problems are, most basically, problems of scale. This very general statement can be explained by a series of analogies. First, consider a generalization of Garrett Hardin's classic argument of the tragedy of common-access resources, which explains by a simple analogy the tragic destruction of common-access resources (1968). Each herder will reason that, by adding an animal to his own herd, he will reap most of the benefits but that the costs in degradation of the pasture will be shared by the whole community. Inexorably, this reasoning leads to the destruction of the pasture and the ruin of all as each individual follows the logic of self-interest without concern for the attendant stress on the communal resource. (I am aware that there have been many discussions of the historical accuracy and the true message of Hardin's elegant analogy. My concern here is so abstract that these fine points are not relevant. As will become clearer below, I am interested in Hardin's analogy as an abstract model – as a concrete guide toward a semantic interpretation of formalistically defined concepts.)

Note that Hardin's schematic analogy, originally formulated to illuminate human population growth, can be generalized to apply to any public good that can be used destructively by individuals. Hardin's tragedy has two scalar aspects. First, the tragedy can be tracked as one of comparative scale – the addition of animals to the collective herd incrementally changes the ratio of animal to vegetable biomass on the range (see Vitouszek et al. 1986). The common pasture analogy represents an ecological relationship between grazing animals and plants in the system. Comparative scale – the size of the herd relative to the volume of vegetation – is a physically describable ratio; at some point in the process of adding animals (and possibly with co-causes such as a drought) the communal grazing resource will predictably be damaged. (This idea of comparative scale, which measures the size [size = throughput] of an economic activity in comparison to its physical resource base, has been developed by Daly and Cobb [1989].) Applying this idea to environmental problems that threaten public goods more generally, comparative scale represents a physical, measurable relationship between trends in individual behavior and the required resource base supporting those behaviors. Eventually these trends, if unchecked, will reveal the fragility of the ecological system in the physical dimension affected.

The increasing comparative scale of activity to resource base is only a symptom of a second, underlying problem of the scale at which decisions are made. The tragedy of the commons dramatizes the fact that self-interested individuals, acting rationally, inexorably destroy a collective good. Note that Hardin's argument involves goods that emerge on separate levels. There is the good of individual herders and the communal good. The tragedy is that unconstrained pursuit of good on one level leads inevitably to the tragedy of destroying another good that exists on the community level and emerges in the long run.

Economists often argue that the problem with open-access resources – as described in Hardin's tragedy of the commons – is a problem of market externalities caused by lack of private ownership of the resource. Based on this analysis, assigning ownership rights to the resource would create incentives to protect the resource. This analysis misses the remarkable robustness of the Hardin model in that the scalar process can be understood under many approaches to ownership of public and private goods. Private goods are myopic in time and parochial across distance; they are allocated according to self-regarding and relatively short-term values. Those self-regarding values are strongly mediated by the constraints and opportunities of the economic system – the rules, subsidies, and constraints imposed by the structure of an economy. The rules of markets shape short-term economic behavior. The herders, acting as individuals, make assumptions that would be quite inappropriate for intergenerational planning.

Even if there is private ownership of a resource, there may be situations (when there are excellent reinvestment opportunities elsewhere, for example) in which individual profits will be maximized by a strategy of destructive over-use of resources accompanied by investment in alternative opportunities (Clark 1974). By their very nature, public goods are nonexclusive and open to all. They emerge on a community level rather than on the individual level. Given that most environmental goods are public goods, these goods can at best be treated as market goods only by counterfactual conditional. The economists' model is, in this sense, based on a necessarily false assumption. It may have heuristic value in some situations, but it can hardly serve as a promising basis for theory-building. One reason to reject the counterfactual market model as a representation of public goods is that environmental public goods, being communal in nature, must be understood on a different time scale than are individual consumer satisfactions; losses to the communal productive capacities will emerge on a slower, community-wide scale, unlike the increments to the incomes of herders as a result from adding more animals. Thus, any model, such as the present preference model of mainstream economists on which all future values must be represented as present preferences, will be nonscalar, and, therefore, unable to capture the important scalar aspects of Hardin's analogy.

The value concepts of mainstream economics are unsuitable for a dynamic analysis of policy for two distinct but related reasons. First, as just noted, preferences are nonscalar, while environmental problems have an unavoidable scalar aspect. Values of the future are to be understood, based on the economists' model, as present values in the sense that a future value is the current willingness of individuals to pay to protect that value for the future. Second, preferences of individuals are conceived by economists as sovereign – they are taken as given for the sake of analysis. This is not to say that economists believe preferences never change but rather that they treat such changes as exogenous to the economic model (Norton 1991b; Chapter 11, this volume). In keeping with their preferences for equilibrium models, economists treat preference change as irrelevant within the duration of their analysis. Demands, and the preferences associated with them, are taken as givens in the static world of equilibrium conditions surrounding economic change. While this assumption may make sense if the goal is to analyze behaviors and values at a given time, it is clearly inadequate if the goal is to understand values, behaviors, and their impacts over longer periods of time (Chapter 11, this volume).

The herders' individual decisions to expand their herds are rational within the frame of time in which individual, economics valuations are expressed and measured, and given the assumptions about markets and current production possibilities. Community values, however, are experienced in connection with a stable and healthy pasture over multiple genrations and cannot be understood within the context of the temporal dynamic of individual choice. Like corporations that keep separate books for operating and capital expenditures, the strategy proposed is to analyze the impact of expenditures on both short-and long-term scales. By treating the destruction of the pasture as a market externality – a cost to individuals who are not involved in the transaction – economists encourage us to consider the two scales and their goods as reducible, one to the other. Given the scalar asymmetry in the decision problem just noted, however, it may be a mistake to attempt to aggregate across these two levels because the two goods are associated with distinct processes that unfold on different scales, and they should be kept in separate accounts. We might say that the individual decision must be understood in the short-term scale of the economy, while the destruction of the pasture emerges on a longer scale of time. Scalar asymmetry, therefore, lends credence to an analysis such as Holling's, which employs two conceptions of resilience applicable at different scales.

Consider a second analogy related to Hardin's but also expanding on the scalar relationships in a new way. Daly has introduced the analogy of a Plimsoll line as a way of understanding the carrying capacity of an ecological or physical system such as the pasture (Daly 1991). The Plimsoll line refers to a line placed on the side of a ship. As long as the Plimsoll line shows above the water, it indicates that the load on the ship does not exceed its capacity. The search for an ecological Plimsoll line provides an illustration of the basic role of carrying capacity in ecological economics, viewed as an expression of separate economic and ecological economics. Economic rationality urges the owner of a boat to add more cargo to increase the efficiency of transport, but physical characteristics of the boat place limits on the volume of the load that can be carried safely. Regulation becomes necessary to protect the lives of sailors from economic forces.

But references to carrying capacity are problematic in their own right and for some of the same reasons that the Plimsoll line analogy is imperfect. The analogy inaccurately suggests that there is a physically definable point at which the boat becomes unsafe. What is safe for a ship depends also on the roughness of the waters it will cross. The same can be applied to resources: Use of a pasture that is acceptable in most years might lead to collapse in a drought year. In extreme cases, for example, aggregated behavior can lead to desertification, a major and irreversible shift into a new, much less desirable steady state. This interscalar dynamic can also enhance productive resources. Today the carrying capacity of the land of Israel is many times greater than it was in 1948. As George Woodwell (1985) says, "The earth is obviously finite; its resources limited. . . . [But] the question of the limits for the support of man, at first sharp and reasonable, loses clarity under scrutiny; what seems finite becomes infinite, and what should be infinite becomes finite. And the entire frame of references shifts with human adaptation to a new circumstance. Nonetheless, there are limits although the limits may not themselves be unitary, stable, and finite."

Daly's Plimsoll analogy can be made more realistic by reference to yet a third analogy based on a tragic incident during U.S. preparations for the Gulf War. Servicemen returning from shore leave all rushed to one side of their smaller transport boat as they approached their home ship, causing the boat to capsize, with much loss of life. This tragic example gives us a simple paradigm case of a communal value that was, through oversight, destroyed by many trivial individual preferences; it may help us to understand how the individual and communal value systems interact in a complex landscape dynamic. Thinking abstractly, this example links two unquestioned values – human, individual freedom of action (represented by the free choices of individual passengers to locate themselves where they wish on the ship) and the shared value of avoiding capsize, a survival value. The example represents the possibility of cross-scale spill-over effects from cumulative individual actions on those very structural stabilities that provide the opportunities for free choice. Within the panarchical model, cross-scale spill-over effects, once a threshold has been passed, express themselves as flips into a new and less attractive steady state.

HOLLING'S PANARCHY AS A SCALE-SENSITIVE MODEL
OF ENVIRONMENTAL PROBLEMS

The capsizing boat example represents the dynamics of economies within a larger ecological system. The safety of the passengers depends on a complex relationship between carrying capacity and trends in individual behaviors. Suppose there are only a few dozen passengers on a boat that has a capacity of 300 passengers. There will be no need for restrictions on passengers' individual activities beyond those necessary to protect other individual passengers from harm or annoyance. On the next trip, suppose that the boat is full to capacity. While the boat can float safely and stay aright with this load, there may need to be special limits on individual behavior. At some point in loading the ship, it begins to matter a great deal where individuals stand or sit.

An increase in the comparative scale of human activities manifests itself in a new value. Constraints on individual freedom serve a social value that emerges only at the community level. The good of not following a trend to congregate on only one side of the boat is, therefore, a clear case of an emergent value – a value that emerges only when the population on the boat nears capacity. This unique characteristic of environmental problems – the fact that they threaten communal, long-term goods as the accumulation of human impacts increases toward maximal capacity – provides an important clue regarding environmental values. Environmental values are inherently and unavoidably scalar, contextual, and emergent.

315

To conclude this section, scale issues are at the heart of environmental problems in two related senses. First, as was learned in our examination of the tragedy of the commons scenario, environmental problems involve an asymmetrical problem in decision scale. Actions that are rational from an individual viewpoint lead inexorably to the destruction of a public, community-level value that emerges on a larger and more long-term scale. In this sense, failures of sustainability are driven by a social trap inherent in a scalar discontinuity between the scale of individual concern and the scale at which landscape-scale environmental problems emerge (Costanza 1987). Second, because environmental problems result from cross-scale impacts of trends in individual decisions, resulting in changes in the larger physical variables that form the environment, these changes are multigenerational impacts. They should be the focal subject of the search for an ecologically sustainable society.

## DEFINING INTEGRITY

Ecological integrity is best thought of as a term in public policy; it has prescriptive as well as descriptive content. Public policy analysis of environmental problems must, therefore, be conducted in a vernacular that is richer than that of either descriptive economics or ecology.

By embedding value formation in a broader search for adaptive management policies and by developing a normative-descriptive language of adaptive management, it is possible to discuss environmental management holistically, rather than treating values and value formation as exogenous to environmental science. The proposed approach, therefore, differs from the treatment of values by both the economists (who treat human values as unquestioned preferences that drive the economic system) and by advocates of nonanthropocentric definitions of integrity that refer to values as being intrinsic to nature. Both of these positions, apparently at loggerheads philosophically, agree that valuation is exogenous to natural science. Values are, in both cases, found outside human experience, exogenous to any process of inquiry that might correct them; both economists and Deep Ecologists insist that values exist independently of the dynamic interaction of humans and their natural environments. (See Callicott [1989] and Westra [1993] for accounts of integrity that assume moral standing for ecosystems – they are in this sense nonanthropocentric, or ecocentric. I have criticized this approach, both as an interpretation of Leopold's idea of integrity and as a guide to management in Norton [1996; Chapter 1, this volume]). Contrary to these separatist approaches, my approach to environmental valuation attempts to make values,

316

and forces leading to their change, endogenous to an adaptive, scale-sensitive model of environmental management.

As noted above, hierarchy theory provides, within panarchical models, a formalization of space-time relations as encountered from within a complex, evolving, multiscalar system. Hierarchy theory was introduced as a useful conceptualization for understanding ecological system, but I propose a broader role for it. Hierarchy theory will provide the basic structure and methodology of a new paradigm of environmental management. As an activist paradigm, it must guide judgments of social goals as well as incorporate the best science of the day into an analysis of environmental problems.

In this application to management, hierarchy theory is really a method for organizing information regarding complex systems into multiple scales rather than a theory. I use it to make explicit the spatio-temporal relationships subsumed within the panarchical model. It rests on two key assumptions/principles: (1) that all observation must be from some point inside the hierarchically organized system that is being measured and (2) that smaller subsystems within the hierarchy change at a slower pace that represents a quantum difference from the pace of change in the larger system in which it is embedded – its environment (Allen & Starr 1982; O'Neill et al. 1986; Norton 1990).

Given the two descriptive axioms of hierarchy theory, it is possible to characterize the scalar aspects of environmental problems quite generally following Holling in treating the object of management as a panarchy observed from within. To hierarchy theory I add Principle (3): all human valuation must be from some point inside the hierarchical organization system that is being measured and evaluated. The ability to reflect important human social values are in this approach as important as descriptive adequacy in determining which models we build for measuring progress toward our goals of environmental protection. In this way, we shape hierarchy theory to the task of scale-sensitive management and away from the task of pure description. Can hierarchical reasoning from within a complex, dynamic system help us also to understand the connection between human-induced changes in the environment and human social values on the community scale?

One problem is that there are many, in fact an infinite number, of descriptively adequate accounts of the complex systems of nature because it is possible to focus on many distinct scales in complex systems, all of them involving dynamics with their own integrity (Allen & Starr 1982; Levin 1992). Further, different descriptors may be used, depending on the level of the system one concentrates upon. There is no amount of data regarding the physical behaviors of the system that will resolve which level is the level of interest to an observer. Only the observer can decide that. Values, either explicitly or implicitly, will determine which scales and which aspects of the system will be described, modeled, and measured. But, within a normative, managerial model, this value aspect is embraced and turned to advantage. If the goal is good management, not just understanding, our analytic language must be powerful enough to model the role of human values, and the factors that cause them to change in response to new information, as part of the system under study.

In the capsizing boat analogy developed above, cumulative changes in the behaviors of passengers represent a trend as all passengers drift toward the railing that is approaching the larger ship. This trend appears, eventually and tragically, on the larger level as a flip into a new steady state resulting from a threshold being passed. This discontinuity results, on the larger scale, in disaster – what might be called a negative cross-scale spill-over – and corresponds to concerns regarding the comparative scale of an economy and its environmental systems. In order to protect the broader interests of the common good or as Dewey (1987) would call it, "The Public," it is necessary to effect changes in the preferences individuals fulfill. In the panarchic model, individual freedom can now be understood as a basic value of the system – a value that emerges over generations as individuals of the future face many, rather than few, options. This value can be represented as the community-level value of maintaining, over many generations, the conditions of choice, a high opportunity-to-constraints index as expressed in the relationship between a culture and its ecological niche. Provided individual freedom is accepted as an unquestioned good, it is, therefore, possible to relate physical changes in comparative scale and impact of an economy to its environment. The value of human freedom (understood as the opportunity of each successive generation to choose among many options) is representable, and in principle measurable, as a function of changes in the comparative scale of an economy to its environment.

## SPATIO-TEMPORAL INTEGRITY AS A GUIDE
## TO ENVIRONMENTAL POLICY

As Aldo Leopold said, the human species differs from historical, natural forces in that technology and conscious planning have created a capacity for unprecedented and rapid alteration of physical processes on a landscape scale. While humans have evolved naturally, human impacts since the industrial revolution have been distinctive in their rapidity, scope, and violence

(Leopold 1939). Leopold spoke of the violence of the treatment of natural systems in the pursuit of economic gain. Cultural evolution has accelerated far beyond the limits inherent in biological evolution. What is also new "under the sun," he noted in his elegy to the passenger pigeon, is a species capable of regret, of mourning the passing of a species (Leopold 1949) or the destruction of the integrity of a place (Ehrenfeld 1993). But not all human actions must extinguish species and destroy the integrity of places. What we need, in developing aspirations for a multigenerational community, is a vision of the future – a vision that is rich in possibilities for human communities to live in harmony with nature (paper 19, this volume).

Given the ability of communities to regret irreversible losses of species or local integrity, it follows that conscious planning and, in some cases, the need for forbearance from certain activities must be considered a political problem with scientific, historical, and cultural aspects. It seems clear that improving standards of living, at least of the poor, is an unquestioned goal; but so is the goal of protecting the integrity of environments for the future. The problem of management is to pursue these twin goals simultaneously and in some reasonable balance. The goal is to define an ideal of a good life in a good environment to guide social policy. What is needed is a language of resource analysis that combines the rigor of scientific discourse with a method for measuring the much less deterministic process of human value formation. The physical aspect of this unified language is represented in the comparative scale of human impacts on productive systems. In its physical aspect, an environmental problem is an interscalar spill-over – an impact on the larger scale value resulting from cumulative changes in individual behavior. Provided that one accepts the value of freedom – the ability of the future to choose among adequate options – there is value in reducing the stress human economies place on natural systems. We can, therefore, define one essential part of the bequest each generation owes the next and following generations. Each generation is required to pass on to the next a set of physical and ecological processes that are capable of supporting many options for future use, understanding, and perhaps worship – a positive opportunities-to-constraints index. A nondeclining opportunities-to-constraints index would mean that the environment future generations inhabits would have roughly the same mix of opportunities and constraints as were encountered by earlier generations. The proposed language is rich enough to express both descriptive and evaluative judgments. It we assume only that individuals in the future will value freedom of action, and if we value the future, we have a rational interest in limiting processes that will have severe impacts on human freedom and options in the multigenerational scale of time. These changes will be experienced by the future as a shift in the ratio of constraints to opportunities facing individuals in their context of action. The question is whether we can determine these physical thresholds and chart a policy course that will avoid approaching the bifurcation points that threaten to irreversibly change the organization of the habitat for future generations and reduce the options available to them.

Destroying resources for short-term gain reduces future options, while saving the resource and the options it embodies requires development that maintains options. Can human societies adjust their preferences in the face of information that unlimited economic growth is incompatible with a future free of constraints on individual action? If not, it is difficult to see how protection of environmental values can occur in a society of free individuals (Ludwig, et al. 1993). If there is to be a feedback loop through which information about the impact of human behaviors and actions will affect those behaviors, it is apparently essential that preferences be considered dynamic and corrigible (Chapter 11, this volume).

## WIN–WIN STRATEGIES IN A MULTISCALAR SYSTEM
## OF MANAGEMENT

So far, the line of argument developed here may sound bleak; growing populations and growing economies will lead to more and more threats to important ecological dynamics and more threats to the options of future generations. Holling recognized the apparent Malthusian consequences of the panarchical model, when applied on all levels up to and including the global scale (Holling 1994). The separation of physical and decisional scales, however, has an optimistic side. This conceptual separation of physical and decision levels opens the possibility of win–win situations in which we devise policies that solve economic problems in ways that also positively affect long-term, larger-scale dynamics that are necessary to support intergenerational human values. In the ship case, a little foresight might have avoided catastrophe; an experienced captain or cook might simply have served lunch on the other side of the boat during the approach. Some passengers would have chosen to eat, others to greet their friends, and the tragedy would have been averted without restrictions on individual freedom. An adjustment of opportunities available to individuals – in this case an expansion of options and opportunities – leads to diversity of responses and to avoidance of catastrophe without regulation of behavior. While restrictive regulations limiting where people can stand provide one means to avoid capsizing, there are other methods

320

that accomplish the larger-scale goal and increase options at the individual level. The adjustment of the opportunities and incentives results in a free-choice solution at the individual level to the risk of capsizing at the level of the ship.

Is there an environmental analogue to serving lunch on the other side of the boat? In a typical tragedy-of-the-common-pasture example, suppose that a local economy has been successful in raising live animals that have been shipped elsewhere to be butchered and processed. Herders are prosperous and are inclined to invest – hence their tendency to buy more animals. A wise policy would be for the local government, or some private group, to begin a herders' cooperative to process meats and tan hides; herders then have expanded opportunities to invest in a profit-making venture by pursuing alternative investment options. These new investments will increase local economic activity and job opportunities for the community and, at the same time, remove pressure from the pasture because economic development can now seek a path that increases local economic activity without increasing the demand on resources, represented as stress on the pasture. (The point, one might say, is that while nature offers no free lunches, there are lunches selling two for the price of one – a policy which simultaneously creates development [through complexification] and encourages the creation of more jobs per unit of resources used, provides new opportunities for investment and consumption, and reduces the stress on resource-producing systems.) This analysis allows us to incorporate the useful distinction of Daly and Cobb (1989) between growth and development into our model. When faced with a threatened tragedy (based on a descriptive model of changing physical scales), the strategy must be to seek developmental opportunities that will encourage greater complexity – adding a layer of processing or recycling – in the economy, rather than simple growth in the throughput of the economy. Development refers to increases in economic activity that does not negatively affect the mix of constraints and opportunities faced in the future; growth, which depends on increased throughput – in comparative scale – irreversibly destroys opportunities and options by skimming opportunities and leaving constraints. This solution embodies the insight that a multiscalar problem, such as the capsizing boat, cannot be solved on the same level. It requires a solution from another level and will normally involve development (complexification) rather than growth on the scale at issue.

To apply this reasoning to a real case, it suggests that the solution to the conflict between industry (loggers) and environmentalists in the Pacific Northwest would have been to encourage diversified industries based on local resources, including old-growth timber, decades ago. While it may be too late to apply this solution given present conditions, the argument is instructive nonetheless. Suppose that, from the perspective of 30 years ago, state and local governments had provided guaranteed, low-interest loans to encourage cooperatively owned sawmills and furniture manufacturing plants and in other ways discouraged the export of whole logs and unprocessed forest materials out of the region. This action, had it created new economic activities and more jobs per unit of resources, might have protected long-term values of the resource base (opportunities) and, at the same time, produced a more diverse economy that generated more jobs per log cut. Examples such as this show how arranging incentives in a panarchy might lead to win–win situations in which development occurs without growth in the throughput of the system and perhaps without tragedy to intergenerational options and freedoms.

Success in this enterprise can be conceptualized as changing incentive structures of individual actors by expanding investment opportunities and encouraging a diversity of responses that may reduce stress on specific resources. This policy can be seen as protecting the ratio of opportunities to constraints as they are exemplified in the interaction of individuals and their resource-producing environment. On this view, the standing timber represents and supports future options, while clear-cut and eroding hillsides represent constraints on these options. Practices such as export of whole logs to other countries or regions may maximize profits of multinational timber companies, but this is a clear case of nonlocal interests skimming the opportunities present in an environment. Because profits can be transferred to investments in other places, while leaving the constraints embodied in that environment for the locals as companies move on to other locales, local and regional economies are victims of cut-and-run business (Lee 1993). This process is simply one of internalizing opportunities and externalizing costs of unwise development onto local communities. These local communities would be better served if the incentive structure for investment opportunities were changed so that the export of whole logs was less profitable and investments in industries that process whole logs into finished lumber or furniture were more profitable. (Somewhat more strongly, one might argue that it would be wise to decrease incentives to export whole logs, perhaps through an extraction tax.) Notice that this line of reasoning cannot be formulated without positing a value that emerges on the local, community scale; a value that cannot be captured in the incomes and profits of individuals and companies. It is a macroscale value placed by the community on a preference for development that will preserve future opportunities and options over the current development that allows the skimming of opportunities for immediate profit at the expense of future options in that local community. . . .

CONCLUSION

I propose that we abandon the attempt to achieve an adequate evaluation of changes in the environment through the serial approach, which first models changes in physical states of natural systems and then evaluates them. I propose an iterative process of interdisciplinary interaction in search of a unified paradigm for management, with valuation and description being pursued in a way that will lead to social learning. The goal is to develop an integrated set of values and policies, a set of economic policies that encourage development paths that reduce disruption of natural systems. This set of values and policies can only occur through a process of social learning. The search for social learning must encourage learning regarding human values and perceptions as well as provide the information base for environmental decision making.

One way to put teeth in the statement that the science of environmental management must be both normative and descriptive is to adopt scale-sensitive, descriptive models that organize information and values into models for environmental problems and associate these values with important ecological processes. This can be accomplished by hypothesizing a relationship between scales of human concern and particular physical dynamics that are associated with broad social values. Then attempts to model these associated physical dynamics can be thought of as a first attempt at determining the scale at which environmental problems should be characterized and addressed. These values can then be associated with different layers of a temporal and spatial panarchy, which is modeled from a perspective within. Human choice is shaped from above by the opportunities and constraints forming the context of individual decisions. The collective impact of human activities/choices can be analyzed, at the individual scale, as individuals reacting to that set of opportunities and constraints.

One advantage of the visualization provided by Figure 17.1 is that it helps us to conceptualize a new criterion for acceptable and appropriate environmental action. I refer to the proposed criterion as advocating actions and policies that conform to the scalar Pareto optimality criterion (see Figure 3.2). The scalar Pareto criterion represents a multiscalar application of the Pareto optimality criterion, which was originally stated on the individual level as the requirement that all actions have a positive impact on some individuals and negative impacts on nobody. The scalar application of the Pareto criterion is stated as follows: choose policies that, from the viewpoint of a representative individual in each community, will have positive (or at least nonnegative) impacts on goals formulated by that person on the individual level, on the community level, and on the global level. While the scalar Pareto approach

| Temporal Horizon of Human Concern | Time Scale | Temporal Dynamics in Nature |
|---|---|---|
| Individual/economic | 0–5 years | Human economies |
| Community, intergenerational bequests | up to 200 years | Ecological dynamics/ Interaction of species in communities |
| Species survival and our genetic successors | indefinite time | Global physical systems |

Figure 17.1. Correlation of human concerns and natural system dynamics at different temporal scales.

retains an individualist perspective (it is human individuals who formulate, discuss, and defend values on all levels), it does not seek reduction of all values to economic preferences. It is pluralistic in the sense that the value of ecosystems is understood on a community, not an individual, scale and no reduction of community-level values to individual values is attempted. But the pluralistic ethic is also integrative in the sense that we choose actions that will have positive (or at least nonnegative) impacts on the relatively distinct dynamics that produce and support human values that are expressed on multiple scales (here hypothesized as three).

The approach to human values outlined here, and applied within a scalar analysis of environmental problems, attempts to avoid reductionism and unidimensionality in valuation and to develop a normative-descriptive language for quantification of the idea of the integrity of a place – the harmonious relating of a culture to its ecological context. Mainstream economic analysis reduces all preferences, including ones that will be experienced in the future, to nonscalar present values in search of policy guidance by computation and aggregation. Put simply, the proposed approach relaxes the assumption that environmental values can only be expressed in a single, unidimensional currency. The outcome of this relaxation leads to a more pluralistic and scalesensitive approach to description, evaluation, and management of human impacts on natural systems.

Intuitively, the system of policy analysis I described represents a search for harmony in resource use. To this end, I avoided developing descriptive accounts and normative accounts of nature separately. Because the goal of management as understood here is to place humans within nature and then to manage human impacts on natural processes, I tried to understand human values within a structurally similar decision space, a space with considerable options, but a space in which exercising some options will close further options in the future and other actions will protect or open options in the future. The key to harmonious solutions to resource-use problems may be to associate social values with distinguishable dynamics on multiple levels and to reduce cross-scale impacts on larger dynamics by encouraging decentralized development. These policy steps would encourage economic and ecological diversity by, for example, maximizing local jobs per unit of timber cut and also by developing a sense of local responsibility for local resources. However, the purpose of this paper has been more conceptual than substantive; the key point made here is that if we are to gain a comprehensive understanding of human impacts on natural systems, a scalar approach to values is as necessary as a scalar understanding of complex physical systems. Furthermore, if we are to model this entire dynamic, the models can only be stated within a language of sufficient semantic richness to express and include normative as well as descriptive content.

## REFERENCES

Allen, T. F. H., Starr, T. B. (1982) *Hierarchy: Perspectives for Ecological Complexity.* University of Chicago Press, Chicago, IL.

Batie, S., Shugart, H. (1989) The biological consequences of climate change: An ecological and economic assessment. In: Rosenberg, N., et al. (eds.), *Greenhouse Warming: Abatement and Adapation,* pp. 121–131. Resources for the Future, Washington, DC.

Bockstael, N., Constanza, R., Strand, I., Boyton, W., Bell, K., Wainger, L. (1995) Ecological economic modeling and valuation of ecosystems. *Ecological Economics* 14: 143–159.

Callicott, J. B. (1989) *In Defense of the Land Ethic.* State University of New York Press, Albany, New York.

Callicott, J. B. (1992) Aldo Leopold's metaphor. In: Costanza, R., Haskell, B., Norton, B. (eds.), *Ecosystem Health: New Goals for Environmental Management.* Island Press, Covelo, CA.

Clark, C. (1974) The economics of overexploitation. *Science* 181, 630–634.

Common, M. S., Blamey, R. K., Norton, T. W. (1993) Sustainability and environmental evaluation. *Environmental Values* 2, 299–334.

Common, M. S., Perrings, C. (1992) Towards an ecological economics of sustainability. *Ecological Economics* 6, 7–34.

Costanza, R. (1987) Social traps and environmental policy. *Bioscience* 37, 407–412.

Costanza, R. (1991) *Ecological Economics: The Science and Management of Sustainability.* Columbia University Press, New York.

Daly, H., (1991) Elements of environmental macroeconomics. In: Costanza, R. (ed.), *Ecological Economics,* pp. 32–46. Columbia University Press, New York.

Daly, H., Cobb, J. (1989) *For the Common Good.* Beacon Press, Boston.

Dewey, J. (1987) The public and its problems. In: Boydson, J. (ed.), *John Dewey: The Later Works, Vol. 2: 1925–27*, pp. 235–372. Southern Illinois University Press, Carbondale.

Ehrenfeld, D. (1993) *Beginning Again: People and Nature in the New Millennium.* Oxford University Press, New York.

Hardin, G. (1968) The tragedy of the commons. *Science* 162, 1243–1248.

Holling, C. S. (1978) *Adaptive Environmental Assessment and Management.* Wiley, London.

Holling, C. S. (1992) Cross-scale morphology, geometry, and dynamics of ecosystems. *Ecological Monographs* 624, 447–502.

Holling, C. S. (1994) An ecologist's view of the Malthusian conflict. In: Lindahl-Kiessling, K., Landburg, H. (eds.), *Population, Economic Development, and the Environment,* pp. 79–103. Oxford University Press, Oxford.

Holling, C. S. (1995) Engineering resilience vs. ecological resilience. In: Schultz, P. (ed.), *Engineering Within Ecological Constraints.* National Academy of Engineering, Washington, DC.

Holling, C. S., Gunderson, L., Light, S. S. (1995) *Barriers and Bridges to the Renewal of Ecosystems.* Columbia University Press, New York.

Holling, C. S., Gunderson, L., Peterson, G. (no date) Comparing ecological and social systems.

Lee, K. N. (1993) *Compass and Gyroscope: Integrating Science and Politics for the Environment.* Island Press, Covelo, CA.

Leopold, A. (1939) A biotic view of land. *Journal of Forestry* 37, 727–730.

Leopold, A. (1949) *A Sand County Almanac and Sketches Here and There.* Oxford University Press, London.

Levin, S. (1992) The problem of pattern and scale in ecology. *Ecology* 73, 1943–1967.

Ludwig, D., Hilburn, R., Walters, C. (1993) Uncertainty, resource exploitation and conservation: Lessons from history. *Science* 260, 17–19.

Norton, B. G. (1988) What is a conservation biologist? *Conservation Biology* 2, 237–238.

Norton, B. G. (1990) Context and hierarchy in Aldo Leopold's theory of environmental management. *Ecological Economics.* 2, 119–127.

Norton, B. G. (1991a) Ecological health and sustainable resource management. In: Costanza, R. (ed.), *Ecological Economics: The Science and Management of Sustainability,* pp. 102–117. Columbia University Press, New York.

Norton, B. G. (1991b) Thoreau's insect analogies: Or, why environmentalists hate mainstream economists. *Environmental Ethics* 13, 235–251.

Norton, B. G. (1992) A new paradigm for environmental management. In: Constanza, R., Norton, B., Haskell, B. (eds.), *Ecosystem Health: New Goals for Environmental Management,* pp. 24–41. Island Press, Covelo, CA.

Norton, B. G. (1996), Reduction integration: Two approaches to environmental values. In: Light, A., Katz, E. (eds.), *Environmental Pragmatism,* pp. 105–138. Routledge Publishers, London.

O'Neill, R. V., DeAngelis, D. L., Waide, J. B., Allen T. F. H. (1986) *A Hierarchical Concept of Ecosystems.* Princeton University Press, Princeton, NJ.

Page, T. (1977) *Conservation and Economic Efficiency.* Johns Hopkins University Press for Resources for the Future, Baltimore, MD.

Pimm, S. L. (1991) *The Balance of Nature? Ecological Issues in the Conservation of Species and Communities.* University of Chicago Press, Chicago, IL.

Rapport, D. (1992) What is clinical ecology? In: Costanza, R., Norton, B., Haskell, B. (eds.), *Ecosystem Health: New Goals for Environmental Management,* pp. 144–156. Island Press, Covelo, CA.

Sagoff, M. (1988) *The Economy of the Earth: Philosophy, Law and the Environment.* Cambridge University Press, Cambridge, U.K.

Toman, M. (1994) Economics and sustainability: Balancing tradeoffs and imperatives. *Land Economics* 70, 399–413.

Vitouszek, P. M., Ehrlich, P. R., Ehrlich, A. H., Matson, P. A. (1986) Human appropriation of the products of photosynthesis. *Bioscience* 36, 368–373.

Walters, C. (1986) *Adaptive Management of Renewable Resources.* Macmillan, New York.

Westra, L. (1993) *An Environmental Proposal for Ethics: The Principle of Integrity.* Rowman and Littlefield, Lanham, MD.

Woodwell, G. M. (1985) On the limits of nature. In: Repetto, R. (ed.), *The Global Possible,* pp. 47–65. Yale University Press, New Haven, CT.

# 18

## Change, Constancy, and Creativity

### The New Ecology and Some Old Problems

The New Ecology emphasizes change and dynamism in ecological systems, claiming that ecology has under-emphasized these features of natural systems and their organizational structures. This emphasis reminds me of a discussion that occurred on the first day of one of my courses in Environmental Ethics. The course mainly covers modern philosophies and attitudes, but I usually spend the first day talking about the ancient background of our modern ideas. I had just spoken of the emphasis in the Hebrew tradition on the eternal nature of Jahweh, and had gone on to expound on the fascination of early Greek philosophers with change and permanence. I noted that the precocious Heraclitus had proclaimed, "All is in flux," but that Parmenides, who denied even the possibility of change, was more representative of Greek thought.[1] I explained how Plato had declared the changing world of the senses illusory because this world lacked the stable and unchanging status of "Ideas" or "Forms." For Plato, only the constant and unchanging could be real. Then, a student asked perhaps the best question I have encountered in over twenty years of teaching: Why do the Judaeo-Christian tradition and the Greek tradition share the same reverence for the fixed and unchanging?

Philosophy, at its best, identifies and questions our deepest assumptions. The student had noticed that both the Hebrews and the Greeks, so different in other respects, apparently gravitated toward static, everlasting, ultimate explanations of the confusing and highly changeable world they encountered experientially. I paused and then gave some answer I do not remember. This question, however, was too good for an off-the-cuff answer. Having thought about the question until the next class meeting, I had a better but still very unsatisfying answer. I had to admit that I could see no philosophical

From Duke *Environmental Law and Policy Forum* 7 (1996): 49–70.

or intellectually defensible principle that could justify such a monumental assumption. My answer was that there seems to be a deep psychological need for constancy and stability in Western cultures, perhaps in all cultures. Somewhat lamely, we left the matter there and proceeded to discuss the rise of modernism. Fortunately, the topic of this special issue, *change* in ecological systems, provides an opportunity to return to this important question: How are we to conceptualize the rich mix of change and constancy that we encounter in the world of experience?

The New Ecology has attacked traditional ecological thought (what I will for convenience call the "Old Ecology") for emphasizing constancy, stasis, and equilibrium in describing ecological systems, and for under-emphasizing the role of change, disturbance, and dynamism. I tell the above anecdote because it occurs to me that the readiness of ecologists to embrace equilibrium theories and to find constancy in ecological events may have deep – perhaps even nonrational – sources. Equilibrium theories may not be *empirical* theories at all, but rather may represent pre-theoretical assumptions, which are perhaps rooted in a deep psychological need for stability in the face of threatening changes. Ecologists, too, are affected by psychological needs. If my speculations about the depth of the Western commitment to stability have any merit, we might acknowledge that we have no choice but to find some level or type of stability. The intellectual question then becomes one of how to characterize stability and how to reconcile it with the empirically obvious change we experience everywhere.

## I. TWO ARGUMENTS AGAINST THE "OLD ECOLOGY"

The question before us is: How will the ideas of the "New Ecology" affect environmental thought and environmental goals? I will begin to answer this question by distinguishing two arguments, both based in the New Ecology, which lead to two distinct criticisms of the Old Ecology.[2]

### A. The Argument from Constant Change

According to the argument from constant change, the Old Ecology was unaware that ecological systems are dynamic, changing systems. The Old Ecologists treated unchanging systems as the norm and therefore assumed that stasis is natural and that change requires explanation. Old Ecologists understood change as ultimately tending toward a climactic state of mature stability.

Disturbance was treated as a temporary derailment of the ecological train; the natural tendency was to build structure and to regain the stable, climactic state.

This emphasis on equilibrium systems negatively affected management practices, according to New Ecologists, because it encouraged managers to assume that they could exploit a particular species or resource while assuming that the system would be unaffected in deep and lasting ways. In the old models of management, exploitation-driven changes in the system represent only temporary deviations from a steady state. Insults to ecological systems can therefore be healed simply by relaxing harvesting pressure or reducing direct damage to resources; the system can be expected to go back to "normal."

According to the New Ecologists, it is dangerous to assume that eco-systems are equilibrium systems capable of absorbing insult and returning to their pre-disturbance state. This line of reasoning is sometimes carried further to suggest that, since the goals of environmental protection have so often been formulated in the vernacular of the Old Ecology, full recognition of the importance of change will necessitate a major re-thinking of environmental policy goals.

## B. *The Argument against Grand Theory*

One can also find a second critique of old-style managers in the writings of New Ecologists. According to this second argument, Old Ecologists over-emphasized grand and speculative theory, while New Ecologists pay less attention to general principles of ecosystem organization and study particular, local ecological interactions and their outcomes. While New Ecologists are careful to deny that they are anti-theory, they insist that theory must be built from the bottom up, by generalization from many specific studies, rather than from the top down, with broadly applicable hypotheses deduced from general principles of structure. According to the argument against grand theory, Old Ecology and old management can be faulted for letting a few grand and psychologically satisfying general theoretical principles guide their activities and for neglecting studies that reveal the special character of particular assemblages of species. If ecology is reformed, according to this second argument of New Ecologists, ecologists and environmental managers should avoid grand generalizations and emphasize local ecological knowledge in support of locally formulated environmental goals.[3] Rather than pontificate about damage to "nature's fabric," ecologists and environmental managers will, once corrected by New Ecology, have to form local coalitions, in which ecologists provide ecological information and support for local efforts to protect systems under stress.

## C. Responses to the Arguments of the New Ecology

I will consider both arguments in light of their potential impacts on the future of environmental policy. The arguments are clearly stated by Daniel Botkin:

> Admitting that change is necessary seems to open a Pandora's box of problems for environmentalists. The fear is simple: Once we have admitted that some kinds of changes are good, how then can we argue against any changes – against any alteration of the environment?[4]

While Botkin goes on to explain that there are more or less comforting answers to this question, and that eventually progress will ensue as we develop new approaches consistent with an ecology of change, his discussion is motivated by two claims: (1) that environmentalists and environmental managers have not been aware of the dynamic nature of ecological systems and (2) that once they do become aware of this dynamism, they will at least initially face a new set of problems in developing and defending their policies.

Botkin is correct to claim that environmental managers have often overlooked the importance of change and that they have often acted as if their exploitation of a species or resource would have no impact on the organization of the larger ecological system. He is also correct in implying that this failure to recognize the importance of change in natural systems has led to tragic failures of management. I disagree, however, with Botkin regarding the exact analysis of what has gone wrong. In order to see the difference between my approach and that of Botkin, I must introduce two qualifications to Botkin's argument: (1) that the idea of developing and using dynamic models has been around for a long time but that problems occur with details of their implementation, and (2) that New Ecologists, acting in reaction to the prior over-emphasis on the grand theory of stability in ecological systems, sometimes over-emphasize the importance of change in ecological systems.

The first qualification is historical in nature. Consider the following quotation:

> To the ecological mind, balance of nature has merits and also defects. Its merits are that it conceives of a collective total, that it imputes some utility to all species, and that it implies oscillations when balance is disturbed. Its defects are that there is only one point at which balance occurs, and that balance is normally static.[5]

According to Professor Botkin's arguments, I assume he would agree with this statement. The interesting thing about this quotation, however, is that the passage was published over fifty years ago *in 1939* by Aldo Leopold. During the 1930's and 1940's, Aldo Leopold, author of *A Sand County Almanac* and a hero to conservationists as the father of the land ethic, developed a general theory of dynamic environmental management based on ecological theory. This quotation represents a step in the evolution of that remarkable theory of management.[6] It could hardly be said that Leopold failed to share his insight with other environmental managers since the paper was published in *The Journal of Forestry*, a leading journal of environmental management. Further, this passage was not unique but represented a fascination with the topic of change and scales of change, to which Leopold returned throughout his career, most notably in the famous simile of "thinking like a mountain."[7] Leopold first became interested in ecology and the study of ecosystems when he attended a conference on natural cycles, which was devoted to the theoretical and practical question of how one can identify changes that are part of natural cycles and distinguish these from changes that result from human stressors.[8] At that conference he met Charles Elton, the distinguished British animal ecologist, and the two became friends and collaborators in theoretical and field studies of ecological systems.[9] Therefore, the idea that natural systems are dynamic and changing and that the use of equilibrium, models in management represents a simplification and even a falsification of natural systems has been present in ecology and environmental management for at least fifty years. Leopold and others recognized that equilibrium assumptions produce, at best, useful models that should always be qualified and supplemented with more dynamic models. Consequently, there can be no doubt that Leopold and his disciples took dynamism very seriously and that they tried, however unsuccessfully, to base their management on this insight.

If Botkin is right that environmental managers, through the present, have continued to act as if they are unaware that natural systems are dynamic, then we are faced with more questions. Why have environmental managers failed, in a half century, to recognize and implement Leopold's clearly-stated insight? Why should we think that Botkin's argument is going to have stunning effect, while Leopold's did not? Surely, we know more about change in ecological systems than we did in Leopold's day, so we can no doubt cite scientific ecological studies describing dynamic systems. But if Botkin's argument is correct, we cannot really trust that science after all – it is tainted because the models used to develop that scientific base were faulty. So again, why should we think Botkin's clear restatement of Leopold's doctrine of dynamism in nature will have more effect this time around? To answer this question fully, we may have to look beyond the failure of *scientific* insight. It may be that the reliance on assumptions of equilibrium and stability is more psychological than rational. While I agree that Botkin is right to point out the failure of both ecologists and environmental managers to develop and use dynamic models, it is important to recognize that the basic idea has been around at least since Leopold. The Devil, it turns out, is in the details.

The second qualification to Botkin's arguments concerns the possible over-reaction of New Ecologists. Because they are writing in reaction to decades of over-emphasis on stasis, stability, and unidirectionality in the development of ecological systems, New Ecologists sometimes over-emphasize the pervasiveness of change. It is not a good idea to pose the question of change versus stability in nature as if there may be an all-or-nothing answer, as if it might turn out that the world is either entirely changing or entirely stable. Both extremes were explored by the Greeks. Heraclitus, as noted, believed that "[a]ll is in flux"; Parmenides, at the other extreme, concluded that all change is illusory.[10] The truth surely is somewhere in between.

Accordingly, I doubt that New Ecologists intend to deny all constancy in nature. For example, they clearly accept the basic assumptions of the evolutionary/ecological worldview – that species, over many generations, adapt to regularities in their environment. Evolutionary theory demands that nature be sufficiently patterned so that a species can be shaped, through many repetitions of births and deaths, by natural selection. New Ecologists do not reject this basic assumption of all evolutionary/ecological theory – there must be enough pattern and predictability in the environment for populations to evolve and adapt to it. Their point, rather, is that change is more important, and constancy less important, than has been assumed by Old Ecologists.

Specifically, this means that disturbances can and should in many situations be thought of as expected and normal, and that many, perhaps most, communities are regulated more by disturbance than by some tendency toward climax. However, it does *not* mean that ecological systems contain nothing that is constant or predictable. The tree species that survive in fire-regulated communties – often cited as examples of disturbance-driven systems – have adapted to a pattern of periodic fires. The point is not to choose between change and constancy, but to achieve a better conceptualization of the confusing mix that we see in ecological communities.

## II. RE-THINKING ECOLOGICAL CONSTANCY

If we are to make sense of New Ecologists' justified emphasis on change, I believe their view must be given a scalar, hierarchical interpretation. We must recognize that every level of nature is constantly changing, but the overall process is driven by an interplay between relative change and relative constancies on different levels of a complex hierarchy. Nature is organized into a scalar hierarchy in which each level reacts to, and is also affected by, levels below and above. These effects decrease in directness and force as one goes up or down the hierarchy.[11]

We have a simple hierarchy of living systems. Each level represents a different dynamic, changing at slower and slower paces as we move out- ward, from cell to community. *Scale* is therefore the key to a hierarchical understanding of ecosystems. In a hierarchical world, scale, not only *tempo- ral scale,* but also *spatial scale*, is crucial to all understanding. Nature can best be understood as an organization of systems and subsystems, with larger systems changing more slowly than their constituent subsystems. The larger systems therefore provide a relatively constant background to which smaller subsystems can adapt. This constant background can be conceptualized as a complex mixture of constraints and as opportunities that present themselves to smaller, adaptive subsystems such as individual organisms.

Stability in nature may be called a *well-founded illusion.* It is an illusion because, as Heraclitus, Leopold, and Botkin tell us, nature changes at every scale. It is, nevertheless, well founded because huge differentials in the scale of Earth's processes are adequate to provide a workable sense of relative stability, stability that is experienced comparatively. Slowly changing back- ground variables are relatively static in contrast to rapidly changing variables, resulting in the appearance of stability. For instance, as a prime example of the pervasiveness of change, Botkin cites the discovery thirty years ago that the Earth's surface is actually composed of dynamic tectonic plates that have literally changed the face of the Earth in the longer scales of geological time.[12] Imagine yourself as a common housefly for a moment, trying to reproduce young and pass on your genes to a future generation. Houseflies go through their whole life cycle in a few days. About 1.03 million generations of house- files will be born and die in the time it takes the San Andreas fault line to move one mile. To a housefly, the chance of being run down by a tectonic plate would not be a major limiting factor in the reproductive process. Rel- ative to housefly time, tectonic change is imperceptible (the location of the continents occurs on a different scale); the continents are stable enough not to matter to individual houseflies. I could have given other examples, such as *Drosophila*, providing even larger numbers or smaller numbers, such as those for generations of humans. The point is that these numbers are all very large. Movements of tectonic plates are not a major determinant of human evolution because the pace of change in tectonic plates is so out of proportion to the scale of events in ordinary human lives. While the tectonic plates are changing, they change so slowly that for purposes affecting survival of human individuals the plates are stable.[13]

The principal point is that, even though everything is constantly changing, not everything changes at the same rate. Differences in temporal rates are so great that from any given perspective there will be environmental factors that are relatively constant, to a degree permitting the necessary adaptation a species requries to maintain its niche. Thus, while it is true that nature changes on every level, it is also true that there are layers upon layers of "relative stability."

I will now explain how my analysis of the failure of environmental management differs from Professor Botkin's. The reason environmental managers have not made use of Leopold's insight is due to the inability to operationalize Leopold's elegant theory of dynamic management. Leopold's theory has not been operationalized because no one has developed adequate conceptual tools for understanding the pace and scale of change in multi-layered ecosystems.[14] Once we follow Leopold and Botkin into the world of dynamic ecological management, we face a hopeless confusion of scalar models, with few rules for organizing them. As Stuart Pimm has recently emphasized, ecologists seldom do studies that can test ecological processes on scales longer than a few years.[15] For example, the National Science Foundation's limit for project funding is usually three years. Much of the theory explaining long-term processes in nature has therefore been based more on analogies and assumptions than on hard ecological evidence.[16] Very little is known about large-scale slow processes in nature; to my knowledge, no one has an algorithm for "scaling up" from small-scale and short-term studies to large-scale and long-term studies.

It is one thing to say, as Leopold clearly did, that nature is composed of many dynamic processes and that scale is crucial; it is quite another thing to furnish a detailed set of concepts for discussing change and its impacts across differing scales of time and space. I believe that hierarchy theory, explicitly stated in the early 1980's, provides a beginning for fashioning such tools.[17] It might be argued that the hierarchy theorists simply codified long-understood facts about differential paces of change. At any rate, the introduction and formalization of hierarchical models defines more precise spatiotemporal relationships, which may prove very helpful in conceptualizing applied problems in environmental policy and analysis.[18]

So while Botkin and I agree that we need to apply dynamic theory to the study of ecosystems and environmental management, I doubt that the problem can be solved simply by asserting that we need dynamic models. We actually need to produce the models, which in turn requires an adequate theoretical understanding of multi-scalar change. I also do not see this recognition and challenge, as Botkin apparently does, as requiring a radical re-thinking of the goals of environmentalism, although it will require a change in some for-mulations of environmental goals. I believe that Leopold's land ethic, which emphasizes the protection of the integrity of multi-scalar ecological systems, should be our basic guide to management.[19] According to this view, humans tend to perceive ecological systems on the shorter time scales of particular plots or of individual useful species, whereas impacts of human management also affect, on longer time scales, the very organization and structure of the larger ecological system. Good management, then, must monitor, in addition to short-term impacts, longer-term impacts on the organization and integrity of larger systems.[20] The puzzle then remains: Why has the health and integrity approach not been implemented in environmental policy?

We can dig deeper for an explanation of the failure of Old Ecology to guide environmental management by looking at the second argument of New Ecologists, the argument against grand theory. According to this second ar-gument, New Ecology rejects grand theory and moves toward more local studies of adaptations. Ecologists, according to this view, should correct Old Ecology by forming a coalition with local environmental managers to study the behavior of species and ecosystems at local scales while trying to under-stand how local systems fit into larger-scale systems.[21] This leads us back to problems of scale: cross-scalar impacts are not well understood because we lack both a theoretical understanding and comprehensive practical solu-tions to the problem of scale in ecology.[22] Aside from the fact that problems of scale in ecology are enormously complex and difficult, I believe there is an intellectual reason why the problem of scale has not been solved – indeed, has hardly been seriously addressed by ecologists and physical sci-entists. That reason, which I think expresses the essence of the second argu-ment of New Ecology, is that ecological theorists (and philosophers as well) have, deep down, been captivated by a particular version of the idea of *or-ganicism.* This idea has caused them to forget the importance of dynamic processes almost as soon as they acknowledge them.[23] To understand how this has occurred, we must look more carefully at the concept of organi-cism because it comes in several varieties with very different philosophical implications.

### III. ORGANICISM: WEAK AND STRONG VERSIONS

The following words were scribbled by John Muir in the margins of a book on evolution: "Every cell, every particle of matter in the world requires a Captain to steer it into its place.... Somewhere, before evolution was, was

336

an intelligence that laid out the plan, and evolution is the process, not the origin, of the harmony."[24] While it may be unfair to hold anyone accountable for what they scribble in margins of books they read, we find Muir adopting an especially strong version of organicism, characterized by two important features. First, he interprets evolution as a process *within a whole being* in the sense that God and the whole of nature are identical, and, consequently, the whole of nature is, or is like, a person (hence the capitalization of *Captain*). Second, Muir attributed *intentions* to this whole Being, crediting it with guiding the dynamic processes in nature according to a divine plan.[25]

Muir's comforting and elegant pantheism is therefore very awkward from a scientific perspective. It apparently recognizes a strategy or goal of ecosystem development, and treats that strategy as if it were both a prior, mentalistic end and a causal force – but a causal force for which there is no clear mechanism. The image of strong organicism, by attributing spiritual personhood to the forces of the whole system, tempted later ecologists such as Frederic Clements and Eugene Odum to over-emphasize the *holistic* characteristic of ecosystems, treating them as acting according to a driving principle.[26] Holism in ecology has therefore been crippled by its flirtations with mysticism.[27]

We have now come full circle, back to the problems arising from Moses, Parmenides, and Plato. Why are Westerners so quick to assume that underneath the constant change we actually encounter, no matter how chaotic the events we observe, there is order, constancy, a plan? This leads to a second question. Are we now in the realm of science, or are we in the realm of the deep unconscious, better studied by psychology, philosophy, or religion?

To understand this Western tendency, it may be necessary to look away from science toward these other disciplines. My discipline, environmental ethics, has also contributed to the failure of policy makers and environmental managers to develop an adequate scalar analysis for addressing environmental policy problems. We, just like the ecologists, have been too quick to personalize nature, to think of ecological systems as identifiable entities with a capability of consciousness and explicit goals, and to attribute to elements of nature a purpose of their own. I am suggesting that Muir's strong organicism underlies and bedevils both ecology and value theory equally. In ecology, it led to Clementsianism and to unfortunate metaphors like the "strategy of ecosystem development." In value theory, it has encouraged an unfortunate tendency to think that ecosystems are sufficiently like persons to justify the extension of inherent value to them.

Why have we been so anxious to consider ecosystems as things at all? Much of the writing in environmental ethics concerns which objects in nature have value of their own, and Leopold and Botkin have shown us that ecosystems are not properly understood as objects at all. They are open systems that unfold on many scales, with changing elements on each level making the best of opportunities gained by reaction to relative stabilities in their larger and slower-scaled environment. Environmental ethics has contributed to the failure to escape strong organicism because members of the field have been too quick to embrace the view that, like human individuals, ecosystems can have inherent values. Behind this quickness to embrace ecocentrism is a deep-seated bias toward things and beings, especially personages, at the expense of open processes.

Let me illustrate my point by reference to perhaps the most important passage in the history of conservation thought, Leopold's famous criterion. "A thing is right when it tends to preserve the integrity, stability and beauty of the biotic community. It is wrong when it tends otherwise."[28] Some advocates of the land ethic, such as J. Baird Callicott and the Deep Ecologists, have taken these two sentences to assert that the community or ecosystem is the object of value which conservationists should be attempting to protect.[29] They assumed, accordingly, that the ecosystem or community must, for Leopold, be *an object of value that exists independent of human values.* Callicott and his followers therefore interpret this passage as Leopold's definitive statement that communities themselves have inherent value, human-independent value that can be considered in competition with – and sometimes override – human values.

Leopold's criterion of good behavior, interpreted as an assertion that ecosystems have human-independent inherent value, has led to a sometimes nasty and wholly misguided discussion of whether Aldo Leopold and his followers are environmental fascists or not. Critics of the land ethic, especially Tom Regan, have argued that if we manage to protect ecosystems because they have inherent value, we will someimes override the rights of individual members (human and nonhuman) of ecological communities for the good of the larger whole. Regan even compared this approach to the way that Adolph Hitler and the Nazis overrode the rights of individuals in their misguided attempt to protect the German state as the embodiment of a master race.[30]

However, if Leopold and the land ethic are interpreted within a multi-scalar system, there is no conflict between individual rights and the protection of ecosystems. Human individuals actually exist within ecosystems. Damage to ecosystems is usually the cumulative damage of whole cultures and civilizations – it is a responsibility at the community level of a multi-scaled, open system, not at the level of individual decisions. Correction of these communal threats need not override the interests of individuals. Individuals, in a properly functioning system, will act in ways that contribute to, rather than destroy, the values that emerge on the larger ecosystem scale.

In the end, it is questionable whether the whole issue of fascism and intrinsic value in nature would arise if we could keep clearly in our minds that ecosystems are open processes. Would interpreters of Leopold be likely to say, without the ghosts of strong organicism influencing their conceptualizations, that ecosystem processes have inherent value? Environmental ethicists have been encouraged to find inherent value in ecological systems because they saw them, in the great tradition of Muir, as personalized organisms.[31]

Being constantly tempted to think of ecosystems as persons – or at least as objects capable of strategizing – philosophers and ecologists have unfortunately failed to confront an ancient philosophical problem: How can an organism behave independently while at the same time functioning as part of a larger organism? In philosophy, it is called the problem of parts and wholes. In ecology, it is called the problem of scale. Because they pay attention only to wholes, as is required by their implicit devotion to personalistic, strong organicism, ecologists and philosophers have had little reason to address the problems of parts and wholes and the related problems of scale, boundaries, and pace of change.

### IV. RESPONDING TO ORGANICISM

Botkin is correct in asserting that dynamism is absolutely necessary to move beyond obsessions with organicism and teleology. This, however, is not enough. We must shift the ground of debate regarding environmental values so that issues of scale become paramount. We must do so by proposing a clear conceptual framework for discussing the scale and pace of change. That ecologists have not solved the problems of scaling up from their small studies to larger and longer-term ones is perhaps excusable. The problem is admittedly a tough one. What is scandalous, given its importance for environmental management, is that ecologists and philosophers have hardly addressed these problems at all.

Emphasizing wholes, rather than the interrelationships of parts, has encouraged an unfortunate ambiguity in Leopold's central term, integrity, and this ambiguity has impeded its application. As Leopold intended the term, the integrity of an ecosystem is the integrity of an open, non-conscious complex of processes unfolding at different paces and scales. All of these ideas can be explained scientifically, without a hint of teleology or mysticism.[32] However, when the term integrity is applied to a whole being, it is natural to also understand integrity as an attribute of a personal being. As a personal attribute, integrity may carry moral weight in opposition to our obligations to human beings. It is a tragedy that Leopold's insights have been unnecessarily linked to unscientific speculations as a consequence of this potentially dual meaning, making them seem unnecessarily radical and open to endless debate. We must get past personalist organicism and address the key issue of system integrity for dynamic, nonlinear systems, and develop concepts for understanding multi-scalar interactions within ecosystems and the processes that constitute them.

The difficulty in all of this, of course, is that organicism does have an important point to make.[33] Organicists are correct that mechanistic models do not explain the ability of ecological processes to create, sustain, and heal themselves. Ecological management cannot begin without accepting two elements of organicism's richer conception of nature – the idea that the whole is more than the sum of its parts and the idea that relationships among multi-scalar processes, not the static characteristics of objects, provide the key to understanding ecosystems as they evolve through time. The problem is to express this idea in a way that does not carry us all the way to teleology and strong organicism. We must emphasize the creative nature of environmental processes and the key role of energy flows in those processes without personalizing them.

Hierarchy theory may provide the middle ground between inadequate forms of mechanism and the unfortunate personalism of strong organicism. We might call this viewpoint *minimalist organicism*. Hierarchy theory views ecological systems as complex, multi-layered systems that are self-organizing. This view does not involve causal mechanisms that work from the top down, but rather posits complex communication both upward and downward across all levels of nature. Creativity is not directed by a unified figure; it emerges from the ability of living things to adapt to relative constancies in their environment. Any system that is self-organizing exhibits this creative force, which we now know depends upon a mix of stable, predictable elements and chaotic, unpredictable ones.[34]

In a hierarchy, behavior of individual organisms responds and adapts to patterns and processes at the system level. This behavior is analogous to one sense of individual freedom – the freedom to act independently of constraint, the ability to choose between available options and to adapt in an effort to survive, prosper, and leave successors. In a dynamic, irreversible system, time is necessarily asymmetrical. Constraints flow down the hierarchy, but information flows upward in the system as well. The aggregated choices of individuals are component processes in the larger landscape-scale environment. Individual behavior is enacted against a stable backdrop, or environment, which appears to individual choosers as a mix of opportunities and constraints. In essence, individuals appear both as individuals on one level and as parts in a larger dynamic system.

In a longer frame of time, the cumulative impacts of individual choices reverberate as changes in the environment, altering the ratio of opportunities to constraints faced by future persons. The intergenerational impacts of a culture on the landscape can therefore be understood in terms of cross-generational exchanges in which the operative currency is freedom (options) expressed as a ratio of opportunities to constraints. These opportunities are stored in the structure of an ecological system. If growth in human impact erodes natural capital stored in ecological systems on larger and slower time scales, future generations will face more constraints and fewer opportunities as the cumulative impact of many individual human actions skims off opportunities (such as capital stored in ancient forests) and leaves constraints (poor and eroding soil exposed by clearcut lumbering). Alternatively, if resources are protected and if important and productive ecosystem processes are maintained, the people of the future will, as we have, face an abundance of opportunities.[35]

Ecological systems thus should be modelled as dynamic, open systems that are organized asymmetrically in space and time. Human economic systems represent dynamics driven by choices of individuals who live, choose, and die in cycles that are rapid in comparison to the rate of change in the surrounding environment. Regularities and predictable patterns on these larger levels provide the opportunity for biological and cultural evolution. Human individuals interact with their environment, usually through the mediation of an economic system. Individual perception, accordingly, is geared to short-term changes that occur in economic time. Large-scale ecological impacts of human activities must be understood both as results of cumulative individual actions on one scale and as spill-over impacts on larger-scale environmental systems that would normally change so slowly that their dynamic is unnoticed by short-sighted humans. In this way, individual decisions made within an economic system can, when acting cumulatively with the impact of many other individuals, affect the breadth of options open to the future.

One prerequisite for charting a rational environmental policy is to analyze the cumulative impact of many individual decisions on these larger and normally slower scales, because accelerated change at these scales may affect the context in which future members of our society will face choices and adapt. Using his own spruce budworm research and supplementing it with several other examples, C. S. Holling argues that when human activities simplify processes by concentrating productivity into a few crops or species, they can make large-scale systems more ecologically "brittle."[36] Because this style of management decreases the redundancy of pathways fulfilling such essential functions as energy transfer, the system becomes more prone to shift into a new steady, functional state, one that is less likely to be supportive of the cultural and economic behavior and adaptations that have emerged in response to the opportunities available in local environments.

Accordingly, Holling has proposed that we must recognize two concepts of resilience.[37] One concept, which operates at the economic or engineering level of the system, assumes a single stable point to which the system, once disturbed, will return. This concept is appropriate for individuals to apply in making day-to-day economic decisions. It assumes that system organization and behavior will be unaffected by an individual decision and develops equilibrium models to understand human behavior on the individual level. The other concept of resilience, ecological resilience, becomes relevant when the concern is to understand whether the cumulative impact of individual decisions may shift the larger ecosystem into a new stable state of functioning. Since such new stable states are almost always less desirable for humans, it is important that environmental management be formulated and evaluated on both the short-term, single-state model and the longer-term model that monitors whether cumulative impacts are threatening to exceed a threshold and cause a flip into a new system state. . . .

. . . Generally speaking, the integrity of a system is maintained if the pace of change in that system is sufficiently slow to allow species and communities to adapt to those changes. If so, we say that the system maintains its health or integrity, despite constant change. While space does not allow us to explore all of the consequences of this conceptualization, I believe it provides a general framework for integrating a dynamic ecology into discussions of environmental policy decisions. Therefore, it also provides the beginnings of an answer to Botkin's concern that we must distinguish acceptable from unacceptable change.

The next step is for ecologists to elaborate the hierarchical concepts of temporal pace and spatial scale and to do so not in terms of general theory, but in the particular local contexts in which particular ecological systems with unique organizational structures are unduly stressed by human activities. New Ecology, which emphasizes dynamism and avoids concepts based in personalistic organicism like the "strategy" of ecosystems, is ideally positioned to contribute to this important task. Still, the New Ecologists' justifiable emphasis on change in ecological systems must be balanced with a concerted effort to understand the pace and scale of that change. We cannot understand the importance of creative, self-organizing activity of ecosystems – or what we mean by ecosystem integrity – unless we also understand the concept of relative stability in dynamic multi-scalar systems.

NOTES

1. *See* G. Kirk and J. Raven, *The Presocratic Philosophers* (1966).
2. My purpose here is not to capture the fine nuances of either argument, but rather to show their skeletal structure and emphasize the differences between them. I have no doubt that the proponents of the arguments would refine them and perhaps even show linkages between them. Because these initial sketches are intended to show the broad outlines of the arguments, they are not attributed to any specific author. More specific attributions, with citations, follow.
3. See generally P. Price, Alternative paradigms in community ecology, *in A New Ecology: Novel Approaches to Interactive Systems* 353 (P. Price et al. eds., 1984); Mark Sagoff, Ethics, ecology, and the environment: Integrating science and law, 56 *Tenn. L. Rev.* 77 (1988).
4. Daniel Botkin, *Discordant Harmonies* 156 (1990).
5. Aldo Leopold, A biotic view of land, 37 J. Forestry 727 (1939).
6. Bryan Norton, *Toward Unity among Environmentalists* 51 (1991).
7. Aldo Leopold, *A Sand County Almanac* 129–33 (1966).
8. Curt Meine, *Aldo Leopold: His Life and Work* 283 (1988).
9. Id.
10. See *Kirk and Raven*, note 1.
11. See Chapter 17, this volume.
12. Botkin, note 4, at 145.
13. For a more technical explanation of this point, see generally Chapter 16, This volume.
14. Norton, note 6, at 39–60.
15. Stuart Pimm, *Balance of Nature?* 1–4 (1991).
16. Id. at 3.
17. See generally T. F. H. Allen and Thomas B. Starr, *Hierarchy: Perspectives for Ecological Complexity* (1982); R. V. O'Neil et al, *A Hierarchical Concept of Ecosystems* (1986). While I think of hierarchy theory as important in its own right as a theory of system organization in ecology, I refer here to its applications in the normative science of environemntal management.
18. See Chapter 16, this volume.
19. Leopold, note 7.
20. Chapter 15, this volume.
21. Bryan Norton and Bruce Hannon, Environmental values: A place-based theory, *19 Envtl. Ethics* (1997).
22. Pimm, note 15.
23. This interpretation is quite consistent with the arguments of New Ecologists; they do not oppose theoretical explantation or generalization, but they oppose the top-down process of filling in the gaps in empirical knowledge with speculative theories about the "strategies" adopted by ecosystems because these general ideas tend to bias specific research and its reporting. See Price, note 3.

24. Stephen Fox, *John Muir and His Legacy: The American Environmental Movement* 82 (1981) (quoting notes of John Muir in his copy of a book by Alfred R. Wallace).

25. These assumptions are not really independent. Throughout the history of Western thought, to attribute a mental life to a being is taken to be tantamount to giving it a spiritual identity.

26. Frederic Edward Clements, *Research Methods in Ecology* (1905). For a more contemporary example, see E. Odum, The strategy of ecosystem development, 164 *Science* 262 (1969).

27. Leopold himself flirted with this seductive, strong version of organicism but always retreated. Yet I have no doubt that this *image* of an ecosystem as a unitary and semiconscious being contributed to Leopold's quickness to adopt the organicist formulations of Clements and some of the surely excessive claims for the unidirectionality of succession that were popular in his day. In this respect, Leopld was – like his contemporaries among pure ecologists – seduced by an image of whole ecosystems as shadowy existents which "strategize" and "optimize." In that way, Leopold may require updating, and purging of the oft-hidden assumption of stasis and predictability.

28. Leopold, note 7.

29. J. Callicott, *In Defense of the Land Ethic* (1989); Bill Devall and George Sessions, *Deep Ecology: Living as if Nature Mattered 86* (1985).

30. Tom Regan, *The Case for Animal Rights 362* (1983).

31. We can also consider another interpretation of Leopold's famous remark, as a comment on the proper focus of conservation management rather than a statement of which objects in nature are of value. Leopold may have been making the point that, because of the complexity of nature's interrelationships and because there are so many different values exemplified in nature on so many scales, the only way to protect these values is to protect the integrity of community processes. Protecting the integrity of biotic systems, taken together with the complementary guidelines – stability and beauty – is the goal of good environmental management. According to this interpretation, Leopold is not telling us what to value in nature, but rather telling us what to protect in our practical environmental management. Leopold is defining the correct action in environmental use and management. Furthermore, he is strongly endorsing a systems approach to environmental management as the only way to encompass our multifarious goals as we mange a multi-level, complex system that is our habitat.

32. See R. Ulanowicz, *Growth and Development* (1986).

33. Bryan Norton, Should environmentalists be organicists?, 12 Topoi 21, 21–30 (1991).

34. See generally Roger Lewin, *Complexity* (1992) (describing how this is possible for a system).

35. The relationship between opportunities/options and our obligations to the future is explored in more detail in Chapter 17, this volume.

36. C. S. Holling, Cross-scale morphology, geometry, and dynamics of ecosystems, 62 *Ecological Monographs* 447, 483–485 (1992).

37. C. S. Holling, Engineering resilience versus ecological resilience, in *Engineering Within Ecological Constraints* 31 (Peter Schulze ed., 1996).

# 19

## Democracy and Sense of Place Values
## in Environmental Policy

### with BRUCE HANNON

An important problem of modern society is to understand how constraints on resource use can be democratically imposed. Recent authors have expressed deep concern about the possibility that resource shortages will lead to totalitarian governments. They hypothesize that such governments would be the only effective means to enforce constraints on resource use and protect the environment from the inevitable consequences of human population growth, overconsumption of natural resources and social chaos (Hannon 1985; Heilbroner 1974; Kennedy 1993; Ophuls 1977, 1992). Ludwig et al. (1993) have provided an elegant argument – based in fisheries management but apparently susceptible to startling generalization – which calls into question the ability of democratic governments with free market economies to protect renewable resources once there has been heavy capitalization and development of exploitative industries. Must societies of the next millennium be undemocratic if they are to protect their natural resources?

In this paper we address one important aspect of the search for a democratically supportable policy that will sustain resources for future generations – the comparative role of national/centralized, regional/state, and local communities in the development of environmental policy. We distinguish two approaches to the evaluation, development, and implementation of policies to protect resources and environments. The two approaches (we will call them "top-down" versus "bottom-up") differ most essentially in that the top-down approach emphasizes centralized control and decision making, while the bottom-up approach attempts to protect, to the extent possible, local prerogatives in setting environmental policy. We will then show that the different approaches would require very different methodologies for gathering and aggregating data regarding environmental values, and that application of the

From *philosophy and geography* 3 (1998): 119–146. Reprinted with permission.

two methodologies results in differing conceptions of the overall good of a society. We conclude with some general applications of our theory to the future of environmentally sensitive planning.

More and more authors and commentators on the environmental crisis endorse sense-of-place values as important components of our enjoyment and valuation of the environment (Ehrenfeld 1993; Norton and Hannon 1997; Sagoff 1993; Sale 1985; Seamon and Mugerauer 1985; Tuan 1977). The idea of sense of place is addressed in the literature of several disciplines. We will mention here key works in geography, anthropology, sociology, environmental ethics and social criticism, and especially ecological history, as well as in architectural and planning theory. We find the concept intriguing, in that the idea/concept is highly praised as important in several subdisciplines (such as phenomenological geography, historic preservation, and environmental ethics), but the concept itself has remained some what peripheral to the main subject matters of each of these disciplines. Perhaps the transdisciplinary nature of the concept, which has prevented its development within a single disciplinary paradigm, explains why the concept has not been given precise operational meaning or application. This lack of a shared operational meaning might at first recommend abandoning the term as too imprecise to serve as the basis for a scientific approach to policy formation and evaluation. We prefer, in contrast to this defeatist approach, to define the term within a transdisciplinary discourse rich enough to express aspects of the idea emphasized in several disciplines.

The concept of place has been most developed by *geographers*, especially by geographers in the phenomenological tradition (outstanding in this tradition is the work of Gould and White 1974, 1986; Relph 1976; Seamon 1979; Tuan 1971, 1974, 1977; also see the essays in Seamon and Mugerauer, 1985). While this work has created interesting conceptual ties with the tradition of philosophical phenomenology – especially the work of the German philosopher Martin Heidegger (1958, 1962; and Vycinas 1961), it has not attracted much attention from more quantitative geographers. Consequently, while this literature has greatly expanded our qualitative understanding of the place-relatedness of humans, their institutions, and their cultures, it unfortunately has not been fully integrated into the broader theoretical and quantitative work of geographers. One promising direction in this area is geographical work that incorporates ecological theory – specifically hierarchy theory (an application of general systems theory to ecological systems) – into large-scale, landscape-level geographical analysis. This work employs the ideas of hierarchy theory to develop more integrated conceptions of scale and interconnectedness in landscapes (Allen and Hoekstra 1992; Allen and Starr 1982; Collins and

Glenn 1990; Holling 1992; Johnson 1993; Lavorel et al. 1993; McMahon et al. 1978; O'Neill et al., 1986). What remains puzzling, here, is how exactly the objectivist, hierarchical models, which seek scientific and culture-free "objective" truths of physical geography, relate to the subjectivist and culturally determined models of phenomenologists (Norton and Hannon, 1997).

*Anthropology, sociology, and related social sciences* have contributed to literature on the relationship of local and regional cultures to their environment and resource base; especially important have been several excellent accounts of successful systems of resource access and protection among indigenous cultures (see, for example, Gadgil and Berkes 1991; Gadgil and Guha 1992; Rappaport 1968). The relationship between societies and their resource base has been effectively studied in a new way by a group of environmental historians – the *ecological* historians – who have used a combination of ecological and social historical data to reconstruct the role of human institutions in the "development" of the new world by colonial and post-colonial societies. The breakthrough book in this area was William Cronon's *Changes in the Land* (1983), which surveyed the changes in the New England countryside as a result of the arrival of European colonists. Crosby (1986) offers a broadly similar account, but Crosby emphasizes introduced species as the agents of "ecological imperialism," while Cronon details the intricate interrelations of ecological processes with the dynamic of human institutions (Cronon 1983, 1991); we find Cronon's institutional/ecological account more illuminating in the context of issues emphasized in this paper. Cronon, Crosby, and also Worster (1979, 1985) have all added to our knowledge of the interrelations of social and natural history. This theoretical work has been followed by numerous studies which have applied similar methods to the social and ecophysical transformation that occurred in the landscapes of every region of the United States and a few foreign countries (see, for example, Gadgil and Guha 1992; Gutierrez 1991; Silver 1990). Many examples of this literature do not consider place relatedness explicitly, but it provides many local and regional examples of how changing social and economic institutions alter, contribute to, and in many cases destroy the distinctiveness and identity of a place.

The idea of place has also been discussed, although not extensively, in *environmental ethics and social criticism.* Murray Bookchin (1965, 1982) articulated the idea that ecology and social criticism both point toward smaller, decentralized communities with strong commitments to place, in opposition to centrally controlled societies and resource management regimes. Other popular authors and essayists, including Wendell Berry (see, especially, 1977) and a whole literary movement sometimes called Southern Agrarians, have been highly critical of top-down management of resources and have advocated a strong commitment to the distinctiveness of local places. Alan Gussow (no date, 1993) has explored sense of place in art and aesthetics. Sale (1985) provides a useful overview of the role of place in understanding human cultures, advocating a "bioregionalist" approach, while Sagoff (1992) explores the cultural and aesthetic aspects of sense of place values. Sagoff, incidentally, argues that Americans generally have a weak sense of place, and attributes this to both the mobility and commercialism of American society. Deep ecologists have advocated more attention to the concept of place, but in our view they have not successfully resolved the apparent conflict between the localism implied by emphasis on place and the centralist, universalist, and Eurocentrist implications of their theory that all life has equal intrinsic value. In general, interest in these universal, extra-cultural values among environmental ethicists has reduced the importance of the concept of place in that discipline (Norton, paper 17, this volume).

Among architects, planners, and designers there was considerable interest in place-relatedness in the late 1960s and 1970s, and there has been a revival of interest in the concept in the 1990s. Important early work ranges from concern for the cultural and physical distinctiveness of local communities (Briggs 1968; Lynch 1960, 1972; Moore et al. 1974) to phenomeno-logical/aesthetic studies in architecture and planning theory (Norberg-Schulz 1979). Norberg-Schulz sharply separates objective from subjective aspects of place awareness, and then argues for a geographical determinism according to which cultural attitudes are determined by the physical structures of the environment (see, especially, chapter III.1). These issues have become intertwined with the question of the nature of the "spirit of place" or "genius loci," as it is sometimes called, as definitions of these ideas can vary according to the mix of physical and cultural factors assumed to determine the spirit of a place. More recent work ranges across the spectrum from approaches mainly descriptive of distinctive features of physical places (e.g., Hough 1990) to work with greater emphasis on cultural and perceptual aspects of place relatedness (e.g., Jackson 1994; Steele 1981). Daniel Kemmis (1990) provides both a provocative discussion of regionalism versus federalism in American politics and a useful application of the idea of place in local politics; also see J. K. Bullard (1991) for an account of professional specialization in a specific local place. Interestingly, Kemmis and Bullard have both been elected mayors of their cities (Missoula, Montana, and New Bedford, Massachusetts, respectively).

There has also been serious study of placelessness and loss of place. Important theoretical work by Relph (1976) was preceded by empirical studies, funded by the National Institute of Mental Health in the 1960s and 1970s, which examined the lives of residents of Boston's West End before and after it was levelled for urban renewal. This work emphasized the psychological importance of attachment to place, and of a sense of grief at loss of one's place, even in an area considered by outsiders to be a "slum" (Brett 1980; Duhl 1963, including especially Fried 1963; Gans, 1962).

One area where consensus has not been reached is in the area of boundary-setting, and the related question of how rigid boundaries between places are assumed to be. This difference spans geography and planning, with physical geographers usually employing physically measured boundaries, while planners and phenomenological geographers follow anthropologists in emphasizing perception and cultural factors in the determination of boundaries between places.

Even if most individuals orient from a place, it is not clear whether this perceptual position translates into "neighborhoods" or whether neighborhoods have more shifting boundaries, depending on human issues and cultural similarities. See Galster (1986) for an incisive empirical approach to these complex questions.

One important agreement among psychologists, geographers, and planners is on the importance of distinguishing *place* from *location*. Despite their local mobility, migratory peoples can have a strong sense of place. This distinction was made sharply by the aesthetician Langer (1953), who argued that a gypsy camp is a *place* which is reestablished in many separate *locations* during migrations. In this case, the placement of wagons, etc. can create a sense of constant "place orientation" at multiple locations. The ideas behind this distinction – that a sense of place emerges from an interaction of cultural and natural settings and that commitment to place is somewhat independent of location – may prove important to planners and environmental managers, because it raises the possibility of consciously building upon and strengthening a community's sense of place identity, even in the face of extraordinary mobility of populations. Of special interest here is the work of Proshansky (1978; Proshansky et al. 1983), who suggested the possibility that place-bonding may be to *types* of settlements as much as to particular locations, and thus may function transspatially. Feldman (1990), working in this tradition, reviews literature on "spatial identity" and presents data establishing loyalty to types of places, such as city, suburb, or small town. Hull (1992) explores Proshansky's hypothesis empirically, developing an operational notion of "place congruity."

If it is possible for individuals and communities to actively promote a positive sense of place, and if sense of place values in a community encourages protection of their local environments, it becomes an important question whether it is possible to adopt a policy of promoting a sense of local re-sponsibility through strengthening the sense of place in a community. Unlike Sagoff, who, as we noted above, explains lack of sense of place by invoking population mobility, this approach treats the relationship between mobility and sense of place commitment as an empirical matter, and as an important area for active participation by local planners, who may function to develop a stronger sense of place commitment in highly mobile communities. We return to this point in the conclusion.

A survey of these varied literatures suggests that the concept of sense of place is much better understood, theoretically and empirically, than it once was, and that the concept may present opportunities for important applications in planning and in environmental management. Despite the increasing interest of environmentalists and planners in the idea of place, however, little has been done so far to operationalize this intriguing but elusive concept, and it has been given little emphasis in actual analyses of environmental values. As a result, important applications have not occurred because of the inability of practitioners of multiple disciplines to settle upon a concept that is operational and sufficiently transdisciplinary to unify insights from the disciplines, all of which address the subject of place from different perspectives and with differing emphases.

This lack of agreement regarding operationalization and application is well illustrated in an exchange published in *Resources*, the newsletter of the economic research institute Resources for the Future. In response to a proposal by Mark Sagoff (1993) that protection of sense of place values be articulated as a criterion of good environmental management, the economist Raymond Kopp (1993) expresses disdain, rejecting the concept as "problematic as a basis for policy. If Congress and the regulatory agencies can figure out how to define environmental policy on [the basis of the concept of place], more power to them."

Kopp's summary dismissal of Sagoff's proposal as unworkable rests upon a key implied premise. Kopp implies that sense of place values must prove themselves useful in guiding *national* environmental policy (through guid-ance to Congress and the regulatory agencies). Notice that this apparently unquestioned assumption functions in effect as a methodological stipulation, requiring that any measure of environmental values must be easily aggre-gable across space. Further, the remark might be taken to imply a preference for centralized control of resource use, a top-down system of environmental management that assumes federal sovereignty over state and local govern-ments in conflicts regarding resource use. We hope we are not imputing too much to Kopp's remark; we worry, however, that a strong methodological preference for cross-scale aggregation of all values up to the national level may well bias environmental economists against sense of place values. Kopp, if we have understood him correctly, argues that sense of place values cannot be aggregated to guide a national environmental policy and, therefore, so much the worse for sense of place values as policy guides. This commitment to aggregating environmental values at the national level quite naturally discriminates against sense of place values, the accounts for which – almost by definition – must be kept local. Setting aside centralist and aggregationist biases, however, Kopp's argument can be reversed. If there is strong evidence for place-relativity of important environmental values, and local values and sovereignty are favored, it might be concluded: So much the worse for measures of value that aggregate at the national level but that cannot account for these locally placed values. Our purpose in this paper is to present a twofold argument. First, there are local, place-relative values which are systematically missed if all environmental accounts must be kept in nationally aggregated accounts (such as market prices recorded as dollar values). And, second, these values have an important role in charting a course toward a democratically acceptable approach to environmental policy and planning.

## AN EXAMPLE

We introduce our argument for sense of place values by discussing a particular example – a farm family who has owned the same farm for several generations – and inquire how this family values its property, hypothesizing that some of these values cannot be captured within a market analysis of social values. We recognize that the example is a favorable one for our case, but we use it simply to establish the existence of a nonmarketable sense of place values in one case. We discuss how widespread and how significant these values are below. For simplicity, we formulate the case as applying to land values, although we believe it applies to environmentally sensitive values more broadly, and perhaps to every "public good" in environmental valuation.

Consider the plight of a fourth-generation farm family who learns that an adjacent property is to be purchased under eminent domain provisions and provided as the site for a toxic waste treatment facility under a contract with the local municipality. Our family is offered a choice. If they wish to stay, they will be compensated for the decrease in their residential property value resulting from the siting or, if they prefer to move, they will be bought out at the estimated market value of their property before the siting. The amount offered in compensation, in case they stay, should in theory represent their property value lost by the change in the adjacent site. Determination of just compensation in the law is usually based on market values (Freeman 1993). It seems fair to ask whether in this case a "fair market value" approach does indeed represent complete compensation. Does fair market value capture all the values that are lost if the family decides that, while they do not want to move, they cannot accept the new risk and disruption and decide to leave?

Apparently not. If the family accepts fair market compensation and then uses the money to purchase a farm elsewhere, they will be compensated for their economic loss, but they will not be compensated for the loss of their "home." Place-relative information such as how to avoid poison ivy on the way to the pond, what time of day to catch the largest fish, and a plethora of other practical and aesthetic details will not be transmitted with the deed to the property and cannot be carried to a new site.

These nontransferrable values are the countless pieces of information and experiences that are entwined with daily life, impossible to separate into fact and value. They represent generations of wisdom accumulated from specific experiences and encoded in the cultural information and attitudes passed from generation to generation. We believe that these place-relative values appear in all cultures, though we expect that their specific form and content will be highly variable across cultures. To see the environmental connection, consider the extent to which loss of vernacular tradition in Florida architecture in favor of generic building techniques has both reduced the cultural distinctiveness of that semi-tropical state and greatly increased the demand for electrical energy.

At least some less technologically advanced cultures, especially those that have survived for many generations, have developed myths and cultural practices – often religious in nature – that have the effect of protecting the resource base (see, for example, Gadgil and Berkes 1991; Gadgil and Guha 1992; Rappaport 1968). These local myths and practices, which are part of the "cultural capital" of a region, are often eroded by "development," leaving local populations with neither traditional practices nor trained scientists who might gauge the impacts of resource use. Because place-relative cultural values must be developed with intimate knowledge of local conditions, migration, especially toward modern cities, diminishes this cultural capital.

Return to the case of our family. If the loss of these values – the loss of a special relationship between people and a place – leaves the family diminished despite having received "fair compensation," what does society risk when these values are ignored in analyzing and shaping our environmental policies? There are specific experiences – a particular granddaughter learning to *like* to fish at her grandfather's side, or an apprentice learning the tricks of the trade for building a more comfortable Florida house, for example – which transmit values and information through experience, practice, and enjoyment that are too personal and place-specific to be exchanged with the sale of property. The sum of these values, practices, and strategies constitutes locally based cultural capital.

Now suppose our family sells and another, nonlocal, family, willing to accept the risks and disutilities of the site, subsequently purchases our family's property at a lower price (determined by its postsiting market value). Most of the place-relative values associated with the first family's long association with the land will not be transferred in the sale. The original family will leave the farm diminished, and the new family will acquire these positive place-oriented values only slowly, if at all. These effects represent a part of what economists have called "adjustment costs" (Hanemann 1993). This concept, which has been developed as a category of easily missed costs in computing the impacts of global warming, may be particularly applicable to locally originating values. Because long-term estimations of changes due to global warning compare two temporally distinct equilibria, Hanemann argues that it is important to consider the disequlibrium costs incurred in adjusting to accelerating environmental change. Values lost during a forced transition may represent a systematically undercounted category of costs, and locally disruptive changes such as sitings of noxious land uses can involve significant costs of this type. If we suppose, alternatively, that the siting goes forward and the land is no longer considered suitable for residential use at all, the land may be rezoned, making it a site for further difficult-to-site industries and facilities. These changes may make it an "attractive" site for further disputed industrial uses. The market value of the farm may then even increase as important place-related values are completely eradicated.

If this argument proves sound, it apparently entails that there exist significant, systematically overlooked externalities associated with large-scale development activities, and with economies and regulations that encourage movement of industries, etc. These losses are losses in the cultural capital of a community, the bits of particular information and experiences that unite a people in a relationship with their resource base or, in ecological terms, their "habitat." At this level, a culture's integration with the larger biotic community – the plants, other animals, and physical forces that make up its habitat – represents both cultural capital and also an important part of the local identity of that culture. Having offered one (admittedly favorable) example to illustrate this local cultural capital, we proceed to examine some of the characteristics of this capital to determine how widespread this phenomenon is, and to inquire whether this cultural capital and the values it embodies could and should have a role in environmental policy analysis.

## CHARACTERISTICS OF SENSE OF PLACE VALUES

It is worth noting several unusual characteristics of these sense of place values. We will describe five features of place-relative values: (1) conditionally transferable, (2) local, (3) culturally constitutive, (4) pervasive, and (5) partially measurable.

### *Conditional Transferability*

First, we must note that nontransferable sense of place values are not the only type of place-relative values. Indeed, many types of place-relative values are easily recognized and readily incorporated into a market-based analysis. A wonderful view across a valley or the sea, for example, is highly relative to a given site, but is readily included in a market analysis in societies in which such value is common; since the view is obvious to any buyer, the added value is reflected in the market price. Similarly, location near an already existing noxious industrial use lowers purchase prices. Our special concern here is with values that are similarly place-relative but that cannot be observed by prospective buyers and that cannot therefore be reflected in the purchase price. Prices in free markets reflect values that are discoverable and exchangeable. The experience of a grandfather teaching his granddaughter to fish in the lake cannot be discovered by potential buyers, nor can it be exchanged. It can have value only to those who experience it and, therefore, if it has value, it is an extra-market value. Both transferable and nontransferable place-relative values affect the character of a place; our interest, here, is in the nontransferable ones, however, because those will be most likely to be missed in a market analysis.

### *Locality*

All sense of place values are place-relative in the sense that they emerge in a specific, local context. Philosophically, this means that actions constitutive of a culture gain their meaning in a specific place, and they are expressive of locally distinctive and highly variable cultures, biogeographies, and physical features. Development by a person of a local sense of place is an important part of developing a sense of personal identity (Fried 1963; Relph 1976).

Place-orientation is complex and many-layered (Norberg-Schulz 1979). To paraphrase the geographer Yi-fu Tuan, "we need a sense of place and a sense of the space around our place" (Norton and Hannon 1997; Tuan 1971). Anthropologically, this means that one should expect as many senses of place as there are unique combinations of culture with varied types of natural communities that form the contexts of those cultures. Practically, it means that no environmental policy that ignores local variation and local experience can be expected to protect biological or cultural diversity. If the only values that get counted in our analyses of policy options are ones that can be aggregated without reference to perspective and scale, then the nontransferable, place-relative values identified here will be "washed out" in the process of aggregation to higher levels.

## Cultural Constitutivity

The contextual knowledge that results from multigenerational interactions between a culture and its distinctive plants and animals ensures that sense of place will be highly variable across cultures and across natural physical systems. This variability expresses the creativity involved in a culture's adapting to varied challenges and opportunities as offered by their biogeographic context. These values can be thought of as representative of "options" – choices to enjoy particular experiences – as viewed from a local perspective. But they are also a series of possible connections to the land, connections that are mediated through patterns of choices. It will be the choices that our generation makes regarding these places, whether to lose or to continue them, that will determine the options of future generations. The fabric woven across generations by this process of interaction with local resources sets the broad outlines of the identity of a culture within its surroundings. It determines the physical context to which subsequent generations will adapt.

In this sense, the choices made by one generation create the context for the next generation's choices; but it is the combination of physical/ecological constraints and the reactions of a culture to them that constitute meaningful activity. Sense of place values, in this sense, capture the "options" open within a community (the patterns of learned behaviors and possibilities for innovation) *and*, at the same time, those behaviors give expression to locally based cultural meanings – they are individualized responses to a particular environment (see Norberg-Schulz 1979; Relph 1976).

## Pervasiveness

While we have developed a quite specific example of private property values associated with experiences that are place-specific, the types of values identified here are by no means limited to this narrow case. We could re-tell the above story of the family, simply moving the experience of a grandparent teaching a grandchild to fish at the public access to the best local fishing lake. The difference in the cases is that, if an undesirable land use is sited near the public fishing site, the experience will be diminished, and the stock of sense-of-place values enjoyed by the family will decrease, but the family will almost certainly not be offered compensation of any kind. Indeed, it appears that there are similar losses in nonmarket values associated with many changes in the context of common access resources such as occur when the Forest Service clear-cuts land near a recreation area, because this action can negatively impact the quality of the experience of visitors to the area.

The types of values introduced in the favorable example can now be broadened. It is as if the fabric of future possibilities for a local culture is woven from choices in the present. Defense of place-relative values reflects the collective will of a people – the common embracing of a cultural identity – a collective decision to maintain an authentic relationship to its past, to its future, and to its natural context. The search for a cultural identity must respect and build upon the natural history of a place, which includes the practices that emerged historically, and it must also project values into the future.

Sense of place values emerge at the local level and are highly dependent on the context at that level; they represent the positive sense of community that, in best cases, arises between a people and the place in which their culture has been defined. These values are therefore "scaled" – they are associated with a particular level of a multiscalar system (see Norberg-Schulz, 1979; also, for some ecological theorizing relevant to this point, see Allen and Hoekstra 1992; Allen and Starr 1982; Collins and Glenn 1990; Holling 1992; Johnson 1993; Lavorel et al. 1993; McMahon et al. 1978; O'Neill et al. 1986). If one attempts to connect local values to a common scale (such as a monetary scale), and then to aggregate the community-level values of many communities to arrive at a "sum" for the nation, all of the richness and context-dependence of these locally originating values will be lost. These sense of place values cannot therefore be meaningfully monetized.

These values are also "scaled" in time – they emerge intergenerationally as the community accommodates itself to its habitat; new practices are adopted and passed from generation to generation in an ongoing process of culture-building. It may be useful to think of these values as "aspirations" or "community preferences" in contrast to economists' "individual preferences" as a second, and essential, aspect of human valuation (Chapter 11, this volume). Whereas preferences exist in the present and are taken as givens, aspirations exist on an intergenerational scale and they represent choices on the part of a society regarding the type of society it will be. The values that emerge on this level require intergenerational continuity, a continuity that should reflect itself across generations on the landscape. Choices each generation makes concerning the landscape govern the range of freedom of the future in the precise sense that the landscape will determine what options are open and what experiences are possible in the future (Chapter 17, this volume). A commitment to sustainable use is a commitment to hold open certain options and the possibility of certain experiences. Because of this dynamic, intergenerational quality, sense of place values cannot be captured in a "snapshot" approach to ascertaining the aggregated preferences of a population at a moment in time.

### *Partial Measurability*

Sense of place values can be expressed as measurable differences in human behavior. We have argued that many sense of place values will not be captured in market transactions because they are so intimately intertwined with bits of knowledge and experience that are too place-specific to transmit at a real estate settlement. Again, our example is instructive. If the family had no positive attachment to the farm – if they were absentee owners or if they had just bought it and had not moved in yet, for example – they would probably be fully compensated for their loss if the price offered is indeed the fair market price. We can hypothesize that, if a family has a large stock of sense of place values, they will be more likely to refuse to accept a fair market price for their property, and their efforts (including, for example, legal fees and volunteer work) expended to protect the place itself provide a rough – or lower-bound – measure of the strength of their commitment to their particular home place. Our emphasis on locally developed values as indicative of commitment to place leads us to take these costs very seriously in evaluating environmental impacts of large development projects and of national management of resources such as national parks and national forests. As an example of the power of local feeling, consider the State of Illinois' eighty million dollar attempt to site a radioactive waste storage facility in Martinsville, Illinois. A local citizen group of the small rural community with relatively high unemployment obtained contributions of nearly one-half million dollars in intervenor funding to organize expertise to defeat the project on several grounds. This remarkable effort was undertaken despite a state offer of over one million dollars per year in compensation to the community.

Willingness to accept the market value indicates an *economically* defensive stance; its motive is to protect the economic investment of the private landowner, but not the place itself, which is treated as having acceptable substitutes in alternative properties. Willingness to fight on against a siting *after fair market compensation has been offered* is indicative that something other than economic interests are being defended. In these cases, the economic model, which assumes that there exists a rate of compensation that will make persons indifferent to loss of their home, seems inapplicable to an analysis of a positive commitment to *this* place, rather than to this and comparable places that could be purchased elsewhere. The values of home are in this sense separable from economic values embodied in the property; indeed, they are in an important sense nonfungible with economic values because they become obvious only when economic interests have been fully protected. When owners reject fair market value for their home, it is always possible to interpret their refusal to sell as economically motivated gaming. The difficulty of controlling for this confounding variable represents one of the major deterrents to the development of a more precise and measurable conception of sense of place values. But sense of place values, however difficult to measure directly because of the gaming effect, are necessary to explain behavior that is likely to be counterproductive economically. Citizen actions to protect parks and preserves are often motivated by noneconomic goals. Another approach to quantifying values associated with particular places might be to pay attention to private landholders gifts and below-value land sales by private landowners to environmental groups for preservation (Robert Mitchell, personal communication). Sense of place values therefore manifest themselves in connection with private goods (as in the case of our farm family) or they can be motivated by a public spirit. The point to be emphasized here is that these are significant and pervasive values, and they are often systematically ignored in market valuations.

We feel justified in calling these place-relative values that cannot be measured in market transfers *positive sense of place values*. These are the values, the positive commitments – the aspirations that are tied up with a particular place – which are not compensable in dollar terms. These values express commitment to a non-interchangeable connection to a particular place. A strong stock of positive sense of place values would represent intimate experience and relationships with the plants, animals, and ecosystems that are distinctive to their region. The idea of aspirations captures the sense of a cultural future, which must of course grow organically from a storied past, and it is these features that give distinctive value to a particular place (Rolston 1988). These aspirations, which correspond to intergenerationally created and protected cultural capital, cannot be understood independently of the resources – the opportunities and options expressed in the local biotic community. Nor can they be separated from the constraints that have been conquered by local technologies and local wisdom. These constraints give meaning to actions of individuals and communities who are integrated into a place (Chapter 17, this volume). They emerge not so much in connection with particular acts and purchases, but rather as expressions of cultural identity and character (Page 1992). A good example of this type of value, from the political realm, would be the types of underlying political ideals (such as rule of law) that express themselves as a community writes a new constitution (Page 1977; Toman 1994). Whereas preferences expressed in markets may represent individual values as felt in the present, aspirations are group values that express themselves on a longer, intergenerational scale.

Notice also that aspirations can be more complex than preferences in the sense that, if one has an aspiration that one's children and grandchildren share and enjoy an experience (such as learning to fish in the family fishing hole) in the future, two conditions are implied: one hopes, first, that the fishing hole will be unspoiled and accessible and, second, that the youngsters will enjoy that experience. So, in addition to the implied value of accessibility to means to fulfill a preference, aspirations represent also second-order preferences regarding what preferences future individuals will have and express. This feature of aspirations (Norton 1987; Sagoff 1986) introduces a level of complexity – they represent preferences over preferences – into the analysis that is impossible to capture in a uniscalar analysis such as that exemplified by microeconomics.

A CLASSIFICATION OF PLACE-RELATIVE VALUES

Drawing on our analysis of the example, we offer the following taxonomy of place-relative values:

1.  Preferences associated with place
    a.  publicly observable benefits, risks, and disutilities associated with a place and adjacent land uses
    b.  recognizable and transferrable amenities (a nice view, fishing opportunities)
2.  Place-oriented aspirations
    a.  enjoyments that depend for their value on the history and future of a place. These intimate values are expressed in a culture's relationship with their physical and ecological context – the geography, plants, and animals that constitute their environment/habitat – and they emerge only through participation in, and acquisition of knowledge about, the natural and cultural history of a place. They are values

that are transmitted and experienced only by intimate contact among people who interact in a culture with a sense of natural as well as social history.

b. the value of the cultural and community continuity conserved when experiences of type (2a) are protected over generations.

While the values in category (1) are presumably captured in market transactions, the values in category (2) cannot be so captured, as argued above. They are, however, the very essence of any positive connection between a human culture and its place.

### MEASURING AND AGGREGATING SENSE OF PLACE VALUES

Nontransferable, place-relative values have measurable behavioral consequences. For example, we predict that communities with a large stock of these values are more likely to continue to contest land use decisions, incurring measurable costs, even after they have been offered compensation. Many of these costs are easily measurable – legal fees, for example; others are difficult to measure but offer no problems in principle – such as donated time and efforts, etc. – while others, such as emotional pain experienced during a long legal battle, will never be measured accurately (Varlamoff 1993). The quantifiable expenses of opponents of a development plan (after having been offered market compensation) would nevertheless appear to provide a lower bound measure of place-relative values that are not transferred in fair-market sales.

The problem with these values is not so much that they are unmeasurable or unquantifiable in particular situations – any method of quantifying social values will include some easily measured and other difficult-to-measure values. But these values are so highly context-dependent that their character is lost when an attempt is made to aggregate them across many communities. We must therefore revisit Kopp's assumption in favor of nationally aggregatable data. Even if we were to develop quite precise measures of positive sense of place values for a given community, Kopp may well be correct in believing it is impossible to aggregate these "values" together to guide federal policy decisions because these values contain a nonfungible commitment to characteristics that are place-specific. These values are not additive with other values to achieve a grand, national total of welfare, because these values exist in so many different, and in some sense incommensurable, contexts. In particular, they embody values that emerge on multiple levels. These values do not aggregate because they are scale-specific (Chapter 17, this volume).

The dispute here is not one about whether to *quantify* values; it is rather about whether to vertically aggregate values across hierarchical levels, once a quantification has been achieved. And it is also about whether it is worthwhile to seek and use information that cannot be monetized and aggregated at the national level. Whereas a decision maker following the top-down approach would aggregate information to the highest decision level, treating decision making as a matter of computation of single-scaled values, a decision maker acting from the bottom-up would first sort decisions into various categories according to a scalar criterion. In the version we have developed, these categorizing decisions are based on an assessment of the scale of the problem and the social scale at which a response is appropriate (Chapters 16 and 17, this volume). Then, the decision maker asks, given the scale and level of dynamics at which we must address this particular environmental problem, what decision rule should apply in this context? Once a decision rule is chosen and goals have been formulated at the relevant governmental level, it should be possible to quantify data about the physical world and to quantify data regarding citizens' values affecting the decision. But on a multiscalar approach, one would not expect that all of the data should be analyzable in terms of a single utility function of individuals and thereby aggregable upward in the decision hierarchy.

To illustrate the importance of the difference, consider two ways of making a decision regarding the placement of a long-term, low-level radioactive waste storage site. Suppose, first, it is noted that the risks and costs to society (considered as the set of all citizens in the nation) attendant upon having many decentralized storage sites are computed to be very large because, for example, security arrangements at many sites are either expensive or risky, or because there will be widespread stress on individuals because far more people will live near multiple sites than would live near a centralized site. A nationally aggregated decision made in the computational style, in other words, would apparently favor the choice of a single storage site in a remote area. But suppose that every local community refuses to accept the site. If we consider the siting to be a matter of the "right" of each local community to self-determination, respecting this right amounts to offering each local community a veto power. With such a multiscalar view of our governing system, in other words, it makes a lot of difference which questions are addressed first. And in this sense the decision systems we offer are not computational, but emphasize careful categorization of risks and the values associated with them, seeking to characterize environmental problems in terms of their appropriate scale, and matching decisions to the appropriate governmental scale.

With a multiscalar view, every decision maker would accept the right to self-determination on the local level and would treat this universal veto as a given. The search for a national policy on radioactive waste storage would now be formulated with a whole different set of policy options. All options considered would be ones that respect the decisions of local governments not to accept the site. Admittedly, this approach may be idealistic in that we already possess – because of past decisions and activities – large stockpiles of waste requiring storage somewhere, which makes the apparent implication of the bottom-up approach inoperative as a complete solution. But this practical complication does not obscure the clear implication for future decisions of a commitment to local self-determination and enforceable local veto powers. Placing a high priority on local self-determination in effect constrains the options for a national policy, perhaps imposing the outcome that there must be a vast reduction in activities that create such waste. The point of such examples is to show that the choice of an aggregative approach to measuring environmental values tends to devalue local considerations that seem persuasive to members of local communities. If all communities reject a particular land use or even if a great number of them do, the message to the higher level would be to stop the imposition and seek another alternative. For example, if no place can be found (due to local vetoes) for storage of low-level radioactive waste, then stop generating it and find substitutes. We could then dispense with cynical reports such as "Building Citizen Support for Responsible Low-Level Radioactive Waste Disposal Solutions: A Handbook for Grass Roots Organizers" (1995), published by the Nuclear Energy Institute, a trade organization consisting mainly of nuclear power utilities and manufacturers.

The special characteristic of hierarchically organized decision processes is that they recognize and meaningfully represent the apparently unavoidable asymmetries in space-time relationships. If actions of individuals are viewed as scale-specific within a multiscalar system, then the larger-scaled systems that provide the context for individual decisions appear as both constraints and opportunities available, locally, to individuals. The amount of forested land in an area offers an opportunity for a timber industry; but a limited quantity of high-quality forest also places a constraint on the development of such an industry. Opportunities provide an explanation of the behaviors of individual actors such as timber companies and their employees, but the constraints embodied in the limited extent of forest cover will determine, on a longer scale of time, the evolving character of the community in relationship to its natural context. Sustainable forestry, thus understood as including the sustenance of local culture and local ecological integrity, would in this sense be a *local* commitment and include local responsibility for forest protection (Norton and Hannon, 1997).

The problem is that in a nationally or internationally organized economy it is possible to "export" opportunities in the form of timber, but the associated constraints – the results of near-total deforestation such as erosion and siltation of streams – necessarily "stay at home." Thus it is "rational" for decisions made by multinational corporations in the context of international markets to change the character of a local community by deforesting it and then moving to a new forested area. Their profits will be exchangeable in the currency of larger-scale systems such as world markets (Clark 1974); the costs of their aggressive exploitation, the degradation of the natural resource, will be felt at the local level and will be lost from computations as the profits are summed at higher levels of geographic, social, and economic organization. Aggregated measures of rationality and single-scaled approaches to decision making therefore tend to over-estimate benefits and under-estimate costs because, in the arena of national economics, wealth will be created by the systematic transference of externalities to local communities as a byproduct of the exportation of their opportunities. Especially when international corporations can command federal subsidies through lobbying at the federal level, local communities will be hard put to resist such exploitation.

We have questioned Kopp's assumption that measures of environmental value must be expressed in terms that are aggregable across space and time, exploring hierarchical systems of analysis which, unlike the single utility measures of mainstream economics, analyze values in a contextual and scale-sensitive manner. Multiscalar systems retain information of local importance in the process of developing a national policy, at the cost of giving up aggregability across scales. Sense of place values exist on many local levels and, within a hierarchical approach, these sense of place values would be satisfied locally before going on to address larger-scale questions of valuation. This creates a burden of proof on the powers of centralization to show why expressions of local goals should not trump centralized interests. If one favors local control and reestablishment of local responsibility, a multiscalar system of analysis which favors a bottom-up approach is more likely to achieve a democratic outcome than a policy process that emphasizes those values that can be aggregated to a national level. We should clarify that our preference for local control is not based on the assertion that all problems are local problems. Indeed, we expect that there will be many cases in which local governments must address large-scale environmental problems, including their contributions to global changes. Our point is that even to address international problems while ensuring democracy at the local level, decision structures must be organized in a bottom-up fashion.

We have sketched two approaches to environmental policy formation and valuation, and have noted that aggregation of values across all levels of environmental policy formation will promote policy goals – and the values that support them – favorable to a centralized decision process. Alternatively, a multiscalar organization of the process can be designed to be scale-sensitive and to encourage processes of valuation that emphasize locally developed environmental values. This approach eschews cross-level aggregation of values and instead recognizes the asymmetry of time-space relations, protecting local values at a local level. While we recognize that local cultures do sometimes destroy their resource base, local control has the advantage that information feedback loops are shorter and local populations may have to live with the consequences of reduced opportunities and economic options more directly than do national and international corporations or national governments.

The evaluative, informational, and decision models we have proposed have been built from the bottom up, in that we have avoided the methodological requirement of cross-scale aggregability of value measures we propose. Our methodological choices have in this sense been guided by a persistent attempt to favor localism and values that are articulated locally, because we believe that centralized goal-setting for economic growth will inevitably favor large economic interests. These large interests will systematically export opportunities, leaving results of ignoring constraints behind as they move to fresher exploitable areas.

In closing, we can return to the very large question with which we began this paper: is there any general approach to environmental policy formation that both protects resources *and* can be established democratically? If we think of democracies, broadly speaking, as governments that are responsive to the will of the people, it is possible to ask "which people?" and "How will their wills be aggregated?" In systems which are multiscalar (such as federal systems), it is possible to identify communities that exist at more than one level. It is comforting to think that, when such a system is working properly, "democracy" occurs at all levels, with the government at each level pursuing policies that are favored by the relevant community. A democratic national government would therefore pursue policies that are acceptable to people at all lower levels.

In the process of analysis, we have uncovered an important ambiguity in this conceptualization of the democratic process. We have shown that aggregating values from smaller to larger communities can bias policy analysis against local communities. We saw that Kopp and other neoclassical microeconomists set, as a methodological constraint/expectation, that all values expressed on all levels must be aggregable in common, fungible terms. On the view of democracy that emerges in this case, a democratic national government would pursue those policies that are indicated by the sum total of the aggregable preferences of all individuals, *qua* individuals, at all lower levels of the society. It allows no statement by a local community of nonfungible, place-relative values. We have seen that, on the system assumed by Kopp, it will often be "rational" to impose a land use on a community despite opposition from that community because the strong local values of the opposition are swamped by the "common good" of millions of people from other communities who will reap the rewards of the government's imposition of a siting on this small community. The harshness of this approach can of course be mitigated by insisting, for example, that the few in the local community should be compensated from the gains of the many, perhaps with direct payments or with new schools, etc. These mitigative "bribes" take the form of subsidies flowing from the federal to the local level; they have the same impact as other federal subsidies for local resource extraction: they confuse the signals so that local governments will not perceive the consequences of nonsustainable use of their resources. This mitigation does not change the fact that the aggregative approach presupposes a "top-down" flow of control in the system. Policy is set on a national level, and local communities are left with no choice but to negotiate the best "compensation package" they can get. The "good" of the society is determined at the national level through an aggregation of individual preferences; any local "goods" such as positive, nontransferable sense of place values will have been lost because they are invisible among the market values that, on the view of mainstream economists, should guide national policies.

If, on the other hand, we relax the aggregability requirement on environmental values and think of environmental policy as hierarchically organized from the bottom up, locally developed values that are highly context-dependent may be instrumental in building many strong local communities in which individuals act upon a strong sense of local place. These local communities might, on this view, reject "fair compensation" for a siting simply because it is not consistent with their conception of their community. Values will be counted, voting will take place, etc., but the political outcomes at local community levels will stand on their own. They cannot be swamped and overridden by values that are aggregated at a larger scale. This difference in the conception of the public good is the operational significance of the difference between a multiscalar and bottom-up approach to conception of environmental values and methods, and an aggregative and computational approach to policy with a national focus only.

Our theory suggests no less than an about-face in current trends toward centralized control in environmental policy formation. To some extent, we are calling for an end to federal and state sovereignty, and an end to top-down thinking in environmental policy. Our approach would, for example, shift the initiative for environmental policy development away from the federal to the local level. But it is important to realize that the approach can work only if devolution of federal control is accompanied by an end to federal subsidies and bail-outs. The point is to reestablish local *responsibility* for resource use and planning decisions, which requires both local control *and* a system in which local populations must accept the consequences of their actions.

One advantage of a compartmentalized, multiscalar, bottom-up system of many local controls is that it will reduce the impact of national economic lobbies. In the current system in which environmental policy goals are set on a national level, it is much easier for national industries to centralize their lobbying efforts. We hypothesize that it will be more difficult for national economic interests to control environmental decision making if it is decentralized.

Our argument also implies that we should provide local activists with resources with which to resist encroachment upon local prerogatives. In particular, new laws are needed to empower local activists to challenge proposed impositions of decisions that are motivated by the concerns of centralized governmental power, often in collusion with large international businesses (Varlamoff 1993). More generally, our argument has several implications for planning and for the planning process.

First, once one draws the distinction between location and place, recognizing that it is possible to create a sense of place even among populations that change locations regularly; it is possible to advocate a strong role for planners as facilitators of a process of building and strengthening a local sense of place. This process of community value articulation should be designed to build a stronger sense of place and a stronger inclination to defend the distinctiveness of local cultures. Planners, on this approach, could become key players in ecosystem management projects, encouraging local communities to articulate and develop their sense of place and local identity as a general guide to setting goals of environmental policy. This approach might also encourage closer cooperation of environmental and cultural/historical preservationists.

A second, related implication is that the process of devolution of responsibilities for environmental quality and regulation from the U.S. federal government to the states – a process that is certain to be accelerated by strong budget cuts at the U.S. Environmental Protection Agency – should be met with a process of social learning (Dewey 1984; Lee 1993). Local communities may, with the help of planner/facilitators, be able to seize as an opportunity the current trends away from federalism, using these events to strengthen local responsibility and control, and to build a healthy sense of the distinctiveness of particular places as broad guides to resource use and environmental protection (see Kemmis 1990). The processes of social learning described here must be iterative; they must encourage ongoing discussion of environmental goals and how those goals interact with socially and culturally expressed statements of the distinctiveness of particular places. This iterative process, well described by Lee (1993) and Gunderson, Holling, and Light (1995), emphasizes the interaction of professional scientists, environmental planners and managers, and the public in an experimental approach to proposing and testing environmental policy goals and policies.

These general implications all follow, directly or indirectly, from the main argument of this paper, that social values can be protected only if local communities can exert self-determination and a veto power in favor of locally originating and locally supportable norms to guide resource use. In general, then, a strengthened and more scientific conception of sense of place values, based on improved understanding of the role of place in developing a sense of community and a willingness to accept responsibility for local resources, may encourage a stronger role for planners in facilitating the articulation, public discussion, and revision of the goals of environmental management.

We recognize that return to local responsibility will be difficult and will require time to develop stronger local senses of place and a commitment to protect the integrity of local places. But this may be the only route toward a democratically supportable approach to sustainable use of resources. We believe that it would also represent an important step toward introducing true democracy, which on the hierarchical bottom-up view of policy involves actions of government that are responsive to the needs of people and communities who live most intimately with their resources and within their habitat.

REFERENCES

Allen, T. F. H., and Hoekstra, T. W. (1992). *Toward A Unified Ecology.* New York: Columbia University Press.
Allen, T. F. H., and Starr, T. B. (1982). *Hierarchy.* Chicago: University of Chicago Press.
Berry, W. (1977). *The Unsettling of America.* New York: Avon Books.

Bookchin, M. (1965). *Crisis in Our Cities.* Englewood Cliffs, N.J.: Prentice-Hall.

————. (1982). *The Ecology of Freedom.* Palo Alto, Calif.: Cheshire Books.

Brett, J. M. (1980). "The Effect of Job Transfer on Employees and Their Families." In *Current Concerns in Occupational Stress,* ed. C. L. Cooper and R. Payne. New York: John Wiley and Sons.

Briggs, A. (1968). "The Sense of Place." In *The Fitness of Man's Environment,* Smithsonian Annual, II. Washington, D.C.: Smithsonian Institution Press.

Bullard, J. K. (1991). "The Specialty of Place." *Places* 7, no. 3: 72–79.

Clark, C. (1974). "The Economics of Over-Exploitation." *Science* 181: 630–634.

Collins, S. L., and Glenn, S. M. (1990). "A Hierarchical Analysis of Species Abundance Patterns in Grassland Vegetation." *American Naturalist* 135: 633–648.

Cronon, W. (1983). *Changes in the Land: Indians, Colonists, and the Ecology of New England.* New York: Hill and Wang.

————. (1991). *Nature's Metropolis: Chicago and the Great West.* New York: W. W. Norton.

Crosby, A. W. (1986). *Ecological Imperialism: The Biological Expansion of Europe, 900–1900.* Cambridge: Cambridge University Press.

Dewey, J. (1984). "The Public and Its Problems." In *The Later Works, 1925–1953, Volume 2.* Carbondale, Ill.: Southern Illinois University Press.

Duhl, L. J., ed. (1963). *The Urban Condition: People and Policy in the Metropolis.* New York: Basic Books.

Ehrenfeld, D. (1993). *Beginning Again: People and Nature in the New Millennium.* New York: Oxford University Press.

Feldman, R. (1990). "Settlement-Identity: Psychological Bonds with Home Places in a Mobile Society." *Environment and Behavior* 22: 183–229.

Freeman, A. M., III. (1993). *The Measurement of Environmental and Resource Values: Theory and Methods.* Washington, D.C.: Resources for the Future.

Fried, M. (1963). "Grieving for a Lost Home." In *The Urban Condition,* ed. L. J. Duhl. New York: Basic Books.

Gadgil, M., and Berkes, F. (1991). "Traditional Resource Management Systems." *Resource Management and Optimization* 18: 127–141.

Gadgil, M., and Guha, R. (1992). *This Fissured Land: An Ecological History of India.* New Delhi and Berkeley: Oxford University Press and University of California Press.

Galster, G. C. (1986). "What Is a Neighborhood? An Externality-Space Approach." *International Journal of Urban and Regional Research* 10: 243–263.

Gans, H. J. (1962). *The Urban Villagers: Group and Class in the Life of Italian Americans.* New York: The Free Press of Glencoe.

Gould, P., and White, R. (1974). *Mental Maps.* New York: Penguin Books.

————. (1986). *Mental Maps.* Second Edition. Boston: Allen & Unwin.

Gunderson, L., Holling, C. S., and Light, S. S. (1995). *Barriers and Bridges to the Renewal of Ecosystems and Institutions.* New York: Columbia University Press.

Gussow, A. (no date). *A Sense of Place: The Artist and the American Land.* San Francisco: Friends of the Earth/Seabury Press.

————. (1993). *The Artist as Native: Reinventing Regionalism.* San Francisco: Pomegranate Artbooks.

Gutierrez, R. (1991). *When Jesus Came, the Corn Mothers Went Away.* Stanford, Calif.: Stanford University Press.

Hanemann, M. (1993). "Assessing Climate Change Risks: Valuation of Effects." In *Assessing Surprises and Nonlinearities in Greenhouse Warming,* ed. J. Darmstadter and M. Toman. Washington, D.C.: Resources for the Future.

Hannon, B. (1985). "World Shogun." *Journal of Social and Biological Structures* 8: 329–341.

———. (1994). "Sense of Place: Geographic Discounting by People, Animals and Plants." *Ecological Economics* 10(2): 157–174.

Heidegger, M. (1958). "An Ontological Consideration of Place." In *The Question of Being.* New York: Twayne Publishers.

———. (1962). *Being and Time.* New York: Harper and Row.

Heilbroner, R. (1974). *An Inquiry into the Human Prospect.* New York: W. W. Norton.

Holling, C. S. (1992). "Cross-Scale Morphology, Geometry, and Dynamics of Ecosystems." *Ecological Monographs* 62(4): 447–502.

Hough, M. (1990). *Out of Place: Restoring Identity to the Regional Landscape.* New Haven: Yale University Press.

Hull, R. B. (1992). "Image Congruity: Place Attachment and Community Design." *Journal of Architecture and Planning Research* 9: 181–191.

Jackson, J. B. (1994). *A Sense of Place, a Sense of Time.* New Haven: Yale University Press.

Johnson, A. R. (1993). "Spatiotemporal Hierarchies in Ecological Theory and Modeling." From *2nd International Conference on Integrating Geographic Information Systems and Environmental Modeling*, Sept. 26–30, Breckenridge, Colorado.

Kemmis, D. (1990). *Community and the Politics of Place.* Norman, Okla.: University of Oklahoma Press.

Kennedy, P. (1993). *Preparing for the Twenty-First Century.* New York: Random House.

Kopp, R. J. (1993). "Environmental Economics: Not Dead but Thriving." *Resources for the Future*, Spring 1993 (111), 7–12.

Labao, L. (1994). "The Place of 'Place' in Current Sociological Research." *Environment and Planning A,* 26: 665–668.

Langer, S. (1953). *Feeling and Form.* New York: Charles Scribner's Sons.

Lavorel, S., Gardner, R. H., and O'Neill, R. V. (1993). "Analysis of Patterns in Hierarchically Structured Landscapes." *Oikos* 67: 521–528.

Lee, K. N. (1993). *Compass and Gyroscope: Integrating Science and Politics for the Environment.* Covelo, Calif.: Island Press.

Ludwig, D., Hilburn, R., and Walters, C. (1993). "Uncertainty, Resource Exploitation and Conservation: Lessons from History." *Science* 260: (April 2): 17–19.

Lynch, K. (1960). *The Image of the City.* Cambridge, Mass.: The MIT Press.

———. (1972). *What Time Is This Place?* Cambridge, Mass.: The MIT Press.

McMahon, J. A., Phillips, D. A., Robinson, J. F., and Schimpf, D. J. (1978). "Levels of Organization: An Organism-Centered Approach." *Bioscience* 28: 700–704.

Moore, C., Allen, G., and Lyndon, D. (1974). *The Place of Houses.* New York: Holt, Rinehart and Winston.

Norberg-Schulz, C. (1979, U.S. Edition, 1980). *Genius Loci: Towards a Phenomenology of Architecture.* New York: Rizzoli.

369

Norton, B. G. (1987). *Why Preserve Natural Variety?* Princeton, N.J.: Princeton University Press.

Norton, B. G. (1995). "Applied Philosophy vs. Practical Philosophy: Toward an Environmental Policy Integrated According to Scale." In *Environmental Philosophy and Environmental Activism*, ed. D. Marietta and L. Embree. Lanham, Md.: Rowman and Littlefield.

Norton, B. G., and Hannon, B. (1997). "Environmental Values: A Place-Based Theory." *Environmental Ethics* 19: 227–245.

Nuclear Energy Institute (1995). "Building Citizen Support for Responsible Low-Level Radioactive Waste Disposal Solutions: A Handbook for Grass Roots Organizers." Nuclear Energy Institute, Washington, D.C.

O'Neill, R. V., DeAngelis, D. L., Waide, J. B., and Allen, T. F. H. (1986). *A Hierarchical Concept of Ecosystems.* Princeton, N.J.: Princeton University Press.

Ophuls, W. (1977). *The Politics of Scarcity: A Prologue to a Political Theory of the Steady State.* San Francisco: Freeman.

———. (1992). *The Politics of Scarcity Revisited: The Unraveling of the American Dream.* New York: Freeman.

Page, T. (1977). *Conservation and Economic Efficiency.* Baltimore: Johns Hopkins University Press.

———. (1992). "Environmental Existentialism." In *Ecosystem Health: New Goals for Environmental Management*, ed. R. Costanza, B. Norton, and B. Haskell. Covelo, Calif.: Island Press.

Proshansky, H. M. (1978). "The City and Self-Identity." *Environment and Behavior* 10: 147–170.

Proshansky, H. M., Favian, A. K., and Kaminoff, R. (1983). "Place Identity: Physical World Socialization of the Self." *Journal of Environmental Psychology*, 3: 57–83.

Rappaport, R. A. (1968). *Pigs for the Ancestors: Ritual in the Ecology of a New Guinea People.* New Haven: Yale University Press.

Relph, E. (1976). *Place and Placelessness.* London: Pion Limited.

Rolston, H. (1988). *Environmental Ethics: Duties to and Values in the Natural World.* Philadelphia: Temple University Press.

Sagoff, M. (1986). "Values and Preferences." *Ethics* 96(2): 301–316.

———. (1992). "Settling America: The Concept of Place in Environmental Ethics." *Journal of Energy, Natural Resorces and Environmental Law* 12: 351–418.

———. (1993). "Environmental Economics: An Epitaph." *Resources (Newsletter of Resources for the Future)* Spring 1993 (111): 2–7.

Sale, K. (1985). *Dwellers in the Land.* San Francisco: Sierra Club Books.

Seamon, D. (1979). *A Geography of the Lifeworld.* London: Croom Helm.

Seamon, D., and Mugerauer, R., eds. (1985). *Dwelling, Place, and Environment.* New York: Columbia University Press.

Silver, T. (1990). *A New Face on the Countryside: Indians, Colonists, and Slaves in South Atlantic Forests, 1500–1900.* New York: Cambridge University Press.

Steele, F. (1981). *The Sense of Place.* Boston: CBI Publishing Company.

Toman, M. (1994). "Economics and 'Sustainability': Balancing Tradeoffs and Imperatives." *Land Economics* 70: 399–413.

Tuan, Y.-F. (1971). *Man and Nature.* Resource Paper No. 10. Commission on College Geography. Washington, D.C.: Association of American Geographers.

_____. (1974). *Topophilia: A Study of Environmentral Perception, Attitudes, and Values.* Englewood Cliffs, N.J.: Prentice-Hall.

_____. (1977). *Space and Place: The Perspective of Experience.* Minneapolis: University of Minnesota Press.

Varlamoff, S. (1993). *The Polluters: A Community Fights Back.* Edna, Minn.: St. John's Publishing.

Vycinas, V. (1961). *Earth and Gods.* The Hague: Martinus Nijhoff.

Walters, C. (1986). *Adaptive Management of Renewable Resources.* New York: Macmillan.

Worster, D. (1979). *Dust Bowl: The Southern Plains in the 1930s.* New York: Oxford University Press.

_____. (1985). *Rivers of Empire: Water, Aridity, and the Growth of the American West.* New York: Pantheon Books.

# V

## Some Elements of a Philosophy of Sustainable Living

In this book the idea of sustainability has been observed from a number of disciplinary perspectives, which have functioned as a sort of intellectual house of mirrors – reflecting and refracting the idea of sustainability through lenses provided by philosophical pragmatism, policy science, economics, and ecology. This part includes three papers that look at the core idea of sustainable living, surveying some of the substantive issues that must be faced if we are to articulate a consistent and practically possible moral guide to sustainable living. Full success at such a task will require solutions to many puzzling moral and conceptual problems; the essays included here address a few of the perplexing questions that have to be clarified and resolved if we are to articulate a coherent approach to sustainable living.

One difficult problem in making sustainable choices is to clarify our moral relations to other species that, whatever else we save, seem deserving of our careful husbandry. Herein lies an apparent paradox, however, having to do with our relations to wild animals. On the one hand, we are responsible for the decimation of historic populations of many species. Recognizing these impacts, thoughtful and caring people – in zoos, aquariums, and in other wildlife facilities – have undertaken heroic efforts to save particular species from extinction through captive breeding programs. These efforts, however, often require serious invasions of the autonomy and well-being of individual members of those species. In the early 1990s, following discussions with Gene Hargrove, Terry Maple, Michael Hutchins, Elizabeth Stevens, and others in the zoo community, we undertook a project – funded by the Ethics and Values in Science Program of the National Science Foundation – to bring thoughtful zoo professionals together with animal welfare activists and animal rights activists. The idea was to have a closed-door airing of issues and an attempt to forge as much compromise as possible, and discuss the remaining issues, in an attempt to save species through breeding of captive populations.

373

A surprising consensus emerged from the meetings, and a detailed set of guidelines were provided to the professional society of zoo administrators, who adopted most of them as guiding principles. The paper "Caring for Nature" is my contribution to this lively discussion and debate, which, along with the other viewpoints from the conference, are all represented in a book, *Ethics on the Ark.*

In the mid-1990s, I attended several meetings and read some material on the Earth Charter, which was billed as a "soft law" document that would express the common values that humans derive from nature. I thought it would be useful to cast a philosophical eye on early versions of the document and the writings that explained its purposes and goals. My examination of an early set of proposed principles led me to some broader considerations about how we value nature and toward skepticism that the dominant general theories of environmental value are adequate to the task of understanding environmental value. This led to speculation about the contents of a theory of environmental value that would be broad enough to encompass the many ways in which humans value nature, and yet emphasize nature's creativity and productivity as common-denominator elements in all valuations of nature. More current versions of the Earth Charter, it turns out, do not raise the philosophical issues so sharply, and the most recent versions I have read avoid apparent contradiction; and yet I think that the paper "Can there be a Universal Ethic?", by raising these issues, provides analysis that remains vital today.

This part ends with a rather long survey paper, "Intergenerational Equity and Sustainability," on the subject of our obligations to future generations, written as my final product of a very productive collaboration with Michael Toman an economist, supported by the EPA. The paper examines, from several perspectives, the problem of intergenerational equity. Although there is much disagreement about the meaning of sustainability, there is broad agreement that it has to do with our moral relations with future generations; the obligation to act fairly to the future must be the touchstone of any theory of sustainability. This paper set out to attack the problem of how to define our obligations to the future head on, asking about the nature of intergenerational obligations to protect the environment, and to do so, as far as possible, by characterizing economists' take on those moral obligations and then surveying the full range of philosophical literature that has dealt with this subject. The resulting paper – more a monograph than a paper – was too long for normal journal publication, although it has spawned several shorter articles and essays. I include it here as my most comprehensive and nuanced treatment of the obligation to be fair to the future, and offer it as a possible moral support for a general theory of sustainability.

# 20

## Caring for Nature

### A Broader Look at Animal Stewardship

The naturalist writer Annie Dillard poignantly poses a central moral issue of our time in a chapter on the fecundity of nature (Dillard 1974). She notes that the rock barnacles on a short stretch of rocky seashore pour forth a million million larvae, of which only a few will survive to maturity. Nature, she notes, apparently cares not a whit about individuals. But all modern Western systems of ethics value individuals above all. "Either this world, my mother, is a monster, or I myself am a freak," Dillard laments. Dillard's point might be stated more calmly as a concern about how broadly to apply traditional concepts of justice when our actions have sometimes terrible impacts on members of other species.

As the crisis of biodiversity unfolds over coming decades, conservationists will advocate more and more invasive methods to save in captivity species that are threatened in the wild. As habitat alteration and fragmentation continue, more and more interventions in the lives of wild animals will be required if their wild populations are to be protected and kept in ecological balance. Actions taken to protect species and ecological communities may not be in the interest of individual members (Hutchins and Wemmer 1987). The recommendations of captive breeders and conservation biologists therefore conflict in many cases with the recommendations of animal welfare advocates (Ehrenfeld 1991).

The excruciating policy questions addressed in this book stem from Dillard's dilemma: committed conservationists tend to concentrate on the protection of populations, species, ecosystems, and processes. When they do so, individuals become the means to carry genetic information from generation to

From B. G. Norton, M. Hutchins, E. F. Stevens, and T. L. Maple, eds., *Ethics on the Ark: Zoos, Annual Welfare, and Wildlife Conservation* (Washington, DC: Smithsonian Press, 1995). Reprinted with permission.

generation, and the commitments of conservation seem to dictate that we follow nature in deemphasizing individuals. But biology has also emphasized the common heritage of human and nonhuman species. As Thoreau noted, "The hare in its extremity cries like a child" (Thoreau 1946), and modern science has in general agreed with modern sensibilities that countless human actions severely affect the interests of individual animals. Hence this book. Our goal is to achieve as much consensus as possible regarding ways to balance the actions required by a concern to save wild species against the equally unquestioned good of reducing animal suffering, discomfort, and invasions of autonomy.

There exists no consensus, among experts or the public, regarding one of the fundamental questions in biological conservation: what should be the central focus (target) of protection efforts? I am assuming that there does exist a broad consensus supporting the goal of protecting biodiversity, but that the meaning of that consensus, especially its meaning for preservation priorities, has not been worked out. Should our goal be to protect genes? Individuals? Populations? Species? Communities? Ecosystems? All of the above? None of the above? To answer these questions intelligently we need some sense of what it is we value in nature. That we cannot answer them in a way that commands wide policy consensus has led to confusion in management practice – confusions that point out just how perplexing are our obligations to protect the natural world. We here seek answers that will help us to avoid these confusions in the future and, we hope, to reduce tension between the well-intentioned persons in the environmental, humane, and zoo communities.

I would like to offer a broader perspective, one that considers not only our obligations as they arise in particular contexts, such as within zoos, in captive breeding programs, on farms, or in environmental restorations – but also one that looks at all of these perspectives simultaneously, seeking general patterns that structure our thinking in all of these areas. My method will be pluralistic in the sense that I do not assume that all values can be measured in a common moral metric such as rights, individual welfare, or ratios of pleasure and pain. In this respect my method differs from that of most attempts to describe an ethic for the treatment of nonhuman animals, in that they usually emphasize the similarities between humans and members of nonhuman species. In these approaches, some characteristic, such as rationality, sentience, or teleological action, is designated as the basis of "intrinsic/inherent value" and "moral considerability" (Goodpaster 1978, Regan 1983). All creatures with this designated characteristic have rights or intrinsic value and their interests must be included in any moral calculations.

In some cases, this approach is coupled with the somewhat dogmatic assertion that whatever has inherent value has it equally, which leaves open the question of how far from humans one should locate the sharp cutoff (Regan 1983; see Callicott 1989 for a criticism of Regan's view). Other theorists conclude that moral-making characteristics emerge gradually from evolutionary complexity; ethical obligations to members of other species therefore ripple outward in fainter and fainter copies of human ethics as the species in question get less humanlike. Whatever the differences in outcomes, these approaches share a conceptional strategy: to reduce all of our obligations regarding nature to as few types as possible in search of a common metric for valuing nature and its elements.

Underlying these reductionistic approaches is a drive toward moral monism, the view that all moral value can and should be traced to a single, foundational principle. It is apparently thought that expressing all of our values in a common metric will make policy decisions less confusing and difficult. While I have learned much from those who demand a monistic criterion, I have become extremely skeptical that these ethical approaches will guide us toward an intelligent protection policy.

I think we need an alternative approach to thinking about our obligations to nature, an approach that emphasizes differences rather than similarities – especially differences in context. Moral pluralism is the view that we value many things in different ways, and that these differing values are sometimes in conflict. Further, these values may be incommensurate, so that they cannot be weighed in a common metric, because they exist at different scales of analysis – individuals, species, ecosystems, etc. The best we can do in these situations is to honestly seek a fair balance point at which we have done our best to minimize harms for which we are responsible (Donnelley 1992). Rather than trying to make these values commensurate, the contextualist emphasizes different obligations according to situation, or context (Norton 1991). The focus then shifts to developing second-order rules determining which moral imperatives should be emphasized in given situations. A direct assault on the problem of protection priorities may now be possible, even though we are not armed with a single moral yardstick for setting our social priorities.

I believe that we have an obligation to minimize the suffering of individual animals in some situations and that we have obligations to emphasize species protection in other situations. The problem is to explain coherently and effectively how to tell the difference between these situations. Let me start by exploring what we value in wild animals and by explaining what I think we should be protecting in wild, natural systems. Then, I will contrast our obligations to wild animals with our obligations to domesticated animals, showing that very different rules apply in these different contexts of interspecific interactions. The key to understanding moral treatment of zoo animals and moral obligations regarding captive breeding programs is to understand how to weave an ethic that respects the integrity of the individual and also recognizes the role of an individual in the broader processes that involve its species. I will then provoke a consideration of the possibility of animal altruism – a concept that challenges our traditional conceptualizations and hastens us toward a more integrated ethic for the treatment of nonhuman animals.

### WILD ANIMALS

I believe that wild animals, considered as individuals, are valuable in an important sense, but that in most situations they are not morally considerable to humans. Individual animals are valuable to humans – they are often valued aesthetically and sentimentally, for example – but they do not exist in a relevant moral sphere for us. Let me explain this assertion, because it is sure to be controversial.

While wild animals *may* be considered morally, this depends on the situation. In the most straightforward situation – when individual wild animals live largely undisturbed by human activities in their natural habitat – humans accept no responsibility for animals as individuals.[1] This very general claim of nonresponsibility is not justified by any absolute claim about the intelligence or sensitivity of those individuals. It is rather a manifestation of a decision to respect the animal individuals *as wild*. By deciding to respect their wildness, we have agreed not to interfere in their daily lives, or deaths. We value them, but we value their wildness more; to respect their wildness is, in effect, to refrain from placing a *moral* value on their welfare or their suffering (see Callicott 1989). It is to treat them as a separate community, one with which we limit our interactions in order to encourage its autonomy from our own society. We also value wild animals as part of natural processes. I believe that our interactions with animals in the wild take on a moral dimension only at the population and species level, not at the individual level.[2]

Strachan Donnelley has introduced a useful distinction between what he calls ontological value and moral value (Donnelley 1992). Because we know that all animals strive to perpetuate their lives and their genetic line, there exists a good that results when these strivings are rewarded and a loss when they are thwarted and an organism dies. We can perhaps determine grades of ontological goodness corresponding to the mental complexity, self-awareness, etc. of the individual organism. Although it is difficult to ascertain the exact level of mental awareness experienced by animal organisms, we recognize it because we have experienced the same struggle between life and death ourselves (Donnelley 1992).

It is clear that animals have varying levels and acuities of experience, including varied levels of self-awareness. But it is equally clear that the experience we share carries with it both a striving to perpetuate that experience and the inevitability of diminution in some situations, and certain death in the long run. Ontological goodness, and grades of it, correspond to the strength and vitality of the experience of individual organisms. But this ontological goodness does not in itself entail moral goodness or moral obligations to interfere to protect it.[3] Humans cannot and ought not to accept responsibility to avoid deaths of individual animals in the wild. Building on Donnelley's distinction between ontological and moral value, I suggest it is not this *content* of animal experience but the *context* in which we encounter it that determines the strength and type of our obligations to animals and other natural objects. The content of the experience of some animals is surely rich enough to make them candidates for moral considerability. Nevertheless, I feel justified in killing feral goats on an island to protect the indigenous plant communities there, even though I have no doubt that the individual goats have a greater ontological value than the plants. Ontological value is morally relevant in some situations – as when animals are in captivity – but it is morally irrelevant in the wild, because the maintenance of the animal's wildness (appropriate behaviors in a wild context) prohibits our manipulating that experience. The forbearance we exercise here is very similar, psychologically and perhaps morally, to the attitude of wise parents who, after the time of maturity, let their children live their own lives.

While it may be necessary to manage wild populations to protect them against acute impacts of human disturbance, the goal of that management should be temporary and intended to remove the need for further interventions. Once an animal is brought into the human context, its ontological value, which corresponds to richness of experience of individual speciments, becomes extremely important in determining our moral responsibilities. I will return to permissible intrusions into the lives of captive and tamed animals below; first, I must say more about obligations that derive from our legitimate concerns for future generations.

## OBLIGATIONS TO SUSTAIN NATURAL PROCESSES

The obligation to sustain biological diversity stretches far into the future, far beyond the point at which we can identify individuals (Norton 1984, 1991) and requires that we shift our sights from individuals and the various elements of biodiversity toward concern for the *processes* of nature (paper 16, this volume; Vrijenhoek 1995).[4] In this larger resolution, individuals can only be seen as parts, functional elements in larger processes (Koestler 1967, Allen and Starr 1982). I will now explore our obligations to wild species in this longer, intergenerational context.

Thermodynamic models provide a better starting point for process-oriented models than do element-oriented models. If we assume that natural processes unfold in hierarchical order, and follow general systems theory in assuming that smaller systems change at a more rapid rate than do the larger systems they partly constitute, we can introduce some conceptual order into the confusing problems of scale in environmental policy. We experience nature from within a complex dynamic in the sense that there exist many dynamic, divergent forces on many scales. Since we are affected by, and affect, many dynamics, any ethic that is to integrate all of our obligations must provide a second-order guide to balancing our varied obligations within these varied dynamics.

When our attention shifts from concern for individuals – whether the legitimate regret we feel when we unavoidably hit an animal on the highway or the sentimental valuing as expressed in the Bambi syndrome – to concern for biodiversity, we shift from an individual to an intergenerational temporal scale and to a landscape-sized spatial scale. In this lengthened and broadened scale, our concern shifts to interactions among populations and species, not individuals. As individuals become rarer, we value individuals more and may accept responsibility to protect individuals as a means to protect species and the processes they perpetuate. The worldwide threat to biological diversity results not so much from the behavior of any individual human but from huge trends in human population and land use. I agree with those who believe that humans have important responsibilities to reverse the trends that are destroying biological diversity,[5] but this is a responsibility that has its natural habitat in discussions of extinction, in concern for long-term biological conservation, and in concerns about intergenerational inequities in the allocation of resources. On this scale, individuals and local populations are regarded as instrumental in perpetuating creative and competitive processes that drive ecology and evolution. This concern for the creative activity of natural processes is not well expressed as entailing obligations to any particular *elements* of the system. It is a process-oriented moral concern, and it directs our attention to the structures and functional interactions that constitute dynamic ecological systems.

Some of us call the creative drive of ecological systems "autopoiesis" (from the Greek: "self-making") – it is the mysterious driving force that creates, through dissipation of energy in open systems, a kind of growth or development, as order is created out of chaos (Maturana and Varela 1980, Ulanowicz 1986, Rees 1990, Costanza et al. 1992, Lewin 1992, paper 16, this volume). As pluralists, we do not assume that this value can be measured on a metric common with human rights or concerns about animal welfare.

Looked at in this larger scale, humans have made mistakes in the past, by intentionally and unintentionally introducing exotic species and by causing extinctions. Among these, I believe it was a mistake to eliminate predators on wilderness ranges – the policy of predator eradication destroyed a crucial (key-stone) process in those ecosystems, and it has saddled subsequent generations of wildlife managers with the onerous task of destroying individuals of prey species who overpopulate and degrade their ranges. A failure to understand a crucial process led our forefathers to destroy that process. The moral responsibility exists not so much on the individual level as on the interpopulation level. I am suggesting, then, that we make moral decisions in different spheres, and that differing considerations should dominate at different scales. Managing to protect biological diversity – and this will more often mean managing humans, not wildlife – occurs on an intergenerational scale on which populations and species interact, not on an individual level (Norton and Hannon 1997). Our moral decision to value wild animals *as wild* isolates us from moral obligations to wild animals as individuals.

It is now possible to split two moral questions that exist into two separate spheres. The question of whether to cull a herd, or to begin a captive breeding program, is a population-level question: are ecological processes as now configured adequate to perpetuate the health and integrity of the essential elements and structures of the system? The question of which method is permissible, or effective, for reducing overpopulations of ungulates, or what treatment is due captive breeding stock of endangered species, can be addressed in a separate context, which might be called interspecific ethics – the ethics of our relations with individuals of other species. The integrated pluralist believes we will judge our actions affecting animal organisms differently, depending on the context of those actions (Midgley 1983, Callicott 1990). Once we have made a decision that we must interfere on the population level in ways that deeply affect individual animal lives, of course, we must accept responsibility for humane implementation of our environmental goals.

This contextual separation of two dynamics and two somewhat independent ethics applying at the populational and individual levels does not, of course, resolve all issues. By separating the question into populational levels, we separate the judgment of whether herd reduction is necessary or whether captive breeding is necessary from the question of what *means* are justified in such interventions. Especially, authors of this book do not agree regarding the means that are permissible on the individual level to achieve objectives that emerge on the population and ecosystem level of the scalar hierarchy. Here, I believe that the long-term value of keeping wild animals wild, especially in national parks and national forests but on private lands as well, dominates our obligation to minimize suffering of individuals, but I expect this position to be controversial. I can understand an alternative position, for example, that would accept the separation of questions but argue that we may never cause unnecessary pain in herd regulation; development of safe and effective birth control devices for wild animals would then become a conservation priority.[6]

Before proceeding further, let me mention several controversial issues that seem to form a coherent class. (1) Should we remove overabundant ungulates from predator-free ranges, and under what situations (to reduce destruction to ecosystems or to reduce competition with domestic grazers, for example)? (2) Suppose that an epidemic of disease were attacking the last wild populations of a primate such as the chimpanzee, and that we desperately needed information that could be obtained only from extremely invasive and painful tests on living chimpanzees. Would we sacrifice several chimpanzee lives to save their species? (3) When, as seems almost inevitable, we get down to the last few mountain gorillas, should we live-capture the last few as breeding stock for a captive population? (4) Should we sacrifice a few Andean condors to gain essential information in order to save the California condor? What is common to all of these cases is that they pose policy choices that set individual members of a species against the well-being of some larger composite such as a species, a community, or an ecosystem. That is, a policy choice that may seem attractive if our goal is to perpetuate species or retain intact systems requires the suffering or death of some individual animals in an effort to save wild species. These issues all, in their own way, highlight the problems of determining what the target of protection efforts should be. And they all turn on the question, under what conditions will we sacrifice individual animals to perpetuate a larger good, either social or ecological? We should be especially interested in these cases because I believe that recommendations on these cases may represent an intellectual watershed in our deliberations. Those who, like Tom Regan, emphasize the similarities between our obligations to animals and our obligations to humans will prohibit any animal altruism toward composites, while those who emphasize that we have obligations, first of all, to protect species and ecological systems will tend to favor animal altruism or sacrifice in at least some cases. We can now characterize, both exactly and quite generally, what is at the heart of disagreements between environmentalists and animal welfare advocates, and especially animal liberationists. Individualists, who predominate in the liberationist tradition (but are much less dominant in the animal welfare tradition), unalterably oppose any nonvoluntary sacrifice of animal individuals for broader goods. Holists, by contrast, are united by their insistence that, at least in many cases, interests of individuals are overridden by the interests of composites; some form of holism is therefore attractive to conservationists.

## ANIMALS IN HUMAN COMMUNITIES

Most of us can agree, I think, that our obligations to domestic animals – farm animals, pets, and other animals that are integrated into our lives on a day-to-day basis – are far more extensive than our obligations to wild animals. The context in which we interact with these animals implies a contract to look after them. By living together with them, we have brought them into our community, and we are obliged to feed them, to care for them if they are injured, and so forth. We also accept responsibility for controlling their populations. My point is that it is the context of our interactions – the responsibilities we have taken – that determines our moral obligations more than the characteristics of the individuals involved; some of these responsibilities rest on us because of unfortunate choices and actions of our predecessors and ancestors. The morally relevant fact is not usually the *content* of experience of an individual creature but the *context* of our interactions with it. If I encounter the neighbor's cat about to get a songbird in my back yard, I would intervene if possible. If I were hiking in the wilderness and I were fortunate to see a wolf pack run down and kill a deer, intervention would be profoundly inappropriate. The crucial moral fact that decides cases like this has nothing to do with the relative mental or moral capacities of songbirds and deer and everything to do with the context of the experiences.

The more we meddle in the daily lives of wild animals, the more we accept moral responsibility for the consequences of our actions in their individual lives. The context in which we interact with domesticated animals implies a contract to look after them. No such contract exists with wild animals; for this reason, we have no moral obligations to individual members of wild species who remain in their natural habitat.

## WILD ANIMALS IN ZOOS

I have argued that it is mainly the context, and not the content, of our interactions with animals that determines our moral obligations to them, and I have argued that our obligations to wild animals generally emerge at the population level, where our policy decisions affect large trends in ecological systems and the processes that sustain them. These obligations are quite distinct and are based on values very different from those underlying our obligations to domestic animals. The obligation to sustain biodiversity intergenerationally is a process-oriented ethic; this process orientation is unavoidable because of the complex, and usually unknown, interactions among species in ecosystems. I do not treat it as commensurate with the interspecific ethic, which is based on the complexity of individual experience, or with ethical systems that state, preemptively, that all objects with independent moral value have that value equally (Regan 1983). A pluralistic, but contextual, ethic has promise to organize our thinking as we struggle to make clear the means by which we decide difficult cases.

But can these distinctions help us to deal with our present quandaries? Zoo animals are neither fully domesticated nor fully wild; they exist in a mixed context (Midgley 1983). Accordingly, humans have some obligations to zoo animals that originate in both contexts. Our taking them from the wild obligates us to protect them in ways they cannot protect themselves, and this clearly involves obligations to individuals. But we also hope and intend that they contribute, in one way or another, to the perpetuation of wild processes, of which their species is a part. We therefore respect their wildness and even hope that they or their descendants can be returned to a wild context free of our manipulation.

Many of the quandaries we face in the treatment of individual animals ultimately depend on which of these two aspects of our interactions with zoo animals we emphasize. I certainly do not have all of the answers, but I want to focus on one crucial flash point where the two moral systems – intergenerational sustainability of natural processes and concern for the autonomy and well-being of individual animals – intersect. Consider the broad class of cases, exemplified above, in which environmental managers believe it is necessary to harm severely one or more individual animals in the interests of protecting their species, some other species, or some ecological system. Can we ever justify what we might call "animal altruism," the sacrifice of significant interests – such as the interest in freedom or life–of individual animals in the interests of a long-term goal such as balancing populations, preserving genetic diversity, or protecting the integrity of an eco-system? In many of the daily decisions of environmental managers, especially those most concerned with biological resources, this is an unavoidable question (Kleiman 1989). If we agree with Dale Jamieson (who argues that the removal of animals from the wild and placing them in zoos harms them) that there is a moral presumption for leaving wild animals wild, it follows that we must provide a countervailing moral principle adequate to override this harm (Jamieson 1995; Regan 1995).

We have therefore arrived at a central issue – the issue of animal altruism – that provides an intellectual map of the confusing territory of ethics between the species. Many well-intentioned people seek a middle-ground position between moral individualism and literal extensionism of human individualistic morality to animals on the one hand and a holistic ethic that values individuals only as repositories of genes and essential parts of system processes on the other hand. But defenders of this middle ground are impaled on the horns of a dilemma: how can they accept the obvious value that exists in the experience of complex animals and at the same time make difficult decisions that will make some animals casualties in the struggle to save biological diversity?

ANIMAL ALTRUISM

It may be useful to contrast these cases in which conservationists believe it necessary to sacrifice animals with cases of human altruism and self-sacrifice. I think most of us believe, with regard to voluntary donors of organs, voluntary participation in the defense of one's country, and countless other selfless acts that we approve and applaud, that acceptance of risk of harm, even predictable loss of life, can be justified in the interest of the life of a loved one or of a higher good such as national defense, provided the action is undertaken and the risk accepted *voluntarily*. The sacrifice of individuals for a higher good is in this case permissible and in many cases laudatory because the requirement of voluntarism guarantees the moral integrity of the individual, rational moral agent and transforms an act that is simply foolish from the perspective of individual survival into an act of heroism when it is viewed in the larger perspective of family or community well-being. What is lacking in the idea of animal altruism, if it is structured to mimic human altruism, is this crucial act of voluntary self-sacrifice that gives nobility to free and selfless community-oriented service.

In individualist approaches to ethics, the key idea is unquestionably consent and voluntary choice, and so we must recognize the systematic ambiguity created by discussing animal altruism or sacrifice. If the moral status of animals strictly parallels the moral status of humans, animal altruism is conceptually and morally impossible. Moral altruism among animals is no more than a euphemism for involuntary animal sacrifice. The sacrifice of individuals to a higher good – such as protecting a crucial process (killing ungulates on predator-free ranges), invasive and risky procedures on individual animals to gain knowledge that will help to save their species, and removal of exotics and feral populations to protect fragile ecological communities – is ruled out as a priori impossible.

I doubt that these issues can be resolved so simply. We should therefore address as directly as we possibly can the following question: are there any conditions under which we should sacrifice individual animals in the interests of protecting a composite of which they are a part? If so, what, exactly are those conditions?

Field studies of social behavior of wild animals of course reveal *behaviors* very analogous to altruism, but our reluctance to say unqualifiedly that such behavior is voluntary (a requirement that is currently at issue) causes us to pause before calling it truly altruistic. But when the dominant male in a gorilla band stays behind and sacrifices himself in order to protect the band from poachers, the analogy seems strong and we have no problem seeing the heroism involved; nobody would quarrel, I think, with calling this animal altruism. Further, analogous behavior can be recognized in far less complex creatures than gorillas; army ants sacrifice themselves by the thousands by marching into a stream to create a bridge of dead bodies for their advancing comrades behind them.

But consider another case: Suppose human keepers of mountain gorillas realize that a healthy adult must be sacrificed to gain knowledge that would further a captive breeding program and perhaps save the species from extinction. The gorilla individuals cannot deliberate and cannot volunteer (even in the anthropomorphized sense of the gorilla hero), because they do not deal in abstractions and they probably cannot consider the importance of a long-range goal such as saving their species. If we think of animal altruism as strictly analogous to human altruism in that it requires unequivocally voluntary compliance with decisions to face risk, pain, or certain death, we can only conclude that no gorilla can be altruistic toward an abstraction such as a species and hence that sacrificing an individual to save its species is without question wrong in every case. If this argument is convincing, then it would seem to follow that other cases of sacrifice of animals (for human health experiments or to benefit another, endangered species) will also be ruled out.

Extensionist approaches to interspecific ethics, which emphasize similarities across species and base all moral decisions on a single value metric, are therefore likely to conclude that sacrifice of individuals for species survival is always wrong because the individuals cannot fulfill the key requirement of voluntary acceptance of risk.

This seems to me paradoxical. An ethic that begins by emphasizing similarities among human and nonhuman individuals ends in greater willingness to sacrifice human individuals than nonhuman individuals because humans have a genetically conditioned and morally developed ability to deliberate and choose heroism. Lacking these genetic and moral prerequisites, animals are disqualified from serving their kind unless they face a palpable and concrete threat to their immediate family and associates. It seems to me that Tom Regan, for example, would be committed by his monistic rights theory to this very general principle, which would have pervasive consequences for decisions by environmental managers (Regan and Francione 1992). If so, I cannot resist asking two more, admittedly rhetorical, questions: Does the conclusion that an individual nonhuman animal can never act heroically in the service of its species not involve a kind of anthropomorphism? Does this conclusion not rest on an implicit requirement that nonhuman animals must in some sense fulfill human criteria in order to extend their own sense of heroism, so evident in the case of the gorilla's self-sacrifice for his group, to include sacrifice for the abstract extension of his commitment to his group, a commitment to perpetuate his genes for many generations?

I am fully aware that there is a certain circularity in the positions we take here. It is perhaps because I side with holistic environmental managers that my intuitions point the way theirs do, but it seems obviously wrong to define altruism in a way that requires any altruist to act voluntarily and then conclude immediately that animals, because they are incapable of voluntarism in service of an abstraction, can never act altruistically. I hope that individualists are not ruling out animal altruism or sacrifice based only on this linguistic sleight of hand. Since I believe that cases such as this are crucial to our discussions, let me explain in general terms how I would decide them.

The goal is to forge a concept of animal altruism that both recognizes the inherent differences between human and nonhuman consciousness and at the same time reflects appropriate levels of respect for individual organisms. I believe we can, but I doubt we can if we conceptually straitjacket ourselves by valuing only individuals and their experiences. I therefore advocate not a middle-of-the-road approach but a both-sides-of-the-road approach based on moral pluralism.

To accomplish this goal, let us return to Donnelley's (1992) distinction between ontological and moral value. Earlier I used this distinction to explain how we do not have obligations to interfere to reduce naturally occurring pain (during acts of predation in the wild, for example), even though we recognize high degrees of ontological value in the prey's struggle to survive. This ontological value makes animals candidates for moral status, depending on context. In the wild, we only express our awe at the power and cruelty of nature, and we avoid intervening whenever possible. But once the individual animal is brought by us into the human community, new constraints are imposed upon human treatment of individuals – we have taken responsibility for the animal's well-being. The humans in question have therefore committed themselves to act as responsible stewards for the now helpless animal, and they must devise a new ethic appropriate to this different situation.

This line of reasoning explains the gravity of our decision to remove wild animals from the wild; it also explains the seriousness of the responsibilities we come to owe individuals once we become their moral guardians. Sometimes this responsibility is thrust upon us, as when a baby animal is orphaned as an unforeseen outcome of human activities; at other times, we grasp this responsibility, recognizing that a given species, which we value, is threatened and that extraordinary interventions are necessary to prevent permanent rents in the fabric of natural systems. In either case we must act responsibly, recognizing to the extent possible the impacts our actions will have on individual lives, and recognizing also the gravity of the situation.

As a moral pluralist, I perceive the ethical problem as one of facing at least three broad moral responsibilities (in addition to the obvious obligation to treat our family and associates justly). One obligation is to our fellow humans of this generation who suffer from famine, war, and poverty. But there is a second obligation, to future generations. In this I count prominently the obligation to sustain nature's bounty and to provide the future with a livable ecological context (niche) that will leave them options for a pleasant life. A third obligation is to respect the other creatures that exist with us on this planet, which is both fragile and resilient. I recognize that intelligent and fair-minded people differ regarding which of these responsibilities is dominant; I therefore advocate what I call a convergence strategy, which places priority on pursuing policies that will simultaneously support our goals in all three areas of obligation. According to the convergence hypothesis, we assume that in the long run, what is good for our species will also be good for other species, taken as species (Norton 1991). We therefore seek policies that protect ecological systems from disruption due to our increasingly ubiquitous activities. Similarly, we place high priority on attempts to raise the standard of living of the poorest humans, recognizing that improvements in standard of living normally lead to reduction of human birth rates and an eventual positive impact on protecting biodiversity.

If human population growth continues and virtually all wildness is wiped out, we will have harmed both future generations of humans and the processes of nature. What we have in common with other animals is the ceaseless striving to exist and reproduce, the source of ontological value. It is essential that we respect that ontological value; it imposes obligations on us when we accept responsibility for the animals we bring into our community. But the community we bring them into is not the human community; it is a *mixed* community (Midgley 1983). And this community cannot be identified with those animals existing now; it is rather a hierarchically organized system that changes on many different scales and according to multiple dynamics. Respect for wild animals as wild requires that we see them as participants in both a personal struggle to survive and in a struggle to perpetuate their kind. The individual drive to survive is of course not just accidentally connected to the individual's contribution to the gene pool of the species. Just as human beings are both social beings who feel sentiment toward others and self-preserving organisms in their own right, we cannot do justice to animal ethics without recognizing that the struggle of all animals to survive has meaning only if it is a part of their correlative struggle to perpetuate their species in the long run.

Now consider again the situation of a captive animal that will be used as a part of a captive breeding program. This animal is a wild animal, but we have accepted responsibility for its care and brought it into the context of the conservation community, and in that context we have obligations to respect its considerable ontological value, which requires that we respect the sentience of the animal by limiting its pain, especially in the extremity of death. But full respect for a truly wild animal, an animal that exists with integrity in its traditional niche and still inhabits the least disturbed areas of its traditional range, requires that we recognize also the intergenerational aspect of the great striving toward life. The struggle of animals to exist in the wild is both a struggle to survive individually and a struggle to perpetuate their species. A reasonable concept of animal altruism must account for the natural instinct of animals toward individual survival and, in addition, their natural instinct to perpetuate their species.

It is an awkward truth that humans must decide which members of the community will be sacrificed in the furtherance of their species and of the ecological community that constitutes their niche. The problem of exploding human population and destruction of habitats for other species is in a profound sense a human problem. It is anthropogenic in its causes, and we must accept responsibility for our past and present actions that inevitably shape the future. But the solution is surely not to domesticate all animals. If we accept responsibility for a wild animal and then reduce that animal to a creature incapable of noble acts in service of its community, this too is an act of disrespect, because in the wild the animal acted both to survive and to perpetuate its species. We can now see the error in individualistic and extensionist positions on interspecific ethics. By denying that animal altruism in the service of higher ideals can ever be justified because the crucial condition of voluntarism cannot be fulfilled, the individualists respect only the drive toward individual preservation and ignore the equally powerful drive of animals to perpetuate their species. Individual members of every social species (which includes at least every sexually reproducing species) live an existential paradox – the drives to survive and to reproduce one's kind, usually mutually reinforcing, can conflict, forcing a choice between individual preservation and contribution to species perpetuation.

This second aspect of the striving of all living things can also require self-sacrifice, as when a parent sacrifices or risks its life to protect its young. I therefore conclude that we can treat animal altruism as a conceptual extension of the striving to live and perpetuate one's species. These animals may therefore be enlisted in a noble cause; they cannot do so voluntarily, but their willingness to sacrifice themselves is implicit (it cannot be explicit because they are incapable of language) in their struggle to perpetuate their species. Following this line of reasoning, individual animals can justifiably be sacrificed, provided (1) the sacrifice contributes toward a goal that is implicit in the life struggle of the animal, (2) the animal is treated appropriately within the context of the struggle to save the natural world community of which humans and all other species are constitutive members, and (3) necessary means are taken to reduce pain and suffering to the extent possible and commensurate with the ontological status of the animal under human care.

Again, context is important, just as it is in human ethics. Suppose my country is at war and I enlist in the defensive forces; I thereby enter a new context. Suppose that the fortunes of war place my regiment at a crucial point in the battlefield, and the general in charge reluctantly concludes that my regiment must be sacrificed to plug a hole in the lines and allow orderly retreat. My voluntary enlistment declared my willingness to be a soldier. It is at this excruciating moment that that voluntary act is consummated. I cannot with honor decline to fight. I have voluntarily entered a context in which I have given up the right to decide whether to retreat or fight. Although his behavior may be more genetically mediated, the dominant male gorilla defends his harem with a similar resignation.

If we think of animals in captive breeding programs as implicitly accepting the goal of perpetuating their species, then they may unfortunately be called upon to make the ultimate sacrifice. Making this decision, however, requires that we, who are capable of seeing the gravity of the situation facing them and other species, respect their striving in all its complexity. We accept the awesome responsibility to manage the lives of other species even to the extent that we accept the responsibility to sacrifice some of those animals as a part of protecting their species and their niche. Our responsible treatment of them then becomes analogous to the decision of the military commander as we reluctantly place individual animals in the breach to hold off the forces that would interrupt its survival as a species in the wild. This is an act not to be taken lightly; the grave responsibility we have accepted by our experiments with change in natural systems should be sobering, and the decisions we make must be sensitive to the complexities of the situation – we must weigh all of the relevant information and we must seek to meet all of the sometimes conflicting responsibilities to the fullest degree possible. The decision maker must accept that the animal in question has great ontological value, value that expresses itself both individually and collectively as the means by which individuals contribute to the perpetuation of their species. As an individual member of a living species, each animal exists on two planes, as a striver to save itself and on the larger scale in which it tries to perpetuate its genes; in this larger scale, the individual life gains existential meaning in a larger context. And it is not simple cruelty to sacrifice an animal for the good of that animal's species.

The question of whether a member of one species can be sacrificed for the survival of another species seems to me more problematic, but I suggest for discussion the following position: a member of one species can be sacrificed for the well-being of another if that sacrifice can be justified as a necessary act to protect natural communities and the habitats that will make possible the perpetuation of all species.

The remaining difficulty lies in the asymmetry between the human role and that of other species. A large part of this asymmetry results from an undeniable fact. Humans, in our own struggle to control nature, are the culprits who cause most of the dilemmas. Then we, in our wisdom, decide to sacrifice individuals to save species and processes that our own excesses have drawn into risk. It is the march of human domination that threatens the habitats of other species, and our own, as well. Can we, as the only species (as far as we know) capable of scientific projections and perception of long-term dangers, inflict upon individual members of other species the ultimate sacrifice in order to avert consequences of our own mistakes? I do not have, I admit, a fully satisfactory answer to this question; I can, however, offer two considerations that put the question in a context in which it is perhaps less perplexing. First, note that viewed in the long run, nature's processes are more basic than individuals. It is processes that create and sustain individuals. If we take seriously the obligation to protect and sustain biological diversity, we must act to protect the wild processes and the species that embody them.

In accepting the responsibility to sustain biodiversity, we ipso facto accept the responsibility to choose, when it is unavoidable, to save those creative and sustaining processes even at the expense of individuals. Second, while it is undeniable that the current crisis in biodiversity is human in origin, our species has, at least until recently, acted without intentional malice in expanding our range. Until we developed, only in the last century or so, an awareness of our ability to destroy natural processes on a huge scale, our striving to perpetuate and expand our own population was morally on a par with the actions of members of other species who strive for analogous goals for their own species. Once one recognizes that sustaining biodiversity will require caring for processes within a dynamic system, and that all species change the larger dynamic as they strive to survive and reproduce their kind, it seems to follow that currently existing humans cannot be held morally responsible for the actions of our forebears. The moral question is, What will we do now that we are beginning to understand the incredible complexity of the factual and moral situation we face? I have argued that we will begin to act in morally acceptable ways only if we develop a complex and pluralistic understanding of our multiple moral obligations in our complex situation.

## SUMMARY AND CONCLUSIONS

I have developed a general approach for balancing our moral responsibilities to animals of other species. This approach is pluralistic in the sense that it recognizes more than one type of value and more than one source of obligations, but it is an integrated pluralism, specifically a contextual pluralism, which manages these multiple obligations by emphasizing them differently in different situations. Wild animals have ontological value, but our decision to leave them wild precludes our accepting moral responsibility for their well-being in the wild. Once they are made a part of the mixed communities of zoos and captive breeding programs, we accept expanded responsibility for their welfare, and our obligations will be governed by the comparative level of self-awareness and complexity that constitutes their ontological good.

I am not sure how far this line of reasoning will carry us, but it does hold promise for developing a more encompassing position – one that places high value on individual wild animals while recognizing that wild animals strive to perpetuate their species, allowing, at least in principle, the possibility of animal altruism in service of the long-term good of wild species and their habitats. This line of reasoning may justify us in enlisting animals in causes that they cannot understand, because the goal of perpetuating their species and other species is implicit in their struggle to perpetuate their genetic line as an element in the fabric of nature, and of the experiment of life more generally.

1. Because I see the problem as one of balance, in pluralistic terms, I do not think this rule should be applied narrowly, independent of other considerations. For example, Alston Chase discusses a case in which the Park Service refused to help a bison that had fallen through ice into the Yellowstone River (1986). Because the bison's life has positive value, I do not take the general preference for wildness to *require* that we let the bison die in these cases. I would reason that while I have no moral obligation to save the bison, a general preference for saving things of ontological value might in this case override the minor cost to the authenticity of the struggle for existence.

2. This is the point of Aldo Leopold's dramatic argument that we must "think like a mountain," which means "thinking on the temporal and spatial scale of the mountain" (Leopold 1949; Norton, papers 1 and 15, this volume). I have argued that it is total diversity over large geographic areas that should be the target of protection efforts (Norton 1987). Emphasis on within-habitat diversity would mandate too much concern for populations (which are naturally ephemeral on some scale of time), whereas concern for cross-habitat diversity would not support enough concern for the genetic diversity that comes from distinctive species and subspecies that evolve in specialized habitats. While I still believe that protecting total diversity over landscape-sized geographical areas is the best general measure of how we are doing over years and decades, I now doubt that any approach to biodiversity policy that emphasizes elements rather than processes will serve to characterize biodiversity in the long run (see Norton and Ulanowicz this volume).

3. Hargrove has insightfully emphasized the importance of this distinction in the naturalist tradition, which has shown little concern for individual specimens of plentiful species but has condemned wanton destruction of individuals – destruction for no good cause (1989). Since what is considered wanton killing of individuals may well depend partly on their rarity, Hargrove links current concern for species and the processes they represent with a long and respected tradition in science, literature, and art. See Hutchins and Wemmer (1987) for a survey of thinking on the justifiability of herd reductions.

4. Ulanowicz and I have provided a precise and potentially quantifiable argument that system characteristics at different scales depend on different dynamics owicz (paper 16, this volume). This independence of dynamics is crucial in understanding environmental policy and management; it implies that policies affecting economics will not all have equivalent impacts on ecological systems that provide their context. It is therefore possible in principle to have development that supports biodiversity.

5. See Jamieson 1985 for a discussion of the moral gravity of the role we have assumed.

6. See Hutchins and Wemmer 1987 for a discussion of the strengths and weaknesses of the case for contraception as opposed to killing.

REFERENCES

Allen, T. F. H., and T. B. Starr. 1982. *Hierarchy: Perspectives for Ecological Complexity.* Chicago: University of Chicago Press.

Callicott, J. B. 1989. *In Defense of the Land Ethic.* Albany: State University of New York Press.

_____. 1990. The case against moral pluralism. *Environmental Ethics* 12:99–124.

Chase, A. 1986. *Playing God in Yellowstone: The Destruction of America's First National Park.* Boston: Atlantic Monthly Press.

Costanza, R., B. G. Norton, and B. D. Haskell, eds. 1992. *Ecosystem Health: New Goals for Environmental Management.* Covelo, Calif.: Island Press.

Dillard, A. 1974. *Pilgrim at Tinker Creek.* New York: Harper & Row.

Donnelley, S. 1992. *Bioethical Troubles: Animal Individuals and Human Organisms.* Jerusalem: Hans Jonas Symposium.

Ehrenfeld, D. 1991. Conservation and the rights of animals. *Conservation Biology* 5:1–3.

Goodpaster, K. 1978. On being morally considerable. *Journal of Philosophy* 75:306–325.

Hargrove, E. 1989. *Foundations of Environmental Ethics.* Englewood Cliffs, N.J.: Prentice-Hall.

Hutchins, M., and C. Wemmer. 1987. Wildlife conservation and animal rights: Are they compatible? In *Advances in Animal Welfare Science 1986–1987,* ed. M. W. Fox and L. D. Mickley, 111–137. Washington, D. C.: Humane Society of the United States.

Jamieson, D. 1985. Against zoos. In *In Defense of Animals,* ed. P. Singer, 108–117. New York: Blackwell.

_____. (1995). Zoos revisited. In *Ethics on the Ark: Zoos, Animal Welfare, and Wildlife Conservation,* ed. B. G. Norton, M. Hutchins, E. F. Stevens, and T. L. Maple, 52–68. Washington D.C.: Smithsonian Institution Press.

Kleiman, D. G. 1989. Reintroduction of captive mammals for conservation. *Bioscience* 39(3):152–161.

Koestler, A. 1967. *The Ghost in the Machine.* New York: Macmillan.

Leopold, A. 1949. *A Sand County Almanac.* New York: Oxford University Press.

Lewin, R. 1992. *Complexity: Life at the Edge of Chaos.* New York: Macmillan.

Maturana, H. R., and F. J. Varela. 1980. Autopoiesis: The organization of the living. In *Autopoiesis and Cognition,* ed. H. R. Maturana and F. J. Varela. Boston: D. Reidel Publishing.

Midgley, M. 1983. *Animals and Why They Matter.* Athens: University of Georgia Press.

Norton, B. G. 1984. Environmental ethics and the rights of future generations. *Environmental Ethics* 4:319–337.

_____. 1987. *Why Preserve Natural Variety?* Princeton, N.J.: Princeton University Press.

_____. 1991. *Toward Unity among Environmentalists.* New York: Oxford University Press.

Norton, B. G., and B. Hannon. 1997. Environmental values: A place based theory. *Environmental Ethics* 19:227–245.

Norton, B. G., M. Hutchims, E. F. Stevens, and T. L. Maple, eds. 1995. *Ethics on the Ark: Zoos, Animal Welfare, and Wildlife Conservation.* Washington D.C.: Smithsonian Institution Press.

Rees, W. R. 1990. The ecology of sustainable development. *Ecologist* 20:18–23.

Regan, T. 1983. *The Case for Animal Rights.* Berkeley: University of California Press.

Regan, T., and G. Francione. 1992. A movement's means create its ends. *Animals' Agenda* 12(1):40–43.

———. (1995). Are zoos morally defensible? In *Ethics on the Ark: Zoos, Animal Welfare, and Wildlife Conservation*, ed. B. G. Norton, M. Hutchins, E. F. Stevens, and T. L. Maple, 38–51. Washington, D.C.: Smithsonian Institution Press.

Thoreau, H. D. 1946. *Walden.* New York: Random House.

Ulanowicz, R. E. 1986. *Growth and Development.* New York: Springer-Verlag.

Vrijenhoek, R. (1995). Natural processes, individuals, and units of conservation. In *Ethics on the Ark: Zoos, Animal Welfare, and Wildlife Conservation*, ed. B. G. Norton, M. Hutchins, E. F. Stevens, and T. L. Maple, 74–92. Washington, D.C.: Smithsonian Institution Press.

# 21

## Can There Be a Universal Earth Ethic?

### A Reflection on Values for the Proposed *Earth Charter*

PART 1. VALUE THEORIES AND BIOLOGICAL DIVERSITY

Recent international discussions of biodiversity policy have established two points: (1) there is a strong and growing international consensus in favor of sustaining/protecting biodiversity, and (2) there is little agreement regarding *why* this should be done. Thus, while a significant international consensus regarding policy has apparently emerged, this consensus is not grounded in a consensually accepted *value theory* to explain why biodiversity protection, however strongly supported, should be a top priority of environmental policy. Lack of agreement on (2) has led to disagreements at the Earth Summit in Rio de Janeiro, for example, where delegates disagreed on whether to emphasize nature's economic, "utilitarian" value or its "intrinsic" value, defined as value that exists independently of human values and motives.

The ambivalence between saving nature for future use and saving nature for its own sake is written into the proposed principles for the new "Earth Charter," which is being urged as a next step in developing a legal and political framework to guide local, regional, national, and international efforts to protect nature.[1] Steven C. Rockefeller, an advocate of the Charter, writes: "In order to address the many interrelated social, economic, and ecological problems that face the world today, humanity must undergo a radical change in its attitudes, values, and behavior.... The purpose of the Earth Charter Project is to create a 'soft law' document that sets forth the fundamental principles of this emerging new ethics, principles that include respect for human rights, peace, economic equity, environmental protection, and sustainable living." He goes on to hope that the Earth Charter "will become a

From *Man & Nature*. Working paper 92 (1997). Reprinted with permission from the Man and Nature Center, Odense, Denmark.

universal code of conduct" that expresses "the shared values of people of all races, cultures, and religions." I laud these goals, and I think the idea of an Earth Charter is a wonderful one. But can we share Rockefeller's optimism about the early arrival, and consensual acceptance, of "this emerging new ethics"?

First, we must ask: *Is* there an overarching ethic that represents the values of all peoples? How might one articulate such an overarching ethic, given the existing tensions surrounding evaluations of nature? In this paper I will explore these questions by surveying the usually cited value theories – utilitarian/economic and nonanthropocentric – and by arguing that neither of these approaches is likely to provide a unifying ethic such as might support a universal Earth Charter. My goal is to point the direction toward a new approach to environmental evaluation, one that is more likely to lead to an inclusivist ethic.

Interestingly, the tension between utilitarian approaches and deep ecological approaches is openly expressed within the "Summary of Principles" now under discussion. One principle of the new Earth Charter, listed in the "Worldview" section, states: "Every life form is unique and possesses intrinsic value independent of its worth to humanity. Nature as a whole and the community of life warrant respect." Meanwhile, the Charter's first principle, on "Sustainable Development," states: "The purpose of development is to meet the basic needs of humanity, improve the quality of life for all, and ensure a secure future."[2] Are these principles consistent? Ultimately, I'm not sure; it depends, obviously, on how the terms are defined.[3] Whether or not these principles are directly contrary to each other, however, they certainly express a *tension* between two broad ways to value nature. The Worldview Principle emphasizes valuing nature in separation from humans and their activities, whereas the Sustainable Development Principle emphasizes the evaluation of nature insofar as it fulfills human needs.

It has also been noted that the tension between these two value theories tends to polarize discussions of international efforts to protect biodiversity. Environmentalists from the United States (and other developed countries) espouse intrinsic values in nature, even though the U.S. government – because of economic concerns – has failed to ratify the Convention on Biodiversity. It is common for spokespersons for the developing world to complain, in international policy forums: "First-World countries have already exploited and converted their forests; now they ask *us* to forgo forest-based development and attendant increases to human welfare." Even as governments of developing countries are attempting to maximize economic development based on exploitation of natural resources, there have emerged minority groups, including minorities from indigenous cultures opposed to capitalistic exploitation, that have attempted to retain or resurrect their animist religions as a counterbalance against economic exploitation. The tension regarding why and how to value nature therefore has practical effects, making it more difficult to forge North–South and other coalitions.

The lack of a consensus regarding foundational values especially affects hopes for an Earth Charter because that effort, by its nature, must be inclusive. So we must also ask: Is it possible to embody both the use-and-development ethic and the utilitarian value concepts that usually come with it, and the save-nature-for-its-own-sake ethic, in one charter? Perhaps it is possible. Given the tension between the two sides in the Great Ethics debate, however, development of a more inclusivist ethic will require both sides to move toward a middle ground. Little of that has happened so far.

I believe that one reason this debate has been so frustrating and polarizing is that we have been asking the wrong questions in the wrong way, given the task at hand, and that *neither* of these theories is a good candidate for providing an inclusivist ethic to guide actions affecting nature, especially including wild life forms. First, I must establish that there is an alternative to those usually cited and discussed. That there is such an alternative can best be explained by showing how the two polar positions regarding values actually share important and highly controversial assumptions; these assumptions will be examined in the remainder of Part 1. The subject of Part 2 is more practical; there I discuss the problem of choosing the most important targets of environmental protectionist policies and priorities, and the role of values in deciding on conservation policies. I conclude that the assumptions informing the traditional theories render us unable to see the most promising solutions to these practical problems. The remainder of the paper, Parts 3 and 4, provides a sketch of one alternative approach, an approach that has some promise to provide an inclusivist theory of environmental values.

So far, most discussions of how to evaluate nature have been based on one or the other of two theories of the value of nature; for brevity, I will call them "Economism" and "Deep Ecology."[4] Although both theories come in multiple variants, for our purposes, we can examine the two theories in their most general forms. First, there is the theory of mainstream Economists that environmental values are economic values. According to this view, elements of nature have *instrumental* value only, and should be valued like other commodities. Economists, of course, recognize that there are no natural markets for many environmental goods and services, so methods other than measuring market behavior must be used if we are to correctly describe human preferences regarding the environment. Deep Ecologists and other nonanthropocentrists directly oppose this viewpoint. They argue, contrary to the instrumentalist, preference-based theories of Economists, that some elements of nature have "inherent" value, and that these elements are therefore deserving of preservation for their own sake. According to this view, human individuals and some other elements of nature, either individuals, species, or ecosystems, have their own values, values that are not dependent upon the preferences of individual human valuers.

But it is important to recognize that these opposed theories rest on a cluster of highly vulnerable assumptions. Both Economists and Deep Ecologists accept a sharp dichotomy between values that are inherent and those that are instrumental; further, both groups proceed to use this sharp dichotomy to separate nature into beings/objects that have, and those that lack, "moral considerability." In a particularly strong version of Economism, for example, Gifford Pinchot (first Head of the U.S. Forest Service) said, "There are just two things on this material earth – people and natural resources."[5] Pinchot was thus enforcing a sharp dichotomy between persons and other things, living or nonliving – the well-being of the former, but not the latter, should be taken into account in our calculations regarding what is acceptable behavior. Interestingly, the position of Pinchot and his Economist allies coincides on a more basic level with that of Immanuel Kant, who is typically cited as opposed to the consequentialist emphasis of utilitarians. For Kant, only rational beings could be "ends-in-themselves." Both Pinchot's utilitarianism and Kant's rights theory are thus based on a sharp distinction among entities – those who are to be regarded as being "ends in themselves" and those objects that can be used, without restriction, in the service of those ends.

Economists and Deep Ecologists, then, agree that there must be some special status for those beings who have noninstrumental value; they simply disagree regarding which objects in nature actually have this special status. For Economists like Pinchot, the special status is co-extensive with humanhood; for the Deep Ecologists, moral considerability is co-extensive with a much larger subset of nature's components. Either way the sharp distinction between "instrumental" and "inherent" values ensures that questions of environmental value are posed in all-or-nothing terms. For the Economist, "Should we protect this river?" becomes "Does this river have net positive economic value (for humans) or not?" For the Deep Ecologist, it becomes, "Does this river have inherent value?" These questions are usually formulated in incommensurable theoretical frameworks; what they share is their tendency to elicit yes-or-no determinations of the value of objects.

Another related consequence of the bipolar formulation of environmental valuations is the apparent bias of both sides in favor of evaluating *objects* or *entities* rather than evaluating *dynamic processes* and *changes in processes*. Protection is assumed to be protection of items in an inventory: should we try hardest to save genes? Individuals? Populations? Species? Ecosystems? This object bias is of course endemic to all of Western culture at least from the classical period; it represents the triumph of Plato's concern for constancy of forms constituting reality over the ideas of Heraclitus, who, around 500 B.C., declared that "All is in flux."[6] This ideological triumph has further been embodied in modern scientific reductionism, which seeks explanation in the motion of elementary particles.

The "atomistic" idea which emphasizes elements is so deeply engrained in Western thinking that alternative conceptualizations of nature have only been considered relatively recently. Since the publication of Charles Darwin's evolutionary theory, however, the importance of systemic change and irreversible developments – of complex, dynamic processes – has asserted itself.[7] This revolution has extended to physics, and physicists are now leaders in an interdisciplinary effort to develop a more dynamic worldview, as is evidenced by the ever-increasing emphasis placed on nonequilibrium dynamics. The full implications of a dynamic worldview are just now being felt. It may be decades before these concepts are well understood, but creative work in nonequilibrium system dynamics is already leading to new insights in ecology, and this direction holds promise for applications to environmental policy.[8] This much we know for sure: full absorption of evolving systems thinking into environmental management will have far-reaching impacts on the policies we advocate, and will almost certainly require more attention to interspecific relationships and system-level characteristics.

Another similarity between Deep Ecologists and Economists is that they are both looking for a universal, "monistic" approach to values. Monism, as defined by Christopher Stone, conceives the ethical enterprise "as aiming to produce, and to defend against all rivals, a single coherent and complete set of principles capable of governing all moral quandaries."[9] Commitment to monism, ultimately, explains the all-or-nothing character of the two competing theories. Both Economists and Deep Ecologists think that there is only one kind of ultimate value; they differ only in how widely they find that value. It seems unlikely to me that either of these all-or-nothing, monolithic theories of value will prove rich enough to guide difficult real-world choices regarding what should be saved, where conservationists should concentrate efforts, and how they should set priorities. But should that not be precisely the role for a theory of environmental value in the conservation policy process?

PART 2. RE-THINKING THE PROBLEM OF CONSERVATION
PRIORITIES/TARGETS

Interestingly, the same entity-oriented concepts that have divided environmentalists regarding value theory have also affected thinking in practical
policy situations. For example, suppose I ask: What should be the highest
priority in protecting biological resources? Typical answers to this question,
given the Western tendency toward entity orientation, might cite "all species"
or "all species and all ecosystems." Other answers might identify particular
types of species, for example "producers," as the most important. Note how
easily we fall into a characterization of the problem as one of providing a list
of entities – of providing an "inventory" of things that should be saved. The
problem of conservation priorities is thus expressed as a "ranking" of various
categories according to their importance.[10]

Biologists, like Economists and Deep Ecologists, tend to talk about conserving inventories of objects and setting priorities among these. But whereas
discussions of value theory emphasize philosophical considerations of what
has ultimate value, in a management context these pure philosophical considerations are inevitably mixed with questions of methods and means, and
even with recognitions of political constraint. For example, a conservation
biologist might believe that *species* are the highest priority, and at the same
time advocate policies to save ecosystems and habitats because the systems
approach is the most efficient means to achieve the goal. Conversely, a policy
analyst might argue that we should save *ecosystems* for the future, but that the
best way to save ecosystems is by legislating protection of species (because
they are fairly easily counted, etc.). The policy analyst and the conservation
biologist, working in the field, differ from the moral theorist in having to take
account of empirical realities imposed by the intricacies of systems, by our
lack of knowledge, and by political constraints. These differences are very
important and are not denied here. The point is that, despite these differences
in the categories of entities and how to sort them, *both* the philosophical,
values debate *and* the practical, policy debate are framed as questions that
can be answered by presenting *lists of entities*. Setting priorities, given this
entity-and-category orientation, is a matter of *excluding* some kinds of objects
from a "preferred" list.

To understand how this entity orientation has affected policy discussions,
it will be useful to briefly explore the history of policy debate in the United
States regarding biological resources. Attempts to protect biological resources
in the United States can be divided – somewhat arbitrarily but usefully – into

three phases. As early as the seventeenth century, local shortages of valued food and game species such as deer led to bag limits and other restrictions on takings of those species.[11] As human populations increased, more and more species suffered population declines and more and more restrictions were contemplated and imposed. These restrictions were sometimes successful and some-times not, but for our purposes we are interested in the *formulation* of the policy issue.[12] In this first phase of protection of biological resources, which we can call the "single-species" phase, the object of protection was always a species or the populations of a species. This first approach continued well into the twentieth century, as the populations of more and more species underwent declines in response to human exploitation and habitat conversion, and culminated in the Endangered Species Act of 1973. This Act, which was unquestionably a remarkable advance in policies to protect nature, nevertheless exemplified the original conceptualization of the problem as a problem that should and could be addressed at the level of species.[13] Accordingly, success in protection efforts was envisioned as protecting an inventory of existing objects.

By the early 1980s, a number of scientists and policy analysts began to question explicitly whether the species-by-species approach is sufficiently comprehensive to protect all biological resources; the term "biological diversity" came into vogue in the early 1980s, and this term was shortened to "biodiversity" in conjunction with an influential Symposium on Biodiversity, sponsored by the Smithsonian Institution and the National Academy of Sciences.[14] This Symposium ushered in the "biodiversity phase" of policy debate regarding biological resources. The new term "biodiversity," as used at the Symposium and subsequently, was defined as "the sum total of distinct species, genetic variation within species, and the variety of habitats and ecological communities."[15] While this new concept no doubt represents a significant advance in the search for more comprehensive policies to protect biological resources, it is nevertheless problematic. Notice that the definition still treats protection of biological resources as a question of protecting objects – an inventory of items. In this sense the biodiversity phase seems simply an extension of the single-species phase; it simply expands the list of items of concern.

But despite this *formal* consonance with older approaches, the definition of biodiversity introduces a more dynamic conception of resources in its *content*. The third element of the definition – "the variety of habitats and ecological communities" in which species exist and adapt – while listed as *elements* in an inventory, refers in actuality to a whole range of ecological *processes*. This change in content, which recognizes that species actually trace multiple and changing trajectories through time, has the subversive effect of undermining the comfortable assumption that protection of biological resources involves protecting a static list of entities. Thus, while the term "biodiversity" is at present a useful one, it embodies the entity orientation so typical of modernist philosophy, science, and policy.

This inadequacy of the biodiversity conception leads me to expect a third phase in the formulation of the goals of biological protection. Indeed, such a third phase is now emerging, but it may be too early to discern its details clearly at this time, and considerable controversy exists. One thing seems certain: in the new phase, processes will be more important and entities less important. So I suggest we call the emerging phase the "ecosystem processes" phase. Since it is not clear how processes will be described in the future, it remains unclear how environmental evaluation will be expressed in a more process-oriented mode. Some of us have suggested that medical analogies are helpful in re-directing our thinking more toward evaluating processes, and have used terms such as "ecological health" or "ecological integrity."[16] Others have argued that these analogically derived terms have no clear scientific meaning, and illegitimately mix description and evaluation.[17] I hope to avoid the purely linguistic issues here by arguing for more linguistically neutral conclusions: in the future, our descriptions of nature will become more dynamic and process-oriented; and our evaluation of biological aspects of these processes will have to change to accommodate these changes in description.[18] In the remainder of this section, I explore what consequences we can draw from these more neutral conclusions regarding the emerging policy phase.

One manifestation of the trend away from static and toward more dynamic models of environmental problems and goals is a corresponding trend toward more "holistic" and more systems-oriented managerial models. In the United States, increasing emphasis has been placed on "ecosystem management plans," often involving also "adaptive management." These represent planning and management processes – often involving multiple agencies of government, all levels of government from local to national, and the public – that are applied to systems bounded by natural features of the ecophysical landscape.[19]

In the United Kingdom, a similar trend toward holism is evidenced by planners and political theorists such as Michael Jacobs, who argue that making economics "ecological" will not be sufficient, and that it is necessary to adopt an even broader analysis of environmental problems, which can be called "socio-ecological economics."[20] Environmental policy experts and advocates in Europe have, using several technical vernaculars, similarly advocated more holistic and integrated environmental management. One group, building on the ideas and practices of Impact Analysis, has advocated "Strategic Environmental (Impact) Assessment."[21] Proponents of this approach expand traditional impact analysis techniques to larger systems, and they advocate applying their tools not just at the project level, but also to impacts of policies and even proposed legislative acts. This approach, still in its infancy, would involve impact assessments at project, local, regional, and perhaps even national and international levels; this embedding of many smaller, simultaneous impact studies in a multilayered, integrated assessment represents one approach to more holistic environmental management.

Another promising approach has been championed by Dutch theorists, modelers, and practitioners, some of whom are developing a model called the TARGET model, which includes both social and ecophysical features. The idea behind this work is that social controversies about new technologies provide informal methods of "technology assessment," and that such informal processes may be integrated into larger, multiscalar models of humans and nature.[22]

These various trends all point toward a more holistic approach to environmental management, and at least implicitly toward a greater respect for dynamic processes. I am a strong supporter of these general directions. In the remainder of this chapter, I will examine in more detail the prospects of constructing a process-oriented theory of values, and a practical, process-oriented method of evaluation as elements of a larger theory of adaptive ecosystem management.


PART 3. ADAPTIVE MANAGEMENT: A PROCESS-ORIENTED
APPROACH


Considerable progress has been made in understanding environmental problems as problems of *adaptation* within complex, multiscalar, dynamic systems, and this approach may emerge as the management approach of the twenty-first century. Writers and practitioners of this tradition have successfully developed an adequate characterization of how environmental problems emerge as problems of adaptation at the individual and cultural levels. This multiscalar system allows conceptualization of an "individual" scale of action, and also a larger community level on which populations interact with their environments. One important consequence of this multiscalar and multivariate formulation is that neither means nor goals of sustainability can be set concretely in the beginning, and the quest for sustainable human communities must involve many individual processes of experimentation, revision of scientific understanding, and reformulation of community goals. This adaptive, experimental approach requires the careful design and nurturing of institutions capable of fostering social learning. C. S. Holling and his colleagues have argued that large, landscape-scaled ecological systems tend to become "brittle" under continuous exploitation and that these large systems can disintegrate and then gradually re-equilibrate at different levels of functioning or with quite different structural organization.[23] On the basis of this hypothesis, and with some success in modelling resource management in the field (including work on spruce budworm outbreaks and fisheries management), Holling and his colleagues have become the champions of dynamic, non-equilibrium modelling in resource management. They argue that environmental management cannot be modelled in single-equilibrium systems, and that impacts on natural systems can approach thresholds which, if exceeded, can cause discontinuous and rapid change into an alternative steady state.[24] It is often noted that such large-scale changes result in systems that are less productive of human services or less attractive to human users. Adaptive management, according to this argument, would choose exploitational patterns that mimic natural processes in order to minimize the likelihood of accelerating system-level change and loss of human use. Speaking theoretically, this insight was embodied by arguing that good management must care for both the *productivity* and the *resilience* of the ecophysical system. "Resilience" is introduced as measure of the magnitude of disturbances that can be absorbed before a system centered on one locally stable equilibrium flips to another.

This hypothesis has also been elaborated by the explicit employment of concepts from "hierarchy theory," an application of general systems theory to ecological modelling. Hierarchy theory models ecological systems according to two assumptions: (1) that all observations and measurements must be taken from some perspective within a hierarchically organized dynamic system, and (2) that the systems, as modelled, exhibit nestedness, with smaller subsystems changing more rapidly than do the larger systems which form their environment. Given this multiscalar framework, it is possible to model environmental problems in a multiscalar system.[25] At the individual level, organisms (including human persons) survive by responding creatively to a set of opportunities and constraints that are presented to them by the environment they inhabit. So the range of options open to the present is a function of the structure, and consequent functioning, of the habitat. Successful choices of individuals are encoded into the system as information about what "works" given current system organization – adaptation is acting successfully on information flowing from the environment to individual actors. Information, however, flows upward in the system also because the collective choices of many individuals, in a nested system, gradually alter the context in which the next generation faces the game of adaptation. The nested subparts also constitute the larger systems, and exhibit cumulative impacts in slower-scaled changes in the structure and function of the systems that support human choice. If generation 1 clear-cuts all of the available forest, opportunities are lost; attention shifts to constraints experienced in generation 2 – how to survive against the constraint of an inadequate wood supply?

The work in adaptive management has, in the past few years, virtually merged with the work of "ecological economists," who have broken with mainstream welfare economics on several important grounds.[26] Ecological economists – to simplify a complex argument – have rejected the idea of "weak sustainability," popular among mainstream economists, that a generation acts responsibly toward the future if it adds to, rather than diminishes, the total stock of human capital. If we accumulate wealth – assets in any form – then the future will have an adequate investment base and an equal opportunity with the present; so they will be as well off as we are, according to mainstream economists and weak sustainability theorists. But ecological economists go beyond this generalized obligation, arguing that there are specific features of the environment – called "natural capital" – that should be preserved for the future in trust.[27] Adaptive managers and ecological economists have thus joined forces, arguing that the resilience of ecological systems is surely one important example of an element of natural capital. These combined forces argue that system resilience is a more useful index of environmental sustainability that alternative measures (such as economic growth measures or simple carrying capacity measures), because economic activities are sustainable only if the life-support ecosystems on which they depend are resilient.[28]

Adaptive management/ecological economics, then, has provided a theoretical model for understanding environmental problems as problems of what might be called "cross-scale spill-overs." If I, as an individual, cut down one tree, or even my whole woodlot, this will have little long-term impact, provided other individuals let their trees grow. If all forest owners in a watershed, however, clear-cut their forest within a few years, a standing resource – an option for use – will have been eliminated for decades; and there may be indirect effects of soil erosion and reduced stream-water quality that may last much longer. Environmental problems can then be understood as multi-scalar, or "cross-scale spill-over problems" – an idea to which we return in the next part. The point I wish to state here is that adaptive managers have provided a model which illuminates the emergence of environmental problems at the systems level. Environmental problems emerge at the systems level with increasing impacts resulting from increased human populations and increased technological power. Cumulative individual choices can accelerate change, sometimes crossing crucial thresholds, and cause systems to undergo rapid structural re-organization. But, if Holling and his colleagues are correct in asserting that "flips" into new states will result in habitats that are less productive of humanly valued goods, then these changes will be experienced by individuals in subsequent generations as a constriction in the options available to them to find means to survive. The mix of opportunities and constraints presented by the habitat will have shifted for the worse.

It is now possible, following Holling and the adaptive managers, to think of values as emerging within a dialectic between culture and nature, with each generation facing a mix of opportunities and constraints, and with the cumulative choices of each generation affecting the landscape in ways that will affect the mixture of opportunities and constraints that will be faced by coming generations. Long-term survival as a community/culture requires, in the short term, adequate economic opportunities and a reasonable pace of economic development. On longer scales, what we worry about is whether future generations will have a roughly equal or superior mix of opportunities to constraints. So, if measures of economic activity, such as the Gross Domestic Product (GDP), are to be used as comparisons of welfare over time in the short run, then what is also needed is an "Opportunities-Constraints Index" (OCI) to measure changes in opportunities available in the future. The OCI would compare rising or falling of options and opportunities that the environment presents to actors across inter-generational frames of time. We can therefore rate development paths, thought of as packages of policies and choices, according to their impact on the range of free choice – opportunities to adapt – that will be open to posterity. So, a good policy or program would be one that, when added to the current economic and ecophysical dynamic, can be expected to (a) increase economic welfare (on the scale of years) and (b) maintain a nondeclining stock of resource-based options for individuals of future generations. It must, that is, perform well on more than one temporal scale. Adaptive management therefore provides a useful, and quite general, representation of many environmental problems as cross-scale spill-overs/impacts of cumulative human activities, and because this representation is expressed in a multi-scalar, dynamic system, it also sets the stage for a more dynamic approach to evaluating changes and processes of change.

Adaptive management has been given a political component by Kai Lee, who accepts the characterization of environmental management as an experimental, community-based search for ways to exploit natural systems without undermining their healthy functioning.[29] Adaptive management, he says, should represent a negotiation within a politically organized ecosystem management process. Scientists, working within a political process in which stakeholder groups express and defend their interests, attempt to develop trust sufficient to undertake "experiments" in management. These experiments are to be designed to produce *both* an "epistemological community" devoted to experimentation and "social learning" *and* a reduction in uncertainty in the present and related situations. Through this process, the community guides scientists to study aspects of the system that are of importance to the community, and scientists respond with studies that will help to reduce negative impacts of valued human activities.

I endorse this general model, but I believe current formulations are lacking in one important respect in that the multi-generational models are put forward as "descriptive" models only. Values, if they are mentioned at all, are treated exogenously.[30] Current versions of Adaptive Management theory therefore assume, with Economism and Deep Ecology, that environmental values have a source and are fully determined outside the policy process. The problem is that, while treated exogenously, valuations by individuals are important system drivers, because individual behaviors express the individual preferences that result in the cumulative impacts that threaten resilience. As long as values are thus maintained as exogenous, and hence independent variables, the system of management can offer no remedy – no informational feedback loop – if individuals in the community are expressing preferences that promote more and more negative cross-scale impacts.

This problem can be perceived if one simply observes the just-mentioned, virtual merger of adaptive management with Ecological Economics. Ecological economists set out to define "natural capital" as those features of an environment that are essential to protect future welfare. Coming from the ecological side, ecologists suggest that changes in ecological organization – "flips" into new states – are socially disvalued. But "resilience" is defined as a descriptive characteristic of ecological systems, so we can ask, "Why *should* we protect the resilience of systems?" We cannot answer, with the welfare economists, that individuals, taken in aggregate, "prefer" to do so if it is the current pattern of preferences which drives economic development and threatens resilience.

What adaptive managers and ecological economists want to say is that people really "should" prefer a higher, or over-riding value – our obligation to protect resilient ecosystems for the future. But this latter alternative cannot be expressed, even within an enriched vocabulary including the concepts of ecology (which describes changes, and thresholds, in the ecological system), and of economics (which describes preferences as they are expressed by present consumers). What is lacking, I would argue, is a theory of *value* that (a) establishes the possibility of articulating values that can compete with currently felt preferences, such as an obligation to sustain opportunities for the future, and (b) some way to link those long-range values to physical features such as "resilience." Lacking such a value theory, adaptive management can simply describe a system going haywire. It can offer no analysis of how the system might right itself by affecting the driving, independent variable – current preferences. Preferences, and the evolution of preferences, must be a part of the adaptational process; adaptive management must test and revise our values as well as our empirical hypotheses. Individual preferences and social values – as well as the institutions that shape them – must be considered, and modelled, as endogenous to the social process of environmental management.

PART 4. ALTERNATIVES TO MONISTIC ASSUMPTIONS
AND THE ENTITY ORIENTATION

I have argued above that the philosophical debate regarding inherent value in nature and the policy debate about how to set conservation priorities have both presupposed an entity evaluation, in keeping with a fundamental form of Western thought. I have also argued, however, that an entity-based evaluative scheme, even if one were available in operational form, would soon be obsolete because of recent developments in environmental management. Management thinking is moving away from single-species management, and I believe it will eventually move away from the inventory-of-objects approach altogether, because environmental problems, as we have just seen, are best understood as problems of adaptation across multiple scales of time. In this part, I examine the prospects for a process-oriented system of environmental evaluation, and explore some of the general features of such a system by discussing the consequences of denying the crucial shared assumptions of the Economists and the Deep Ecologists. Once freed from the the shared assumptions that bind *both* Economists *and* Deep Ecologists in polar opposition over classifying objects of value, it is possible to look with fresh eyes at questions of value and policy; I will discuss the effects of rejecting some of the central assumptions shared by these opposed groups.

1.   We have seen above that Economists and Deep Ecologists share a complex web of beliefs about the nature of environmental value; the core of that complex web is the belief that a sharp distinction must be drawn between two kinds of value, intrinsic and instrumental. If we reject this sharp dichotomy between instrumental and intrinsic values, a pluralist and integrative position

emerges as a possibility: there are many ways in which humans value nature. These ways range along a continuum from entirely self-directed and consumptive uses, and includes also human spiritual values and aesthetic values and other forms of noninstrumental valuations. If one forgoes a sharp definitional distinction between these two opposed types of valuing, the moral task of sorting entities into those that have and those that lack this special feature of "noninstrumental" value becomes a nonproblem. The sorting question, that is, has interest only after one enters the bipolarized conceptualization of environmental values that comes with the web of assumptions shared by Economists and Deep Ecologists.

2. Rejecting the Entity Orientation. Suppose that we, following this more pluralistic approach, stop thinking of environmental evaluation as an exercise in categorizing objects at all. Rather, the goal is to choose indicators of the adaptability of various technologies and policies. Attention would then turn to impacts of existing and proposed technologies and policies on ecophysical and social processes; the task is to develop an indicator, or suite of indicators, that would allow the ranking of "development paths." A development path would then be thought of as a scenario that can be projected to unfold under a given policy or set of policies. The task of evaluation will then be one of ranking various development processes that might unfold from the present into the future. We hope, in the end, to be able to say, "Development Path A is more (less) likely to fulfill social values, V1, V2, V3, . . . than Development Path B." So the process approach advocated here simply ignores the problems and possibilities of entification and sets out to evaluate processes of development and change as they play out on a landscape.

Rejection of the entity bias has an even more profound implication for the theory of environmental value. If we reject the assumption that environmental evaluation is basically a matter of sorting entities, and focus instead on evaluating processes and paths of change, it is possible to recognize a deeper source of value in nature, what might be called "nature's creativity." Ilya Prigogine and his co-author, Isabelle Stengers, have argued persuasively that Western thought has for too long emphasized "Being" at the expense of "Becoming" and entities at the expense of processes.[31] Prigogine and other leaders of the emerging science of chaos and complexity have set out to repair this imbalance, arguing that change, process, and becoming are more basic than being – that the world of objects we see is simply our stilted perception of a rich, multi-scalar, evolving system.[32]

If we were to apply this kind of thinking to biodiversity policy, we would focus on the processes that have created and sustained the species/elements that currently exist and populate the world rather than on the species/elements themselves. Indeed, emphasis on the value of creative processes in nature may go a long way toward expressing the common denominator in most people's valuing of nature. When the native animist worships or respects trees or animals, it is their activity and presumed potency, their ability to affect processes that entwine with human life, that excite religious impulses. When the agriculturalist or the forester values nature, it is the ongoing processes of productivity, the ability to provide a *flow* of useful products, that is the essence of the value perceived. Similarly, when the Deep Ecologist says that elements of nature have intrinsic or inherent value, one might express this object-oriented statement in a more process-oriented vernacular as an insistence that there is majesty and meaning to be found in the evolving processes of life. The common element of these different object-oriented statements of value is a correspondence, in a more process-oriented vernacular, to an important aspect of nature's ability to create, and to an important human impulse to value that creativity. Similarly, it is reasonable to interpret the advocate of biodiversity protection as valuing natural processes for their capacity to maintain, support, and repair damage to its parts.

Perhaps the impulse to value nature's creativity – an impulse that, thankfully, has been expressed in a multitude of ways in different persons and cultures – comes closer than the theories of either the Deep Ecologists or of the Economists to capturing the deep and universal value that could unite all peoples behind an Earth Charter. Those theories, one might say, are directed at the specific *content* of people's values, rather than the real and shared *source* of those specific values in nature. Emphasis of one type of value at the expense of others can only lead to conflict and divisiveness, because humans – struggling to survive in many local situations with differing constraints and opportunities, and different natural and cultural histories – will have different needs. Some humans are hunters, some are bird watchers, some are shamans, others are developers and capitalists. The common denominator of all of these types of value derived from nature – when expressed in a dynamic, process model – is nature valued as a multi-scaled system of creative processes. This creativity is exhibited on many scales of nature. On the paleontological scale, it has resulted in diversity of all kinds; and on the shortest scale, it gives hope of the next harvest to the faithful peasant who plants seeds. These creative processes, we can further say, are valued by humans because a creative and building nature provides options and opportunities to fulfill human values *whatever those human values are*. These values *emerge* from the human–nature dialectic of co-evolution; they do not exist in *either* the humans-only world of economists *or* in the independent realm of non-human values envisaged by the Deep Ecologists.

The point of an Earth Charter should not be to tell the many peoples and cultures of the world *how* they should value nature; it should rather express the underlying value placed on nature's creativity and the opportunities this creativity offers humans to choose, to adapt, and to "value." It is this underlying creative and sustaining force that allows species to reproduce and maintain themselves, and to create new adaptive responses to changing ecophysical processes that form their environment. This value, it can be argued, exists at a deeper level than do the values of Economists and Deep Ecologists. Creativity – nature as a source – is the sine qua non for all of these, and other, more specific values.

Following our excursion into foundational values, we have now circled back to the position of the adaptive managers and their concept of resilience. This concept can be thought of as a promising attempt to identify and operationalize a characteristic of natural systems that is essential to their continued creativity. What has been added by our expansion of adaptive management to include a corrective to destructive human preferences within the management model is a capacity to close the valuational loop. It is now possible to explain *why* individuals who value the future – those who are committed to living sustainably – *should* care about resilience. We value resilience because resilience allows a system to remain productive, to dissipate energy and maintain structure, and to heal wounds and repair stresses; these are the essential features of a system that maintains its creative force by maintaining its self-organizing structure.

Nature's creativity is valued both in the present and for the future because it is the very basis of human opportunity. At the same time, making value analysis endogenous to policy process allows us to explain why individuals, who value natural products for personal consumption, might also come to see how certain consumption patterns in the present – and the preferences that drive them – are inconsistent with maintaining opportunities and a range of free choices – opportunities to adapt – for future generations. Just as the smoker who realizes that continued smoking is inconsistent with the longer-term value of good health, the driver might someday realize that excessive use of fossil fuels is inconsistent with maintaining opportunities for the future. It is granted that the latter case differs from the smoking case in involving also an element of altruism, but my point is that both individuals face the need to adjust their behaviors, and in doing so they will likely also alter their preferences, at least eventually. To carry the analogy one step further, science contributes to change in the individual smoker by providing physiological models showing that continued smoking threatens the future value of health later in life. Ecology and other adaptive managers do experiments and, likewise, can show an inconsistency between the consumptive behavior of a person or culture and the value of sustaining opportunities for their descendants. Science in both cases provides data and models that alert consumers to a conflict between their behavior in fulfilling one value and maintenance of another value. It should be possible for social scientists to model resulting changes in preference. So I expect that cognitive psychology and related disciplines will become increasingly important as aspects of community-based adaptive management.[33]

The creation of a stronger values component for adaptive management can emerge from the confluence of two initially separable, but ultimately unifying, forces. The first force comes from the side of nature and must be understood by the natural sciences – it is the creative force that spun out our communities as a part of a far more vast creation.[34] The second force is the striving of individual human beings to survive, to reproduce, and to perpetuate their kind. This striving gives rise to many diverse goals in many different evolutionary contexts; it also gives rise to the ingenuity which is so valuable to humans. Nature's creativity is experienced by us as a set of opportunities. The range of individual choices – an important pre-requisite of human freedom – is thus affected by the range of nature's creativity. For example, the opportunities of a whittler or a wood sculptor are provided, but also limited, by the range of tree species at his or her disposal. Similarly, the opportunities – and range of choice – available to a house builder or developer is a reflection of the types and variety of landscapes and settings available.

The confluence of these intellectual forces, then, affirms a crucial and deep connection between human choice, freedom, and creativity on the one hand and nature's creativity on the other. The human choice and freedom which expresses itself in different values and in different behaviors expressed by different persons and different cultures – from the use value of the hunter-gatherer to the inherent value attributed by Deep Ecologists – finds its ground in the creativity of nature. More practically, I am suggesting that, when we measure environmental values, we should measure the extent to which the creativity of a natural system serves, and is served by, human creativity. What we need, on this way of thinking, is a measure of how well a system is maintaining those forms of creativity that support a wide range of human opportunities, both in the present and in the future. As noted above, such a measure might be called an OCI.[35]

3. Avoiding Reductionism and Monism. One reason such sweeping consequences follow from rejecting the entity orientation is that, on this deeper level, monism becomes irrelevant. Since we no longer need to sort entities into those that are instrumentally and those that are inherently valued, who

cares? Our evaluations are no longer constrained by the requirement that environmental values must be commensurable and measurable within a unified system of evaluation, with a single moral or evaluative "currency." If we reject the monistic assumption, according to which all value must be explained according to a single principle, it is possible to start from the pluralistic viewpoint that all cultures value nature and natural processes in many ways. We should, as a first step, develop a vocabulary and operational measurements that are rich enough to express these multiple values. We thus embrace pluralism as a working hypothesis, setting out to characterize and operationalize as many values and types of values as possible. This leaves, for subsequent discussion, the question of whether some of these types of values can be usefully "reduced" to other types, assuming that some level of consolidation of multiple frameworks will eventually emerge.[36]

One reason that monism and reductionism have been popular in environmental ethics and moral philosophy is that a monistic system, which has only one principle to apply in each case, avoids any fear of relativism or subjectivism. Monists reason that, if there are multiple principles, theories, and rules in the moral arena, and some of these rules yield different conclusions in the same situation, moral evaluation will collapse into relativism.[37] While I agree that moral relativism must be avoided, there are a number of strategies available for avoiding relativism within pluralistic systems; it is not necessary to embrace monism in order to avoid relativism. For example, it is possible to develop a two-tiered system of analysis in which the "action" tier includes multiple rules for choosing acceptable behaviors and a second "meta" tier contains rational principles for deciding which of the action rules is appropriate in various situations.[38]

4. Rejecting the Assumption of Placeless Evaluation. Evaluation models like Economism and Deep Ecology are constrained by their monism to express all value in a common currency, so their accounts of value tend to lose, in the process of aggregation, the place-relative knowledge and value that emerges within a specific dialectic between a human culture and its physical and ecological setting, or context.[39] One implication of the adaptational model for understanding environmental problems is to emphasize the importance of localism; as we know from evolution, all adaptation takes place at the local level, as individuals survive, or fail to, as they "experiment" with various adaptations to local conditions. As one relaxes the assumption that we need a single, universally aggregable accounting system for all environmental values, it becomes more possible to hear, and register, the very real concerns of local cultures trapped between the hard realities of international economic forces beyond their control and the equally real limits and constraints that

manifest themselves at the local and regional levels. Localism, as a replacement for universalism, leads to an emphasis on local variation, on diversity from locale to locale and from region to region, and to many local "senses of place," each of which expresses a unique outcome, at each particular place of the infinitely variable dialectics between local cultures and their habitats. Development, and various development paths, can therefore represent differing trajectories created by the nature–culture dialectic in a specific culturally evolved place. This trajectory, given the above conceptualizations, can be measured at an economic scale and also – using an OCI that measures the extent to which the development trajectory maintains and increases ecophysically based opportunities – at the multi-generational scale required to judge the true sustainability of a culture.

CONCLUSION

The attempt to develop an Earth Charter, a document that expresses "the shared values of people of all races, cultures, and religions," may provide an occasion for re-examining current approaches to environmental valuation. In this paper it has been argued that neither the narrow, human-centered utilitarianism of Economism nor the assertion by Deep Ecologists of human-independent values in nature are able to characterize a universal value that can unite humankind behind an effort to protect biological diversity. It was first shown that both of these widely espoused, opposed theories share a set of assumptions about the nature of environmental value – monism, a sharp separation of intrinsic and instrumental values, and an object orientation. Further, it was shown that these assumptions are highly vulnerable when examined objectively. Then it was shown that discussions of environmental values in practical contexts such as discussions of conservation priorities, no less than discussions of value theories, usually employ the entity orientation, and formulate problems of conservation as problems of sorting objects and saving inventories. But it was also argued that the entity orientation seems increasingly obsolete as environmental management everywhere moves toward more holistic and process-oriented management models.

Adopting an attitude of skepticism toward entity-oriented concepts and the value theories of the Economists and the Deep ecologists, we then examined in particular the adaptive management model as a way of understanding human–nature interactions from the viewpoint of a community adapting to a larger, changing ecophysical system. This approach, it was found, provides a simple and plausible model of how environmental problems emerge at the interface of human technological change and environmental change: collective individual choices, in response to the opportunities and constraints offered by the environment of individuals in Generation 1, can change the environment so that individuals of Generation 2 face a differing mix of opportunities and constraints. Environmental problems, on this view, emerge as cross-generational spill-over effects, effects that reduce the range of choices that will be faced in the future. Conversely, successful restoration efforts can be interpreted as positive cross-scale impacts/spill-overs. Thus, while all peoples must derive economic sustenance from their environment, a concern for the future demands that one also monitor the impacts of current actions on the future mix of opportunities and constraints. Humans, as communities, that is, can accept responsibility to maintain a non-declining set of opportunities based on possible uses of the environment. This responsibility is based on a sense of community with the future and on a sense of fairness – the future ought not to face, as a result of our actions today, a seriously reduced range of options and choices as they try to adapt to the environment that they face. Acceptance of this responsibility as an important aspect of an adaptive management model, however, would require that the adaptive management model allow the reconsideration of values and preferences, if evidence suggests that current values and behaviors are likely to reduce the amount of opportunities, and increase the magnitude of constraints, that will be faced in the future.

Building on this adaptational model, and trying to avoid reducing the many values humans derive from nature to a single type, it is possible to see that what is valued in common by persons with diverse relationships to nature is its *creativity*. The creativity of nature provides "options" which are the basis for human opportunity. Here, then, we have located a level of environmental value that may be universal. While the hunter, the developer, the shaman, and the bird watcher all exercise very different individual values and options, some self-oriented, some not, what they all share is a powerful dependence on the creative aspect of nature. If, then, we can avoid the assumptions – of monism, of a sharp dichotomy between intrinsic and instrumental values, of the entity orientation – that bind Deep Ecologists and Economists in a polarized opposition, it may be possible to find, in a celebration of nature's ongoing creativity, a universal value capable of supporting a truly unifying Earth Charter.

NOTES

1.  The effort is being organized by the Earth Council and Green Cross International with support from the government of the Netherlands. For more information, see the special issue of *Earth Ethics* 7, No. 3&4 (Spring–Summer, 1996).

2. Steven C. Rockefeller, "Global Ethics, International Law, and the Earth Charter," *Earth Ethics* 7, pp. 3–5.

3. At first glance, it would appear that these two principles are inconsistent, as stated, assuming that "development" means "development of natural resources, including biological resources," and assuming, as the syntax of the Sustainable Development principle implies, that "all" refers to "all humans."

4. I here use these terms, capitalized, as labels for advocates of the respective theories of value as explained. It should not be inferred, of course, that all economists would fit the label "Economist" as defined and used here; similarly, the capitalized phrase "Deep Ecologist" refers more to a caricature based on sharing the characteristic belief that nature has value independent of humans and their motives. Since the purpose of this paper is to look at the bigger picture, details and possible differences within the large categories are not important here.

   For brief statements of the Economist position, see William Baxter, "People or Penguins," and A. Myrick Freeman, "The Ethical Basis of the Economic View of the Environment," which are both reprinted in D. VanDeVeer and C. Pierce, *The Environmental Ethics and Policy Book* (Belmont, CA: Wadsworth Publishing Company, 1994). For a statement of the position I have called Deep Ecologism see, for example, Holmes Rolston III, *Conserving Natural Value* (New York: Columbia University Press, 1994), especially Chapter 6.

5. Gifford Pinchot, *Breaking New Ground* (Covelo, CA: Island Press, 1987 reprint of the 1947 edition), p. 323.

6. Heraclitus, it appears, was the first "New Ecologist," in emphasizing the importance of change in nature. See paper 18, this volume.

7. See John Dewey, "The Influence of Darwinism on Philosophy," in *The Influence of Darwin on Philosophy and Other Essays in Contemporary Thought* (New York: Henry Holt and Company, Inc., 1910).

8. See Stuart Pimm, *Balance of Nature?* (Chicago: The University of Chicago Press, 1991; and Bryan Norton, "A New Paradigm for Environmental Management," in R. Costanza, B. Norton, and B. Haskell, Eds., *Ecosystem Health: New Goals for Environmental Management* (Covelo, CA: Island Press, 1992).

9. Christopher Stone, *Earth and Other Ethics* (New York: Harper and Row, Publishers, 1988), p. 116.

10. See Bryan Norton, *Why Preserve Natural Variety?* (Princeton, NJ: Princeton University Press, 1987), Chs. 12 and 13, for a history and criticism of attempts to "prioritize" taxonomic categories as a basis for setting policy in the United States Fish and Wildlife Service.

11. William Cronon, *Changes in the Land* (New York: Hill and Wang, 1983).

12. The three phases should be understood as *policy* phases; I recognize that particular scientists and many groups of scientists developed alternative formulations of the underlying biological relationships, but the focus of this discussion is the formulation of biological protection issues in policy and legislative contexts.

13. See paper 7, this volume, for a more detailed discussion of the value and implications of the Endangered Species Act.

14. The Symposium was held in Washington, D. C., in September 1986. See E. O. Wilson, Ed., *Biodiversity* (Washington, DC: National Academy Press, 1988), for proceedings of the Forum.

15. This is the working definition that was used at the National Forum on Biodiversity.
16. See, for example, the essays in Costanza, Norton, and Haskell, *Ecosystem Health.*
17. Note that the issue of inherent value emerges again in this context as some authors, such as J. Baird Callicott, in *In Defense of the Land Ethic* (Albany, N. Y.: State University of New York Press, 1989) and Laura Westra, in *An Environmental Proposal for Ethics: The Principle of Integrity* (Lanham, MD: Rowman and Littlefield Publishers, 1994), argue that use of the term "integrity" signals an acceptance of some form of nonanthropocentric evaluation. Other authors, including most of them in Costanza et al., *Ecosystem Health,* use the terms analogically, and draw no ontological or deontological conclusions from attributions to "health" or "integrity" to ecosystems.
18. I have discussed these linguistic aspects in more detail in paper 9, this volume.
19. See Kai Lee, *Compass and Gyroscope* (Covelo, CA: Island Press, 1993), and the essays in L. H. Gunderson, C. S. Holling, and S. S. Light, Eds., *Barriers and Bridges* (New York: Columbia University Press, 1995).
20. See Michael Jacobs, "What Is Socio-Ecological Economics?" *Ecological Economics Bulletin* 1:2 (April, 1996) and *The Green Economy* (London: The Pluto Press, 1991).
21. See R. Therivel, E. Wilson, S. Thompson, D. Heaney, and D. Pritchard, *Strategic Environmental Assessment* (London: Earthscan, 1992); R. Therivel and M. R. Partidario, Eds., *The Practice of Strategic Environmental Assessment* (London: Earthscan, 1996); and, for a review and application to urban planning, A. Shepherd and L. Ortolano, "Strategic Environmental Assessment for Sustainable Urban Development," *Environmental Impact Assessment Review,* 16 (1996), 321–335.
22. See, for example, Sym Van Der Ryn and Stuart Cowan, "Nature's Geometry," *Whole Earth Review* (Fall, 1995), pp. 9–14; Arie Rip, "Controversies as Informal Technology Assessment," *Knowledge Creation, Diffusion, Utilization,* 8 (1986), pp. 349–371.
23. C. S. Holling, *Adaptive Environmental Assessment and Management* (London: John Wiley, 1977); Carl J. Walters, *Adaptive Management of Natural Resources* (New York: Macmillan, 1986); Lee, *Compass and Gyroscope*; Gunderson, Holling, and Light, *Barriers and Bridges,* op. cit.
24. Mick Common and Charles Perrings, "Towards an Ecological Economics of Sustainability," *Ecological Economics* 6 (1992): pp. 7–34; C. S. Holling, "Engineering Resilience versus Ecological Resilience," in P. C. Schulze, Ed., *Engineering within Ecological Constraints* (Washington, DC: National Academy Press, 1996).
25. Lance Gunderson, C. S. Holling, and Stephen Light, "Barriers Broken and Bridges Built: A Synthesis," in Gunderson, Holling, and Light, *Barriers and Bridges,* op. cit.
26. See H. Daly and J. Cobb, *For the Common Good* (Boston: Beacon Press, 1989); R. Costanza, Ed., *Ecological Economics: The Science and Management of Sustainability* (New York: Columbia University Press. 1991.
27. See paper 12, this volume, for a more detailed analysis of this debate.
28. K. Arrow, B. Bolin, R. Costanza, P. Dasgupta, C. Folke, C. S. Holling, B.-O. Jansson, S. Levin, K.-G Maler, C. Perrings, and D. Pimentel, "Economic Growth, Carrying Capacity, and the Environment," *Science* 268 (1995), pp. 520–521.
29. Lee, *Compass and Gyroscope,* op. cit.

30. Ibid., p. 192.
31. I. Prigogine and I. Stengers, *Order Out of Chaos: Man's New Dialogue with Nature* (New York: Bantam, 1984).
32. Leading scientists have thus joined Heraclitus, Henri Bergson, and A. N. Whitehead in embracing a process-oriented understanding of natural systems. See Henri Bergson, *Creative Evolution*, A. Mitchell. trans. (New York: Holt, Rinehart & Winston, 1911), and A. N. Whitehead, *Process and Reality* (New York: The Macmillan Co., 1929). Space does not permit a full discussion of these topics here, of course. The reader is referred to any one of a number of popular and semi-popular discussions of complexity and chaos.
33. See Bryan Norton, "Thoreau's Insect Analogies: Or, Why Environmentalists Hate Mainstream Economists," *Environmental Ethics*, 13 (1991), pp. 235–251; papers 11 and 14, this volume, for further discussion of how preferences change across time and in response to information regarding impacts of current preferences and actions.
34. There is no intent to suggest that the "creative force" mentioned here involves *intentional* creation. Unfortunately, all terms and phrases that capture the creative aspect of natural systems seem also to include an implication of intentional creation. This is a defect in our unfortunately dualistic everyday language. See Prigogine and Stengers, *Order Out of Chaos*, for an explanation of these points.
35. While space does not permit a full development of these ideas here, I hope I have said enough to exhibit the intellectual and methodological advantages that emerge when the entity orientation is abandoned. Paper 17, this volume.
36. For a description of such a process as located within a broader "multicriteria analysis," see Harold Glasser, "Towards a Descriptive, Participatory Theory of Environmental Policy Analysis and Project Evaluation," Ph.D. dissertation, Department of Civil and Environmental Engineering, University of California, Davis, 1995.
37. See, for example, J. Baird Callicott, "The Case against Moral Pluralism," *Environmental Ethics* (12) 1990, pp. 9–24.
38. See Bryan Norton, *Toward Unity among Environmentalists* (New York: Oxford University Press, 1991), p. 200, for an outline of such a system and paper 13, this volume.
39. See Bryan Norton and Bruce Hannon, "Environmental Values: A Place-Based Theory," *Environmental Ethics*, 19 (1997), pp. 227–245 and Norton and Hannon, (this volume).

# 22

# Intergenerational Equity and Sustainability

INTRODUCTION: WHAT DO WE OWE THE FUTURE?

The concepts of "sustainability" and "sustainable development" have become the central concepts – indeed, shibboleths – in today's environmental policy discussions. This popularity has persisted, despite almost universal complaints – even by those who use them regularly and approvingly – that the concepts are vague and lack consensually accepted meanings. While it seems clear that calls for sustainable activities and policies must rest on an obligation of current people to future generations, philosophers have contributed little to the ongoing policy debates regarding how to define and measure these key terms.

There is of course a significant philosophical literature on the fascinating, and puzzling, subject of the nature of our obligations to the future; the problem is that little attempt has been made to relate the abstract philosophical arguments of the 1970s and 1980s to the more practical problems of stating operational criteria for sustainable living in the 1990s. Especially, philosophers have had little to say about specifying a *metric* by which progress – or lack thereof – toward sustainability can be measured. Perhaps this is true because philosophical treatments of intergenerational obligation have dealt mainly with a host of puzzles and paradoxes having to do with the inevitable asymmetry of intertemporal moral relationships;[1] in general, attention to these

The research for this paper was partially supported by a grant from the Environmental Protection Agency, Office of Exploratory Research, and by a grant from the National Science Foundation for the study of ecosystem valuation, awarded under the NSF/EPA partnership for environmental research. I was also much helped by comments from colleagues who read and commented on earlier versions of the paper, including Andrew Dobson, Richard Howarth, Leslie Jacobs, Michael Toman, and Clark Wolf.

420

puzzles and paradoxes has not led to a simple and compelling *positive* theory of intergenerational obligations.[2] Indeed, most philosophical writers have been dismissive of any longer-term obligations, such as obligations that extend beyond one's grandchildren.[3] Lacking such a compelling general theory of intergenerational morality, philosophers have assumed that, despite problems of concept and measurement, any theory of what we owe the future will be based upon a comparison, by some means, of the well-being of subsequent generations with the well-being of earlier generations. While this formulation may ultimately be correct, the problem in policy discussions has been that – since the central assumption has never been given a definitive, or even plausible, positive philosophical interpretation – economic models of valuation have been enlisted to operationalize the key idea of comparative "well-being" across time. Having fallen into the assumption that intergenerational morality must involve comparisons of individual well-being, and lacking a convincing account, in moral terms, of how to compare well-being across time, welfare economics has become, by default, the metric for measuring it. I will argue, however, that the substitution of "economic welfare" for "well-being" in the general formula represents a dangerous oversimplification of obligations to live sustainably. In particular, this conceptual capitulation – by virtue of the uncompromising individualism of mainstream welfare economics – leaves no role within the sustainability debates for community-based obligations.

I begin by decomposing the question of intergenerational equity into four separable aspects in Part 1. In Part 2, I will show how this complex of philosophical problems have been subject, in environmental policy debate, to a Grand Simplification, a simplification that has resulted in an overly simple treatment of intergenerational fairness in sustainability policy discussions. In Parts 3–6, I will propose and begin to defend a more complex understanding of the problem of intergenerational assessments of environmental value and argue that, once the problem is adequately framed, it may be possible to provide a more comprehensive, and more finely nuanced, conceptual framework for discussing environmental policy proposals with long-term consequences. The framework proposed will be multigenerational and pluralistic, and will be communitarian and "place-based." This approach differs, then, from the approach of welfare economists in positing goods beyond individual ones, but it differs also from nonanthropocentric approaches to environmental valuation by locating environmental values squarely within a multigenerational human community that exists in a place.

PART 1. FOUR PROBLEMS CONCERNING
INTERGENERATIONAL EQUITY

The problem of what we owe the future is not a single monolithic problem, but rather a cluster of interrelated problems. For convenience, I have grouped these problems into four categories and given them somewhat descriptive names. They are (1) *the problem of intergenerational trade-offs* – How should an earlier generation balance concern for future generations against its own moral and prudential concerns? (2) *the distance problem* – How far into the future do our moral obligations extend? (3) *the ignorance problem* – Who will future people be and how can we identify them? How can we know what they will want or need, what rights they will insist upon, or what they will blame us for? And (4) *the typology of effects problem* – How can we determine which of our actions truly have moral implications for the future? I begin by briefly discussing these four aspects of intergenerational ethics.

## 1. The Trade-Offs Problem[4]

In most discussions of intergenerational fairness – which, as noted, are carried out against a backdrop of utilitarian and often economistic assumptions – the major focus is on the trade-offs problem – the problem of how we should weigh the demands of the future against the undeniable and more palpable needs of present people, many of whom live in abject poverty. The trade-offs problem appears to be the most important aspect of intergenerational ethics and, eventually, it must be addressed: How much sacrifice on the part of the present can be justified or required on the basis of obligations to the future? I will not directly answer this question in this paper, however; my goal, rather, is to alter the conditions and assumptions under which the trade-offs question is addressed by examining problems 2–4, which, in my view, set the "boundary conditions" for any adequate solution to the trade-offs question. By employing their available methods to address the problem of intergenerational trade-offs straightaway, without first examining the conditions that determine the terms of such trade-offs, economists and other utilitarians have severely constrained answers to the less synoptic problems that determine the moral terrain on which trade-offs must be negotiated. In this paper, I concentrate on problems 2–4, seeking answers that make sense in terms of these less synoptic questions themselves, in order to set the stage for a new, and less constrained, examination of the issue of fair trade-offs.

## 2. The Distance Problem

There is an important philosophical puzzle about the "horizon" of ethical concern. One might, on the one hand, insist that our obligations are only to the next generation. According to this view, which we can call "presentist," each generation should accept considerable responsibility for the impacts of its actions on its children and their children's cohorts, because these persons are identifiable moral patients, known to us, in some cases loved by us, and worthy of our full moral concern and commitment.

A "one-generational presentist" might argue that all intertemporal obligations should be handled as simple bequests from one generation to the next. We accept responsibility for our children, both singly and as members of a cohort of persons we care about, and we leave the caring for the subsequent generation to our children and their cohorts, who will in turn pass on responsibility to their children. In this view, while our children's children are identifiable persons whom we may care for and love, we can simplify the problem of intergenerational bequests by concentrating on obligations that stretch only from one generation to the next. A less presentist position – "two-generational presentism" – would argue also that our grandchildren, as identifiable individuals with whom we have concrete moral relations, should also be moral patients for us, but that our obligations do not extend beyond those we can know and love.

In contrast to these forms of presentism, one can also imagine an opposed position that we might call "futurist." The futurist – impressed with the fact that there will probably be many persons who exist in the future and that these persons will, like us, no doubt require inputs of natural resources if they are to achieve a reasonable level of welfare – would attempt to balance present entitlements to use resources against the demands of an ever-expanding class of future claimants. Especially with regard to nonrenewable resources such as stocks of fossilized energy, strong concern for the entitlements of the future – because there are so many potential claimants on this finite resource – might lead to paralysis or to severe limits on resource use today. At any rate, it seems essential to provide an answer to the distance problem if we are to know how to address the trade-offs question – the future extent of our moral obligations delimits the set of individuals whose interests are to be considered in making any trade-off. So, one cluster of interesting normative and conceptual problems surrounds the question of determining how far into the future we must project our values in order to fulfill all of our obligations to the future.

### 3. The Ignorance Problem

The ignorance problem includes at least two variants. Both variants depend upon the difficulty of foreseeing the future. (3a) One variant – the ignorance-of-values variant, which will be very important in our discussions here – refers to the difficulty of knowing, today, what people in the future will want or need. This variant, which may be formulated somewhat differently, depending on which moral theory one assumes, raises questions about whether and how we can know what future individuals will *want or desire* (if one is a preference utilitarian), what future people will *truly need* (if one's moral theory emphasizes a distinction between simple preferences and basic needs), or what the future has a *right* or *entitlement* to (in the vernacular of rights theory). There is also the related problem that technological change can render many resources unnecessary while creating a demand for others (3b). Another variant of the ignorance problem, which will not be discussed in detail here, has to do with the difficulty of knowing, in the present, which individuals, and how many of them, will be born in the future. This problem, which has been developed interestingly by Derek Parfit, has to do not just with ignorance about the future, but also with special problems of individual identity that occur in nonpresentist positions. If I accept obligations to people who will live several generations hence, I seem to face a sort of paradox: how can I determine my obligations regarding future people, and formulate actions and policies to fulfill these obligations in reference to their welfare, when my decisions may determine which people are born and even how many of them can survive?[5]

### 4. The Typology of Effects Problem

If I cut down a mature tree and plant a seedling of the same species, it seems unlikely that I have significantly harmed people of the future, though there will be a period of "recovery." As long as there are many trees left for others, my seedling will replace my consumption, and no real harm is done. If, on the other hand, I clear-cut a whole watershed, thereby setting in motion severe and irreversible erosion, siltation of streams, and so on, it is arguable that I have significantly harmed future generations, having irreversibly limited the resources available to them, and that I have restricted their options for pursuing their own well-being. However intuitive this distinction seems in particular cases such as this, however, it is extremely difficult to provide a general definition for characterizing, and to offer a theoretically justifiable practical criterion for separating, cases of these two types.[6] One position, discussed in more detail later, argues that all resources are intersubstitutable ("fungible") in an important sense, and that the intuitive distinction between inherently benevolent or neutral uses of resources on the one hand and destructive ones on the other proves untenable. Another approach to the typology of effects question is to step outside the intertemporal framework and to argue that certain elements of nature have "inherent value" – value that is independent, in some important sense, of what any human being desires or values. Members of this diverse group of theorists – called "nonanthropocentrists" – argue that we should extend something like the special moral status we attribute to human beings to other elements of nature.[7] This formulation of the basic value issue in environmental policy has led to a polarized situation in environmental value theory, with nonanthropocentrists faced off against welfare economists.

I believe these two polar positions are almost surely both wrong.[8] Worse, the polarized rhetoric they foster is unlikely to bring us closer to a consensually accepted, practical environmental ethic, an ethic that would help us to sort actions according to a morally significant typology. I therefore seek a theory of environmental values that (1) recognizes obligations *regarding the environment* to future generations; (2) describes those obligations in a way that illuminates the complexities of living in a society in which decisions made today may determine the nature of the world that future generations will encounter; and (3) encourages us to think of these values within a communitarian, rather than within a purely individualistic, conception of ethical obligations.

Current discussions of environmental policy – especially interdisciplinary disagreements regarding how to define "sustainable" – reflect differing approaches to specifying what one generation actually owes the next; these differing approaches, in turn, rest on varied answers in response to the four problem areas sketched above. It turns out that theoretical positions in the second area – problems of ignorance, especially as formulated in (2a), ignorance of future values – have proven dominant in policy discussions.

## PART 2. SOLOW AND THE GRAND SIMPLIFICATION

In a series of lectures and papers, Robert Solow, winner of a Nobel Prize in Economics for his work in growth theory, has defended the view that sustainability can be fully defined, characterized, and measured within the neoclassical theory that shapes the mainstream economic tradition of resource analysis.[9] Solow's basic idea is that the obligation to sustainability "is an obligation to conduct ourselves so that we leave to the future the option or the capacity to be as well off as we are."[10] He doubts that "one can be more precise than that." A central implication of Solow's view is that, while to talk about sustainability is "not empty, . . . there is no specific object that the goal of sustainability, the obligation of sustainability, requires us to leave untouched" (p. 181). These theoretically argued positions, of course, directly challenge environmentalists' whole program, which includes many more specific items of obligation. Solow's argument is worth examining in detail.

First, Solow dismisses a "straw man" (sometimes referred to as "absurdly strong sustainability") the theory that we should leave the world completely unchanged for the future. "But you can't be obligated to do something that is not feasible" (p. 180), he argues. Solow therefore concludes that, since we cannot leave nature exactly as it is, there are no limits to the substitution of wealth for natural resources at all. We should simply try to maintain an expanding stock of total capital. Sustainability, within this complex of principles and assumptions, is a matter of balancing consumption with adequate investment so that future generations will have enough wealth to invest in efficient production and support their desires to consume. Note that, on this view, monetary capital, technology, labor, and natural resources are interchangeable elements of general capital. Within this set of definitions, the future cannot fault us as long as we leave the next generation as able to fulfill their needs and desires as we have been in our generation.

Having dismissed the straw man, Solow asserts our total ignorance regarding the preferences of future people: "we realize that the tastes, the preferences, of future generations are something that we don't know about" (p. 181). So, he argues, the best that we can do is to maintain a nondiminishing stock of capital in the form of wealth for investment and in the form of productive capacity and technological knowledge. Solow says: "Resources are, to use a favorite word of economists, fungible in a certain sense. They can take the place of each other" (p. 181). Because we do not know what people in the future will want, and because resources are intersubstitutable anyway, all we can be expected to do is to avoid impoverishing the future by overconsuming and undersaving. Provided we maintain capital stocks across time, efficient production and real income will be maintained, and each generation will have the opportunity to achieve as high a standard of living as its predecessors.[11]

Notice that, in this argument, which concludes with a particular solution to the trade-offs problem, ignorance problems completely overshadow the distance problem and the typology of effects problem. Ignorance of what people will want effectively wipes away obligations beyond the next generation;[12] and in a system of dynamic technological change we cannot identify any resources that will be crucial to tomorrow's production possibilities, so it makes no sense to distinguish between unacceptable and acceptable use of resources – there are only "efficient" and "inefficient" ways to generate human welfare. We need only maintain an adequate savings rate in order to rule out spending down the general stock of capital, and that will take care of our obligations to the future. As long as future generations are richer, they will have no right to complain that they were treated unfairly.[13]

I call this the "Grand Simplification" of the problem of intergenerational equity, and this simplification provides the theoretical foundation for the treatments of sustainability advocated by most mainstream welfare economists.[14] Interestingly, the philosopher John Passmore offers somewhat different, but parallel, arguments for a very similar simplification, to which I shall return in Part 5.[15]

Note that the Grand Simplification has three elements: (1) it effectively foreshortens our obligations to only one generation, the next;[16] (2) it assumes the fungibility of resources across uses and across time; and (3) it implies that there are reasonable intergenerational comparisons of the availability of a society's opportunities to enjoy welfare across time. The Grand Simplification is so Grand because it resolves the seemingly perplexing distance question, erasing any possible specific concerns for distant generations, even as it sidesteps the typology of effects problem by assuming the fungibility of resources. All we need do is to avoid impoverishing the future by overspending and undersaving, which can be achieved simply by maintaining a fair savings rate.[17] The Grand Simplification therefore simplifies the question of intergenerational obligations to one of maintaining a nondeclining stock of general capital (which is taken to ensure nondeclining real income). This simplified reasoning therefore allows a direct comparison of welfare opportunities available to members of subsequent generations; advocates of this approach to intergenerational obligations prefer to compare "unstructured bequest packages" across generations. As long as subsequent generations have an opportunity to be as wealthy as we are, and to consume as much as we do, we cannot be faulted.

What is remarkable is that this simplification is achieved on no better foundation than a simple declaration of an implausibly strong, even extreme, statement of the ignorance problem,[18] coupled with the unargued assertion that economists know that resources "can take the place of each other" – the fungibility assumption. I have argued elsewhere that this set of assumptions and beliefs, so presented, should not be considered an empirical theory, but rather a proposed conceptual – eventually, perhaps, an operational – model for judging the sustainability of proposed policies and activities.[19] Solow's approach to intergenerational equity must therefore be examined not just for

427

the verifiability of assertions made within the theory, but also with respect to the appropriateness of its assumptions and conceptual commitments to the task of understanding what we owe the future.

PART 3. THE SEARCH FOR STRONGER ECONOMIC SUSTAINABILITY

Ecological economists, noticing that the assumptions of mainstream welfare economists greatly restrict our ability to analyze impacts of specific policies on the future, have challenged the heart of this complex of assumptions.[20] In particular, they have questioned Solow's principle of unlimited fungibility. They argue that certain elements, relationships, or processes of nature represent irreplaceable resources, and that these resources constitute a scientifically separable and normatively significant category of capital – natural capital. This position apparently contradicts Solow's central conclusion – that sustainability is achieved provided simply that the *total* stock of capital is not declining. We therefore face what is described by Herman Daly and John Cobb as the clash between "strong" and "weak" senses of sustainability.[21] Strong sustainability requires not just adequate savings from generation to generation but, in addition, protection of special features of the environment and natural resources that can be designated as natural capital.[22]

Ecological economists such as Daly argue that human-created capital and natural capital are complements, and that they are imperfect substitutes for one another.[23] Historically, when natural resources were plentiful and human technology was limited, we worried most about human capital; as the proportions shift, we must worry more about the impacts of future resource shortages on prices and income. Daly and Cobb propose that how that certain resources have been identified as natural capital, these particular assets should be protected because they are essential to the welfare of future generations. We can say, then, that whereas the weak economic sustainability theorists believe we owe to the future an *unstructured bequest package*, the strong economic sustainability theorists *structure their bequest package*, differentiating special elements of capital-in-general that must be included in the capital base passed on to coming generations.[24] The language of "natural capital" allows ecological economists to insist on what seem to be somewhat stronger sustainability principles and also to retain many of the evaluative assumptions of welfare economics. In particular, this group never questions the weak sustainability theorists' unitary conception of value – all values must be measured in present dollars, the common currency of human welfare. And this reduction of all values to a single measure allows them to protect the marginalist assumptions of welfare economics. Strong sustainability theorists never question the comparability of natural and other forms of capital with respect to their impacts on welfare as experienced at different times. Their analysis retains the main features of marginalism and compensability so familiar in welfare economics. They never question, therefore, that losses to the future can be represented in terms of present values, and they expect to represent losses of natural capital within a synoptic computation of present welfare values.

As an example of this viewpoint, consider the "green" accounting system proposed by Daly and Cobb. They note that national accounts, even when measured inadequately as Gross National Product (GNP), are usefully corrected for depreciation of capital and that, at least over time, consumption must be limited by the Net National Product (NNP) for that year. Daly and Cobb then argue that it makes equal sense to correct national accounts for "depreciation of natural capital" (along with other corrections that need not concern us here). So, by analogy to the adjustment account kept for depreciation of capital, Daly and Cobb suggest another adjustment account to measure loss of natural capital as losses to be deducted from general accounts of all wealth, including technological development, factories, and so on. This deduction from national accounts, they conclude, is justified "simply to gain a better approximation to the central and well-established meaning of income" (p. 71). To follow up on this conclusion, they operationalize this notion in their "Index of Sustainable Economic Welfare," which is presented as an extensive Appendix for the Common Good. In their argument, they dispute the conclusion of economists William Nordhaus and James Tobin on the point of substitutability, concluding that resource depletion is already affecting income, and stating that "since 1972, the stagnation of productivity for about a decade are signs [sic] of the effect of rising real resource costs, particularly energy resources" (p. 410). They say, "We have thus deducted an estimate of the amount that would need to be set aside in a perpetual income stream to compensate future generations for the loss of services from nonrenewable energy resources (as well as other exhaustible mineral resources). In addition, we have deducted for the loss of resources such as wetlands and croplands.... This may be thought of as an accounting device for the depreciation of 'natural capital...'" (p. 411). The value of resources for the future is equated to an "income stream" from a trust fund that will compensate the future for losses in income due to our destruction of natural capital today.[25]

At this point in the argument, however, the differences between strong and weak economic sustainability become less and less tangible, even as

429

their similarities – especially the willingness to compensate the future at the present exchange rate for a lost future opportunity – become more evident.[26] In this sense, Daly and Cobb have *backed into* the Grand Simplification – what we owe the future, in toto, is to protect natural capital *or* to compensate the future for its loss if we fail to do so. While they do not embrace presentism, it is difficult to see how this concern with *income* will translate into specific requirements affecting distant generations. No less than for Solow, one acceptable outcome is for people of the present to destroy or use up elements of natural capital; such destruction is to be repaired by investing in a trust fund to compensate for income lost as a result. But it now appears that the distinction between Solow, on the one hand, and Daly and Cobb, on the other, comes to very little, conceptually and morally. They simply disagree about whether losses of natural resource shortages will have a significant impact on future income. Solow doubts that they will; Daly and Cobb claim that they already have. But that is an empirical disagreement, not a conceptual or moral one.[27] Both approaches accept the Grand Simplification to the extent that both see resource degradation and reduction of resource availability in the future as compensable by replacement with investments equal to the current exchange value of any future option; Daly and Cobb, no less than Solow, have bought into the Grand Simplification by effectively reducing the question of fairness across generations to one of equality of opportunity to consume, measured in terms of income or wealth.

Does this result mean that strong sustainability is a sham – that the only kind of sustainability is weak sustainability? No, it does not, although it apparently does call into question the use of a marginalist/individualist approach to comparing welfare across generations and to specifying the requirements of sustainability.[28] As long as the system of analysis allows compensability across generations for any losses, it is difficult to see how one might generate specific many-generation protections for prized resources. Fortunately, there is another position available, a position that differs from both Solow's and that of Daly and Cobb. To see the remaining possible view, return to the position of Solow's straw man – absurdly strong sustainability – which hopes to save everything exactly as it is for the future, so rightfully dismissed out of hand. Solow goes straight from this dismissal to the conclusion that there is *nothing* in particular that we owe to the future. But this reasoning represents a fallacy of false alternatives: what about "less than absurdly strong sustainability" – the apparently sensible view that there are *some things* about nature that we may change at our will and other things that we ought not to change? We can state this position as follows: There are some elements/processes in nature that are so important to the future that no generation is permitted to destroy them. If these resources are lost, the future will be worse off than it would have been had they been protected – persons in the future will be made worse off by this destruction *even if they are more wealthy than their ancestors.* They have suffered an uncompensable harm. A theory of value with these features would differ from *both* the Solow view *and* the Daly–Cobb view, because it recognizes uncompensable harms, based on a belief that some items of value are in fact priceless. They cannot be assigned an economic value; there is no dollar amount that, if placed in trust as repayment to the future for harms perpetrated in the present, would represent adequate compensation. The nature of such uncompensable harms will be further explored later.

This position, which cannot even be conceived within the marginalist, evaluative language of economic welfare calculations, represents the most promising direction for developing a new approach to environmental evaluation and to defining a stronger sense of sustainability. The purpose of this position, of course, is to identify aspects of nature that are priceless, and to provide a strong moral argument that these aspects must be protected if we are to be fair to the future. This is a formidable challenge but, for those who are convinced that present obligations to the future require more than weak sustainability, attacking this task may turn out to be a more productive strategy than accepting the burden – inherent in any economic approach to strong sustainability – of defining natural capital.

If the preceding analysis is correct, the task of defining natural capital in economic terms was a fool's errand from the start. If the theory we adopt, like that of Daly and Cobb, allows *compensation* across generations for any type of lost natural capital, then what we owe the future is simply the present value of "options" to use resources in the future. If natural capital can be replaced with payments from a trust fund, and if fair compensation payments can be ascertained in the present, then fungibility has simply come in the back door – the resources the future needs must be given a present economic value – in order to calculate how large the trust fund must be to compensate the future adequately.[29] But to accomplish the calculation, it will be necessary to know which resources are necessary to maintain nondeclining income in the future. If one knew that, one would have no need, economically, for a specification of natural capital. To determine what we owe the future economically, we must know the present economic value of natural capital; and once we know the present economic value of natural capital, the next step – taken fearlessly in the appendix on accounting by Daly and Cobb – is to set up a trust fund to compensate for lost income as a result of that harm. Since the compensation is acceptable only if the trust fund payments are acceptable substitutes for despoiled natural capital, the practical effects of Solow's substitutability assumption, and of Daly and Cobb's compensation provision, are apparently identical. Future people will be without natural capital, and they will have been offered monetary value in its place. The moral of this story is that fungibility of natural and human-made capital is made unavoidable by the adoption of the economic paradigm, not simply by the explicit stipulation of Solow. The much-vaunted concept of natural capital comes to very little: consistent welfare economists are, for all practical purposes, weak sustainability theorists, who accept the reasoning of the Grand Simplification – explicitly, as Solow does, or implicitly, as Daly and Cobb do.[30]

PART 4. GRANDLY OVER-SIMPLIFIED?

For all its Grandness – and regardless of its acceptability to thinkers as diverse as Solow, Passmore, Daly and Cobb, and Barry – the Grand Simplification has serious problems both in theory and in practice. In this part, practical implications of the Grand Simplification are compared with ordinary intuitions as they emerge in concrete, possible situations. When one applies the simplified reasoning of the Grand Simplification to real cases of environmental decisions with long-term consequences, one finds that it supports highly questionable moral outcomes and contradicts reasonable moral intuitions. In this part, I sketch four classes of cases that apparently call the Grand Simplification into question, and suggest a need for a more complex analysis of our obligations to the future. In Part 5, the theoretical problems raised by these cases will be explored.

### 1. Toxic Time Bomb Cases

Suppose it were proposed that the most dangerous toxic wastes from technological processes of production be stored in a new type of container. Manufacturers guarantee that these containers will reliably isolate toxic substances for at least 150 years, eliminating any chance of exposure to people of the present and next generations. It is also known that these containers can explode unpredictably, spreading their contents throughout the environment, at some time after 150 years. Suppose policymakers nevertheless pursue this policy, because it offers an affordable solution to the hitherto insoluble problem of disposing of toxic wastes. If anyone objects to the effects of this wanton proposal, apparently Solow and other advocates of the Grand Simplification will reassure them by pointing out that we know nothing regarding what future people will want. People of the future may prefer to bathe in toxic wastes to bathing in clean water, so we may be doing them a favor by providing them a free chemical bath unexpectedly![31]

## 2. *Greenhouse Warming and Gradual Climate Change*

Economists who have examined the costs and benefits of programs to limit the emission of greenhouse gases have argued that economic impacts on industrialized economies will not be severe, and that the climate changes likely to occur in the first generation or two (after anthropogenic changes in climate are detected) will have neutral, and perhaps positive, impacts on economic growth, as lengthening growing seasons boost agricultural production, for example.[32] Add the Grand Simplification to this line of reasoning and it seems to follow that we would do no harm by setting in motion irreversible climate change, because anyone who might be harmed by it is distant from us in time and beyond the purview of our ethical obligations.[33] Persons within our ethical purview, according to the presentist assumptions, will be affected neutrally or positively, so we may even have a moral obligation to *ensure* global warming (even if it has disastrous consequences three generations hence).

## 3. *Old-Growth and Wilderness Conversion*

Suppose our generation systematically converts all old-growth forests and wilderness areas to productive uses such as farming and mining, producing wealth but making it impossible for future persons to experience unspoiled wilderness or other natural places. Since they are wealthy, they can perhaps provide themselves with Disney-type facsimiles of wilderness. As long as they have adequate income to be able to afford such substitutes, the economists tell us, they will have been adequately compensated for their unavailability in reality.[34]

## 4. *Severe but Gradual Ecological Declines*

According to the Grand Simplification, there would be no harm in degrading the Chesapeake Bay into an irretrievably polluted anaerobic slime pond, provided we do so slowly enough that the most negative impacts will occur a couple of generations hence. In the meantime, people will be well compensated for the loss of the Chesapeake, provided that an adequate portion of the profits from its degradation is invested so that the intervening generations retain an undiminished opportunity to achieve welfare comparable to ours.

If we accept the Grand Simplification, consequences to the third generation, however severe, would simply not count in our decision.

These four families of apparent counterexamples, if taken seriously and not ruled out by theoretical fiat, point to at least two serious weaknesses of the Grand Simplification. Our intuition that we would do something wrong by creating toxic time bombs suggests, first of all, that our intuitions strongly favor, at least in some cases, a less than narrowly presentist viewpoint. These cases also seem to call into question Solow's broad claim that we do not know what people of the future will want. It is simply ludicrous to suggest that we do not know whether or not people of the future will want to be doused unexpectedly with toxic chemicals. Indeed, to the extent that one has intuitions that any of the four cases represent moral harm to the future, one's intuitions run counter to Solow's unqualified claims of ignorance, and these intuitions seem, at least, to demand a more nuanced presentism. The following question is unavoidable: Given what we reasonably believe about future tastes, which effects of our activities can be predicted to be benign and which are likely to be harmful? Once we raise the question of what we do in fact know, or justifiably believe, and which of our actions threaten reasonably expectable future values, however, the typology of effects question immediately returns to center stage. And then the Grand Simplification entirely unravels. We are back to trying to figure out what we owe, to how many generations, with a knowledge base containing some near certainties and a great deal of uncertainty.

The Grand Simplification is appealing because it apparently allows the formulation of all problems of intergenerational morality as questions of "just" trade-offs accounted for in terms of present exchange values. But if the Grand Simplification fails philosophically, the best conceptual formulation of questions of intergenerational morality, including the question of trade-offs, must be reconsidered. Further, once one calls into question whether economic calculations and utilitarian comparisons are the only ways to think about intergenerational obligations, more nuanced solutions to Problems 2–4 may lead to new insights concerning intergenerational trade-offs.

Once we reject the extreme ignorance claim and recognize that we have a convincing basis for some expectations about what the future will want, it is possible to reestablish a closer relationship between distance questions and typology of effects questions. It seems reasonable, once we have a typology of effects based on what we are sure will be harmful to the future, to limit consideration of some risks over shorter time scales, while other issues (such as storage of nuclear wastes, for example) have a longer horizon. And while the problem of knowledge and ignorance still drives any solution, a recognition of gradations in our knowledge and ignorance makes possible a more nuanced response to distance questions. Once the question of what seems almost certain to harm people in the future is on the table, a hard-edged answer to the distance question is simply insensitive to key aspects of the choice problem. To answer it would require much more specific information than would be available in Solow's data on economic growth and savings rates. If we desert the Grand Simplification, however, specification of a typology of effects would be a major step toward defining a "stronger" sense of sustainability. Unfortunately, the prevalence of the Grand Simplification has virtually blocked – or totally confused – discussions of possible typologies of effects.

The acceleration of greenhouse gases is similar to the toxic time bomb case in providing an example of present activities that cause neutral or even beneficial changes for a short time, only to be followed by a risk of irreversible and cumulatively disastrous outcomes beyond the horizon of one or two generations. While there has been much uncertainty about what the impacts of anthropogenic increases in greenhouse gases will be, few discussants have taken the view that (1) impacts on the third generation hence will be huge and disastrously negative but that (2) such an outcome would be morally irrelevant to our current choices. Again, it seems that many environmentalists and scientists believe there is a range of outcomes that the future will surely wish to avoid if possible, and that this commonsense position apparently implies that we need a typology of benign, neutral, and damaging effects if we are to address the complex policy cases faced today. Using the ignorance-of-preferences and the unlimited substitutability assumptions to justify presentism runs roughshod over common sense and over important moral intuitions.

All four cases, as noted, suggest that we are not comfortable with Solow's unqualified claims of ignorance or with the inference he draws regarding our obligations to the future. But cases 3 and 4 also raise another set of puzzles having to do with what might be called "intergenerational paternalism." Solow, at a different point in the essay quoted previously, claims that not only do we not know what the future will want, but that "to be honest, it is none of our business" (p. 182). The cases of wilderness protection and protection of special features such as Chesapeake Bay (what for simplicity of reference we can call "protection of special places") highlight another disagreement between Solow's theory and the intuitions of environmentalists.[35]

If people assert an obligation, contrary to Solow, of members of the present generation to protect special places from severe degradation, they are apparently making some assumptions about what will be valued in the future. They might, for example, be assuming that people in the future will, in fact, greatly value the special place in question. But the environmental protectionists also believe that people in the future *should* value these special places. Imagine that our generation, through conscientious effort and some sacrifice, succeeds in protecting many of these special places; and further suppose that our children's generation continues the protection, but that the subsequent generation declares its preference for development everywhere and for degraded bays, and sets out systematically to destroy the natural legacy we have left them. If members of our generation could somehow learn that their grandchildren or great-grandchildren will desecrate the heritage our generation so carefully preserved for them, I submit that they would not, as passively as Solow says, accept this change in preferences as none of their business. On the contrary, their reaction would be to double and redouble their efforts to educate today's population and to build lasting institutions that will perpetuate their deeply held values and ideals. Natural protectionists accept responsibility to protect places – and in doing so, they also accept responsibility to avert, by any means at their disposal, a loss of commitment to protection. The environmentalist accepts, as part of the obligation to save special places, an obligation to perpetuate the values and mindset that finds those special places worthy of respect and protection within our society and culture.[36] On this view, there is a paternalistic streak in protectionist thought. Protectionists hope to save the wonders of nature, but they also accept a responsibility to perpetuate, in their society and among their offspring, a love and respect for the natural places they have loved enough to protect.[37]

It was noted earlier that the natural protectionists would balk at Solow's unqualified claim of ignorance. But now it becomes clear that their disagreement with Solow may not be based mainly on a belief that they, as present protectors of future values, are infallible *predictors* of what people in the future will in fact prefer. It may rather be that present protectors accept responsibility for inculcating certain values and for ensuring that those values are perpetuated in future generations. This analysis of the preservationist mentality suggests – as might be argued on a number of other grounds as well – that wilderness areas and other natural wonders are not valued by preservationists simply as opportunities for preference satisfaction and for welfare gain. To reduce the question of fairness across generations to comparisons of opportunities to consume across generations is to miss an essential part of the protectionists' program and commitment. Protectionists, in contrast to Solow's declaration, *do* care about the preferences of future generations and their preservation efforts are a part of a larger project, that of shaping the values of the culture to include love and respect for natural things and to perpetuate these ideals for the future. The preservationist acts to create a communal sense of caring, and does so by creating or contributing to a community that expresses a deep and abiding value for nature – a community, a kind of organic unity across generations.

Our analysis of the wilderness and special places examples identified a special problem for the Grand Simplification. The concern that is felt to protect wilderness and special places apparently involves a special type of complex intention: the protectionist in the present generation will not be satisfied merely to pass on to the next generation, or even to several generations, intact wilderness or special places.[38] Successful protection of wilderness and special places such as the Chesapeake Bay requires not only provision of these elements of nature as objects available for future enjoyment and consumption, but also successful transmission of an attitude of love, respect, and caring for these places to the persons of subsequent generations – including a sense of moral obligation to continue protectionist policies and ideas. Solow, who is committed to the individualistic, utilitarian view of mainstream welfare economics, sees the value of an object as identical to its ability to fulfill preferences that people, understood as individual consumers, actually have. Rejection of this purely individualistic value theory is at the very heart of the environmentalists' message. They not only work to protect natural areas and other resources in the present, they also attempt to project their values into the future. The questions of how people come to have, and to change, their preferences, and of whether we can judge some preference sets as morally superior to others, are interesting and important questions for understanding intergenerational equity issues, given the central commitments of most environmentalists. But these questions have not been given attention by economists because they are totally exogenous to the discourse of welfare economics.[39] Accordingly, the value system and vocabulary the economist employs to analyze our obligations to future generations, which models all values as fulfillments of individual preferences, cannot express this added aspect of the protectionists' concern for the future. Similarly, philosophers' tendency to approach the problem of intergenerational fairness as one of comparing aggregated individual well-being also obscures this commitment of many protectionists.

If we are to understand the moral commitments of the nature protectionist, we must express them in a richer and more nuanced vocabulary than that of Solow's welfare economics or other comparisons of individual well-being. This vocabulary must be strong enough to express commitment of present people to value beyond economic and consumptive values. Protecting special places almost always involves attributing noneconomic value to them. So, our argument leads back to the idea, introduced at the end of Part 3, that truly strong sustainability will involve a commitment to protect certain places or processes of nature, and that that commitment must be a commitment to a priceless value. The commitment of the protectionist is to protect something that, if lost, would be an uncompensable loss to people of the future. Truly strong sustainability, of the sort explored here, asserts that if we destroy important aspects of natural systems, then we may harm future generations, making them worse off than they would have been, even if they are able to be as well off, in terms of comparative welfare, as we are.

Viewing value from within the economists' perspective of comparing welfare across generations, moral judgments about the preferences that future people express and act upon are simply irrelevant. But for the nature protectionist, it makes sense to say that those people in the future who have lost all interest in nature are worse off in ways that have little to do with their ability to fulfill their actual preferences. Clearly, this additional claim by protectionists is controversial. Solow would no doubt argue that it is a meaningless question and that an advantage of his value calculus is that questions such as this fall by the wayside.[40] But the fact remains that there is a clear commitment – at least initially intelligible – of nature protectionists regarding the values that future people express and live according to, and this commitment cannot be expressed in the utilitarian calculus of economists. Can we help the nature protectionists to make sense of their claim as a *moral* claim?

Speaking logically, there are at least two – perhaps several – possible assumptions/premises that might be introduced to create the richer vocabulary we need to go beyond Solow's preference-based value theory and to express the more complex moral commitments the protectionist feels toward protection of special places. One option, mentioned in the text in note 7 at the end of this chapter and favored by many environmental ethicists, is to claim that the elements of nature in question have value in their own right, "inherently." On this view, the nature protectionists could argue that wilderness areas and special places such as the Chesapeake have a good of their own, and that this good is independent of individual preferences or culturally developed values. They could then reply to Solow that the objects of their protectionist efforts are simply good, and that they must be protected for their own inherent values.[41] I will not pursue the intrinsic-value-in-nature option in this paper for two reasons: (1) I have argued in note 7 that this approach is unlikely to succeed in providing a workable theory of environmental values and valuation; (2) this theory does not relate directly to the topic of this paper, intergenerational obligations, because attribution of inherent value to special places would justify protectionism, quite independently of intergenerational obligations. But since attributions of intrinsic value to nature are both confusing and difficult to defend, it makes sense to see whether protectionism can be supported and guided by a commitment to human intergenerational values.

A second possibility, as mentioned earlier, is to base this additional commitment on a more communitarian/cultural ethic, accepting responsibilities to the future as an obligation to the community and to cultural ideals, rather than to individual consumers of the future. The communitarian option is explored in Parts 5 and 6.

### PART 5. PASSMORE AND SHARED MORAL COMMUNITIES

It was noted earlier that the philosopher John Passmore embraces a conclusion similar to Solow's Grand Simplification. Passmore states that since "our obligations are to *immediate* posterity, we ought to try to improve the world so that we shall be able to hand it over to our immediate successors in a better condition, and that is all."[42] Passmore's argument, while sharing its conclusion with Solow's, differs in that he employs a lemma regarding membership in moral communities, so his argument must be addressed as we progress toward a more communitarian conception of intergenerational obligations. Passmore, following the legal theorist M. P. Golding, reasons that care for immediate posterity is based on "love" of a kind that justifies equal treatment with our contemporaries.[43] Still following Golding, Passmore then asks whether the present generation should sacrifice to protect resources for generations beyond its grandchildren. His answer is unyieldingly presentist: "When the posterity in question is remote, we can have no assurance whatsoever that they will form with us a single moral community, that what they take as good we should also take as good. . . . The man of the future may well be 'Programmed Man, fabricated to order, with his finger constantly on the Delgado button that stimulates the pleasure centers of the brain.' Towards such a being we should have no obligations."[44]

While Passmore includes a reference to community membership, the ignorance premise again drives the argument toward presentism and very nonspecific obligations. For Passmore, obligations are owed to other members of our moral community. But future people may have very different values, values foreign to us – and we cannot know what their values will be. Therefore, he reasons, they cannot be members of our moral community and we cannot owe them anything. Starting from ignorance as a premise, he uses this premise to question whether we share values and a moral community with people of the future, and he cancels all obligations to descendants in the "remote" future, settling for presentism.

A second look at Passmore's argument, however, cannot ignore an odd assumption/inference at work in it. By implication, he understands a "moral community" to be a group of people who share the same values and ideals. Suppose that the people who live two or three generations hence turn out in fact to be very traditional in their values and share most of their values and ideals with Passmore and the traditions he reveres. One would think, if such were to be the case, and Passmore *could* somehow know this, he would conclude that these people *do* form with him and his cohorts a moral community stretched across several generations. In his argument, however, Passmore seems uninterested in what values people of the future will actually hold, or in puzzling about how to decide who, in the future, will be a member of our community. Instead, he cites one possible worst-case scenario, reasons that we cannot know for certain that future humans will *not* prefer self-stimulation to ideals, and banishes all remote generations from his moral community. But this reasoning is remarkable: remote persons of the future are banished from our moral community on mere suspicion of possible – and unknowable – infidelity to our values and ideals!

More pertinent to our present explorations, however, there is also an assumption, implicit in Passmore's unqualified ignorance claim, of our inability to affect the values and ideals of the future. Its not just that Passmore would drum future people out of the community because they *may* fail to embrace his values; it is also implied that the values they will in fact accept are totally contingent, unconnected to our activities or to the activities of institutions we create and nurture. So, Passmore, no less than Solow, is missing a central aspect of the nature protectionists' moral commitment. Far from accepting our inability to affect future values, the nature protectionist accepts responsibility both to protect special places and to develop ideas, cultural ideals, and institutions that will ensure that people, even in the more distant future, share our ideas and ideals. The protectionist sets out to ensure, to the extent possible, that people of the future will share with us a love and caring respect for these same special places, and for other places that become special, for natural or cultural reasons, in their time. The nature protectionist, in short, sees the protectionist effort as a process of community-building, and transmission of values as much as a process of actually saving the places from destruction and degradation. This added commitment includes both the development of institutions and the development of narratives and artistic traditions. At the heart of this value articulation and transmission process is a particular dialectic between nature and culture, a dialectic that is unfolding in that place.

Looked at from this perspective, Passmore's argument represents a self-fulfilling prophecy of disaster. If we believe we are powerless to affect the values of the future, and if we expect that people in the future will be Delgado button freaks instead of nature freaks, then these beliefs would indeed count against our devoting efforts, especially sacrificial efforts, to protecting special natural places for them. And if we follow Passmore in assuming that the people of the future will be monsters or freaks, then it does indeed make sense to act on shortsighted, selfish motives – failing to protect special places that express our ideals and, to some extent, our identity as a culture. If we pay no heed to the legacy we leave, then we may in fact ensure that future persons are Delgado persons and share no moral ideals or community with us. Indeed, it could be charged that acting on Passmore's pessimism is the most likely way to deprive members of future generations of their link, through us, to a rich, nature-supported and nature-loving tradition in our culture. If we, at Passmore's urging, break that link, we may by our very action doom them to a meaningless life at the Delgado button.

Nature protectionists should not, given their ideology, accept Passmore's passivity and pessimism. Environmentalists since Thoreau – including especially John Muir and Aldo Leopold, both of whom elaborated Thoreau's idea in detail – have believed that experience of nature affects moral commitments, builds moral character, and is increasingly important for the self-perception and moral development of modern industrialized humans.[45] Nature protectionists, in short, reject Passmore's pessimism. They believe instead that if the present generation is diligent in protecting special places, in shaping the ideals and values of the future, and in building effective institutions for the protection of nature, then the people of the future *will* favor nature and special places over pointless self-stimulation, and that these future people, too, will project their values to *their* offspring. Given this set of beliefs, Passmore's argument is simply irrelevant. When nature protectionists build private institutions and land trusts, and when they lobby politicians to pass protectionist legislation, they are expressing and perpetuating their values; they are also doing their best to avert the intellectual and moral disaster Passmore passively accepts. If our generation and successive generations act on these beliefs, it is reasonable to hope that humans of the future *will* share a community with us, and that the special places that are preserved may remain for them shrines to cultural, intellectual, and moral ideals that unify and give meaning to our culture.

To summarize the argument so far, reasons have been given to doubt that a purely economic analysis and comparison of welfare across generations – even when supplemented with a concept of natural capital – can do justice to the commitment of many environmentalists, and in particular to those environmentalists who attempt to save wilderness and other special places. We are left, then, with theories that share a rejection of economic analysis as the *sole* basis of moral obligations across generations, in that all theories capable of expressing the protectionists' values must recognize the possibility of uncompensable losses to the future. All of them, I think it is safe to say, believe weak sustainability is a necessary but not sufficient condition for treating the future well. It is also required that we not deprive future people of the opportunity to experience, to value, and to share with us a commitment to our common ideals. This remaining group of theories, which can be called "theories of uncompensable harm," includes inherent value theories, but we have dismissed this approach as unproductive, leaving theories that recognize uncompensable harms to *humans* of the future.

One way to ground the idea of uncompensable harm is to understand obligations to future generations as obligations to a community – obligations not to destroy the natural and cultural history of a "place" in which humans and nature have interacted to create an organic process, a process that can be understood multigenerationally.[46] The communitarian, unlike the welfare economist, sees goods beyond individual ones, rejecting the economists' model of decision making as based solely on aggregation of the individualistic values of *Homo economicus*. Nature protectionists and their program can best be understood within the broadly communitarian political ontology of the conservative philosopher Edmund Burke, who defined a society as "a partnership not only between those who are living, but between those who are living, those who are dead and those to be born."[47] What nature protectionists need to add to Burke's version of political community is a stronger sense of human territoriality and a more explicit recognition that both our past and our future are entwined with the broader community of living things, the living things and ecophysical systems that form the habitat and the context of multigenerational human communities. Nature protectionists do not, as citizens in Burkean communities, evaluate special places as possibilities for present or future consumption, but rather as shrines, as occasions for present and future people to recollect, and stay in touch with, their authentic natural and cultural history. This is a history of humans as evolved animals, and also as cultural beings who have evolved culturally within a particular natural setting. These are cultural beings who cannot also deny their wild origins.[48]

While I believe this communitarian approach is most likely to support a rich and nuanced understanding of our intergenerational obligations, an exploration of these ideas would require another paper.[49] Here we must explore another avenue of argument, which will lead us back, as promised in the beginning, to the centrality of the trade-offs question, and the light to be shed on that problem from our analysis of the boundary-setting problems that form its conceptual context.

PART 6. WHAT DO WE OWE THE FUTURE (II)?

Let us now reexamine the trade-offs problem, which, as noted in Part 1, has been at the center of the public debate over how to define and measure sustainability. I have shown that the very formulation of this problem in terms of intergenerational welfare comparisons presupposes implausible answers to three less synoptic intergenerational moral problems and, since these less synoptic problems set the boundary conditions for determining what would represent fair trade-offs across generations, the attempt to characterize sustainability in terms of utility comparisons necessarily begs important moral questions. In particular, the formulation of the intergenerational moral problem as a problem of utility trade-offs, and as necessarily responding – in the manner of Solow, Passmore, and others – to the values/preferences of future persons, dooms any hopes of specifying stronger sustainability principles. If specifying our obligations to the future depends upon predicting in detail what individuals in the future will want or need, then assertions of obligations to the future will, at best, be plagued by unavoidable uncertainties; if we can be fair to future generations only if we can predict their needs in detail, then there will always be an impossible task at the heart of all specific (strong) sustainability requirements. From this perspective, the reduction of sustainability to weak sustainability – of the problem of future obligations to that of determining a fair savings rate – is simply a figment of the assumptions introduced in order to characterize the moral problem in utilitarian and economistic terms. Having seen this connection, however, and the way the Grand Simplification sweeps away, in the face of ignorance, crucial distinctions essential to the protectionists' case, it becomes clear that prior theoretical commitments – to utilitarianism and to economic operationalizations of intertemporal welfare comparisons – have determined the contours of the playing field on which intergenerational obligations are discussed and determined. By insisting that intergenerational moral obligations be measured in terms of comparisons of aggregated welfare, utilitarians have formulated the problem of intergenerational fairness so as to require information that cannot be available at the time crucial decisions must be made. The collapse of sustainability into weak sustainability on the basis of ignorance is therefore preordained by the theoretical scaffolding chosen to express the trade-offs problem.

This outcome, in turn, can be traced to the utilitarian dogma that normative questions must be construed as empirical questions with empirical answers. Ultimately, it is this dogma that brings the question of *prediction* center stage in discussions of our obligations to the future; and it is this dogma that undermines any attempt to specify stronger sustainability requirements within

443

the broadly utilitarian framework of analysis. At its deepest level, the Grand Simplification rests not upon the *fact* of our ignorance about future values, but rather on a deep and unquestioned commitment to reduce all moral questions to descriptive questions, to questions that can be fully resolved on an empirical basis. The commitment of economists to the empirical resolubility of moral questions pushes them toward a commitment to measuring and comparing quantities of welfare across time. This tendency puts extraordinary weight on our ability to *predict* future values and preferences. It is, furthermore, this commitment that renders the analytic framework of welfare economics unable to express the core ideas of environmentalists who act to protect special places.

There is a name for the mistake committed by economists and many other utilitarians: it has been called the "discriptivist fallacy" by J. L. Austin, the Oxian analyst. In his book *How to Do Things with Words*, Austin argues that many of our sentences that look like ordinary statements have purposes other than to describe. As examples and illustrations, Austin mentions "I do" (when uttered in the context of a marriage ceremony), "I name this ship the *Queen Elizabeth*" (while striking the ship's bow with a bottle of champagne), and "I bequeath this watch to my brother" (in the context of a will). He then says that, in these examples, to utter the sentence in question (in the appropriate circumstances) is not to *describe* the doing of what is being done, but rather to *do* it. Austin proposes that we characterize such uses of language as "performatives" and mentions that they can be of many types, including "contractual," "declaratory," and so on. [50] Later, Austin says: "A great many of the acts which fall within the province of Ethics . . . have the general character, in whole or in part, of conventional or ritual acts."[51]

Reflecting on the views of environmental protectionists – those who try to save special places – we found that they embrace a *commitment*, not only to save these places, but also to create and sustain institutions and traditions necessary to carry on the commitment indefinitely. These acts include the creation of a place-based literature and narratives, as well as public and private trusts set up to secure, for example, habitats for indigenous species. All of these actions signal commitments to continuity between the past and the future; they are best understood in Austin's sense as performatives. They are founded on the commitments that a community makes to continuity with its past, to its natural and cultural histories, and to a future in which its roots in nature are revered, protected, learned from, and cared for.

Applying Austin's idea, and based on the analysis of this paper, a new way of thinking about intergenerational morality emerges. If we see the problem as one of a community making choices and articulating moral principles – a question of which moral values the community is willing to commit itself to – the values they express in the footprint they leave on their place, then the problems of ignorance about the future become less obtrusive. The question at issue is a question about the present; it is a question of whether the community will or will not take responsibility for the long-term impacts of its actions, and whether the community has the collective moral will to create a community that represents a distinct expression of the nature–culture dialectic as it emerges in a place. Will they rationally choose and implement a bequest package – a trust or legacy – that they will pass on to future generations? We do not then ask what the future will want or need; we ask by what process a community might specify its legacy for the future. If one wishes to study such questions empirically, there is important information available. One might, for example, study how communities engaged in watershed management processes, or community-based watershed management plans, achieve or fail to achieve consensus on environmental goals and policies. While empirical studies such as these may contribute to the process of community-based environmental management, I am suggesting that the foundations of a stronger sustainability commitment lie more in the community's articulated moral commitments to the past and to the future, more than to any *description* of welfare outcomes.

This basic point makes all the difference in the way information is used in defining sustainability, and it changes the way we should think about environmental values and valuation. If the argument of this paper is correct, then the problem of how to measure sustainability, while important, is logically subsequent to the prior question of commitment to preserving a natural and cultural legacy. So, we face the prior task – and I admit it is a difficult and complex one – of developing community processes by which democratic communities can, through the voices of their members, explore their common values and their differences and choose which places and traditions will be saved, achieving as much consensus as possible and continuing the debate about their differences. These commitments, made by earlier generations, represent the voluntary, morally motivated contribution of the earlier generation to the ongoing community. While choosing measurable indicators is logically subsequent to commitment to moral goals, the task of choosing measurable indicators can, and must, proceed simultaneously with the articulation of long-term environmental goals. It cannot be otherwise because the choices that are made by real communities regarding which indicators are relevant to their moral commitments represent, in effect, an operationalization of moral commitments. The task of choosing community values similarly cannot be sharply separated from the specification of certain indicators that would track the extent to which actual choices and practices achieve those commitments.

The specification of a legacy or bequest for the future must ultimately be a political problem, to be determined in political arenas. The best way to achieve consensus in such arenas is to involve real communities in an articulation of values and in a search for common management goals, and to include in that process a publicly accountable search for accurate indicators to correspond to proposed management goals.

The advantages of the shift in perspective on the trade-offs question are now evident: this approach suggests that the key terms "sustainable" and "sustainable development" are not themselves abstract *descriptors* of states of societies or cultures, in general, but rather refer to many sets of commitments of specific societies, communities, and cultures to perpetuate certain values and to project them into the future, one that includes a strong sense of community and a respect for the "place" of that community. The problem of how to measure success and failure in attempts at living sustainably is now the problem, for each community, of choosing, in a fair and democratic way, a fair natural legacy for the future, and then operationalizing these commitments as concrete goals to be measured by democratically agreed-upon indicators. The problem of trade-offs is still a key issue, but it is more manageable because it is no longer dominated by the constraints imposed by our ignorance about the future. The trade-offs problem no longer appears as a problem of comparing aggregated welfare at different times, but rather as a problem of allocating resources to various, sometimes competing, social goals.

Here, it is undeniable – as the economists will be quick to point out – that, ultimately, people in the present must balance their concern and investments in the future against real needs today – the trade-offs question. There will be situations in which setting aside special places, or protecting traditional relationships between cultures and their natural settings, will compete with other values. But now the question is transformed. If we see the problem as one of commitment of people in the present not to see certain of their values and commitments eroded, the fact of our (partial) ignorance of their desires and needs – while a limitation in some ways – is not really relevant to the protectionists' case. The case the protectionist must make is that, to the extent that the community has committed itself to certain values and associated management goals, these goals are deserving of social resources and investments in the future. Certainly, the task for the protectionist is a daunting one, given the competing demands upon society's limited resources. To the extent that a community and its members see the creation of a legacy for the future as a contribution to an ongoing dialectic between their culture and its natural context, and to the extent that they accept responsibility for their legacy to the future, they have embraced a commitment that gives meaning and continuity to their lives that, in some deep sense, affects their sense of self and community.[52]

One aspect of the daunting task of protectionists, of course, remains the unavoidable present uncertainty about what will be important to the future. But that should not keep us from acting in cases where uncertainty is low (as in our prediction that people in the future will prefer that we not store toxic wastes in insecure containers), nor does it keep protectionists from advocating more positive goals and values that they wish and intend to per-petuate. On the contrary, having shifted from seeing the trade-off as one of comparing welfare across generations to seeing trade-offs as a set of difficult choices about how to allocate social resources among competing social goals, it is now possible to envision an approach to environmental management that need not be paralyzed by lack of information about future preferences. While this reformulation of the trade-offs problem leaves many difficult conceptual issues unresolved, its main value may be that it opens up for discussion the problems of choosing a reasonable distance, or "scale," for the moral horizons applicable in various problems, with respect to dealing with inevitable uncer-tainty of saving things for a cohort of people who do not yet exist, and with the problem of deciding what losses would be unacceptable. While we may, for the present, be able only to muddle through with temporary solutions to these three boundary-setting problems of moral trade-offs across time, it can only be for the better if these problems are discussed openly by philosophers, policy analysts, and policy makers. Rejecting the Grand Simplification, in this sense, is only a prerequisite for a more rational formulation of the complicated and difficult nest of problems faced when a community attempts, conscientiously, to determine its obligations to the future, and to choose a fair trade-off be-tween supporting the ideals its members would like to project into the future and the very real and present needs of cohorts in the present generation.

NOTES

1. The essays published in R. I. Sikora and Brian Barry, eds., *Obligations to Future Generations* (Philadelphia: Temple University Press, 1978) typify philosophical discussions. Most of these discussions assume (a) either utilitarianism or some alternative way of comparing aggregated well-being of people across generational times and (b) a form of moral individualism that says that all obligations must be obligations to specifiable individuals who might be harmed. Further, in my opinion, these papers are inordinately concerned with the population problem. I say "inordinately" concerned with the population problem because, while I ac-cept the importance of the population problem for environmental sustainability, the elevation of this problem to primary status is an artifact of individualism. If, as is argued later, obligations to the future are more community-based than

447

individual-regarding, then maintaining an acceptable population level will be implied as a *means* to avoid uncompensable harms to future communities of people. Derek Parfit, in his essay in the Sikora and Barry volume, rejects (b), in that he takes seriously the possibility that there are real harms that are not person-regarding harms; he nevertheless seems reluctant to give up (a), and he continues to see the issue of how to count and value individuals as supremely puzzling (see note 2).

While fascinating, most of these philosophical discussions are theory-driven and approach the problem of intergenerational obligations from the viewpoint of theories of social justice, which encourages participants to emphasize distributive concepts as they apply to interpersonal justice, and have usually taken the assumption of a universal (sometimes called "monistic") ethical theory as a part of the context of the discussion. It is assumed, that is, that the relevant problem is to state a single universal principle to resolve all questions of moral dispute. As Mary Midgley points out, this "conceptual monoculture" has the effect of limiting flexibility in moral inquiry. "Sustainability and Moral Pluralism," *Ethics and the Environment* I, 1 (1996): 41–54, especially 51–3. Also see Midgley, "Duties to Islands," in D. VanDe Veer and C. Pieree, eds., *People, Penguins, and Plastic Trees*, (Belmont, CA: Wadsworth Publishing Company, 1986): 156–65.

This paper, following, Midgley, is premised on the belief that monistic and individualistic assumptions unduly constrain the discussion of obligations to the future. My approach differs from theoretically driven arguments in that I begin with a pluralistic conception, and multiple expressions of environmental value, and then struggle toward integration and theory, rather than assuming a single, complete theory of morality, such as utilitarianism, and attempting to enforce a single-yardstick measurement of value. See paper 3, this volume, for a discussion of the disadvantages of assuming monism in environmental values.

2. I take this to be the main lesson of Parfit's creatively destructive arguments detailing the many puzzles and paradoxes resulting from attempts to find theory to support widely held intuitions that we have obligations regarding the future. See "Future Generations: Further Problems," *Philosophy and Public Affairs*, 11 (1982): 113–72, and *Reasons and Persons* (Oxford: Clarendon Press, 1984), which extends the 1982 arguments and concludes that we need a new "Theory X" in order to support our intuitively felt obligations to the future. See especially pp. 449–54. Also see James S. Fishkin, "The Limits of Intergenerational Justice," in Peter Laslett and James Fishkin, eds., *Justice between Age Groups and Generations* (New Haven, CT: Yale University Press, 1992).

3. See, for example, John Passmore, *Man's Responsibility for Nature* (New York: Charles Scribner's Sons, 1974); Martin Golding, "Obligations to Future Generations," in Ernest Partridge, ed., *Responsibilities to Future Generations* (Buffalo, NY: Prometheus Books, 1981); Jan Narveson, "Future People and Us," in Sikora and Barry, *Obligations to Future Generations*, p. 60; and Hillel Stiner, "The Rights of Future Generations," in ed. Douglas MacLean and Peter Brown, eds., *Energy and the Future* (Totowa, NJ: Rowman and Littlefield, 1983). For exceptions, see Ronald M. Green, "Intergenerational Distributive Justice and Environmental Responsibility," in Sikora and Barry, *Obligations to Future Generations*; Clark Wolf, "Markets, Justice, and the Interests of Future Generations, *Ethics and*

*the Environment* 1, 2 (1996): 153–75; Annette Baier, "For the Sake of Future Generations," in Tom Regan, ed., *Earthbound* (New York: Random House, 1984); and Daniel Callahan, "What Obligations Do We Have to Future Generations?" in Partridge, *Responsibilities to Future Generations*.

4. I am indebted to Leslie Jacobs, who helped me see the centrality of the trade-offs question with a particularly acute criticism of a prior version of this paper.

5. See note 2 for references; and Derek Parfit, "Energy Policy and the Further Future," in Maclean and Brown, *Energy and the Future*. I have discussed this paradox and its importance in "Environmental Ethics and the Rights of Future Generations, "*Environmental Ethics* 4 (1982): 319–37, and in "Future Generations, Obligations to," *Encyclopedia of Bioethics*, 2nd ed. (New York: Macmillan Publishers, 1995), Vol. 2, pp. 892–99, and will not discuss it further here. Also see Gregory Kavka, "The Futurity Problem" in Partridge, *Responsibilities to Future Generations*, pp. 113–15.

6. See Talbot Page, "Intergenerational Justice as Opportunity," in MacLean and Brown, *Energy and the Future*, for an excellent discussion of this problem and a demonstration that criteria of economic efficiency cannot solve it. Also see Page, *Conservation and Economic Efficiency: An Approach to Materials Policy* (Baltimore: Johns Hopkins University Press, 1977), for a more detailed treatment in terms of material policy.

7. See, for example, J. Baird Callicott, "On the Intrinsic Value of Nonhuman Species," in B. G. Norton, ed., *The Preservation of Species* (Princeton, NJ: Princeton University Press, 1986), esp. p. 157F; Callicott, *In Defense of the Land Ethic* (Albany: State University of New York Press, 1988).

8. As I have argued at length elsewhere. See, for example, papers 3 and 12, this volume. For a summary of both arguments, see paper 21, this volume.

9. Robert M. Solow, "The Economics of Resources or the Resources of Economics," *American Economic Review Proceedings* 64 (1974): 1–14; Solow, "On the Intergenerational Allocation of Natural Resources," *Scandinavian Journal of Economics* 88 (1986): 141–9; Solow, "An Almost Practical Step toward Sustainability," invited lecture on the occasion of the fortieth anniversary of Resources for the Future, Washington, DC, October 8, 1992; Solow, "Sustainability: An Economist's Perspective," in Robert and Nancy Dorfman, eds., *Economics of the Environment: Selected Readings* (New York: W. W. Norton and Company, 1993). Page numbers in the text refer to the last item.

10. This approach is also supported by Brian Barry, the philosopher. See "Circumstances of Justice and Future Generations" in Sikora and Barry, *Obligations to Future Generations*, p. 24, and "The Ethics of Resource Depletion" in Barry, *Democracy, Power, and Justice* (Oxford: Clarendon Press, 1989).

11. See Wolf, "Markets, Justice, and the Interests of Future Generations," for an argument that this conceptual model fails, even within the system of assumptions constituting the economic study of welfare.

12. Solow does not explicitly endorse presentism and, indeed, advocated a zero social discount rate in one of his early papers, "The Economics of Resources or the Resources of Economics." It is difficult, however, once one has reduced all concern for the future to a matter of maintaining general capital over time, to specify any particular concerns for the distant future. See Barry, *Theories of Justice* (Oxford: Clarendon Press, 1989), p. 193.

13. One might think that this generalization overstates the case, as was argued by Alan Gibbard and Stephen Darwall at a presentation of an earlier version of this paper. Economists accept a responsibility to choose "good investments," so, if natural areas and other natural amenities are in danger of becoming scarce, economists would counsel that we invest in protection of those resources and that future generations might fault us if we fail to do so. On these grounds, it was argued, economists will reach at least many of the same conclusions that environmentalists do, and economists accomplish much the same goals as environmentalists by insisting on good investments in the resource area.

I do not believe, however, that this argument significantly alters the generic nature of the weak sustainability approach to savings. It is true that economists would insist on well-defined property rights, and they believe that ownership in a resource is likely to promote protection of that resource for economic reasons. The mathematical biologist Colin Clark, in "The Economics of Over-Exploitation," *Science* 182 (1974): 630–4, has shown, however, that under not unusual conditions, it will be conducive to greater profits on investment, and therefore economically rational, for agents who own a resource to extinguish the resource. To simplify a complex argument, Clark showed that if one distinguishes the "rent" derivable from sustainable use of a resource such as a fishery in which exploiters have ownership (in the form of equipment, licenses, and exclusive rights) from maximization of profits from that resource by a particular owner, the two goals diverge. While the total income *from the resource itself* might be maximized by taking only sustainable yield, total income on the investment may be higher if the agent maximizes income in the early years, exploits the resource to (economic) extinction, and then reinvests these large early profits in other high-return industries. The key variable, in other words, will be the "opportunity cost of money." If there are other, equal or greater opportunities to invest in other areas of the economy, then the best strategy may be to maximize profits in early years, exploit the resource to extinction, and reinvest in other regions or in other sectors of the economy.

This result from bioeconomics is relevant to the issue at hand because Clark's result shows that as long as there are adequate investment opportunities elsewhere in an economy (and today this refers to most of the world), property owners who act on profit motives alone (the pursuit of which will maximize the *total* stock of capital over the short run) will act to extinguish a resource by rapid early exploitation. This tendency is greatly exacerbated in less developed countries (where much of the biodiversity of the world exists) by the typically high discount rates applied to investments in those countries. Clark concludes his argument: "In view of the likelihood of private firms adopting high rates of discount, the conservation of renewable resources would appear to require continual public surveillance and control of the physical yield and the condition of the stocks" (p. 634). In other words, pursuit of maximal profit for agents – even for owner-agents – diverges from policies that protect resources.

14. This definition is apparently also equivalent, or nearly so, to the widely cited "Brundtland" definition. World Commission on Environment and Development, *Our Common Future* (Oxford: Oxford University Press, 1987), p. 43.

15. See John Passmore, "Conservation," in *Man's Responsibility for Nature* (New York: Charles Scribner's Sons, 1974), pp. 88–92.

16. Presentism stated as a moral principle might be thought to justify discounting. But many authors have argued that applying discounting across generations begs important moral questions. See Frank Ramsey, "A Mathematical Theory of Saving," *Economic Journal* 38 (1928): 543–59; A. C. Pigou, *The Economics of Welfare* (London: Macmillan, 1932); Talbot Page, "Intergenerational Equity and the Social Rate of Discount," in V. K. Smith and John Krutilla, eds., *Environmental Resource and Applied Welfare Economics* (Washington, DC: Resources for the Future, 1988); and Derek Parfit, "Energy Policy and the Further Future: The Social Discount Rate," in MacLean and Brown, *Energy and the Future*. Also see Daniel Farber and Paul A. Hemmersbaugh, "The Shadow of the Future: Discount Rates, Later Generations, and the Environment," *Vanderbilt Law Review* 46 (1993): 267–304, where it is argued that the social rate of discount should not exceed 1 percent.

17. Again, philosophers have largely agreed with economists' optimistic view, assuming that welfare will expand indefinitely. John Rawls, whose *Theory of Justice* (Cambridge, MA: Harvard University Press, 1971) has shaped many discussions of both intragenerational and intergenerational justice, for example, assumes that each generation will be better off than its predecessors, and considers only the question of choosing a fair savings rate in that optimistic context.

18. See Kavka, "The Futurity Problem," pp. 111–13; Page, "Intergenerational Justice as Opportunity"; Callahan, "What Obligations Do We Have to Future Generations?"; and Barry, "Justice between Generations," in *Democracy, Power, and Justice*, p. 500, for convincing reasons that this strong ignorance premise is implausible.

19. Chapter 3, this volume.

20. See especially Herman Daly and John Cobb, *For the Common Good* (Boston: Beacon Press, 1989); Robert Costanza, *Ecological Economics: The Science and Management of Sustainability* (New York: Columbia University Press, 1991).

21. Daly and Cobb, *For the Common Good*, pp. 72f.

22. In fact, there are several ways of defining the concept of strong sustainability. We can use this definition, which reflects Daly's position, for general purposes. For a more detailed discussion, see Norton and Toman, this volume.

23. Herman Daly, "On Wilfred Beckerman's Critique of Sustainable Development," *Environmental Values* 4 (1995): 49–55.

24. Chapter 3, this volume.

25. Barry, like Daly and Cobb, emphasizes income comparisons, and apparently allows widespread fungibility and unlimited compensability across time. See especially "The Ethics of Resource Depletion," p. 117, where Barry treats the problem of defining and comparing levels of opportunity as a problem of choosing just compensation for destroyed resources.

26. See Alan Holland, "Sustainability: Should We Start from Here?" in Andrew Dobson, ed., *Fairness and Futurity* (Oxford: Oxford University Press, 1999). I credit Holland for convincing me how difficult it would be conceptually to maintain the idea of "capital" – as in "natural capital" – and at the same time to provide a sharp and operational notion of strong sustainability.

27. While economists usually simply assume that resource inputs do not constrain economic growth or raise prices, when pressed they cite a 100-year study by

Howard J. Barnett and Chandler Morse, *Scarcity and Growth: The Economics of Natural Resource Availability* (Baltimore: Johns Hopkins University Press, for Resources for the Future, 1963), which found no increase in the proportion of inflation-controlled prices that goes to resource inputs in three of four natural resource sectors. There have also been several partial updates of Barnett and Morse's work, with mixed findings. See, for example, V. K. Smith, ed., *Scarcity and Growth Reconsidered* (Baltimore: Johns Hopkins University Press, 1979). For a new look and a somewhat different set of expectations, see D. C. Hall and J. V. Hall, "Concepts and Measures of Natural Resource Scarcity with a Summary of Recent Trends," *Journal of Environmental Economics and Management* 11(1984): 363–79; Hall and Hall expect resource prices to follow a U-shaped curve, falling at first because of technological developments but eventually rising due to shortages.

Price-based studies of this sort, whatever their outcome, are convincing, however, only if the prices are right – if, that is, the prices include the full social costs of resource development and extraction. Given that the discipline of environmental and resource economics is based on the assumption that prices for environmental and resource goods are usually not right, because these prices ignore uninternalized costs, it is unclear why economists give any weight to this evidence at all. Arguments such as those of Barnett and Morse are especially unjustifiable in sectors of the economy such as minerals, where prices have seldom included the huge environmental damages resulting from open pit mining and other aggressive extraction techniques, and where lax regulation has allowed companies to avoid the costs of environmental clean-ups, as corporations abandon mines and move to greener pastures or to bankruptcy court, thereby imposing the losses on local communities. Worse, as Richard Norgaard has argued effectively, using prices as indicators predicting scarcity requires that actors in markets be well informed regarding the impacts of scarcity in the future ("Economic Indicators of Resource Scarcity: A Critical Essay," *Journal of Environmental Economics and Management* 19 (1990): 19–25). But if they were well informed, then they would already have the answer toward which the indicators point. The empirical work on "scarcity," in other words, is a big muddle. Assuming that market actors are well informed about future shortages in the present should be especially embarassing to economists such as Solow, who doubt that we even know what future consumers will prefer!

28. Whether strong sustainability theorists can avoid the collapse of their position into weak sustainability, and yet retain the main features of economic analysis, remains to be seen. There are a few economists, including Talbot Page and Richard Howarth, who assert entitlements of the future to set constraints on today's economic maximizing: Howarth, personal communication; see Page, "Intergenerational Justice as Opportunity," and Howarth, "Sustainability as Opportunity," *Land Economics* 73 (1997): 569–79; also see D. W. Bromley, "Entitlements, Missing Markets, and Environmental Uncertainty," *Journal of Environmental Economics and Management* 17 (1989): 181–94. What remains uncertain is whether approaches such as this – which deny universal fungibility and insist that the burden of proof is on degraders of resources to show that adequate substitutes exist or are forthcoming – can specify measurable sustainability requirements while retaining the main features of an economic system of analysis.

29. See Edith Brown Weiss, *In Fairness to Future Generations* (Tokyo and Dobbs Ferry, NY: The United Nations University and Transnational Publishers, Inc., 1989), and Peter Brown, *Restoring the Public Trust: A Fresh Vision for Progressive Government in America* (Boston: Beacon Press, 1994), for alternative conceptions of intergenerational trusts. They differ from Daly and Cobb in that they do not think of the intergenerational trust as a fund, with compensation being an adequate substitute for actual protection of the resources that are needed. Both are vulnerable, however, to the criticism that they provide inadequate guidance for actual choices as to what must be protected; they offer, that is, no definitive or practical solution to the typology of effects problem. Also see Page, "Intergenerational Justice as Opportunity," for a particularly clear explanation of how one must go beyond economic analysis to formulate the important questions of environmental and energy policy, because the assumptions of economic analysis are insensitive to crucial questions affecting intergenerational justice.

30. See Holland, "Sustainability: Should We Start from Here?" for a somewhat different, but complementary, argument for this conclusion.

31. This point is well made by Callahan, "What Obligations Do We Have to Future Generations?"

32. See, for example, Thomas C. Schelling, "Some Economics of Global Warming," *American Economic Review* 82 (1992): 1–15; and Robert Mendelsohn, William Nordhaus, and Daigee Shaw, "The Impact of Global Warming on Agriculture: A Ricardian Analysis," *American Economic Review* 84 (1994): 753–71.

33. There are a number of conceptual and moral perplexities associated with global and long-term problems, among them being that people distant in time (in the future) include both descendants of us and our families (who are strangers to us in the sense that we will never be acquainted with each other) and also "strangers" who are strangers both in that sense and also in the sense that they will exist far away, on the other side of the world. Later in this chapter, briefly, and in other writings, I have discussed the need to understand environmental problems as existing on multiple scales of space and time, and have advocated a priority for local environmental problems. But it is clear that local communities eventually must come together to face larger-scale and, eventually, global environmental problems. See Chapters 17 and 19, this volume.

34. See Chapters 11 and 14, this volume.

35. It is of course something of a simplification to refer to "environmentalists" as if they represent a monolithic group, especially given the many discussions of dissensions between the conservationist and preservationist wings of the movement. See B. G. Norton, "Conservation and Preservation: A Conceptual Rehabilitation," *Environmental Ethics* 8 (Fall 1986): 195–220. Since I wish here to avoid these disagreements, I will refer to the activity of saving areas for future enjoyment, use, or contemplation as "nature protection" or "protection of natural places."

     It should also be noted that Solow also recognizes that people may wish to protect special places. See note 41 for further discussion.

36. This approach does not require recognition of inherent values in the object or place itself, provided one believes in the value of cultural and community values. See Norton, *Why Preserve Natural Variety?* (Princeton, NJ: Princeton University Press, 1987), Ch. 10.

37. This point is eloquently made by Joseph Sax, *Mountains without Handrails* (Ann Arbor: University of Michigan Press, 1980), and by Mark Sagoff, "On Preserving the Natural Environment," *Yale Law Journal* 81 (1974): 205–67.

38. While such complex intentions are given little credence by utilitarians and others who attempt to compare well-being across generations, they have a plausible precedent in the process of constitution-writing. Authors of a constitution, when acting appropriately to complete their task, devise a system of rules that will protect – and foster respect for – key principles such as the rule of law. I argue in Chapter 17 in this volume that because of the multigenerational nature of the questions addressed, the best analogy for understanding a community-constituting commitment to live sustainably is the political process of creating and ratifying a constitution. It must be a political *act* that expresses shared values and that binds the future, if not unalterably, to follow certain rules and procedures and to uphold certain values and ideals. Also see Page, *Conservation and Economic Efficiency*, especially pp. 200f; Michael A. Toman, "Economics and 'Sustainability': Balancing Trade-offs and Imperatives," *Land Economics* 70 (1994): 399–413.

39. I have argued elsewhere that the economists' theoretical/methodological commitment to "consumer sovereignty" is a key source of disagreement between environmentalists and economists. Consumer sovereignty is given several formulations and a variety of justifications in the theoretical literature of economics, but its practical effect is to banish from economics any study of how preferences are formed and reformed over time. See note 30 for references; also see Norton, "Thoreau's Insect Analogies: Or, Why Environmentalists Hate Mainstream Economists," *Environmental Ethics* 13 (Fall 1991): 235–51.

40. See Page, "Intergenerational Justice as Opportunity," in MacLean and Brown, *Energy and the Future*, for a demonstration that neoclassical economics is committed to a simplistic behaviorist conception of the mind. This conception would not be hospitable to the ideas expressed here. See Norton, Costanza, and Bishop, this volume.

41. Here it is important to note that Solow also states that, to the extent that we save special places such as parks or redwood forests, it is because "we love them and value them for their own sakes" (p. 181). But this viewpoint should be carefully distinguished from the position being developed in this paper and from at least many philosophical versions of nonanthropocentric intrinsic value theory. While both Solow's and my version of intertemporal obligations are based on commitments that are clearly culturally affected, I offer a context-based approach to determining which special places should be protected, rather than a purely preference-based test such as Solow suggests. Even more decisively, Solow's position should not be confused with strong versions of the theory that some objects, including ecosystems, have inherent value – value that inheres in them quite independently of human valuation or even existence. See, for example, Holmes Rolston III, *Environmental Ethics: Duties to and Values in the Natural World* (Philadelphia: Temple University Press, 1988); *Conserving Natural Value* (New York: Columbia University Press, 1994); and Paul Taylor, *Respect for Nature* (Princeton, NJ: Princeton University Press, 1986). Solow is simply saying that we value such places noninstrumentally (but anthropocentrically), as we would a great painting or another aesthetic object. This difference is clinched, of course, by Solow's laissez-faire attitude toward future

generations: if we were to believe that nature has value in its own right, then we should, contra Solow, care that people in the future will perceive and protect that value.

42. Passmore, *Man's Responsibility for Nature*, p. 91.

43. Martin P. Golding, "Obligations to Future Generations," in Ernest Partridge, ed., *Responsibilities to Future Generations* (Buffalo, NY: Prometheus Books, 1981).

44. Passmore, *Man's Responsibility for Nature*, p. 90. The quotation is from ibid., p. 71.

45. See Norton, "Thoreau's Insect Analogies," *Why Preserve Natural Variety?* (especially Ch. 10), and *Toward Unity Among Environmentalists* (New York: Oxford University Press, 1991).

46. Avner de Shalit, in *Why Posterity Matters* (London: Routledge, 1995), has also offered a broadly communitarian theory of intergenerational morality, which emphasizes the community itself as the basis for concern for the future. The theory I am suggesting differs from de Shalit's in that the human communities to which I refer are to some degree place-based. Thus, whereas de Shalit emphasizes the historical development of human communities, I emphasize the dialectical, historical relationship that evolves between nature and a culture in a specific geographic location.

47. Edmund Burke, *Reflections on the Revolution in France* (London: Dent, 1910), pp. 93–4.

48. See Henry David Thoreau, *Walden* (New York: The New American Library, 1960), p. 144; and Aldo Leopold, *A Sand County Almanac* (Oxford: Oxford University Press, 1949), pp. 199–201.

49. In particular, not nearly enough has been said here about the nature of community-based obligations and the source of their moral force. A bit more will be said about the nature of these obligations later.

50. J. L. Austin, *How to Do Things with Words* (New York: Oxford University Press, 1962), pp. 5–7.

51. Ibid., pp. 19–20.

52. See Alan Holland and John O'Neill, "The Integrity of Nature Over Time," Thingmount Working Paper TWP 96-08 (1966), Department of Philosophy, Lancaster University, Lancaster, U.K.; and Holland and Kate Rawls, "Values in Conservation," *Ecos* 14 (1993): 14–19, for useful discussions of the inseparability of cultural and ecological ideals. Also see Per Ariansen, "The Non-utility Value of Nature. An Investigation into Biodiversity and the Value of Natural Wholes," in Skogforsk, *Communications of the Norwegian Forest Research Institute*, (Aas, Norway: Agricultural University of Norway, 1997). Ariansen suggests that some choices we make with respect to protecting our environment represent "constitutive" values. Loving and protecting special places and special features of a place, on this view, may be constitutive of a person's sense of self and of community membership. Careless destruction of these special features, correlatively, might be considered a kind of "cultural suicide." Ariansen's insight provides one interesting direction for the explication of what was called earlier "uncompensable harms."

# VI

## Valuing Sustainability: Toward a More Comprehensive Approach to Environmental Evaluation

So, what is the upshot of my multidisciplinary experiment? The shortest answer is that I learned that the idea of sustainability cannot be fully captured in the theories and concepts of any one of the diverse disciplines that contribute to environmental science. In particular, the idea cannot be captured by any science that is understood as an exemplar of objective, descriptive, and value-neutral science, whether natural or social. The understanding of science as value neutral, it is now agreed, is at best an abstraction – an ideal that is never achieved by any real science, much less by an openly normative and committed, mission-oriented science such as conservation biology, conservation ecology, or environmental management.

One central theme of these papers has been that the "serial" model of doing and gathering science – which begins by describing the changes and impacts of human actions and proceeds to evaluate outcomes only after scientific description is complete – is a pipe dream based on arbitrary and artificial disciplinary boundaries. The upshot of my experiment is that sustainability has an inevitable normative aspect, which cannot be fully appreciated unless it is *contextualized* within an action-oriented situation in which real people compete, conflict, and deliberate about what to do in response to real environmental problems. What we need is mission-oriented science, and the mission is to find creative, cooperative solutions to environmental problems. Achieving rational solutions to real problems, however, requires a rational approach to assessing, comparing, and measuring human values in real contexts in which communities face real management choices, contexts in which varied, multilayered, cross-cutting human values interact.

This final part includes five papers addressing the question of how to evaluate environmental change due to human choices and impacts. The papers, taken together, examine theoretical and practical problems in stating and estimating the value of environmental goods. "Commodity, Amenity, and

Morality" was prepared for the Smithsonian/National Academy of Sciences Symposium of Biodiversity held in 1986. This paper, which summarized briefly the arguments I had developed in an extended project on endangered species and priorities in species protection, surveys the usual territory of environmental values by economists and social scientists. But it also casts doubt on the idea that we can ever measure the value of even one species by assigning values to discrete uses or enjoyments and then by aggregating toward a total value for that species. Species, I argued, are too entwined in ecological processes to allow the valuation of species separately from the interrelated functions they express through varied relationships with elements and processes of continuing and evolving systems. The second paper, "The Cultural Approach to Conservation Biology," explores the inevitable cultural element in any evaluation of biological resources, extending the idea that valuing resources is far more complicated than counting and placing monetary values on particular uses. My evolving approach to valuation led me, in "Evaluation and Ecosystem Management," to state as precisely as I could why economic methods of valuation are, however useful in some contexts, simply not flexible enough to apply to emerging, more holistic and complex management situations in which ecosystems and watersheds are chosen as the proper unit of management.

"What Do We Owe the Future? How Should We Decide?", which was prepared for a symposium on the feasibility and desirability of reintroducing wolves into Adirondack Park, provides a case study of how one community might work through the complex issues involved in articulating the long-term multigenerational values involved in restoring wolves to an area from which they have been intentionally extirpated. This ecological restoration proposal is especially interesting because it will not be adopted unless it is embraced by the local, regional, and state governments of New York; the case therefore provides a laboratory for observing democratic deliberation at work. Finally, in "Environmental Values and Adaptive Management," with my colleague Anne Steinemann, I propose a general approach to environmental valuation that can serve as the starting point for developing an iterative, pluralistic method for evaluating *development paths* as an essential element in a public process of adaptive management.

Taken together, these papers are representative of my gradual reformulation of the problem of loss of biological resources over a period of almost fifteen years. My disenchantment with the approach that tries to evaluate species by counting and monetarily valuing their uses first expressed itself as a recognition that the real problem is broader than species protection; in the last paper, this line of thinking is carried forward and broader questions of ecologically sensitive valuation are posed. These broader questions, and their social and political context, prompt a shift of focus from narrow questions of placing a value on individual species to the broader questions of evaluating the contribution of wild species, populations, and ecosystems to communities on multiple levels within an ongoing, adaptive management process. Evaluation of environmental change, as understood within this context, demands a deliberative political approach to deciding what is important enough for interested and committed communities to monitor, measure, protect, and restore, which connects back to the idea, which culminated in Part V, that the first step toward sustainability is the formation of a hypothesis about what we believe we *should* save for the future.

# 23

# Commodity, Amenity, and Morality

## The Limits of Quantification in Valuing Biodiversity

What is the value of the biological diversity of the planet? That question reminds me of a game we used to play at ice cream socials and church picnics when I was growing up in the Midwest. Someone on the entertainment committee would count an assortment of screws and gimcracks, or nuts and bolts, and put them into a mason jar. At the Christmas party, it was pecans, walnuts, and hickory nuts. Everybody else had to guess: How many whatchamacallits are in the jar?

Pretend we're having an ice cream social on an improved version of the space shuttle. Someone looks down and says, "What's the value of the life on that planet down there?" The closest guess wins a door prize.

But our question is tougher than nuts and bolts. Recently, scientists discovered bones from a dinosaur they have called *seismosaurus*. That animal was 18 feet tall, more than 100 feet long, and weighed 80 tons. The diversity in size between a seismosaurus and the smallest microbe is staggering. And I used to be thrown off when they put washers of two different sizes in the mason jar! Given the diversity in size among species, not to mention the fact that many species live inside others, it is not surprising that scientists have left themselves some latitude in their guesses as to how many species there are: they estimate that there are between 5 and 30 million species.

That's O.K. I never did very well at the guessing game myself. One time I guessed that a jar contained 452 nuts and bolts. The correct answer was more than 2,000. I won the booby prize for being the furthest off; my prize was the jar and its contents.

But again, I can't help mentioning how much more difficult our current task is. We would hardly have begun to place a value on biodiversity if we had known how many species there are. We're supposed to put a *value* on them. In what terms?

When I looked into a jar, I was always a bit overwhelmed at first, so let's not give up yet. Eventually, I'd decide to be systematic about my guess. I'd divide the jar into somewhat equal sections and try to do a rough count for one of them. Then, I'd multiply by the number of sections. Despite my lack of success, it's a reasonable approach; we can call it the divide-and-conquer method. Economists and other policy analysts have adopted a similar method for valuing biotic resources. They usually try to estimate, however roughly, a value for one species (Fisher, 1981; Fisher and Hanemann, 1985). If they could assign a value to a few species, such as the snail darter, the Furbish lousewort, and the California condor, then we might average the values of those species and then multiply that average value by the number of species there are, if we only knew how many species there are.

All this averaging and multiplying will require that we use numerical values, so we might as well follow economists in trying to use present dollars as the unit of value. Before introducing the technical terms used by economists, let's start with some ordinary concepts: species can have value as commodities and as amenities, and they can have moral value.

We'll say that a species has *commodity value* if it can be made into a product that can be bought or sold in the marketplace. In this category, alligators have potential value in the manufacture of shoes, but they may also have indirect commodity value if it turns out that vinyl shoes stamped in an alligator pattern sell for more than plain vinyl shoes. Indirect value of this sort is especially important in the pharmaceutical industry, since many of our most valuable medicines are synthetic copies of biologically produced chemicals (Lewis and Elvin-Lewis, 1977; Myers, 1983).

A species has *amenity value* if its existence improves our lives in some nonmaterial way, e.g., when we experience joy at sighting a hummingbird or when we enjoy walks in the forest more when we sight a ladyslipper. Hiking, fishing, hunting, bird-watching, and other pursuits have a huge market value as recreation, and wild species contribute, as amenities, to these activities. Bald eagles, for example, have not only inspired the production of millions of dollars worth of Americana, but they also generate aesthetic excitement through a whole area that is blessed with a nesting pair of them.

Finally, species have *moral value*. Here, we begin to encounter controversy. Some philosophers would say that species have moral value on their own. They are, according to this view, valuable in themselves, and their value is not dependent on any uses to which we put them (Regan, 1981; Taylor, 1986). We will not be able to settle this issue. Suffice it to say that species have moral value even if that moral value depends on us. Here, Thoreau comes to mind. He believed that his careful observation of other species helped him to live a better life (Thoreau, 1942). I believe this also. So there are at least two people, and perhaps many others, who believe that species have value as a moral resource to humans, as a chance for humans to form, re-form, and improve their own value systems (Norton, 1984; Norton, 1987).

Moral values that people attach to species are quite high. Responses to questionnaires have indicated that people place a surprisingly high value on just the knowledge that a thing exists independent of any use (Randall, 1986). Economists, using a method called *contingent valuation*, create shadow markets in which they can ask people how much they would be willing to pay to protect a species, quite independent of any use of the species. If existence values can be thought of as a rough indicator of moral values for present purposes, we can say that species also have considerable moral value, measurable in dollars.

So, we can say with some confidence that some species have considerable commodity, amenity, and moral value. The problem that economists have encountered is that these values are distributed very unevenly among species, at least given our current knowledge. For example, Fisher and Hanemann (1985) have surmised that, under certain assumptions, a wild grass recently discovered in Mexico, a perennial related to corn, may prove to have a value of $6.82 billion annually, and they calculated its value for only one possible use – the creation of a perennial hybrid of corn (Fisher and Hanemann, 1985).

At present, however, we do not have sufficient knowledge to calculate the value of most species. Consequently, in addition to the known values that economists note with respect to some small number of species, they also calculate an *option value* for species of unknown worth, i.e., the value we should place on the possibility that a future discovery will make useful a species that we currently think useless (Fisher and Hanemann, 1985). If we extinguish a species now, such discoveries are precluded. Fisher and Hanemann therefore define option value as the present benefit of holding open the possibility that some species we might eradicate today may prove valuable in the future. They would ask people how much they are willing to pay to retain the option of saving the species, given the possibility that new knowledge indicating its value may be discovered in the future.

One important aspect of option value is that it applies equally to commodity, amenity, and morality. As time passes, we gain knowledge in all of these areas, and new knowledge may lead to new commodity uses for a species or to a new level of aesthetic appreciation, or our moral values may change and some species will, in the future, prove to have moral value that we cannot now recognize.

If placing a dollar figure on these option values seems a daunting task, the situation is actually far worse than it first seems. Calculations of option value can be begun only after we identify a species, guess what uses that species might have, place some dollar value on those uses, and estimate the likelihood of such discoveries occurring at any future date (so that we can discount the values across time). Once we've done all that, we can try to figure out how to translate those future, possible values into present dollars. I think it is safe to say that despite the great theoretical interest in assigning use and option values to species, and some impressive strides in modeling these formally, it may be a long time before the total value of even one species can be stated in terms of present dollars (Norton, 1987).

It is worth stepping back to look at the most difficult problems faced by the divide-and-conquer method. First, there is the problem of irreversibility. In general, economists have trouble with decisions where one of the options cannot be reversed. This is an especially important problem for biodiversity. If we decide to have a dam and give up a species, blowing up the dam won't bring the species back.

Second, we are forced to make present decisions under conditions of uncertainty – another problem for assigning present values. Our ignorance of species is mind-boggling. Suppose you're walking on a hillside in Mexico. Your eyes fall on a few tufts of nondescript grass. Would you guess that grass is worth $6.82 billion annually? Only if you knew that it was a member of the corn family, that it is a perennial, that . . . , and so on. Scientists believe that they have identified and named approximately 15% of the species on Earth (Myers, 1979), and we have rudimentary knowledge of the life characteristics of only a few of them. It is an understatement to refer to this level of ignorance as mere "uncertainty."

A third problem with the divide-and-conquer method derives from ecological knowledge. Species do not exist independently; they have coevolved in ecosystems on which they depend. This means that each individual species depends on some set of other species for its continued existence. A species may depend on just one other species for food, or it may depend on an entire complex of interrelated species. This seems to imply that if we now take actions that cause the extinction of any species, then the loss in future benefits should include losses accruing if any other dependent species succumbs as well. Species on which others depend therefore have contributory value in addition to their direct uses (Norton, 1987). To extinguish a species on which two other species depend is to extinguish three species. Thus, to get the full value of a species, we would somehow have to determine the values of all the other species that depend on it.

It also appears that some species are keystones in their ecosystems. For example, when the Florida alligator populations dipped dangerously low about 15 years ago, wildlife biologists noticed that many populations of other species also declined. During the dry winters in the Florida Everglades, other species depended on alligator wallows as their source of water (Taylor, 1986). Must we say, then, that the value of the alligator includes the value of most of the wildlife in the Everglades?

In principle, these ecological facts add no complication. We need only factor in the ecological information regarding the interdependencies among species in ecosystems. Then, we could tally the direct uses and option values of a species and add to this the uses and option values of all dependent species, and so forth. But, of all the areas of biology and ecology, few are less understood than interspecific dependencies. Ecologists cannot even identify all the interdependencies in the systems they understand best. There is no hope that sufficient information will become available for us to determine the interdependencies in tropical forest ecosystems before the forests are destroyed.

Aside from all these problems, the divide-and-conquer method is not even asking the right question. The value of biological diversity is more than the sum of its parts. Even if we could place a value on the biological diversity represented by all species, we would be only partway to an answer to the question "What is the value of biodiversity?" To answer that question, we would also have to include the genetic variation within species across populations and the variety of interrelationships in which species exist in different ecosystems.

The reason my guesses on nuts and bolts were often very far off, even with my divide-and-conquer method, was that I never completed my calculations before an answer was required. I was always overcome by the uncertainties involved. Did the little area I counted represent one-twentieth or one-twenty-fifth of the jar? Is it representative? In order to answer that, I'd shake the jar, only to discover that all the small washers were at the bottom. So, I'd have to count again and recalculate. "Time's up. Turn in the scrap of paper with your name and number. The game's over." I'd end up writing down a random number and suffering the embarrassment attendant thereto. As species become extinct at an ever-increasing rate, resulting in the loss of a fifth or a fourth of all species in the next two decades, according to various estimates, I fear economists and biologists are in a similar situation.

Rather than continuing my attempt to answer this difficult question on the value of diversity, it may make more sense to take a careful look at the question itself and why we are trying to answer it. The question says a lot about us, the questioners. It is a measure of our unique arrogance that we are the only species that calls symposia and writes books to address that question. The sense of arrogance is hardly diminished when we note our usual reasons for asking it. Why are some people so insistent that we put dollar values on species diversity? Because, we are told, important decisions are being made that may extinguish other species. These decisions must be based on some kind of analytic framework (which means each species must be given value in our economic system). If we do not put some dollar value on a species, it will get left out altogether. In other words, they want us to put dollar values on species so that they can compare these to the value of real estate around reservoirs and to kilowatt-hours of hydroelectric power.

Suddenly, the fun goes out of our guessing game. A new analogy seems more apt: I have been in a terrible accident, and I wake up in a hospital bed on a life-support system. The hospital is short on funds, and the hospital administrators are having a meeting at my bedside. They say they have examined all the other methods to raise the necessary money, and they are proposing to sell a few spare parts from my life-support system at a yard sale. One of them says, "This equipment is so complicated that a few parts won't be missed." "How much do you think this part is worth?" asks another, pointing toward a piece of shiny metal. I try to see what the part is connected to, but it is screwed into a big metal box that looks important. "Or that one over there? It looks like it's just cosmetic," another of them suggests. I almost agree, and then I notice that a main power line passes through it. "Stop! Not that one," I say. Just in time.

It is one thing to treat the valuation of biodiversity as a guessing game or as a set of very interesting theoretical problems in welfare economics. It is quite another thing to suggest that the guesses we make are to be the basis of decision making that will affect the functioning of the ecosystems on which we and our children will depend for life.

If we are not taken seriously unless we quantify our answer, I would like to suggest some new units of measurement. An *oops* is the smallest unit of chagrin that we would feel if we willfully extinguish a species we need later on. A *boggle* is the amount of ignorance encountered when an economist asks a biologist a question about species and ecosystems and the biologist answers: "I don't know, and I'm so far from knowing, it boggles the mind." If I understand what the economists are saying, irreversible oopses and boggles of uncertainty are the main factors in decisions affecting biodiversity. In the passion to express the values of a species in dollar figures, it will be unfortunate if we forget to count oopses and boggles as well.

I believe that we should abandon the divide-and-conquer approach. I suggest we use the big picture method instead. Now the question is easier. The value of biodiversity is the value of everything there is. It is the summed value of all the GNPs of all countries from now until the end of the world. We know that, because our very lives and our economies are dependent upon biodiversity. If biodiversity is reduced sufficiently, and we do not know the disaster point, there will no longer be any conscious beings. With them will go all value – economic and otherwise.

I am afraid this answer will not be useful to those who want to know the value lost when they act to extinguish a species, but it seems a better answer than a guess, even a guess that counts oopses and boggles as well as dollars.

One thing we know: if we lose enough species, we will be sorry. The guessing game is really Russian roulette. Each species lost without serious consequences has been a blank in the chamber. But how can we know before we pull the trigger? That is the question we should be asking (Ehrlich and Ehrlich, 1981).

## REFERENCES

Ehrlich, P. R., and A. Ehrlich. 1981. *Extinction. The Causes and Consequences of the Disappearance of Species*. Random House, New York.

Fisher, A. C., 1981. *Economic Analysis and the Extinction of Species*. Report No. ERG-WP-81-4. Energy and Resources Group, Berkeley, Calif.

Fisher, A. C., and W. M. Hanemann. 1985. *Option Value and the Extinction of Species*. California Agricultural Experiment Station, Berkeley.

Lewis, W. H., and M. P. F. Elvin-Lewis. 1977. *Medical Botany*. John Wiley & Sons, New York.

Myers, N. 1979. *The Sinking Ark*. Pergamon, Oxford.

_____1983. *A Wealth of Wild Species: Storehouse for Human Welfare*. Westview Press, Boulder, Colo.

Norton, B. G. 1984. Environmental ethics and weak anthropocentrism. *Environ. Ethics* 6: 131–148.

_____1987. *Why Save Natural Variety?* Princeton University Press, Princeton, N. J.

Randall, A. 1986, Human preferences, economics, and the preservation of species. pp. 79–109 in B. G. Norton, ed. *The Preservation of Species*. Princeton University Press, Princeton, N. J.

Regan, T. 1981. The nature and possibility of an environmental ethic. *Environ. Ethics* 3: 19–34.

Taylor, P. W. 1986. *Respect for Nature*. Princeton University Press, Princeton, N. J.

Thoreau, H. D. 1942. *Walden*. New American Library, New York.

# 24

# The Cultural Approach to Conservation Biology

One lazy Saturday afternoon I was walking on the beach on the North end of Longboat Key, Florida – the last unspoiled strip of beach on that once-beautiful island. The currents in Longboat Pass had shifted and were dumping sand in a crescent spit out into the Gulf of Mexico. The new sandbar was forming a tidal lagoon. As I walked along on the sandbar, I came face to face with an eight-year-old girl as she clambered from the lagoon onto the ledge of the sanbar. She was cradling a dozen fresh sand dollars in her arms. Looking past her, I saw her mother and older sister dredging sand dollars from the shallow lagoon. They walked back and forth, systematically scuffing their feet through the soft sand. As they dislodged the sand dollars, they picked them up and held them for the little girl, who transported and piled them by their powerboat that was beached on the sandbar. A pile of several hundred had accumulated on the sand by the boat.

"You know, they're alive," I said indignantly.

"We can bleach 'em at home and they'll turn white," the little girl informed me. I could hardly argue with that.

"Do you need so many?" I asked.

"My momma makes 'em outta things," she explained.

I pressed my case: "How many does she need to make things?"

"We can get a nickel apiece for the extras at the craft store," the little girl replied.

Our brief conversation ended, as suddenly as it had begun, in ideological impasse. As I wandered away, muttering to myself, I turned over and over in my mind what I should have said to the little girl. Our discussion on the sand had posed, in microcosm, the problem of environmental values.

From M. Pearl and D. Western, eds., *Conservation for the Twenty-First Century* (New York: Oxford University Press, 1989). Reprinted with permission.

I might have talked to the little girl about sustainable yields and tried to get her to worry about whether there would be sand dollars the next time she came to the beach. That's what I was trying to get at when I asked if she needed so many. But, when I asked how many she needed, I'd already granted the utilitarian value of sand dollars. If one is worth a nickel, more will be worth dollars. I didn't want to object to the exploitation of the sand dollars merely on conservationist grounds; I wanted to say more, that the sand dollars are more than mere commodities to be measured in nickels.

I had encountered what David Ehrenfeld has aptly described as the conservationist's dilemma (Ehrenfeld 1976, 1978). To give "economic" arguments in terms of sustainable yields is to admit that species have value as commodities. And it is difficult to rein in exploitation once it is admitted there are profits to be made, for, in our society at least, the value of commodities is described by the adage "More is better."

Ehrenfeld has accordingly chosen the other horn of the conservationist's dilemma. He says that, by reason of their long-standing existence in nature, other species have an "unimpeachable right to continued existence" (Ehrenfeld 1978:208). This is an appealing position. It expresses the deeply moral intuition I felt when I saw the pile of green disks drying in the Florida sun and evokes the revulsion we feel in response to apartheid and other abominable discriminations against races of humans. Philosophically, however, this approach is beset with enormous problems. Rights have, historically and semantically, been ascribed to individuals. Could I, with a straight face, say to the little girl, "Put them all back, each and every one of them; they have a right to live"? I'd have felt silly because I knew this industrious family would not be moved by speeches for sand dollar liberation, however eloquent. Worse, I'd have been a hypocrite. These discoid echinoderms surely have no greater rights than the red snapper I had enjoyed for lunch.

Nor is it easy to transfer the concept of individual rights to species and ecosystems. Individual rights surely have something to do with the fact that individual organisms have a "good of their own" (Taylor 1986). But what is the corresponding "good" of a species? Is it good for the species of sand dollars to be left alone in the lagoon or is it better for little girls to exert adaptational pressure so that the species will be prepared for greater exploitive challenges in the future? Should we put a fence around the lagoon to keep out all predators, including little girls? Or should we let sand dollars face, as all truly wild species must, selectional pressures (Norton 1987)?

While references to rights of wild species are appealing and may be rhetorically useful as propaganda, most philosophers now agree that appeals to rights of nonhuman species do not provide a coherent and adequate basis for protecting biological diversity. Many philosophers have therefore concluded that when environmentalists speak of rights, they really mean that wild species have *intrinsic value* in some broader sense, a sense that does not require that we attribute rights to them (Callicott 1986). What is important, on this view, is that wild species are valued *for themselves*, and not as *mere instruments* for the fulfillment of human needs and desires. This view is attractive in many ways, but it is extremely difficult to explain clearly. Does it mean that other species had value even before any conscious valuers emerged on the evolutionary scene? Could species be valued by no valuers? (See, for example, Rolston 1986, chap. 22; Taylor 1986.)

Or does this view that species have intrinsic value mean only that all conscious valuers should be able to perceive it (Callicott 1986)? But what if some people, such as the little girl and her family, fail to see it? Do we simply accuse them, metaphorically, of moral blindness? Or can we somehow explain to them what this intrinsic value is and why they should perceive it (Norton 1987)?

My point in asking all of these questions is not to prove that intrinsic value in other species does not exist. It may. I hope that someday we will be able to show that it does. In asking these questions, and emphasizing their difficulty, I am trying to refocus attention on a related but importantly different question: Can appeals to intrinsic value in nonhuman species be made with sufficient clarity and persuasiveness to effect new policies adequate to protect biological diversity before it is too late? With some experts projecting that a fourth of all species could be lost in the next two decades, I fear not. As a philosopher, I can perhaps say in modesty what would appear to be carping criticism if said by someone else: philosophers seldom resolve big issues quickly. We are still, for example, struggling with a number of questions posed by Socrates. It seems unlikely that the issue of whether wild species have intrinsic value will be decided before the question of saving wild nature has become moot. There are, then, two separate debates about environmental values. One debate is *intellectual*, the other is *strategic*. The first debate concerns the *correct moral stance* toward nature. The second debate concerns which moral stance, or rationale, is likely to be *effective* in saving wild species and natural ecosystems.

Important environmentalists have addressed the second debate and taken a pragmatic approach to rationales for environmental policy. For example, Aldo Leopold advocated a biocentric ethical system, hoping to undermine the human arrogance that reduces nature to an instrument for human satisfactions (Leopold 1949). But he decided, when entering the public policy arena, to rely on human-centered reasons. In one of his early essays on conservation policy, Leopold outlined a non-human-centered outlook on nature, but then said, " . . . to most men of affairs, this reason is too intangible to either accept or reject as a guide to human conduct." He proceeded to argue for conservation on the basis of concern for future generations of humans (Leopold 1923; Norton 1986).

Similarly, Rachel Carson's career as a writer began with beautiful books on the sea. There, she used the ecological idea that all things are interrelated and interdependent to question human-centered attitudes and human arrogance. But, when she saw that the wild world she loved was threatened by persistent pesticides, she wrote *Silent Spring* (Carson 1962). In that book, she used the same ideas of interconnectedness to argue that spreading persistent pesticides indiscriminately in the environment places *humans* at risk of cancer and other debilitating illnesses. William Butler, who was the Environmental Defense Fund's counsel when the DDT case was argued in hearings before the Environmental Protection Agency (EPA), told me in an interview that they first presented evidence that wildlife was being killed by DDT, and that it was by no means clear that the case to ban DDT would prevail. But when they introduced into the hearings evidence of human effects, Butler reports that their arguments were attended to, and DDT was banned (Butler 1986).

To those who are uncommitted to environmentalism (and this includes many important decision makers), appeals to intrinsic values in nature and to rights of nonhumans appear "soft," "subjective," and "speculative." We can accept this fact of political life without agreeing with it. Whatever the answer to the intellectual question of whether nonhuman species have intrinsic value, I agree with Leopold, Carson, and Butler that human-oriented reasons carry more weight in current policy debates. Given the urgency of environmental degradation and the irreversibility of losses in biodiversity, it would be equivalent to fiddling while Rome burns to delay action until the achievement of a positive social consensus attributing rights and intrinsic value to nonhuman species.

In the remainder of this chapter, I shall therefore concentrate on the human, cultural reasons to preserve species. I shall emphasize two points about these human reasons. First, these reasons are far broader, deeper, and more powerful than is usually recognized. Both environmentalists and their critics tend to emphasize, when focusing on human reasons for preservation, the commercial and commodity-oriented concerns about saving species (see, for example, Myers 1983; Prescott-Allen and Prescott-Allen 1985). But this approach ignores a vast range of important cultural, aesthetic, and social reasons for a preservationist policy. Second, I would like to emphasize that, at least in the short run, there is hardly any difference in the public policies that would be advocated under a broadly cultural, human-centered value system and the policies, advocated by a biocentric value system. Both world views imply that we should do all we can do to save species and representative ecosystems.

Before I explain why I say the cultural and biocentric views have apparently identical policy implications, I must emphasize the breadth of human values that are served by wild nature. To do that, I want to return to the scene on the sandbar. There I was, facing Ehrenfeld's conservationist's dilemma while talking to an eight-year-old. I tried the respect-for-life approach as well as the conservation-for-future-use approach. The little girl wasn't buying the right-to-life approach: and the conservationist approach, by admitting that sand dollars are commodities to be exploited, admitted too much. It may be useful to set aside philosophical abstractions for a moment and consider the microcosmic situation on the sandbar strategically. I wanted the little girl to put most of the living sand dollars back. How could I accomplish that end?

If I have the chance again, if I run into another little girl with too many sand dollars, I think I'll become a labor organizer on the beach. I'll raise the question of child labor. I'll ask the little girl if she's having fun. "Wouldn't you rather build sand castles?" I'd say. Or, better yet, I'd ask her if she'd like to get to know a sand dollar. I'd pick one up and show her its tiny sucker feet and let her feel them knead her hand, almost imperceptibly. I'd explain how, with those tiny feet, the sand dollar pulls itself through the sand by tugging on many individual grains. I'd tell her how the sand dollar picks up particles of sand with its teeth and digests the tiny diatoms that cling to the particles as they pass through its alimentary canal.

Sand dollars and many other echinoderms, such as sea urchins and starfish, have five sections. The little girl might be surprised to note that they share their pentagonal structure with us, in that we have one head, two arms, and two legs. But I'd point out that the sand dollar's five sections are all the same. They have evolved a less differentiated nervous system. Their behavior is consequently less complex than ours, and I hope the little girl will see that different species have evolved different adaptations to deal with quite different situations. The ancestors of sand dollars, beset with many predators in lagoons, invested in armor rather than mobility. But I'd try to get her to see that our mobility isn't necessarily better; what works depends on the situation.

I hope that by the time we talk about all those things, the little girl will be distracted from her task of bleaching sand dollars for nickels and that she will prefer live sand dollars to dead ones. Ehrenfeld's conservationist's dilemma encourages us to face a bipolar choice: Will we value sand dollars as commodities to be used or will we attribute to them rights and intrinsic value and thereby respect them as worthy of our moral concern? In fact, these are not the only two alternatives, and the dilemma posed is a false one. By focusing on the little girl's attitudes and behavior, we can see a third alternative: The environmentalist's case for preserving wild nature can be expressed in terms of human culture.

The beach was being used that afternoon not just as an opportunity for exploitation; it was also being used as a schoolroom – the little girl was being taught that the value of beaches is a commodity value. In the process of teaching children that sand dollars are mere commodities to be exploited, we also teach them that children are mere consumers. The reduction of little girls to mere consumers follows inevitably upon the reduction of sand dollars and other wildlife to mere commodities. When all of nature is subdued and made mere commodities in our economy, we will have destroyed our only symbols of freedom. When the commercial world view entirely takes over, sand dollars will be mere commodities, and little girls in search of nickels will be mere consumers. Like other domesticated species, they will obediently seek the products that are glamorized in media advertising. The manipulative culture of capitalistic commerce will have become all-pervasive, and the greatest losers will be children who have lost the ability to wonder at wild, living nature.

And the reduction of children to mere consumers is unfortunate for *human* reasons – it represents a contraction of the child's value system. Whatever we believe about the intrinsic value of sand dollars, we can say that, when the little girl lost her chance to *wonder* at the living world of sand dollars at the bottom of lagoons, she was impoverished. The little girl's experience at the beach that Saturday was merely a manifestation of a broader cultural phenomenon. When little girls look at sand dollars and see nickels, they are expressing the same attitude as developers who look at unspoiled beaches and see condominium sites. The beach is shelled, fished, sand-dollared, and used as a tanning salon. But is it appreciated for its real value? Do we cherish its ability to contrast with, and call into question, our everyday world of commerce and profit-making? Hardly at all. Here, we can apply Roderick Nash's point that recreation and tourism involve a search for contrasts (Nash 1986). If all of our beaches are sea-walled, boardwalked, and lined with pizza stands, how will they contrast with our roadways?

When we teach little girls to encounter sand dollars as mere opportunities to make nickels, we impoverish childhood experience. And, as that process is repeated throughout our culture, we impoverish our culture. We begin to see nature as a mere storehouse, rather than as the context in which our culture, our struggle to maintain a niche in nature, has developed and will develop. And we forget that our future, as well as our past, depends on our integration into the broader processes of life-building and niche-carving.

The struggle of sand dollars to survive in newly formed lagoons provides a symbol of our struggle to survive as self-determining creatures. The sand dollars must, in the face of hostile and predatory forces in the lagoon, negotiate a niche. In doing so, they create a tiny colony of life, a community that is supported by, and in turn supports, other forms of life. While the Darwinian revolution in biological science has undermined many of our older metaphysical, religious, and spiritual beliefs, it has also pointed out a new direction for understanding human life and the place of humans in the great experiment of evolution. Our species has emerged from the same processes that created sand dollars. Our natural history is, in a sense, contemporary with us; every species illustrates an alternative means of sustaining life. Every wild species is a repository of analogies that inform our ongoing struggle to survive. In the same way that we might learn to avoid errors of past cultures by studying cultural history, we can learn to avoid ecological disasters by studying the natural history of other living things.

The little girl was being cheated out of valuable experiences. She exchanged an afternoon of discovering nature for a few nickels. On this point, I'm sure. Ehrenfeld and I are in agreement. But I do not accept the implication that the only alternative to valuing wild species as nickels is to say that they have intrinsic value independent of us; they can also be valued as occasions for expanding and uplifting human experience.

If our culture proves incapable of preserving other forms of life in the wild, we will lose our only means to understand the great mystery of life's emergence and diversification. We will doom ourselves and our descendants to ignorance of the roots of our existence. The reduction of human valuing of wild nature to mere commodities, to mere objects to be exploited, doubly undermines human values. First, it narrows and cheapens human experience. The little girl entirely missed the mystery and wonder of living beaches. Second, that attitude is destroying the possibilities for future children to experience wild beaches and to expand their horizons of valuing. If we do not change this attitude, and the trends it promotes, there will be no more beaches where little girls might learn the wonder and value of living sand dollars. Those trends will have, and already have had, dreadful consequences for our culture. The valuing of solitude, the value of experiencing unmanipulated ecosystems, the value of seeing nature as a larger enterprise than our search for economic gain are all threatened by the attitude that sand dollars are mere commodities.

I've tried, then, to explain why the cultural approach to valuing biological diversity is broad, deep, and powerful. It is broad, because it appeals to the values of diversity, contrast, solitude, and so forth that commercial culture is in danger of forgetting. It is deep because it digs below the shallow tendency of our culture to reduce all things to the mere value they will bring in a marketplace. Finally, it is powerful because it addresses the question. What sort of world do we want our children and grand-children to experience? If we extinguish wild nature, we will extinguish with it our ability to wonder; we will have extinguished a part of our consciousness. We might, of course, decry the passing of wild nature because we will have failed to protect intrinsic values residing in other species. But we can as well, and less controversially, decry the loss of richness from the human experience.

Now I would like to apply this cultural rationale for saving species to policy issues. What policy would be pursued if one believes in the cultural value of wildlife and natural ecosystems?

I believe it is important to emphasize the *contributory* value of wild species. Economists who try to measure commercial and other economic values for species and environmentalists who emphasize the aesthetic-cultural value of species both tend to underemphasize the contributory value of species. Species do not exist in isolation. Most species function as important parts of ecosystems and therefore contribute to the support and sustenance of other species (Norton 1987). If one values a wild grass because it provides an opportunity to develop a perennial hybrid of corn, or if one values a sighting of a rare whooping crane, one must value not just those species, but the entire fabric of life on which those species depend. Similarly, we cannot place a value on a little girl discovering the wonder of living sand dollars without valuing the beach where sand dollars live. Economists and ecologists cannot tell us how to save economically and aesthetically important species without saving the ecosystems on which they depend. This is especially true of culturally valuable species. A dead sand dollar on a dissecting tray cannot provide the same experience as a living sand dollar with sucker feet kneading a little girl's hand. If we are to save the experience of nature, we must save that experience whole (Fowles 1979). The species we never notice, like the tiny diatom that nourishes the sand dollar, must be appreciated along with the sand dollar and the red snapper. If we are to save the experiences that enliven and enlighten our culture, then we must save whole ecosystems and all the species that contribute to them. It is difficult to see what believers in the intrinsic value of species would advocate, in addition to this policy.

The debate about policy regarding biological diversity, the debate about what ought to be done, is best seen as a *strategic* debate, largely independent of the *intellectual* debate about whether wild species have intrinsic value. If we recognize the full range of human values served by wild species, we will adopt as a policy goal an attempt to save all wild nature. Admittedly, we will fail in this idealistic goal – the task is too large and our culture has evolved a

474

set of commodity-oriented values that push us inexorably to alter habitats and to threaten wild species. But pious assertions that species, dying out around us, have intrinsic value will make them no less extinct. My point is that a policy of saving as many species as possible is the logical implication of either a non-human-centered value system *or* a human-centered value system which recognizes the full range of human values. It is true that if our culture perceived species as intrinsically valuable, we would do more to save wild species. But it is also true that if our culture perceived the breadth, depth, and power of human-oriented values of species, we would do more to save them. Given this context, the debate about intrinsic values in nature should be seen as an intellectual one and the task of saving species should be addressed strategically. We need not first prove that wild species have intrinsic value and *then* begin working to save them.

An analogy may illustrate this point. Suppose a family must move across country and fit all their belongings in their station wagon. In the end, they find they cannot fit both the television and the large, old family Bible in the space available. One family member suggests they go to the pawn shop and determine which is more valuable and take the commercially valuable item. Another member might be affronted by the treatment of the family Bible, with its record of births and deaths recorded in the hands of patriarchal ancestors, as a mere commodity. This family member might insist that the Bible has intrinsic value and should therefore be saved. A debate ensues and the trip is in danger of being delayed. But a wise member of the family cuts through the problem by pointing out that the Bible is valuable as a symbol of the family's struggle and unity, and that seeing and touching the ancestral record will be valuable *for the family and its offspring*. If this point is made, that the Bible has irreplaceable value for building the family's character, the question of what to do need not wait upon agreement concerning its intrinsic value. Rather than standing on the sidewalk discussing the intrinsic value of family Bibles, the family should leave the television, pack up the Bible, and discuss intrinsic value en route. Similarly for saving wild species. Given the powerful case for saving species as a cultural need, it makes very little difference to policy whether those species are also attributed intrinsic value.

Cultural attitudes determine the values we express and pursue. If we develop a healthy attitude toward nature and experiences of nature, we will act to save wild species and natural ecosystems. I have tried to emphasize the depth, the breadth, and the power of these attitudes and rationales for saving nature based on them.

I suspect that there is an underlying fear among American conservationists that if conservation goals are based on cultural attitudes, we have no right to work to save wildlife in other lands inhabited by other cultures. If the cultural value of wildlife were limited to American values, this limitation might pose a problem (see Regan 1981; Sagoff 1974). But all cultures relate to nature and derive important symbols from the unique communities of wildlife that grace their countryside. Further, if we believe in evolutionary theory, all cultures have evolved in a context of natural systems and can learn their place in the natural world by paying attention to their natural heritage.

A possible corollary to the concern about international action to save wildlife is the thought that perhaps we Americans, as citizens of an advanced industrial nation, have no right to impose what we have learned about evolution and the interdependence of living things upon other nations with different ideas. It is correct to say that we have no right to *impose* our knowledge and related attitudes, just as we have no right to impose our commercial attitudes on them, and just as we have no right to export our idea that true civilization requires a fast-food restaurant on every corner and a herd of cattle on every hectare of formerly forested tropics. But at the same time, we cannot deny our role as world leaders in cultural opinions as well as in economic organization. What we can do is act responsibly as leaders of world opinion to make available our science, our understanding of evolution, and the importance of that understanding in achieving a modern, but healthy, attitude toward human culture and its context in natural ecosystems. If we do that, if we develop a healthy attitude toward the natural world at home and share that attitude with people of other cultures, we will have gone a long way toward preserving the natural world and, more than incidentally, a place for a truly human and humane existence in that world.

### REFERENCES

Butler, W. (1986). Interview with Bryan Norton, April 4, Audubon Society, New York.
Callicott, J. B. (1986). The intrinsic value of non-human species. In *The preservation of Species: The value of Biological Diversity*, ed. B. G. Norton. Princeton, NJ: Princeton University Press.
Carson, R. (1962). *Silent Spring*. Boston, Mass: Houghton Mifflin.
Ehrenfeld, D. (1976). The conservation of non-resources. *Am. Scientist* 64:648–656.
_____.(1978). *The Arrogance of Humanism*. New York: Oxford University Press.
Fowles, J. (1979). Seeing nature whole. *Harper's Magazine*, November: 49–68.
Leopold, A. (1923). Some fundamentals in conservation in the southwest. *Environmental Ethics* (1979) 1:131–141.
_____.(1949). *Sand County Almanac*. New York: Oxford University Press.
Meyers, N. (1983) *A Wealth of Wild Species*. Boulder, Colo.: Westview Press.
Nash, R. (1986). The future of nature tourism Unpublished remarks, Conservation 2100 Conference, New York Zoological Society and the Rockefeller University, New York.

Norton, B. G. (1986). Conservation and preservation of species: A conceptual rehabilitation. *Environmental Ethics* 8:195–220.

———.(1987). *Why Preserve Natural Variety?* Princeton, NJ: Princeton University Press.

Prescott-Allen, R. and C. Prescott-Allen (1985). *What's Wildlife Worth?* Washington, D.C.: Earthscan Publications.

Regan, T. (1981). The nature and possibility of an environmental Ethic. *Environmental Ethics* 3:19–34.

Rolston, H. (1986). *Values Gone Wild: Essays in Environmental Ethics.* Buffalo, NY: Prometheus Books.

Sagoff, M. (1974). On preserving the natural environment. *Yale Law Journal* 84:205–267.

Taylor, P. (1986). *Respect for Nature.* Princeton, NJ: Princeton University Press.

# 25

## Evaluation and Ecosystem Management

### New Directions Needed?

1. INTRODUCTION

There is a pervasive trend toward ecosystem management today, a trend which is evident in federal, state, and regional agency actions and in countless publications. This is an important and laudable development; the purpose of this paper, however, will be to show how this pervasive trend creates an important gap in our ability to evaluate environmental policies, and to suggest a general direction that may prove useful in filling this gap.

Environmental management has traditionally been atomistic in the sense that it has addressed particular problems with particular legislation, and in the sense that it has usually addressed problems of wildlife management on a species-by-species basis (Norton, 1991). Accordingly, the methods of valuation that have been developed, and with which managers have most experience, are almost all scaled at the level of particular populations or species, and are designed to measure specific and identifiable changes in one aspect of the environment. There have been, for example, studies of the recreational use values of game species; there have been hypotheses about the genetic value of particular species for pharmaceuticals and as breeding stock for crop species; and attempts to determine by questionnaires how much consumers are willing to pay to retain this or that population or species for non-use, or 'existence' value (see Freeman (1993) for a survey of these methods). However imperfect these methods are, the really bad news is that even these limited methods are apparently unavailable to evaluate alternative ecosystem management plans because the features protected in ecosystem management are features exhibited in the processes and structures of ecological communities viewed not

Reprinted from *Landscape and Urban Planning* 40 (1998): 185–194, copyright 1998, with permission from Elsevier Science.

atomically, but as holistic, functioning communities. In Section 2, I will provide as precise as possible a proof of this negative conclusion. Reacting to this result in the remainder of the paper, I then propose a new approach to environmental valuation, an approach that evaluates policies on multiple scales and emphasizes the importance of protecting options for future generations as an important goal of long-term environmental management.

## 2. A DILEMMA

I am forced to the conclusion that the need to evaluate ecosystem management plans and strategies will require a new approach to environmental valuation. My reasoning can be summarized in the following argument, in which "D" is a descriptive measure of changes in a *physical* system and "V" is a change in a measure of value.

1.  ASSUME: All social values are (in principle) measurable as units of individual welfare (the Welfare Assumption).
2.  It apparently follows that we can evaluate changes in ecosystem states only if either A or B is fulfilled:
    A.  There is a method by which it is possible to correlate, by using a physical, causal model that relates changes in the descriptive state of systems with changes in aggregated individual welfare, OR:
    B.  There is a method by which it is possible to associate changes in welfare with changes in ecosystem states without employing a physical, causal model correlating changes in D with changes in V, directly. But
3.  Ecologists doubt that a physical, causal model of this degree of resolution is possible, at least for the foreseeable future, making A an unlikely solution. And
4.  Economists have not yet devised a method which fulfils condition B (Freeman, 1993). Therefore:
5.  There currently exist no methods by which to quantify changes in V over time as a result of changes in states of ecological systems.

Possibility A is unlikely on its face. How could a physical model evaluate changes in ecosystem states? As Page (1992, pp. 111–112) has pointed out, ecologists, who are trained to approach science in a value-free manner, have not developed an evaluative vocabulary, and to them the problem of evaluation is seen as a task to be left to others. But this quick rejection of possibility A only pushes us over onto the other horn of the dilemma, where we must

479

determine which economic or other evaluative methods might be appropriate in the ecosystem context.

Certainly, there are interesting welfare approaches that could be explored if one chooses possibility B. For example, it might be posited that changes in large-scale ecological systems might be considered positive or negative 'externalities' of the market system of production and consumption and that impacts on ecological communities might impact future opportunities for consumption. At present, however, there exist no methods to measure ecosystem-level impacts as effects on economic well-being of individuals. Even if such methods were developed, they could be applied only if a rate of time preference applicable over long temporal durations could be agreed upon.

Worse, it seems likely that large-scale changes in ecological structure and functioning would also have impacts on non-use values, and here the problems are even more daunting. For example, we heard one activist group, organizing to oppose chip mills in the southern Appalachian Mountains, say: "We like our hardwood forests; if the chip mills come in, they will strip the hardwoods and, at best, plant pines. We don't want to change the ecological character of our region." Assertions such as this would apparently involve significant commitments to non-use values. It is widely agreed that contingent valuation questionnaires, the only method available for measuring non-use values, must include a careful characterization of what the consumer will get as a purchased 'commodity' (Mitchell and Carson, 1989). Since the information does not exist to inform consumers of likely impacts on their welfare of any unit of ecosystem protection, the condition of economic modeling that assumes that the consumer has full, or adequate, information cannot be fulfilled. To simply ask consumers what they are willing to pay to retain ecosystematic features without explanation of likely impacts is implicitly to expect consumers to do the very analysis of impacts of degradation and protection on their well-being that is impossible for experts.

We therefore must agree with the economist Freeman (1993, p. 485) that at least at present, no economic methodologies exist to measure welfare impacts of changes in ecological function and in ecosystem character. This conclusion forces us – assuming we believe it is necessary to evaluate ecosystem management activities in some manner – to call into question the welfare assumption itself, and consider evaluative measures that are defined independently of individual welfare measures. So the general approach outlined here ascribes social value to characteristics of ecosystems as complexes of processes directly, without claiming any ability to measure the impacts of these changes on individuals and their welfare. Note that this is a methodological decision and does not deny that there would be welfare impacts resulting from ecosystematic changes. Indeed, it may be assumed that welfare impacts are in many cases substantial – but it is decided for methodological reasons that these large-scale impacts will not be measured in terms of individual welfare. These social values will instead be expressed on a second level, or scale, and we will not attempt to aggregate across scales. This approach strives for quantification of values on all levels but does not attempt aggregation of values across scales of time. In the remainder of this paper, I introduce a scale-sensitive approach to environmental values, by defining normative–descriptive terms such as 'ecosystem health', 'ecological resilience', or 'ecosystem integrity' as indicators of values states of ecological systems.

### 3. OPTIONS VERSUS CONSTRAINTS AS A GUIDE TO LONGTERM MANAGEMENT

Theories of environmental value employing terms such as 'health', 'integrity', and 'resilience' fall in the general category of 'strong sustainability' theories. Sustainability theories are usually advanced as accounts of what we owe the future and why. 'Weak sustainability' refers to the maintenance, into the future, of a non-declining stock of aggregated capital; according to this definition, a culture is acting sustainably if each generation passes on to the next as much capital in the form of natural resources, wealth, technological capabilities, laboring power, knowledge, etc., as they inherited from their predecessors (see, for example, Solow, 1993). Weak sustainability is built on the assumption of unlimited substitution among resources; it can accordingly be defined within the marginalist, single-equilibrium models of mainstream welfare economics. Weak sustainability is achieved provided each generation devotes an adequate proportion of income to capital investments, and thereby offers future generations economic opportunities equal to those encountered by individuals of earlier generations. Strong sustainability proposes a more stringent requirement: in addition to weak sustainability, strong sustainability requires that each generation protect certain specified processes and features of natural systems as essential elements of their bequest to future generations. Strong sustainability theorists believe that some processes and features of ecosystems – what might be called 'health', 'integrity', or 'resilience' – must be a part of any morally acceptable bequest package to future generations. 'Health' and 'integrity' are not simple descriptive terms in ecological science. They are rather, terms in public policy discourse, and their purpose is to articulate characteristics of systems that are associated with long-term social values and goals (paper 9, this volume). Admitting that these terms are evaluative as well as descriptive, the task of developing an integrated approach to ecosystem evaluation requires that we can associate long-term and widespread human values with specifiable and in principle measurable states of ecological systems.

In a Policy Forum article in *Science*, a prestigious multi-disciplinary group of scholars urged that environmental policy set the protection of the resilience of ecological systems as an important social goal (Arrow et al., 1995). I would follow these authors in arguing that true sustainability requires resilient ecosystems, but I would go further and attempt to relate resilience more explicitly to important social values, especially to the value of main-taining options that depend upon ecological processes and features. Ecolog-ical systems will continue to respond and adapt to both natural and human-caused disturbances – some change is therefore inevitable. Humans cannot protect every process just as it is without freezing nature, which would be the ultimate, and self-defeating, outcome of overdoing 'preservation'. But ecosystem management must not go to the other extreme either; it should not assume that ecosystems are unlimited in their plasticity and that they can really be 'managed', controlled, and manipulated at all levels for human ends. Ecosystem management is understanding human communities as ecological elements in larger, and longer-term, ecological communities and physical systems. Once we fully accept that humans are a part of natural systems, ecosystem 'management' loses its taint of hubris; more often than not, in ecosystem management, the unruly element in biotic communities – what requires 'management' – is the human community and its impacts. What is needed, given these arguments, is a suite of characteristics – such as 'resilience' or 'integrity' – which are sufficiently flexible to avoid 'freezing' ecosystems and stopping their natural development, but which are neverthe-less essential to supporting future well-being and cultural development.

Members of every culture encounter their 'environment' as a mixture of opportunities and constraints. This mix is partly based on characteristics of the environment itself and partly based on what goals are being pursued. Explorers searching for gold encounter a paucity of gold ore as a constraint, whereas this lack is no constraint to the agriculturalist. lack of gold ore is a state of reality, but the evaluation of that state is also a function of the goals and purposes of the explorers. The concepts of 'opportunities', 'options for free choice,' and 'constraints on free choice', therefore, represent an im-plicit 'negotiation' between the 'hard' facts of physical reality and the values and goals of individuals and cultures. This cluster of terms, then, represents an attractive approach to defining intergenerational obligations, because the terms 'options' and 'opportunities' imply both a physical state of the world and a positive judgment of its value. These words are 'morally thick' terms. Like 'stalwart' or 'honorable' in ordinary discourse, they embody both descriptive and prescriptive content (Williams, 1986; Callicott, 1995; Nelson, 1995).

I suggest that we attempt to operationalize, as the basis for a new approach to evaluation of ecosystem management plans, a physical, measurable index that tracks the degree to which ecological, as well as economic, options are protected for future generations. A process or feature of an ecological system will then be understood to have value to the extent that it is associated with economically or culturally important options that should be held open for future generations. The idea of options can play an important role because of its dual nature. The options available to members of a local culture in the future will be dependent on the land, on the physical and ecological characteristics of the landscape that are passed on to them by the present, and also by the goals, values, and aspirations of people in the future. The non-reductionistic approach suggested above now comes into play. Just as firms often keep separate accounts, with different time preference assumptions for operating and for capital budgets, the approach proposed here would keep separate accounts for economic well-being and for inter-generational values, understood as stored options.

Once we are resigned to keeping separate accounts for short- and long-term values, we can continue to use willingness-to-pay as a guide in the short term. Economic well-being – having economic resources to purchase needed and desired services – can be measured in dollars, which can be thought of as 'options' that can be exercised through exchange in 'markets'. These markets model individual behavior against a backdrop of assumptions about market rules and trends, and also against a backdrop of assumptions about the constancy of the available resource base and the quality of the functioning of ecological systems and processes. Economic, cost-benefit models, therefore, continue to have an important role in policy analysis, especially regarding policies with short temporal horizons usually associated with economic planning and decision making.

When concern shifts to multi-generational frames of time, we can retain the goal of measuring value in terms of options and free choices available, but we will not assume constancy of economic and background ecological conditions on this longer temporal scale. So we must replace dollars with another currency – a measure of ecologically sustained options maintained as a culture adapts to the opportunities and constraints embodied in the habitat of a human community over decades and generations. Whereas both operating and capital budgets are normally kept in monetary units, our system of evaluation for ecosystem management policies goes one step further – the longer-term analysis is expressed in different currency and recognizes a different criterion for success. To keep matters simple, assume that short-term impacts of human choices can be measured in terms of dollars representing individual welfare, and that the decision criterion applicable to decisions with short-term impacts is the highest possible benefits-to-costs ratio (BCA criterion). Accounts constructed to calculate benefits of sustaining ecologically based options, however, should be measured in terms of an Opportunities/Constraints Index (OCI). The OCI would be designed to track those particular physical characteristics of ecological and physical systems that would indicate the presence, in the physical environment of a culture, of healthy ecological processes and structures that will maintain, into the indefinite future, culturally valued options and opportunities. Because this value is not assumed to be reducible to immediate and individual economic welfare, it can be interpreted as a more holistic, communitarian, and ecosystematic characteristic.

The evaluative system proposed is pluralistic in the sense that it keeps at least two sets of books and applies different criteria of acceptable action within distinct systems of analysis. Further, the metrics employed to measure success in the two sets of books refer to different units of value, one individualistic and one holistic. This dualistic system requires that we resolve two formidable conceptual problems. First, we must operationalize the OCI in such a way that it (a) designates physical states of ecological systems that are measurable, or at least operationalizable, and at the same time (b) represents a legitimate social value that could be supported democratically by concerned and informed citizens. Second, the employment of two decision criteria raises the problem of possible conflict: What happens if the two criteria point toward different policies? We can call the first problem that of operationalization and the second the problem of reconciliation. We will deal consecutively with these two problems in Sections 4 and 5.

### 4. OPERATIONALIZING THE OPPORTUNITIES/CONSTRAINTS INDEX

Setting aside for the moment the problem that our criteria might point in different directions for policy and action, in this part we push the concept of an OCI as far as possible toward operationalization. It should be noted that, because of the high degree of local determinism regarding ecological conditions, opportunities, and constraints, it is questionable whether it is possible to define a single OCI as a general concept applicable everywhere. We may only be able to give general guidelines for developing many locally defined indices of the integrity of particular places. My goal here is to establish that it is in principle possible to define a measurable characteristic of human and natural systems that are likely to maintain their OCI, and to propose this index as an operationalizable measure of intergenerational equity. Again, we interpret our problem as that of defining a fair bequest package for future generations, and we assume our definition will express the idea of strong sustainability: that there are some structures and processes of nature – natural capital – that are essential elements in any fair bequest package to future generations. The next step is to show how certain interactions of economic and ecological forces can result in increased options, and how others can result in reduced options for a community as it develops over decades and generations.

As a first step in explaining the social value of maintaining options, consider an actual case, that of the rapid deforestation of the ancient temperate rain forests of the U.S. Pacific Northwest over the last century. Can we – with hindsight, but from the perspective of 50 years ago – define an alternative development path for the Northwest that would have (a) used the magnificent resource of forests to build a strong regional economy and (b) resulted in a landscape that included a sustainable source of timber and sufficient old growth to protect the biological diversity and key ecological processes in the region? In considering this example, it may be difficult to avoid well-publicized disagreements about what has happened, and is happening, in this area, but the example is a good one because it is well known and it helps to be as concrete as possible. I will characterize the present outcome in the most sketchy terms, because it is my intention to avoid controversies about specific factual statements regarding the actual situation. The development of the Pacific Northwest since the 1800s has been characterized by (a) rapid development of resources such as old-growth forests, fisheries, and water power resources, (b) relative independence of planning and development across sectors (for example, the power resource was developed without much thought regarding impact on the salmon fishery), (c) rapidly escalating exhaustion of resources, and conflicts among resource users across sectors and within sectors in the late twentieth century, resulting in serious political conflict, and (d) overall acceptable rates of regional economic growth, as the economy makes up for losses in timber and fishery jobs with high-tech development in the larger cities, but localized pockets of extreme hardship emerge as resource-dependent industries shut down in areas where resources are exhausted.

There is no question that there are job losses and general decline in the importance of the resource-exploitation sector in the Pacific Northwest, but there are several important differences in the analyses offered of these changes. For example, one analysis says that there is a real shortage of timber and that the timber industry is entering an inevitable decline, becoming a less productive sector in the economy. Many economists, however, deny that there is a real shortage of timber. There is plenty of harvestable new growth, this analysis argues, but it is not in the right place to provide jobs for existing timber-based communities, and because it is much smaller in stem size, it does not provide appropriate inputs for the technologies developed to exploit old growth.

Leaving aside these disputes, let us suppose that 50 years ago farsighted state governments in the Pacific Northwest had set up a revolving fund of low-interest loans to encourage local entrepreneurs to form milling and furniture-building cooperatives, diverting some investment from expansion of logging operations into the wood processing industry. Suppose also that the program was successful and the Pacific Northwest developed so that most cities or regions had a timber extraction industry, cooperatives to mill raw timber, and other businesses that produce wood products such as pre-fabricated elements of homes or outdoor furniture. The economy, rather than cutting and exporting whole logs, with these added incentives, might have organized to maximize value added near the site of timber extraction. It seems reasonable to believe that, since there would have been more value added per log and more jobs generated per log cut, this alternative development path would have resulted in a more varied and diverse economy, with more options for careers and investments, and also with retention of more old growth than in actuality has been retained. To define the difference between these two development paths in a general and operationalizable way – to distinguish economic policies that protect and expand opportunities from those that destroy and limit future options – would be to provide a more positive characterization of the values and goals embodied in the search for sustainable institutions and policies. This solution, that is, would provide a measure not of individual welfare, but of a set of characteristics observable at the community / ecosystem level; it would be to define strong sustainability as an intergenerationally measurable index of opportunities embodied in resilient ecological systems, as viewed from local perspectives.

It is now possible to understand how the current movement toward ecosystem planning and evaluation represents an important new direction in environmental management. Whereas traditional atomistic environmental management has focused mainly on commodity production and on the economic impacts of such production on individual consumers, ecosystem management layers a second scale of value on top of short-term economic measurement of social values. Economic valuation, based on supply and demand assumptions, models the relationships between the economic system of production and the freedom of consumers to choose affordable products generated by that system. Ecosystem valuation and management, by contrast, focuses on the larger- and longer-scaled relationships that develop between a human population and its habitat over generations. These levels of activities can be modeled independently, because many individual choices of producers and consumers will be cancelled out and have no significant impact on the larger system. For example, if one farmer chooses to cut down his woodlot and plant wheat, this may have no long-term impact on ecological features of the system if the farmer's neighbor simultaneously chooses to let his wheat field go fallow. Ecosystem management can build upon this independence between the short-term effects of individual decisions and long-term impacts, and set out to model and evaluate these two relationships as distinct dynamic systems.

Traditional economic evaluation models attempt to represent future values as social values expressed in the present – as the willingness of present consumers to pay to protect future options. Valuation of ecosystem protection efforts, by contrast, can be envisaged as occurring on a larger scale – the multi-generational scale on which collective individual decisions (trends in an economy or a culture) impact the processes and structure of large scale ecological and physical systems on the landscape level. Imagine an ecosystem management project undertaken in the Pacific Northwest 50 years ago. Such a project might have involved a careful inventory of physical resources and productive processes. But such an inventory would have meaning only if it were accompanied by a social valuation process in which options and opportunities the community values are identified. For example, a social consensus that favored a wooded landscape with a mix of old-growth and new-growth forests could be expressed as a commitment to hold open the option of remaining a culture that is based upon a timber production base. This recognition of a valued option might have led to a commitment to sustainable use of forests, and to the system of subsidies and incentives for investment in timber processing and building of components of houses mentioned above.

The most important step in such a process would have been the explicit self-definition of the community as having a particular cultural identity, and that step can be understood as a choice of a set of 'core values' that provide, in general outlines, the goals and direction of future development. We can think of these values as defining the boundaries between culture and nature – as defining the shape of relationships that guide the intertwining of local cultures / communities with the specific, particular habitat that forms the context of their future adaptation. The imaginary example is the more poignant because it presses upon us the irreversibility of bad decisions in contexts such as these. Failure, 50 years ago, to address issues of the impacts of unrestrained forest extraction as the dominant development path in the Pacific Northwest has led to the current sad outcome. The extraction industry, having stripped the countryside of its options and opportunities (stored as centuries' old trees in ancient forests), has now noted the comparative lack of sunshine to accelerate tree growth as a constraint on second-growth profitability, and is divesting itself of holdings in the Pacific Northwest and moving to the Southeast. Since the forest landscape has been pushed to its limit, any pattern of cutting the remaining scraps of old-growth forest entails the virtual end of the forest-based economy and the culture of the region. Having turned the ancient forests into economic profits, the timber industry can move its investments (the fruits of exploiting the Northwest's resources) elsewhere, leaving residents mainly with the constraints of local ecophysical limitations. Nor can timber culture be made whole again by replacing the option of being a lumberjack with an equivalent number of jobs in high-tech industry. The loss of a timber industry in the Pacific Northwest expresses itself in countless experiences, losses of meaning and value as the children of timber workers are denied the option of following in their parents' footsteps. Ecosystem management projects, at their best, can avert tragedies such as this if they help communities and regions to articulate not just their economic goals, but also their multi-generational aspirations – the values that give meaning and distinctiveness to their culture. Once these values are articulated, ecologists, ecosystem managers, and the concerned public must undertake many experiments by which one hopes to discover development opportunities that build upon, and protect, core values as a guide to decisions that shape the landscape of the future.

But a democratic process such as an ecosystem management project, involving communication of scientists and the public, could be undertaken rationally only if there were some means to (a) determine which culture and economic options are of lasting social value and (b) relate those socially valued options to measurable characteristics of ecological systems. These steps in the ecosystem management process, I am suggesting, would best be undertaken with a system of analytic concepts that are scaled, embodying descriptive and evaluative concepts that would apply at different physical scales.

Consider, by analogy, the way a citizen who is asked to serve as a delegate to a constitutional convention will be expected to evaluate proposals on a scale of multiple generations, whereas most consumer decisions are evaluated according to a relatively shorter scale of time (Page, 1997). We can think of these different evaluation processes as taking place within more than one distinct temporal horizon – our citizen is understood as evaluating changes as part of more than just one dynamic (paper 16, this volume). Consumer choices are understood and modeled in short temporal frames – from zero to three to five years. On this scale, it makes sense to use willingness-to-pay as a reasonable measure of value. But most individuals think of themselves also as members of an ongoing and developing community. On this level, citizens share a love of culture and natural heritage, and they share hopes and aspirations to see future cultural adaptations – institutions and practices – as having continuity with the present and the past. Continuity of a culture as it unfolds within a place, what might be thought of as the 'natural history' of the evolution of a culture, is in this sense an expression of a hard-won cultural self-identity.

I am proposing that we undertake the evaluation of ecosystem management plans by, first, opening the possibility that important decisions made by a community should be evaluated on at least two scales, which embody two separate accounting systems and two different approaches to time preference. The short-term accounting system, which should be dominant in decisions with impacts of up to 5 years or so, can be very similar to current cost-benefit methods. Decisions with a longer frame of time, the consequences of which might last for decades, will also be evaluated according to an evaluation system with a horizon of many generations.

## 5. THE RECONCILIATION PROBLEM

But what of the reconciliation problem? The system proposed here is plu-ralistic in the sense of both employing multiple action criteria and invoking different sources for, and interpretations of, the values that motivate those criteria. Indeed, as just noted, the evaluation system proposed here implies that some decisions, at least those that have long-term impacts as well as shorter term economic impacts, will be evaluated according to two separate criteria. It is possible that, given two accounting systems, we will get conflict-ing evaluations from our two criteria, which may lead to arbitrary choice of one action rather than another (Callicott, 1990). But a pluralistic system need not justify arbitrarily different actions in the same situation. Our approach is, rather, to seek an 'integrated pluralism', an approach that recognizes multi-ple values, and multiple action criteria, and then to show how these multiple criteria interact to yield a single policy direction in each particular situation (Norton, 1991).

Our approach to integration is to assume that our system of policy analysis is 'two-tiered' (Page, 1997; paper 12, this volume). A two-tiered system includes, in addition to the tier on which we apply a given action criterion to a particular problem in search of a policy decision, a prior decision tier in which

| Tier II: The Meta-Criterion: (A Categorization of Problems According to Spatial-Temporal Scale of Impacts and the Social Values Threatened) | | |
|---|---|---|
| A. Classification of scale of potential impacts of a practice/policy B. Judge likelihood of impacts at levels that affect social values C. Choose appropriate criterion from Tier I. | | |
| **Tier I: Action Criteria: (Several Criteria Are Offered to Guide Practice/Policy in Situations as Determined in Tier II)** | | |
| A. Benefit/Cost (BCA) | B. Safe Minimum Standard (SMS) | C. Precautionary Principle (PP) |
| A good policy is one that maximizes the ratio of B/C | Save the resource, provided that social costs are bearable | Take affordable steps today to avoid catastrophe tomorrow |
| Applies to decisions with small-scale impacts and short time horizons | Applies to decisions with ecosystem/landscape level with more than one lifetime reversal time | Applies to possibly catastrophic outcomes in the distant future |
| ECONOMIC SUSTAINABILITY | ECOLOGICAL SUSTAINABILITY | GLOBAL SUSTAINABILITY |

Figure 25.1. A two-tier decision process.

there is a procedure by which one classifies and categorizes a problem. Once a given problem is analyzed and categorized, it is then possible to choose, non-arbitrarily and for good reason, an action rule that is appropriate to the type of risk entailed in the decision (paper 12, this volume; see Fig. 25.1).

A two-tier system is especially helpful when it is elaborated so that the second tier, on which it is decided which action criterion to apply, embodies a system of categories based on the scale of potential problems resulting from a proposed action or policy. If an action or policy entails the risk of irreversible harm over a very large area, an appropriate decision rule would be the precautionary principle – better safe than sorry. If, on the other hand, there is a risk of impacts that are local and reversible, it makes sense to apply the benefit-cost test. The two-tier approach builds scale into the system for analysis of impacts: as a part of the formulation of the problem and the choice of indices to monitor, an environmental problem is assigned a scale that is appropriate given the social value that is to be protected (paper 12, this volume). On an economic scale, policies should encourage individual opportunity. On the scale of multiple generations and large landscapes, our management criteria, and the scientific testing that is done, however, relate to a broader, longer-term goal – that of protecting the integrity of a place. This management goal can be thought of as protecting socially important options, options that give continuity and meaning to social life, and that establish our connection as individuals with the larger ecological system on which we directly and indirectly depend. The bequest from one generation to the next, according to the strong sustainability approach outlined here, requires not just equality of economic opportunity across generations, but also equality of ecological and cultural opportunity – the opportunity to build upon past natural and cultural history, and to contribute to an ongoing culture with economic, institutional, and ecological integrity expressed in personal and cultural ties to a particular place.

We finally have all of the elements necessary to define the ecological integrity of a place – which can be defined as a state of multi-generational harmony between the economic / cultural activities of a human community and its ecophysical context, such that each generation achieves economic well-being through activities and policies that do not cause a cross-generational decline in the OCI. Or, to use our multi-scaled analysis more explicitly, maintaining the integrity of a place is to (a) maintain an expanding set of economic options within each generation while (b) ensuring that future generations will encounter a constant or expanding OCI, which measures the mix of opportunities and constraints that determine the range of ecophysically supported options available to each generation. The term, 'integrity of place' therefore acts as a dual filter, selecting policies that protect both economic and ecological opportunities, and is therefore a useful term to serve as a guide in defining, at many and diverse local places, the proper goal of an integrated ecosystem management plan.

## 6. CONCLUSION

Ecosystem management is a social process, a process that is cognizant of, and interested in, good ecological science, but that is driven by a search for deeply held, culturally rich connections between local communities and their place. One of the most important steps in any ecosystem management plan is early public involvement, involvement in a process of value articulation and in the development of a shared sense of community identity. Scientists must participate in this process, because the public needs to know its natural history and to understand the ecological scenarios that may unfold under various management plans and policies. But scientists can also learn which ecological features are important to the identity of a culture, and in this way receive guidance about what local cultures value in their interactions with the natural communities they inhabit.

Terms such as 'integrity' and 'resilience' – once they are defined as associated with maintaining human options – embody sufficient semantic richness to connect discourse about values, especially the value of future freedom of choice, with physical discourse about ecological systems experienced as a mixture of opportunities and constraints. These terms therefore provide a new alternative for evaluation methodologies. Admitting that we still have much to learn about how to value options that will be faced in the future, and admitting that we need a lot more work to operationalize physical features that can be expected to be associated with important human option values, we at least have a bridge for connecting these two bodies of information, a bridge that will allow communication back and forth between the social and natural sciences, and new opportunities to correct our beliefs and our valuations.

#### REFERENCES

Arrow, K., Bolin, B., Costanza, R., Dasgupta, P., Folke, C., Holling, C. S., Jansson, B., Levin, S., Maler, K., Perrings, C., Pimentel, D., 1995. Economic growth, carrying capacity, and the environment. *Science* 268, 520–521.

Callicott, J. B., 1990. The case against moral pluralism. *Environ. Ethics* 12, 9–24.

_____1995. A review of some problems with the concept of ecosystem health. *Ecosystem Health* 1, 101–112.

Freeman, A. M., 1993. *The Measurement of Environmental and Resource Values: Theory and Methods*. Resources for the Future, Washington, DC.

Mitchell, R. C., Carson, R. T., 1989. *Using Surveys to Value Public Goods: The Contingent Valuation Method*. Resources for the Future, Washington, DC.

Nelson, J. L., 1995. Health and disease as morally thick concepts in ecosystem contexts. *Environ. Values* 4, 287–310.

Norton, B. G., 1991. *Toward Unity among Environmentalists*. Oxford Univ. Press, New York, NY.

Page. T., 1992. Environmental existentialism. In: Costanza, R., Norton, B., Haskell, B. (Eds.), *Ecosystem Health: New Goals for Environmental Management*. Island Press, Covelo, CA, pp. 95–123.

_____1997. On the problem of achieving efficiency and equity, intergenerationally. *Land Economics* 73, 580–596.

Solow, R. M., 1993. Sustainability: an economist's perspective. In: Dorfman, R., Dorfman, N. (Eds.), *Selected Readings in Environmental Economics*, 3rd edn. W. W. Norton, New York, NY.

Williams, B. A. O., 1986. *Ethics and the Limits of Philosophy*. Harvard Univ. Press, Cambridge, MA.

# 26

# What Do We Owe the Future? How Should We Decide?

Each generation, either thoughtlessly or thoughtfully, leaves a bequest to generations that follow it. One of the purposes of this book is to ask, Should a restored and flourishing population of wolves in the Adirondack mountains be a part of our generation's bequest to future generations? As my contribution to the analysis of this case study, I intend to focus on how best to understand, formulate, and analyze this question as a question of public values, which, like all important aspects of ecological restoration, must be addressed in an open and democratic public process.

For most of our history as a transplanted civilization, North Americans of European descent have concentrated on two aspects of the bequest they would leave future generations: the moral, spiritual, and cultural aspect and the economic aspect. Early in our history, colonists explicitly stated that it was their purpose to "build a city on a hill" and to set an example of spirituality and righteousness. Today, it seems, most concern for the future centers on building wealth, great cities, and technological prowess – in other words, on maximizing economic growth. Both of these two broad aspects relate to the human heritage – the spiritual and the economic – portion of our bequest. North American nations have, in my view correctly, tried to minimize the role of their government and public actions in shaping the spiritual aspect of intergenerational bequests. To the extent that they have succeeded, the religious bequest has become mainly a matter for the private sector. Accordingly, most writers who have examined the question of the nature of our publicly provided

From V. A. Sharpe, B. Norton, and S. D. Donnelly, eds., *Wolves and Human Communities* (Covelo, CA: Island Press, 2001). Reprinted with permission.

493

bequest – what we should empower our governments to help us accomplish as a society – have focused mainly on the economic and utilitarian aspects of the bequest. That is, they have asked, What represents a fair savings rate? Each generation should add to, or at least protect, the economic capital base it inherited, it is argued. One view, widely accepted among mainstream welfare economists, claims that this requirement exhausts our obligations to the future – that as long as we leave it as well off as we are, economically, we will have fulfilled all of our obligations to the future.

But one of the questions posed in this book – whether our generation might have an obligation, or accept an obligation, to reintroduce wolves into the Adirondack Mountains – may not have an easily calculable economic effect. It has to do with not the economic but rather the biological bequest we leave for future generations. Unlike the methodology of economics, which measures all impacts of decisions in the single currency of units of utility, biology is a science of the particular. In the first section of this chapter, to pursue the idea that our bequest to the future may need to include some specific biological entities, I take issue with economists who believe that the question of a fair bequest can be resolved at the general, nonspecific level of calculating a fair savings rate. In the second section, I seek a more biologically sensitive criterion for determining what our biological bequest should be, and in the third section I consider the exact nature of the resource that is referred to when we place a high value on biodiversity. The fourth section is devoted to clarifying the question of costs and how we should think about determining when the costs of an ecological restoration are bearable. Finally, I discuss how the decision of whether to reintroduce wolves into the Adirondacks can be addressed within an open, democratic process, a process in which free people might rationally decide to accept a commitment to reestablish a lost population as an important expression of their legacy to the future.

### A QUESTION OF ECONOMICS?

Let us start by asking how far we can go by casting the problem of wolf reintroduction in economic terms. As noted earlier, most economists and many philosophers believe that the question of what we owe the future is reducible to the question of what constitutes a fair savings rate. That is, each generation must invest enough and wisely enough to expand, or at least not reduce, the total capital of the society. There are some obvious economic impacts of wolf reintroduction. Yellowstone Park and surrounding tourist facilities have experienced a surge of tourists interested in wolves; in addition, there are monetary costs associated with managing the wolf population. The point is not to deny an economic aspect of the decision but rather to examine whether an economic analysis can claim to be reasonably comprehensive with respect to the social values at stake when one generation chooses a bequest for its successors.

The key issue facing economic analysts is that of substitutability between natural and human-built resources (Chapter 13, this volume). In general, mainstream welfare economists have adopted a strong substitutability principle, assuming or arguing that resource shortages are unlikely to affect prices because new technologies allow the development of adequate substitutes for depleted resources. Furthermore, they have taken this high degree of substitutability to imply that our decisions about what we owe the future can be guided by principles of fair savings and prudent investment (Solow 1993; Beckerman 1994). They have not been persuaded by arguments that we should differentiate our bequest to ensure that natural as well as human-created capital exists in sufficient quantities. For them, creating a fair bequest package is not a question of *what* to save (because resources are largely interchangeable) but *how much* to save (with the caveat that we make good investments and not squander the capital accumulated by our ancestors).

According to the methods and assumptions of welfare economics, biodiversity should be valued much as we would value any other commodity, but in this case we are purchasing a commodity in the present that will deliver much of its value in the future. To the extent that having wolves in the Northeast represents a significant biodiversity value, we must evaluate its impact on future generations as well as the present generation. The economist will tell us that we should protect as much biodiversity as consumers in the present are willing to pay for, and no more. Furthermore, because the basic unit of value in the economic system is welfare, represented as consumers' willingness-to-pay (WTP) for a given product, any obligation we have to future generations is at most an obligation not to reduce their income (their ability to purchase commodities they judge to improve their welfare). In this economic approach to valuing biodiversity, reintroducing the wolf may prove to have social value if, for example, it leads (as in the Yellowstone reintroduction) to an increase in tourism and economic activity in the area. Or if the wolf were to reduce problems with coyotes, driving them out of some of their recently colonized habitats and reducing their populations, or reestablish some other important ecosystem function, it would be possible to attribute to the newly reestablished wolf population a value sometimes called an ecosystem service (Dailey 1997).

Economists also recognize another type of value that they call, somewhat nebulously, existence value. This is the value a person expresses through WTP to maintain or reestablish a species simply to know that it exists. In their research, economists have found that respondents express significant WTP to protect such existence values, and this WTP is widely distributed in the American public. So one more reason persons might state a WTP to reintroduce wolves, even in competition with other goods they might seek, would be to ensure that future generations will be able to experience wolves and the more whole ecological system that includes wolves. Taking these less obvious values into account as affecting present consumers' WTP for wolf reintroduction, economists in effect reduce concern for the future to a matter of a voluntary, altruistic consumer choice to invest some present income for the benefit of the future (Norton 1991a; Chapters 11 and 13, this volume).

This approach to sustainability reduces to what is sometimes called weak sustainability, the doctrine that all we owe the future is the opportunity to be as well off in achieving welfare as we are. In a sense, weak sustainability is the conceptual flip side of strong substitutability. The more one emphasizes strong substitutability between resources as one structures the bequest, the less one will put values on particular items and the more emphasis can be placed on simply making wise investments in purely economic terms. On this thinking, it can be no affront to the future if we use up particular resources, provided we save a sufficient portion of the profits to maintain a nondeclining stock of general capital. According to weak sustainability theory, then, we are obligated only to avoid improverishing the future economically. Because of fungibility of resources and changing technologies, we cannot identify any particular elements of nature that we are obligated to save. If weak sustainability represents a complete and adequate picture of intergenerational obligations, then it follows that the bequest package one generation owes the next is nonspecific and unstructured. Economic analysis, so used, only measures and compares aggregated wealth acquired and maintained across time, reducing obligations across generations to one of maintaining a fair savings rate.

Why are economists and many philosophical utilitarians convinced by this simplification of the question of what we owe the future? The most common argument in the literature (see, for example, Passmore, 1974; Solow 1993); for reducing obligations to the future to determining a fair savings rate rests on ignorance of future preferences (henceforth called the argument from ignorance). It is argued that if we are to determine what, exactly, we should save for the future, we would have to predict or correctly guess what people of the future will want and need. In the words of Nobel laureate economist Robert Solow, "If we try to look far ahead, as presumably we ought to do if we are to obey the injunction to sustainability, we realize the tastes, the preferences, of future generations are something we don't know about" (Solow 1993, 181). Invoking an unquestioned axiom of ethical reasoning – that one cannot be obligated to do the impossible – Solow concludes, because of the ignorance argument, that we cannot be obligated to save particular things that the future might want or need because we cannot predict what they will want or need. This outcome does not bother Solow, however; after noting that economists believe that resources can "take the place of each other," he boldly concludes, "There is no specific object that the goal of sustainability, the obligation of sustainability, requires us to leave [for the future]" (Solow 1993, 181). Maintaining a fair savings rate, he believes, will fulfill any obligations to the future.

In general, economic methods of valuation do not have obvious applications to choices designed to affect ecological function. In his authoritative survey of economic valuation theories and models, Myrick Freeman (1993) agrees with this assessment. At least as currently used, economic methods of valuation have no available methods for evaluating contributory and other ecosystem values. Worse, for reasons explained later in this chapter, there is no apparent extension of standard economic methods to encompass these values. It therefore misses the very important contributory value: the value they confer in keeping the system functioning and contributing the support of other necessary species. The economic approach apparently places no special value on species or ecosystems, trusting their protection or restoration to be a matter for consumer choice. As Freeman says, "If there is no link between [an] organism and human production or consumption activity, there is no basis for establishing an economic value. Those species that lie completely outside of the economic system also are beyond the reach of the economic rubric for establishing value. . . . Rather than introduce some arbitrary or biased method for imputing a value to such organisms, I prefer to be honest about the limitations of the economic approach to determining values. This means that we should acknowledge that certain ecological effects are not commensurable with economic effects measured in dollars. Where trade-offs between commensurable magnitudes are involved, choices must be made through the political system" (1993, 300).

Every species has a value not just as an entity that might be exploited or worshipped; it also has a value that is important to, but lost in the functioning of the system at a larger scale (Norton 1987; Chapter 23, this volume). Aldo Leopold said it best 60 years ago:

The emergence of ecology has placed the economic biologist in a peculiar dilemma: with one hand he points out the accumulated findings of his search for utility, or lack of utility, in this or that species: with the other he lifts the veil from a biota so complex, so conditioned by interwoven cooperations and competitions, that no man can say where utility begins or ends.... The only sure conclusion is that the biota is useful and biota includes not only plants and animals, but soils and waters as well. (1939, 727)

It is perhaps useful to dwell briefly on this limitation of economic valuation because it illustrates the quandary facing us in evaluating the proposed wolf reintroduction. The limitation in question is that, as noted earlier, economists have assumed a very liberal approach to substitutability between resources and substitutability between natural and human-made capital. This is a substitutability of commodities for each other, however, and the economic conceptualization of environmental value as a collection of commodities suggests no obvious method for counting contributory values of the sort Leopold identified. These contributory values seem to be very important in the long term, affecting future generations significantly, insofar as indirect advantages (such as the improved fitness of deer populations as a result of predation by wolves on the weaker members of the herds) appear and have significance only over multiple generations. This and many other contributory values are important to future generations, but it is difficult to see how they can be measured and accounted for in a model of intergenerational fairness that considers only values reflected in current prices or even in current expressions of WTP.

Solow believes that we have an obligation to be fair to future generations and that we fulfill this obligation by maintaining a fair savings rate, avoiding the impoverishment of the future. How far can this model of intergenerational obligations carry us in specifying what we should save for the future? To maintain a fair savings rate efficiently, Solow could argue, we are implicitly obligated to make good investments with the portion of our income that is set aside as a contribution to fair savings for the future. And if any given element or process of biodiversity has high contributory value, it seems to follow that we should invest in it. Solow thus accepts, by implication, an obligation of present consumers to invest wisely for the future, and if a wolf reintroduction is a wise investment, people in the present should make this investment as a contribution to savings – to wealth creation. If economic actors in the present choose their investments wisely, Solow could say, then today's prices should reflect in present dollars the future value of specific objects such as elements and processes of biodiversity. As an economist who relies on market analysis, Solow seems on strong ground here because it is an assumption of market-based economic analysis that consumers act with full knowledge of the consequences of their actions, so we should be able to expect today's consumers to make the investments that will maintain capital efficiently and avoid impoverishing the future.

So far, this position seems compatible with a protectionist attitude towards biodiversity and nature: Present markets reflect future values, so our obligation is to protect for the future the elements and processes that hold open the most and the best options for future consumers. Solow faces a serious quandary, however, that is exacerbated by his appeal to ignorance as a reason to limit our concerns to maintaining a fair savings rate in the first place (Norgaard 1990). He cannot, on the one hand, claim that prices correctly reflect the future value of increments in biodiversity protection (on the grounds that today's informed consumers' choices reflect future values) and, on the other hand, claim that we are too ignorant of future needs and wants (demands) to define specific obligations to save specific elements and processes of biodiversity. In doing so, he would be implicitly imputing to today's consumers knowledge that he has claimed nobody can have. Either present prices reflect future demands and, therefore, prices provide a good guide to investments in our bequest for the future, or they do not. If they do, then Solow is necessarily wrong in saying that we do not, and cannot, know what the future will want or need because it is precisely this knowledge consumers will need to predict what will be good investments if they are to maintain a fair savings rate for the benefit of the future. If today's prices do not reflect future values, however, then today's prices cannot be considered a good guide in the present to investments for the long-term future.[1]

Either way, we must conclude that there is information that is essential to know, information that is not reflected in analyses of present prices and fair savings rates. Weak sustainability cannot, without supplementation, without biological information and information about future values, guide us toward good investments in the well-being of the future in policy areas such as biodiversity protection and restoration.

To summarize the argument so far, on the economists' assumption of ignorance of future demands and full substitutability between resources, there is one and only one way in which the present can harm the future, and that is by impoverishing future consumers by spending more than we produce. But this seems too narrow a concern in cases such as restoring particular species because it seems reasonable to posit a closer relationship between maintaining a diverse habitat as our environment and maintaining real options that will be valued and valuable in the future. If we want to protect the future from loss of indirect but highly significant values such as the contributory values of species and processes, we must supplement economic valuation with noneconomic principles and reasoning. Despite the apparent inadequacy of weak sustainability to provide a comprehensive accounting of the requirements for fairness to future generations, it nevertheless seems reasonable to consider weak, or economic, sustainability to be a necessary condition for sustainable living. But the argument of this section shows that if we want to articulate a more structured and specific set of obligations to guide the formation of a more structured and biologically sensitive bequest package, we must consider weak sustainability as only a necessary but not sufficient conceptualization of the intergenerational trust.

## AN ECOLOGICALLY INFORMED SAFE MINIMUM STANDARD CRITERION

Some environmental economists have recognized that weak sustainability based on rational, individual consumer choice does not go far enough in providing a presumption in favor of saving specific productive resources such as biodiversity. A number of respected economists and others have advocated an important modification of cost-benefit accounting (CBA) by applying the safe minimum standard (SMS) rule of conservation. This rule says that a productive resource should be saved if the costs are bearable (Ciriacy-Wantrup 1968; Bishop 1978; Norton 1991b; Farmer and Randall 1997). In effect, this modification assumes that the resource is worth saving (thereby avoiding the need for a benefits analysis) and focuses analytic attention on the cost side of the ledger, encouraging a search for affordable means to an unquestioned goal. In effect, this modification shifts the burden of proof to those who would eliminate a productive resource. To sacrifice a species, for example, SMS advocates believe one must make a case that protecting the species will impose unbearable costs on society. This modification is an important step in the right direction, as I have argued elsewhere (Norton 1987, 1991b).

One advantage of the principle is that although the particular term *SMS* has emerged from the discipline of resource economics, it has an important rough analogue in political jargon, especially in Europe and Australia, in the idea of a precautionary principle. The precautionary principle states that in situations of high risk and high uncertainty, always choose the lowest-risk option. I prefer the SMS criterion to the precautionary principle because the SMS provides explicit direction in determining when risks should be avoided. SMS instructs us that in situations of high uncertainty and risk of irreversible impacts such as species extinctions, adequate steps should be taken to protect the vulnerable resource if the costs of protection are bearable.

It cannot be denied, however, that the SMS criterion suffers from vagueness in both of its crucial terms: *resource* and *bearable costs*. How can we decide what we should save or restore? Of course, one can say, "Save everything! Restore everything!"[2] But that policy is not possible without unbearable costs, not to mention the fact that, taken literally, this injunction would freeze nature unnaturally. So, if we are to apply the promising SMS criterion, we must specify which particular resources should be given priority in protection and recovery programs.

The rule that we should save a resource is especially vague in the case of biodiversity because one could emphasize individuals, populations, species, and ecosystems at several scales.

Can we give biological or ethical reasons to determine what to save and what to restore? Can we give reasons that are clear and convincing enough to persuade the people of New York that they should take steps to reestablish populations of wolves in feasible wilderness areas? Can biology provide some general guidelines as to what should be in our bequest? The question of what to save as a biological bequest is addressed in the next section; the question of how to clarify "unbearable costs" is the subject of the final section. Taken together, the arguments of these two sections should allow us to apply the SMS criterion with greater precision.

## BIODIVERSITY AND RESOURCES

If the reader will permit the interjection of a personal note, I have been puz-zling over, reading a lot about, attending conferences on, and writing a good bit about the value of biodiversity since 1980. I thought I had heard it all. About 2 years ago I read an article, before its publication, that changed the way I think about this much-discussed topic. The author of the article, Paul Wood, first departs from usual practice and distinguishes sharply between biodiversity and biological resources. According to Wood, biodiversity sim-ply equals "the differences among biological entities" (Wood 1997, 254). He then makes a persuasive case that it leads only to confusion to consider bio-diversity to be a resource among other resources and to trade off elements of biodiversity against other resources. Biodiversity, he argues, is not a re-source among others, but a generator – a source – of biological resources. As such, biodiversity is a necessary condition of enjoying biological resources, especially over extended time. To use an analogy from the field of economics, biodiversity should be valued on the higher logical plane on which economists and some policy analysts value free markets as generators of value. Just as economists would find it odd to be asked, "What is the market value of a free market?", ecologists may find it odd to think of biodiversity as a commodity. Free markets, on the theory of the famous invisible hand by which individuals seeking their own interest create an efficient system for all, generate value. Similarly, advocates of biodiversity do not conceive biodiversity as one resource among others, to be traded off to get an "optimal mix" of biological and other resources, but rather as a source that, in the long run, is a necessary precondition for the system to generate more biological resources. This generative process exists on at least two levels. On one level, diverse ecological processes provide diverse outputs, including standing resources of many types and a variety of ecological services. But on a longer scale, diversity of biological processes, especially at the ecosystem scale, is also responsible for generating and maintaining diversity itself.

To see the importance of diverse biological processes, it is useful to examine, briefly, an often-used working definition of biodiversity. Biodiversity is often defined as the sum total of all diversity, including the genetic diversity within species, the diversity of species, and the diversity of habitats within which species live. Though perhaps useful as a simple definition for the purposes of "accounting," this definition, by listing the diversity of entities at important biological levels, does not sufficiently emphasize the importance of diverse processes because it concentrates on the collection of diversity at a single point in time. If one thinks of biodiversity not as a simple sum of existing variety but as a process that generates and sustains multiple evolutionary regimens, and hence creates greater variety across time, biodiversity is a self-generating and self-sustaining force, as close as nature has come to the much-sought *perpetuum mobile*. Systems that are diverse provide more opportunities for species to create and evolve into new niches. Diversity creates more diversity (Whittaker 1969; Levin 1981; Norton 1987; Wood 1997). Conversely, losses in diversity eventually lead to more losses, as species that depend on lost species are themselves extirpated, and a cascade of losses is entrained. Adding a time element to our thinking about biodiversity, in this sense, shifts us away from thinking about biodiversity as a list of entities or resources and toward an emphasis on the creative, self-sustaining nature of diverse ecological systems.

To continue with this level-oriented or scale-sensitive treatment of biodiversity, we can now examine the relationship between multiple evolutionary processes at different scales and the traditional understanding of types of biodiversity. Writing in the 1960s and 1970s, R. H. Whittaker (1969) usefully separated three types of diversity in ecological systems. First, there is the diversity of species, sometimes called richness, in any given habitat. On this level, for example, we know that tropical rainforest habitats are much more diverse than are tundra or even temperate habitats. But the diversity within habitats probably is less important to generating and maintaining diversity than is the diversity of habitats that exist in an area. This cross-habitat type of variety is important in providing multiple opportunities for species to adapt and create new niches by providing in close proximity a variety of alternative habitats, which encourages opportunities and thus more variance. If a species can develop a new survival trait, it may be able to colonize a new type of habitat and, once it has colonized a new habitat, the species is submitted to a whole new set of evolutionary pressures and may adapt in quite new directions. In the exceptional case, which usually occurs when the new population is largely isolated from its earlier habitat, a new species may emerge. The total diversity of an area at any given time is a product of within-habitat diversity and cross-habitat diversity. But total diversity over time is a function of the diverse processes at all scales because these diverse processes provide and maintain opportunities for diverse species and populations to survive and adapt across time. Habitat diversity provides diverse options for survival, and each of these diverse options may represent an opportunity for some organism with a genetic or behavioral quirk. Similarly, this occurs for populations.

It is now possible to explain, in broad biological terms, why a wolf reintroduction in the Adirondacks would represent a high-impact effect on the biodiversity of North America. Because most areas of the continental United States no longer have wolves as their top predator, the Adirondack system would be unusual in this respect; because the top predator exerts evolutionary pressure that is felt throughout the community, a variety of old and new relationships will be reestablished or established. At least in Yellowstone, the return of wolves apparently has caused a greater diversity of wildlife (Robbins 1997), and a similar impact may occur in the Adirondacks. However that comes out, the Adirondacks take on added biological interest in their own right in that they increase their difference from other systems of the northeastern United States and thus contribute to the total diversity of the region.

I believe that the goal of biodiversity management at the continental level should be to maintain and restore the greatest authentic total diversity possible (Norton 1987). Calling the goal "authentic" total diversity adds to the idea of maximizing total diversity the additional requirement that species counted must either have existed historically or could reasonably be expected to colonize without the aid of human managers. This requirement rules out augmenting diversity by importing exotics, emphasizing that the maintenance of diversity should favor traditional species because these have evolved to have competitors and symbionts in the habitat.

To be biologically precise, the resource to be protected should be the source of biological resources lodged in the differences between biological entities at all levels of the biological and evolutionary hierarchies. The policy goal should be to protect biodiversity, understood as the sum of all biological differences at all levels and scales, including especially the diverse processes that underlie, maintain, and augment the diversity spiral. These differences between entities, including differences in the evolutionary trajectories they embody over time, sustain diverse processes and diverse entities that, as Wood argues, are the source of biological resources.

Reintroducing wolves, the authentic top predator for the region, to a special area such as the Adirondacks would reestablish processes that support differences between entities in the Adirondack system. Looking at the larger scale of the eastern United States, total diversity likewise will be augmented and sustained. The existence of a more complex and authentic area within the larger region is sure to expand the number of evolutionary trajectories for many species, creating more diversity at every scale of the system. This increase in diverse evolutionary trajectories leads to greater and greater cross-habitat diversity, including especially an augmentation of cross-habitat diversity as a potent driver of the processes creating more total diversity. Diversity creates more diversity. Losses of diversity lead to future losses. This self-sustaining and self-creating process is ultimately the source of all biological resources.

I began this section by asking what exactly is the resource in question in the SMS criterion "save the resource, provided the costs are bearable," as applied to biodiversity. The answer, I have argued, is that the resource in question is the sum total of differences between biological entities at all levels and that the sum includes, very importantly, differences between ecosystem processes because these ultimately are the sources that must be saved. These processes create and sustain both the differences themselves and the flow of biological resources that are necessary to fulfill human needs. But even this may command us to save too much; it is no doubt impossible to save or restore all biological differences. Why wolves? Why in the Adirondacks?

My answer is that reintroducing an authentic top predator to a unique habitat such as the Adirondack Mountains will greatly multiply cross-habitat differences in the larger region. This particular reintroduction would recreate an old evolutionary dynamic, diversify the evolutionary regimens faced by many species in the eastern United States, and create a habitat that exists almost nowhere else in the region. If one wants to maximize our bequest of biodiversity, understood as the biological differences that are the source of biological resources, the best investment available would be to restore the authentic top predator to an area within a larger region where it once existed, regaining lost cross-habitat diversity and spinning the flywheel that creates and maintains diversity over time.

INTERPRETING "UNBEARABLE COSTS"

Though helpful in determining the target of the effort of SMS advocates to save the resource, these considerations do not fully resolve the ambiguity of the other key phrase of the SMS criterion, "unbearable cost to society." The argument that wolves can recreate lost cross-habitat diversity shows that a great social value is at issue, but it remains a difficult question: How might the supporters of a reintroduction in the Adirondacks make a case to the residents of the park that diversity is valuable to them and to all humanity and that the costs of such a reintroduction are bearable, given the benefits to be expected?

The first and in some ways most important answer to these questions is that they will demand political answers. In a democracy, the political will necessary to undertake a restoration must emerge as a legitimate expression of the community's will. In the case in point, the decision to reintroduce wolves must have strong support at every political scale, from local to statewide, especially including residents of the park, if it is to succeed. We have recognized a significant gain to the future – the augmentation of cross-scale diversity – but we can also recognize that protecting this increased value (which will emerge over generations) might be costly to at least some members of the present generation. But now the advantages of using a scale-sensitive model for expressing values may help. It may be possible that the biological bequest can in this case be protected with little cost to the future if we can identify independent dynamics that generate these values (Chapter 16, this volume). It may be possible, with creative planning and policy, to create a win–win situation by finding a path toward community development that will generate short-term economic growth in a way that is compatible with protecting and augmenting biodiversity.

I do not know whether the political will to reintroduce wolves exists at these various levels, nor do I know whether it could be created or developed. This is yet another fascinating question, but one that can be answered only by observing the political process unfold over the next years and decades. Rather than speculate about what will happen, a question that can only be answered empirically as the future unfolds, I will ask another question: On what scientific and moral basis might a human community, through a process of deliberation and participation, decide to accept a moral obligation to protect

505

and augment authentic biodiversity? In answering this question, I will use my remaining space to explain why diversity can be expected to provide social value, even if uncertainty and ignorance of future preferences make it difficult to identify particular uses or commodities of units of biodiversity. If the residents of the park agree that wolves, as top predators, are an important aspect of the diversity of wild places, then my addition of an explanation of how diversity can be expected to have social value should strengthen the argument that a reintroduction, if biologically feasible, should be a social priority.

Again, it is helpful to keep in mind the multiscalar nature of biodiversity and the values it supports. By arguing that reintroducing wolves to the Adirondacks increases the cross-habitat and thus the total diversity of the eastern United States, it becomes clear that one important social value affected by the proposed reintroduction – protecting total diversity for future generations – is manifest at a larger, regional and national scale. This recognition also helps us understand the political complexities of the wolf reintroduction process. Local people, who look mainly at values to their community, might conclude that from their point of view, the reintroduction is not worth the costs and risks involved. But when the multiscalar nature of biodiversity is understood, it is also possible to see that from the larger scale of the region or nation, existence of wolves in this unique habitat may have enough value to outweigh the loss and risks to the local community.

This conceptualization of the problem suggests that, to the extent that augmenting diversity is a social or public good, it ought not to be imposed on individual property owners, and the society may have an obligation, based on intragenerational fairness, to compensate property owners and the communities that may suffer losses as a result of the restoration effort. The Defenders of Wildlife compensation fund is one such method. If the greatest values of biodiversity are experienced over generations and at a national scale, the costs ought not to be inordinately heavy at the local scale. It may be necessary to provide compensation in some form to local communities or some groups that may suffer inordinate losses. Let us then assume that for the most serious losses suffered at a local level, a compensation plan that is fair and acceptable to local communities could be worked out in a process of negotiation.

Having acknowledged that some compensation may be due landowners and residents if they are supporting a public value, this places the local communities at the center of the decision process, a process that is embedded in a comprehensive, multiscale examination and evaluation of the reintroduction at regional and state levels. Assuming that intragenerational equity concerns can be addressed and that the feasibility studies are positive, we can now concentrate on characterizing the long-term benefits of reintroducing wolves to the Adirondacks and the costs to the future if we fail to make this investment when we still have a chance for it to succeed. We can at last ask whether the biological importance to future generations of augmented diversity, understood as total diversity, justifies our generation in including restored wolves in the Adirondacks as an important part of our bequest to future generations.

But immediately upon asking this central question, we encounter our old nemesis, ignorance about the preference of future people. The question of specifying bearable costs turns our attention to an analysis of benefits, but how can we understand benefits to future people if we don't know what they will want or need? One thing we can say, based simply on our analysis that favored the SMS over CBA as an appropriate decision tool, is that we are not as ignorant of the preferences of the future as Solow suggests. It seems obvious that people of the future will not want to discover toxic time bombs in their waste repositories, and they will prefer a diverse and opulent world of living things to a sterile, dead planet. In fact, we have a pretty good idea what will be valuable – and what disastrous – for future people and future communities.

Once again, our understanding of biodiversity as the differences between biological entities and processes provides guidance. We value differences and the varied biological entities and processes that embody and support them because these differences support and enhance human options in the future. Thus, although we probably lack specific guidance about wants and needs, we do know that people of the future – assuming they will value free choice as we do – will want to have options and opportunities available. In addition to impoverishing the future, we could be unfair to them if we unjustifiably reduced their options, their range of choices. We should protect differences between biological entities when, and to the extent to which, they are necessary to hold open options that may prove important to people in the future. Although much more must be said to clarify this point, I am suggesting that the SMS can be supplemented and made more biologically specific by relating biological differences to the social value of maintaining freedom, which is accomplished by protecting resources associated with important options and opportunities (Norton 1999). The SMS might stand in addition to the requirements of weak sustainability as a second necessary condition of sustainable living, especially applicable to decisions that strongly affect the future in irreversible ways.

We have articulated, in broad outlines, a two-criteria basis on which we can evaluate possible options and policies with respect to their sustainability and fairness to future generations. To be fair to the future, our actions today must avoid impoverishing the future (as measured by economists' weak sustainability notion of simply maintaining a fair savings rate) and avoid destroying options that may prove important in the future. I am hypothesizing, then, that a general and conceptual connection holds between protecting biological differences across generations and maintaining freedom of choice.

It is tempting to say that saving options for the future is valuable no matter what the future turns out to want or need, and in one sense this is true (to the extent that options support freedom of choice, they are generally valued); however, we may still not know which options are worth investing in, given that we probably will not be able to protect or reestablish all authentic biological differences. This limitation is not debilitating to our use of the biologically informed SMS in the case of the wolf reintroduction, however, because we have a strong biological reason to favor this particular reintroduction. This reintroduction would recreate a process difference that supports and reestablishes many other differences in competitive regimes between species and hence would have a large net positive impact on differences and, accordingly, on options and choices of future humans.

However, it may be more difficult to provide a general account of which differences to maximize if the goal is to maintain important specific options for the future. Here, our theory may be silent. Beyond emphasizing the importance of protecting sources of biological differences – differences that clearly have high contributory value – it cannot provide specific guidance about all the options that should or must be saved for future generations if we are to be fair to them. Beyond a general value placed on contributory values, it would still be helpful to know which options we should protect for the future.

Once the obligations to protect major sources of differences is fulfilled, it will be a matter of deciding, in each specific geographic place, which options and which associated biological differences should receive priority efforts. To decide specifically which biological differences to protect in a particular place, we seem once again to be stymied by the problem of ignorance. What will future residents of this place want? What should we save for them? As a way through the ignorance-of-future-preferences impasse, I propose a bold solution, a solution that shifts the local decision context radically. Let us not ask, What will the people in the future prefer? Let us instead ask, What values and opportunities are community members in this place willing to commit to and work to protect? In addition to the weak sustainability obligation to not impoverish the future, I am suggesting that the second requirement of intergenerational fairness is to identify and protect the differences and processes that represent the values of the community in question.

In effect, this solution is to identify a sociocultural process by which communities might identify important options, options associated with their very sense of the place they call home. We can call these values constitutive values, and the options and opportunities associated with them are constitutive options. Constitutive values for people are values such that, if they were lost, the person in question would no longer call that place home and would judge that its integrity had been compromised. Suppose we could describe a public process that truly engages the public, at a given geographic place, in an iterative and ongoing search for a sense of place and associated values, and that a consensus emerges about certain values as constitutive values. Suppose, further, that this process included all involved stakeholders, allowed open discussion, and resulted in a set of values that are widely thougtht to be constitutive of the community and its character. Accordingly, there could be public discussion of the particular options and the associated biological differences that give distinctiveness and meaning to the place. If these steps were all completed, then I think one might consider these constitutive values, accepted through democratic processes, as the democratically expressed positive values that tie the community to the specific place.

All of these exercises, I submit, can be thought of as exercises in community self-definition. This approach owes much to the conservative philosophy of Edmund Burke (1790, 93–94), who stated that a society is "a partnership not only between those who are living, but between those who are living, those who are dead, and those to be born." What we add to Burke's historical conservatism with respect to institutions is an ecological sense of the ongoing, multigenerational community as anchored in a physical place. If a community accepts responsibility for a place and for the values that serve as threads weaving the multigenerational tapestry of life and community into a single distinctive culture and community in that place, then it would be reasonable for the people of that community to protect and preserve the constitutive links between their culture and their physical place.

To operationalize this advice, a community, mindful of its natural and cultural history, might pose the question as one of choosing a development path that is fair to the future. Here, we understand a development path as a way in which a community would develop from a given point in time. So far, we have proposed two general criteria by which we can judge the fairness of the proposed development paths. We can then evaluate development paths according to these general criteria: a criterion of economic growth and fair savings (measured as short-term economic impacts on the total capital base of the society, or weak sustainability) and a second indicator or suite of indicators that are designed, after public discussions of values and associated options, to track democratically accepted and supported constitutive values. Because the weak sustainability criterion tracks individual decisions conditioned by today's market, this criterion reflects today's prices as they are set in today's markets. The second criterion, which demands that we protect and enhance authentic biological and ecological differences, applies at a multigenerational scale and tracks the ecological and cultural character of the place over the longer run. I have argued that the added diversity and unique wildness reestablished by a wolf reintroduction would maintain and enhance important opportunities for future generations. These long-term values should be treated on the model of a trust by which each generation invests in a series of community-defining decisions, decisions that will determine the identity and character of the place the community inhabits. Again, the emphasis on protecting options applies at multiple levels, and some system-level decisions and impacts (such as extirpations and reintroductions of top predators) may be very important to the future.

Communities in the Adirondacks, the State of New York, and the larger communities of New England, on this line of reasoning, are obligated to reintroduce wolves if and only if this policy emerges from a process of community self-definition. I see the role of the Adirondack Citizen's Advisory Committee as exemplifying and perhaps directing this process.

Much has been written about the "science" of sustainability – and surely science has a role in the search for sustainability – but my line of reasoning suggests that the most important aspect of identifying a sustainable development path is an act of community commitment. One might say, to borrow a useful term from philosopher J. L. Austin, that what is involved is a performative act, an act of commitment by a community that is willing to embrace and project certain values into the future. Austin notes that many apparent acts of description actually function as performatives. As examples, Austin mentions "I do" (when uttered in the context of a marriage ceremony), "I name this ship the *Queen Elizabeth*" (while striking the ship's bow with a bottle of champagne), and "I bequeath this watch to my brother" (in the context of a will). He then says that in these examples, to utter the sentence in question (in the appropriate circumstances) is not to describe the doing of what is being done, but rather to do it (Austin 1962, 5–7). He proposes that we characterize such uses of language as performatives and mentions that they can be of many types, including contractual and declaratory. Later, Austin says, "A great many of the acts which fall within the province of Ethics . . . have the general character, in whole or in part, of conventional or ritual acts" (Austin 1962, 19–20).

Applying Austin's idea, and based on the analysis of this paper, a new way of thinking about intergenerational morality and sustainable living emerges. If we see the problem as one of a community making choices and articulating moral principles – a question of which moral values the community is willing to commit itself to – then the problems of ignorance about the future become less obtrusive. The question at issue is a question about the present; it is a question of whether the community will take responsibility for the long-term impacts of its actions and whether the community has the collective moral will to create a community that represents a distinct expression of the nature–culture dialectic as it emerges in a place. Will they rationally choose and implement a bequest – a trust or legacy – that they will pass on to future generations? This approach bypasses the problem of ignorance because we need not ask what the future will want or need. Rather, we ask by what process, and on what grounds, a community might specify its legacy for the future. If one wants to study such questions empirically, important information is available. For example, one might study how communities engaged in watershed management processes or community-based watershed management plans achieve (or fail to achieve) consensus on environmental goals and policies. Although empirical studies such as these may contribute to the process of community-based environmental management, I am suggesting that the foundations of a stronger sustainability commitment lie more in the community's articulated moral commitments to the past and to the future than in any description of welfare expectations or outcomes.

This basic point makes all the difference in the way information is used to define sustainability, and it changes the way we should think about environmental values and valuation. So we face the prior task – and I admit it is a difficult and complex one – of developing community processes by which democratically governed communities can, through the voices of their members, explore their common values and differences and choose which places and key values will be saved, achieving as much consensus as possible, and continuing debate about differences. These commitments, made by earlier generations, represent the voluntary, morally motivated contribution of the earlier generation to the ongoing community. The specification of a legacy, or bequest, for the future must ultimately be a political problem, to be determined in political arenas. To choose to protect wolves, on this view, would be to articulate and, through a group process, affirm certain values as constitutive of the distinct and unusual wildness of the Adirondacks.

We must also realize that this choice need not conflict with good policies for economic development. If my argument that specific dynamics are unfolding at different scales is correct, it may be possible to turn controversy into consensus by showing that a reintroduction policy will have positive economic impacts as tourists visit the area and create economic opportunities for entrepreneurs, providing short-term benefits that increase the standard of living of citizens of the Adirondacks and at the same time set in motion the reestablishment of processes that will augment and sustain diversity. The policy would then yield benefits on economic, utilitarian criteria (manifest in the short run) while representing an investment in the options and opportunities that will be available to future people. I conclude that if all of these conditions were fulfilled and if a multilevel political process were to lead to a consensus to protect the future's diverse biological heritage and the opportunities it supports, then our bequest to the future should include a restored population of eastern timber wolves in the Adirondacks.

### NOTES

1. This argument is an elaboration of reasoning first developed by Norgaard (1990).
2. See Howarth (1995) for an argument that the future has proprietary, moral rights in resources, which may imply a policy close to that of total preservation.

### REFERENCES

Austin, J. L. *How to Do Things with Words*. New York: Oxford University Press, 1962.
Bishop, R. "Endangered species and uncertainty: the economics of a safe minimum standard." *American Journal of Agricultural Economics* 60 (1978): 10–18.
Beckerman, W. "'Sustainable development': is it a useful concept?" *Environmental Values* 3 (1994): 191–209.
Burke, E. *Reflections on the Revolution in France*. London: Dent, 1790.
Ciriacy-Wantrup, S. V. *Resource Conservation: Economics and Politics*. Berkeley: University of California Division of Agricultural Sciences, 1968.
Dailey, G., ed. *Nature's Services*. Covelo, CA: Island Press, 1997.
Farmer, M. C. and A. Randall. "Policies for sustainability." *Land Economics* 73 (1997): 608.
Howarth, R. B. "Sustainability under uncertainty: A Kantian approach. *Land Economics* 71 (1995): 417–427.
Freeman, A. M., III. *The Measurement of Environmental and Resource Values: Theory and Methods*. Washington, DC: Resources for the Future, 1993.
Leopold, A. "A biotic view of land." *Journal of Forestry* 37 (1939): 727–730.
Levin, S. A. "Mechanisms for the generation and maintenance of diversity." In R. W. Hiorns and D. Cooke, eds., *The Mathematical Theory of the Dynamics of Biological Populations*. London: Academic Press, 1981, pp. 173–194.
Norgaard, R. "Economic indicators of resource scarcity: a critical essay." *Journal of Environmental Economics and Management* 19 (1990): 19–25.
Norton, B. G. *Why Preserve Natural Variety?* Princeton, NJ: Princeton University Press, 1987.
_____ "Thoreau's insect analogies: or, why environmentalists hate mainstream economists." *Environmental Ethics* 13 (1991a): 234–251.
_____ *Toward Unity among Environmentalists*. New York: Oxford University Press, 1991b.

_____ "Ecology and opportunity: intergenerational equity and sustainable options." In A. Dobson, ed. *Fairness and Futurity*. Oxford: Oxford University Press, 1999, pp. 118–150.

Passmore, J. *Mans Responsibility for Nature*. New York: Scribner, 1974.

Robbins, J. "In two years, wolves reshaped Yellowstone." *New York Times*, December 30, 1997, D1.

Solow, R. M. "Sustainability: an economist's perspective." In R. Dorfman and N. Dorfman, eds., *Economics of the Environment*, 3rd ed. W. W. Norton, 1993, pp. 179–187.

Whittaker, R. H. "Evolution of diversity in plant communities." In G. M. Woodwell and H. H. Smith, eds., *Diversity and Sustainability in Ecological Systems*. Brookhaven National Laboratory Publication No. 22. Springfield. VA: Clearinghouse for Federal Scientific and Technical Information, 1969, pp. 178–196.

Wood, P. "Biodiversity as the source of biological resources." *Environmental Values* 6 (1997): 251–268.

# 27

# Environmental Values and Adaptive Management

## with ANNE STEINEMANN

### INTRODUCTION

While it is a truism that environmental policy is ultimately driven by "social values," there is currently considerable confusion regarding how to understand and assess social values, and corresponding confusion regarding the role of values in the broader process of environmental policy formation, implementation, and management. In this paper, we set out to describe an alternative direction for environmental value studies, a process that emphasizes pluralism, participation, and iteration rather than just simple elicitation of preferences and preference aggregation. Our goal will be to provide an approach that will help to better understand environmental values, especially in public processes for environmental management.

Our approach to environmental values is based on theory, but our theory is not one of the usual ones, such as utilitarianism, or a theory that nature or its elements have "rights" or something like that. Ours is a theory about *process* rather than a theory about *ultimate values*. We take the view that, since we live in a diverse society – and neither hope nor expect that this will change – the problem is not to decide which theory of ultimate value is correct, but rather to design a process by which diverse societies – with many voices expressing many worldviews and ultimate values – can act in a way that will tend toward a working consensus on environmental policy decisions (Norton, 1991). In the context of our theory, it is useful to have a variety of ways of expressing values and a variety of ways of measuring values. We do not, initially, seek a universal currency, such as dollars or units of happiness by which all values can be expressed; instead, we seek a set of

From *Environmental Values* 10 (2001): 473–506. Reprinted by permission of the White Horse Press, Cambridge, UK.

indicators that expresses the values of the community as directly, clearly, and precisely as possible. In order to accomplish this goal, it will be necessary to go beyond one-time elicitations of the preferences of individual consumers, and to engage community members in a process of further clarification and integration of these values as a part of the search for democratically accepted management goals.

The goal of this paper is to add one important piece to a very complex puzzle – the gradual emergence of a new, more holistic understanding of environmental management, what is sometimes called "ecosystem management" or "adaptive management." We offer a general approach that encourages the development of a more comprehensive and systematic approach to identifying and measuring social values, an approach that is pluralistic in the modes of expression and measurement of environmental values. We seek, that is, a way to use the social sciences – including economics but not limited to economics – in a broader public discourse about goals and objectives of management. At present, there seem to be only two ways to talk about environmental values – the relatively undisciplined discourse of ordinary language or the algorithmic (but incomplete) models of technical policy analysts such as risk assessors or microeconomists. We seek a more formal method for ascertaining public values than that of common, everyday discourse; we do not try to treat decisions about how to manage resources holistically as decidable within technical models such as cost–benefit analysis. We believe that it will be helpful to introduce some common conventions and procedures that will guide the evaluation of environmental changes – thereby improving upon public discourse – without shifting completely out of public discourse and into technical, computational approaches to counting value. Even the advocates of such technical, computational approaches admit that such approaches are unable to capture large-scale ecological values (Freeman, 1993), the very values adaptive managers and other holists should embrace as key to their management process. The middle ground we seek is that of a pluralistic theory that can be supplemented with a process heuristic intended to focus diverse communities on the right questions and an evaluative heuristic that guides communities to discuss various indicators. These heuristics can, we believe, guide a forum of people with diverse values to focus on what to measure and on what to protect – by appealing to their values – but to do so in a way that may allow people with differing values to choose mutually acceptable indicators and goals regarding those indicators. The heuristics, if successful, provide a link between pluralist theory and various procedural practices that may encourage the development of consensus and cooperative behaviors in community-based management processes.

Some background may help to place our goal in the larger context of management studies. More or less independently, advocates of ecological, holistic approaches to environmental management from several countries have developed local institutions and public processes to address local and regional environmental problems, problems that emerge at the level of larger ecological systems that function as habitats for human settlements and activities. Some theory, based in ecology and in the Leopoldian simile of learning to "think like a mountain" (Leopold, 1949), has been developed. Adaptive monitoring and management, developed in western Canada and incorporated into many management efforts in the United States, represents one "package" of theory and and suggested practices and guidelines (see, for example, Lee, 1993; Gunderson, Holling, and Light, 1995). Other practitioners employ similar or overlapping methods, without adopting the label "adaptive manager," so our emphasis on adaptive management can be thought of as more or less representative of a range of holistic, community-based environmental management projects. The premise of our paper is that these new approaches to management will require a new, more systematic approach to the evaluation of ecosystem-level environmental change.

We are aware that some of these projects have been studied by social scientists and that evaluations of such projects have been mixed (for reviews, see Cortner and Moote, 1994; McClain and Lee, 1996; also see Innes and Booher, 1999a, 1999b; Walters, 1997; Sabatier, 1998). Our purpose, however, is not to assess the success and failure of such projects empirically, but rather to take tentative steps toward a new approach to environmental values, an approach that is appropriate to public discourse in the context of an adaptive, ecosystem management process. Our approach combines elements of the two existing approaches. Evaluation is undertaken in ordinary public discourse, using a suite of technical devices – multiple measurable indicators – that employ measures that are not necessarily inter-definable or technically comparable. Given this understanding, environmental values and evaluations will be summarized and balanced in ordinary discourse, but the balancing will include a careful look at specific measurable indicators as useful technical guidance in the more political decision as to what to do. This careful look, however, will *not* be represented as an algorithmic aggregation of the multiple indicators and measures. This ordinary-discourse summation and political discussion concerns how to weigh and prioritize multiple measures as guided by heuristics which, while incapable of resolving substantive value questions by themselves, can guide an orderly public process of deliberation, summation, experiment, and revision. This pluralistic approach to ways of measuring environmental values simply recognizes that, in diverse modern democracies, multiple values are expressed in multiple vernaculars. So, public deliberation, while carried on in public, ordinary discourse, is gradually studded with technical devices that prove themselves useful in measuring and evaluating environmental change. Balancing of technical measures is itself not a technical measure; integration of these plural values is carried out in public discourse, but public discourse provides some guidance through heuristics.

The pluralism proposed here is motivated by methodological considerations (paper 14, this volume) and need not be understood as a doctrine about ultimate values. It is part of a broader experimental strategy that seeks first to express diverse values in multiple and perhaps incommensurable ways, and then seeks ways to organize and present those diverse goals as a starting point for a more holistic analysis. Here, the various "reductionist" ideas of moral and economic theory can contribute to the process; they serve as guides toward systematization and integration of values, and help us to formulate both consensus positions and disagreements more clearly. The search for successful environmental policies, on our broader approach, however, becomes a search for specific policies and practices that support multiple values, rather than an attempt to maximize a single variable such as economic efficiency or ecosystem preservation. This pluralistic, integrative approach focuses attention on the process whereby communities with diverse values articulate, discuss, revise, and reconcile competing values. In this way, it may be possible to create an environmental policy that protects many or most of the values that are articulated by community members, and to do so democratically (Burgess et al., 1988a; 1988b; 1998; Kemmis, 1990; Gundersen, 1995; Kempton et al., 1995; Morrison, 1995; Harrison et al., 1996; paper 19, this volume). Since this process approach encourages the expression of multiple values, and does not insist that these diverse values be expressed in a single measure, we advocate the use of more than one criterion applied within an iterative, adaptive system of management (see Glasser, 1995, for a review of multi-criteria systems).

Looking forward, in Part I, we consider the current trend toward what is called "adaptive management," exploring the following three core principles that articulate the basic approach of adaptive managers:

1. Experimentalism. Adaptive managers emphasize experimentalism within a dynamic system, recognizing that an ongoing search for knowledge is necessary to set and achieve environmental goals.
2. Multi-Scalar Analysis. Adaptive managers model and monitor natural systems on multiple scales of space and time.

3.  Place Sensitivity. Adaptive managers adopt local places, understood as humanly occupied geographic places, as the perspective from which multi-scalar management orients.

Having stated and explained these core principles in Part I, which is intended to set the broader context for the introduction of the new approach to valuation in adaptive management, we ask: What value theory, and what general approach to valuation studies, fit appropriately into adaptive management processes, such as watershed management plans or ecosystem management projects? In Part II, we relax some of the assumptions of single-criterion analysis and compare our approach to existing alternatives for characterizing and analyzing environmental values, and propose that the object of efforts at evaluation in an adaptive management context should be *various possible development paths*. Development paths can be judged according to multiple criteria; it is helpful to think of one category of long-term concerns as whether policies are likely to hold open valued options for the future. In Part III, we show how focusing on a choice of indicators for successful management within an adaptive management context can create a locally appropriate set of measurable indicators that "stand in for" important and widely shared social values. We offer two heuristics that encourage communities involved in adaptive management processes to propose and discuss multiple criteria and to "try out" many indicators that might separately track important social values. Finally, in Part IV, we illustrate our approach by applying our system of valuation to a real discussion of environmental and development goals in the southern Appalachians.

PART I. ADAPTIVE MANAGEMENT: AN EMERGING PARADIGM?

We propose a more comprehensive, process-oriented approach to valuation and we suggest that this can be embedded within the tradition, and growing practice, of adaptive management. In this part, we introduce adaptive management by associating this trend with three core principles, which will set the stage for asking what types of valuation and public participation processes can be expected to be successful in the management context.

Our hypothesis is that, if the new, adaptive management processes being proposed today are to be successful, they will require new ways of involving the public in environmental decision making (Shepherd and Bowler, 1997). Decades of experience with public involvement in traditional processes, such as environmental impact assessment, have revealed systematic limitations.

First, public involvement is often a discrete event or events, a snapshot of pre-project conditions, rather than a dynamic, adaptive process that considers changes over time, especially changes after project implementation (Shepherd, 1998). Our approach recognizes that individuals' preferences and perceptions can and do change, particularly in response to new information and changing environmental conditions, and that ongoing community involvement is an important part of the overall dynamic of adaptive management. Second, traditional methods tend to emphasize two-directional, but mainly episodic, information exchanges between decision makers and the public, rather than social learning and communication among individuals. "Social learning" refers to changes in the social conditions that occur when individuals learn from one another and their environment, including how individuals see their private interests linked with the shared interests of other citizens (Gunderson, 1995; Gunderson et al., 1995; Webler et al., 1995; Daniels and Walker, 1996). Our approach builds upon the concept of social learning to include values and goals associated with place-based features of a community. Third, usual methods often treat the public interest as a one-time accommodation or aggregation of individual interests (Reich, 1998), rather than preserve the plurality of values in an ongoing process of decision making. Our approach encourages individuals to express such multiple values, without requiring that they be measured according to a single criterion. This approach could, in principle, permit communities to better examine trade-offs and choose among alternative development paths in order to preserve valued place-based features. Adopting a place-based approach does not imply that only local values count – to take a place-based approach is to look at environmental problems, as they emerge on multiple scales, from specific, local perspectives. Local perspective is thus not inconsistent with development of regional, national, or global policies; it is simply a recognition that most people and most communities look at larger-scale problems *from* their local viewpoints. As attention is turned to larger-scale environmental problems which affect larger and larger communities, these locally oriented individuals and communities must – if they are to act effectively in this larger arena – form larger communities and develop policies that are also adaptive at regional and larger scales.

As noted in the Introduction, adaptive management is used here as representative of a variety of holistic, community-based environmental management processes. Speaking specifically, however, for concreteness, the adaptive management tradition has roots in the ideas of Aldo Leopold (1949; also see Norton, 1990; Lee, 1993; Chapter 3, this volume) and even earlier in the philosophical tradition of American pragmatism (Lee, 1993; Chapters 1 and 3, this volume); it was christened and given prominence by C. S. Holling and associated scientists in the late 1970s (Holling, 1978; Walters, 1986; Lee, 1993; Gunderson et al., 1995; Chapter 3, this volume). One also hears many references to "ecosystem management" (Agee and Johnson, 1988; Grumbine, 1994; Samson and Knopf, 1996). We see these trends as complementary, with "ecosystem management" being a term that relates to choosing physical boundaries of the management unit, while "adaptive management" refers to the methods and processes often favored once an ecologically delineated management unit is identified.

In this paper, we emphasize adaptive management and the methods available to adaptive managers, while recognizing that other, emerging traditions share its basic ideas, discuss adaptive management as representative of holistic, community-based environmental management more generally, and argue that these three axioms can be thought of as representative of a generic notion of holistic, ecological management. We believe that this widely shared core of ideas and axioms sets significant constraints on the type of valuation approach that will be appropriate in the day-to-day practice of such management.

Adaptive management is, above all, *experimental* management, and this represents its first core principle. Adaptive management assumes a dynamic system as the context of management – surprise is to be expected – but management methods should be designed, along with other primary goals, to reduce uncertainty through conscious study of management practice. Calling this style of management "adaptive" links the tradition to the evolutionary ideas of Charles Darwin and his successors, since communities as well as organisms "experiment" with various survival strategies.

This first core idea of adaptive management, then, entails that management actions, whenever possible, should test hypotheses about natural systems, and that controls should be designed so as to learn from pilot projects and other isolable experiments. Adaptive management is self-consciously experimental scientific management in a dynamic system. As will be emphasized below, this same experimental spirit can be applied in the search for environmental values and goals.

The second core principle of adaptive management is that the dynamic systems of nature must be modeled as *multi-scalar*. These are complex, dynamic systems which unfold at multiple scales of time and space. This insight was articulated by Leopold (1949; Norton, 1990), who in the 1940s encouraged managers and citizens not just to think as individuals, but also, metaphorically, to "think like a mountain" – to think, that is, on the temporal and spatial scales of a mountain and the ecological and geological processes going on there. This multi-scalar insight has been applied more technically in adaptive management today by the incorporation of the principles of hierarchy theory into multi-scalar management models (Holling, 1992; 1996; Gunderson et al., 1995; Norton, 1991; Chapters 12 and 16, this volume), as will be discussed after the third core principle.

The third core principle of adaptive management, "place sensitivity," has both a physical and a social aspect. Physically, place-based management is very aware of the particularities of local conditions and the function of local subsystems in larger systems. It emphasizes the particularity of complex local processes and emphasizes information derived locally. This place-based anchoring therefore encourages perspectival and case-based science (Sagoff, 1988; 1998; Shrader-Frechette and McCoy, 1994); while theory is not eschewed, it is generalized from specific local cases rather than spun out from top-down reasoning and "applied" to local situations. Socially, adaptive management recognizes the importance of local communities and the ways they use their physical resource base and, accordingly, adaptive managers emphasize public involvement and social learning in the management process. While not assuming that local people always know best, adaptive managers are respectful of public inputs from local groups and residents, taking their hopes, concerns, and values as a starting point in the search for management goals. Local habitation of a place, one might say, forms an integral part of the dialectic between nature and culture that has evolved in a place, and should be taken into account in forming management goals and plans.

One might ask, Why these three principles? The first principle is usually considered to be the defining attribute of adaptive management – it is what distinguishes it as a movement or tradition. Adaptive management is management designed to use the experimental method – in juxtaposition with public involvement and stakeholder advocacy – to reduce uncertainty in environmental decisions making (Lee, 1993). How, then, do the other two principles gain special, or core, status? The answer, we believe, is found in the central role of hierarchy theory in adaptive management.

Hierarchy theory, which emerged roughly synchronously with adaptive management, has been incorporated into the thinking of adaptive managers. It is so central to their conceptualizations of the management problem because it provides adaptive managers with a means to organize the spatial and temporal relationships that are so important in multi-scalar management. It thereby functions as a general guide to operationalizing the second core principle of adaptive management, and embodies this principle in the structure of more complex, scale-sensitive models of management. Hierarchy theory can be summarized in only two "axioms" (Allen and Starr, 1982; O'Neill et al., 1986; O'Neill, Allen and Hoekstra, 1992): (A1) all observation and measurement must be oriented from some point within the system (more on this axiom below), and (A2) smaller sub-systems change at a more rapid rate than the slower-changing larger systems which provide their environment.

Hierarchy theory, then, especially in its second axiom – to take their significance in reverse of their natural order – operationalizes the multi-scalar nature of adaptive management by modeling natural systems on spatio-temporal scales that differ by at least an order of magnitude. This multi-scalar approach also operationalizes the idea that cultural evolution proceeds at a much more rapid pace than did purely genetic evolution because of the ability of cultures to store and pass on information to their successors, rather than having that information passed on by processes of natural selection. But culture becomes ever more essential, so human communities must survive if the individuals who compose them are to succeed in perpetuating their genes and their practices. Hierarchically organized models are rich enough, conceptually, to model both processes and to relate these processes, which unfold at different temporal scales, to each other. Applying this framework to adaptive management models, in particular, we can say that struggles for individual and community survival unfold at different spatio-temporal scales. Hierarchy theory can also provide opportunities for operationalizing choices, based on the expected scale of impacts of a policy, as to which criteria should be emphasized in various situations (Chapters 12, 14, and 16, this volume).

Since the first axiom of hierarchy theory treats all observation and policy discussion as orienting from some location in a complex dynamic system, it encourages – in the study of social values, as well as in descriptive modeling – the involvement of local communities in the articulation of management goals and in the design of management experiments. The two axioms of hierarchy theory therefore corresponds (in reverse order) to, and in a broad sense operationalize, the second and third core principles of adaptive management. The first axiom is important scientifically because it operationalizes a post-Newtonian, participatory notion of observation in contrast to the traditional scientific view of the world as observed by an outside observer.

This first axiom of hierarchy theory is also important in the tradition of adaptive management, however, for the role of local communities in management activities. Adaptive managers, committed to experimentation and to the ongoing formulation and re-formulation of both management models and management goals, believe that involvement of affected stakeholders is essential if they are to develop the necessary relationship of mutual trust with local communities and to aid in the development of larger-scale regional and national communities devoted to better management at larger scales. This trust is essential if communities are to "buy into" ongoing adaptive management processes, and to remain sufficiently involved to allow social learning to occur at community and regional levels (Lee, 1993; Gunderson et al., 1995). The first axiom, which orients adaptive management practice, as well as science, from a specific place within a larger multi-scaled system, thus operationalizes both a scientific and a political focus *from a specific locale, which represents a point within a complex, dynamic, and multi-scalar system*. This axiom of hierarchy theory supports the adaptive managers' commitment to a place-based approach to communities and their resource use.

The incorporation of these two axioms of hierarchy theory into adaptive management creates a conceptual model in which environmental problems are understood from particular local places within a complex, multi-scaled system in which small, fast-changing components – both physical and social – behave against the backdrop of larger-scaled and slower-cycling super-systems that serve as their environments. Environmental problems might, of course, arise at larger as well as smaller scales, but adaptive management conceptualizes problems *from* a given local place and *within* a multi-scaled system.

A useful way of thinking about these formal assumptions is to see "choices" of individuals in the more complex system of hierarchy theory, which roughly represents the individual as being in a "place" within a complex, dynamic system in which change occurs on multiple scales according to significantly different dynamics. This location and complexity are expressed in Figure 27.1, in which individuals at one point in time face a mixture of opportunities and constraints that reflect resources available at that time. Alternatively, these can also be thought of as representing various strategies for survival open to them at the time in question. Certain patterns of individual choices in an earlier generation can, when taken in the aggregate, change the environment in ways that decrease the opportunities available to persons who live in the future, making the range of choices they face poorer than the ones found by the prior generation. If this occurs as a result of conscious choices and policies of the earlier generation, then the earlier generation can be blamed for reducing the opportunities of future people, who will also be struggling to survive given the resources available to them. This conceptual model, then, incorporates both aspects of hierarchy theory in its structure and, as a corollary, provides a schematic definition of failures of sustainability. A community is not living sustainably if the development path it is following will lead to a situation in which future individuals lack crucial opportunities that will, once lost, irreversibly diminish their life choices. Correspondingly, a positive, but still schematic, definition of sustainability would require the maintenance, over future generations, of options and opportunities essential to the ecological integrity and social identity of a given community.

## A. At a Given Time:

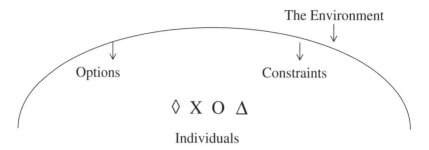

Individuals face their environment as a complex mix of
<u>options</u> and <u>constraints</u> as they <u>adapt</u> to their environment at
any given time.

## B. The cross-scale dynamic across time:

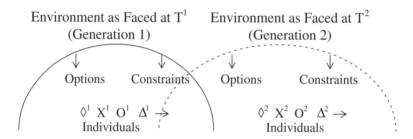

Environment and resource use problems now appear as
<u>cross-scale spill-over effects</u> as collective impacts of individuals
in <u>Generation 1</u> alter the large environmental system, creating
a changed environment for individuals in <u>Generation 2</u>
(causing them to face a new mix of option and constraints).

Figure 27.1. The basic panarchical model (Holling), hierarchically organized.

In this part, we have characterized adaptive management – characteristic
of holistic, community-based management – as a tradition that is unified by at
least three core principles and a schematic definition of "sustainability." In this
sense, the core principles might be thought of as constituting a "paradigm"
or a broad "conceptual model" for thinking about environmental problems
and solutions. We are suggesting at least that these three ideas hang together

as more than three random beliefs of adaptive managers. The first defines their distinctive approach to management, and the other two simply embody and elaborate the physical and social consequences of their formal modeling decision to use the assumptions of hierarchy theory to organize space–time relationships. Having defined sustainability by embedding the concept within an adaptive management model, we can now proceed to draw out some of the consequences that would seem to follow for environmental valuation.


PART II. AN EVALUATIVE APPROACH FOR ADAPTIVE MANAGEMENT

Since the main focus of this paper is environmental values and valuation, we can now ask: What approach to the study of values is appropriate for adaptive management, given that it must be guided by, or at least consistent with, these three core principles of adaptive management? Implicit in this question is the suggestion that different frameworks of evaluation might be compared and chosen according to their appropriateness for particular tasks. We thus understand the question of choosing an appropriate approach to environmental valuation as sorting through available and possible paradigms for the articulation, interpretation, and measurement of values. As has been made clear by philosophers of science (Kuhn, 1996; Toulmin, 1972), broad scientific approaches such as welfare economics embody constellations of important assumptions, norms, principles, and definitions, sometimes referred to as a "paradigm." A paradigm is characterized by the assumptions it makes in constituting its subject matter as a topic for research; these assumptions can make a paradigm exquisitely suited for a specific task – and these assumptions can also make a paradigm unsuitable for other tasks with different demands.

Valuational approaches can be divided into two broad categories: *single-criterion* and *multi-criteria* systems. In the first category, several general approaches have been prominently suggested in the literature; welfare economics and intrinsic value theory provide divergent examples of single-criterion systems of evaluation. There are even more types of multi-criteria systems, but most of these are not characterized by a significant body of consensually accepted principles or theory, and they are distinguished more by their intent to achieve value inclusiveness than by adherence to strict theoretical assumptions. Most multi-criteria systems are partial, indeterminate, and generally unable to provide comprehensive and non-arbitrary guidance in decision making (Glasser, 1995), so examination of them requires considerable speculation. Despite this, we propose a multi-criteria system and accompany it with a means to make it more comprehensive and less arbitrary through an iterative public process.

Theoretically, the system we propose is best thought of, initially, as resulting from the relaxation of various methodological assumptions of single-valued, economic-style environmental valuation according to which preferences, represented as individual willingness of consumers to pay for changes in their environment, are aggregated to arrive at the economic value of an outcome. In relaxing these conditions, it is not necessary to repudiate economic valuation, provided we are pluralistic. Rather, we recognize that maintenance and growth of a healthy economy may be one important and necessary condition for sustainable living and that economic valuation is therefore important, but there may well be social goals that cannot at this time be well represented in terms of willingness of consumers to pay for isolable "environmental commodities" (Vatn and Bromley, 1994).

One important difference between our approach and that of economists is that we are interested in the ways that the values of respondents change over time. Since economists emphasize the stability of preferences, they often assume that respondents' preferences are stable for the period of a given study.[1] Since we expect to be involved in studying, and contributing to, an ongoing process, it seems more appropriate to assume that participants' values will change over time. Once the evaluative task is conceived in this way, contingent valuation methods and other economic measures become useful tools among others, and in some cases contingent valuation methodologies could be used to register changes in preferences in longitudinal studies of preferences expressed at different times. How participants' preferences and values change across time therefore becomes an interesting subject of empirical study, and a number of methods are available to begin such studies. This alteration from a static to a dynamic viewpoint on preferences apparently requires more than a simple one-time elicitation of preferences, and thereby encourages the development of new evaluative tools, especially ones that can be applied iteratively and over time. These changes, in turn, suggest a somewhat different role for social scientists in the process of evaluation and of goal-setting. By introducing process-related techniques that are hypothesized to encourage consensus, the social scientist has admittedly become a part of the process, having assumed an expanded role that is likely to be controversial. Again, we simply respond that our approach is experimental and iterative; our commitment to social learning requires experiments to expand our ways of evaluating environmental policies; we can learn by doing. The experimental attitude can be maintained, methods tried out, and hypotheses tested, provided there is healthy skepticism regarding all assumptions and a commitment to improve our decision process in the next application.

Increasing evidence from cognitive psychology and related fields contradicts economists' assumption of stable preferences if it is taken as an empirical generalization that preferences are in fact stable. This evidence is of two types. One type of evidence, referring to what are called "preference reversals" (Grether and Plott, 1979; Slovic et al., 1990; Tversky et al., 1990) shows that respondents, when they respond variously to equivalent inquiries, apparently do not express pre-existing preferences, but rather "construct" preferences on demand (Gregory et al., 1993; Slovik, 1995). A second type of evidence shows that the context in which a question regarding preferences is posed seriously affects the answers elicited (Kahneman et al., 1982; Sagoff, 1988; Blamey et al., 1993). Again, it is not necessary to challenge economists' decision to elicit preferences as one-time snapshots. It may be that decisions having mostly economic impacts – decisions made in competition with other opportunities for consumption – usually are made against a backdrop of accepted current market conditions and can be viewed as relatively constant over the relevant time periods. Because of the multi-scalar nature of adaptive management and the multi-generational implications of the sustainability concept, it makes sense to view preferences as changeable across time, and to treat their change as an object of social science study when we look at long-range directions of environmental policy (Chapter 13, this volume). A multi-criteria system of evaluation can in this way supplement economic valuation with longer-term indicators associated with community values. Given the multi-scalar nature of adaptive monitoring and management models, these longer-horizon values are particularly appropriate.

While economists and decision scientists assume that discounting will eventually provide a solution to balancing social values across time, its use remains suspect and controversial outside those disciplines. In the Environmental Protection Agency's much-cited *Reducing Risk* report (USEPA, 1990), the Ecology and Welfare subcommittee rejected the use of discounting to compare present with future risks. We believe that understanding how to evaluate long-term impacts of environmental policy will require a more dynamic approach to environmental valuation, one in which the articulation of community goals is considered an ongoing and creative process. We also believe that, in improving our understanding of long-term evaluation, we must recognize the potential for major shifts in the preferences and values people express (Chapters 11 and 13, this volume).

Our approach, which emphasizes iterative public participation across time, can respond to the lability and the contextuality of preferences in two related ways. First, our approach addresses the problem of arbitrariness among criteria *theoretically*, by offering a theory of environmental values that explains how, and on what basis, communities may pursue multiple goals, even goals that are associated with quite different scales and dynamics (Chapters 12 and 16, this volume). Figure 27.1 illustrates these goals as represented in a single multi-scalar system. Second, our approach addresses the problem of weighting criteria in actual decision processes *empirically*. By working with stakeholder groups and other participants in a particular community over a period of time, we can help the participants to articulate multiple independent criteria, making it possible for the community members themselves to debate and balance competing goals.

Our idea is neither to assume that there are many incommensurable values, nor that there is a single measure of value. We simply note that discussion begins with the expression of multiple values. Our process is then designed to articulate and make these initially independent values more precise by encouraging articulation of independent criteria, and by experimenting with multiple criteria and associated measurable indicators. Since we begin with no commitment to a given number or type of indicators, we enter the process of articulating the social values of a community with an experimental spirit. Throughout the process, we can be watchful for ways to integrate, systematize, and simplify diverse values and goals. We seek to accomplish these systematizations through experimentation and interaction within a process of public discussion, rather than by definitional fiat.

Another difference between our approach and economic valuation methodologies is that we focus on a different, and more holistic, "object" of valuation. Economists attempt, to the extent possible, to construe changes in the environment as discrete "commodities" – commodities that could (at least hypothetically) be available for "purchase" in a market situation. This creates an atomistic approach to valuation, with the environment understood as many discrete elements. Values of actual changes in the environment – which will usually involve changes in the availability or price of several or many such "commodities" – are then aggregated from distinct elements. Our approach, by contrast, is to evaluate "development paths" more holistically. A development path can be thought of as direction by which the community could proceed into the future, a direction that will be significantly affected by the policies and decisions the community makes.

By considering the object of evaluation to be development paths, we can see a given community's problem as that of choosing, among the acceptable paths to economic development, that path that also holds open the most important options for future generations. On this view, individuals in the present

528

must, as in Leopold's (1949) simile, "think like a mountain"; this requries thinking about the long-term as well as the short-term impacts of decisions and thoughtful attempts to integrate these. We are proposing that we operationalize Leopold's idea as an explicit element of the adaptive management process by evaluating development paths according to multiple criteria, recognizing with Leopold that different management criteria are applicable to dynamics that unfold on distinct temporal scales. For example, the rate of erosion of mountainsides is not normally a major factor in our economic decisions; but when we think of our bequest to future generations, erosion rates might represent important evidence about how we are doing. The object of evaluation would thus be alternative development paths, which can be thought of as coherent scenarios – ways that development of a community could go from a given point. These might include actual projections based on no-intervention assumptions, but could also include coherent alternative development scenarios, paths that would unfold over several scales of time, given various policy interventions.

Further, our approach differs from that of economic valuation in that, since we place less emphasis on aggregation of values across geographic space, we emphasize and encourage place-based and local values, and we expect that the scientific data sought, and the management experiments undertaken, will often be tailored to local ecological conditions and local social concerns. We recognize that some communities will place a very high value on certain local features of their environment, features that lend distinctiveness to local places. Thus we are not surprised if, in the process of discussing values, goals, and indicators, a community adopts somewhat idiosyncratic indicators associated with its special sense of its local place. We believe this emphasis on local features and on local distinctiveness, while not necessarily inconsistent with the economists' goal of creating a single criterion employing a single currency, fits especially well with the basic tenets of an adaptive management system as described above. It allows each community to choose indicators without restricting available measures to those that can be translated into the universal vocabulary of "willingness-to-pay," and encourages cross-community variation in process and outcome.

So, in a variety of ways, our approach diverges from the assumptions and day-to-day practices of economic valuation; but our approach also differs from extant multi-criteria approaches. The approach described in the literature on multi-attribute utility theory (MAUT), including value trees and value-focused thinking (Gregory et al., 1992; Gregory et al., 1993; Gregory and Keeney, 1994; Keeney, 1996), appears initially similar to ours in that these methods undertake valuation with an eye on social context. Generally, these methods use stakeholder objectives to frame decision problems: they identify and structure attributes, elicit stakeholder values, assign weights to attributes based on those values, and then mathematically combine values and facts to obtain a summary measure. Like our approach, these methods seek to incorporate the multi-dimensional aspects of value into decision making. Neither approach has a simple algorithm for weighting various values, and it is possible for management efforts to be stalled if powerful interest groups steadfastly advocate opposed values. The goal of both approaches is to continue the dialogue, creating and nurturing community and a sense of trust, even as differing policy mixes are advocated. Our work differs from that of MAUT advocates in several important respects, however.

One primary difference is that these other methods place numbers on values and then structure values into an algorithmic system, such as a value tree. Values, on this approach, are inferred from public behavior or elicited in discourse with participants, given a numeric value within a particular theory of value, and then analyzed within a technically defined system of analysis. This approach assumes that individual and social values can be quantified, organized, and then combined mathematically. Our approach, by contrast, is engaged in ordinary discourse, with technical calculations, indicators, and measures serving as aids to a more open and deliberative public discourse. Our approach seeks to incorporate diverse and perhaps qualitative expressions of values into decision making, without requiring that all values be quantified and modeled, or that all stakeholders agree on the hierarchy of attributes and the values of those attributes. Whereas advocates of MAUT elicit preferences and then use a multi-criteria system to analyze and aggregate those preferences within a technical model, we view preferences as changing as a result of public deliberation and new information, and we embed multi-criteria analysis within the public process, refining and changing the criteria in response to changing public knowledge and values.

Another difference is that many of these other methods ultimately reduce diverse and multiple values to a single summary number. In this sense, multi-criteria decision making becomes single-criterion decision making. While this approach is analytically convenient, it loses crucial aspects of context by collapsing multiple scales into a single dimension. Questions remain regarding whose values count, how much they count, and how to combine those values. In contrast, our approach preserves the plurality of values, and encourages expression of multiple values as part of the public process, without requiring that diverse and perhaps incommensurate values be combined into a single measure.

Additionally, MAUT assumes the explicit separation of facts and values (Gregory et al., 1993). We, on the other hand, assume that facts and values are often linked, and that it may not always be possible to separate the two or the effects of one on the other. What people believe can affect what they prefer and vice versa. So, our approach explicitly considers and supports the effects of social learning on the articulation and reconsideration of values. For instance, through a public participation process, stakeholders' understanding of the causes and consequences of environmental degradation may change, thereby influencing their preferences for one policy alternative over another. By evaluating development paths more holistically, we avoid the arbitrary separation of respondents' *perceptions* of a good and their *preference* for it. We propose, one might say, endogenizing the development, analysis, and weighting of competing values into a broader adaptive management process.

In summary, our approach focuses on adaptation rather than algorithm, on plurality rather than combination, and on participation rather than quantification. Even supporters of MAUT note that "Very few arenas can accommodate this type of rational display of facts, values, and conflicts" (von Winterfeldt and Edwards 1986:379). By concentrating on development paths throughout an ongoing process, we shift the emphasis away from discretizing and quantifying particular values placed on singular "commodities," and toward a public process that evaluates development paths holistically and continues also to discuss whether chosen criteria are adequate to capture the community's values.

## PART III. SOME HEURISTICS FOR PARTICIPANTS IN ADAPTIVE MANAGEMENT PROCESSES

Development of multiple criteria within an ongoing public process may suggest to some a chaotic approach to policy formation; but we believe it is possible to design a process that can bring some semblance of order to a public process of setting environmental goals and deciding on the priorities among them. One crucial aspect of this process is to choose management goals and criteria of success that reflect broad social values, social values that are reflected in the fondest hopes and the greatest fears of the public. Here, one might expect considerable diversity in people's statements of goals at the beginning of the process. But as participation continues, the group may embrace integrative indicators as ways of stating goals that would protect many values simultaneously.

While we believe that the process must involve a serious discussion of the values held by various community members and stakeholders, we do not recommend that the search for consensus start with discussions of the "ultimate value" of nature or with the articulation of very general values. An advantage of a circular, interative process is that we can choose to "begin" our interventions at any point in the ongoing process. We suggest that public discussions of management goals begin with an examination of the environmental indicators that will be used to measure "success" in management. Values will be relevant and will enter the discussion, because social values held by individuals will be invoked as reasons to choose, or give weight to, a particular criterion or indicator. But starting with the problem of choosing an initial set of rough-and-ready indicators – which will then be submitted to further discussion, refinement, and revision – allows us to make the problem a concrete issue about what we should measure and monitor, and it also leaves open the possibility that some specific indicator will be supported by people with quite different values. In this way, it may be possible to integrate many social values on a quite practical level by agreeing on a suite of indicators that support several social values simultaneously.

The history of environmental policy has made it clear that, in many cases, advocates of quite diverse values – bird hunters and bird watchers, for example – can unite behind shared goals, such as maintaining or creating habitats for migratory birds, without resolving their underlying differences in the way they value birds. Much can sometimes be gained, then, by postponing direct confrontations over ultimate values, or at least pushing these into the background, allowing stakeholders with diverse values to seek concrete goals that will further their quite different values (Norton, 1991). This outcome can be encouraged by actively seeking "integrative" indicators, ones that track a variety of values and that are acceptable to participants with diverse moral viewpoints.

A good example of an integrative indicator for regional planning is a "percentage-of-impervious-surfaces-measure." This indicator, which can be fairly accurately measured by satellite imagery, also has arguably scientific connections to important social values including a clean water supply, the amount of wildlife habitat, and other management objectives that may be favored by a patchwork constituency. This constituency, though embracing diverse values, may thus support a goal of minimizing impervious surfaces. This goal, accompanied with a means to measure success in reducing pervious surfaces over time through satellite imagery, may be a useful guide to decision makers because it can serve as a stand-in for some pretty important – and widely held – social values. Shared management foci such as this can also create a public context in which management experiments are undertaken, measurements are carefully recorded, and management options are explored through pilot projects designed to reduce uncertainty about the outcomes of various, proposed management options. The important thing is that, in the meantime, the community goes forward to discuss *both* the question of how they are doing in achieving stated goals *and also* the question of how well our chosen indicators and measures seem to be tracking socially important variables. Again, the central concept is the Deweyan idea of social learning, which can occur when communities commit themselves to an ongoing process of participation in setting management goals and priorities. Adaptive management, when it incorporates ongoing public involvement through stakeholder groups and interactions between managers, scientists, and the public, can provide a context, and help to create a trusting, experimentally minded community that encourages social learning and the gradual adoption of shared criteria by which to measure how the society is doing in protecting social values.

Given our emphasis on local participation in defining management goals, it is impossible of course to provide anything like a complete list of sustainability indicators in a theoretical paper such as this. Choosing and weighting these indicators will require, we believe, many local processes that will no doubt lead to many and diverse outcomes. The most we can do in this paper, then, is to sketch some characteristics of a process, including some tools for evaluating environmental change, that might help communities to develop a set of indicators that will define, for them, the goal of sustainable living in their place. Our contribution to the process is to offer some heuristics that might guide participants in locally based adaptive management processes to ask, and to answer, the right questions on the way to this result. Expecting diversity of viewpoints, we seek a process that can develop trust and cooperation and allow social learning, even within a diverse community. How can we improve the likelihood that communities engaged in these processes will tend toward consensus in the choice of goals and of policies to pursue those goals?

Since we focus on the task of choosing measurable indicators as a goal of public participation, a task that will require unusual attention of participants to scientific and political aspects of the management process, it would be unrealistic to hope that the task of choosing indicators could be accomplished by the general public, through direct democracy. Adaptive managers have instead advocated an inclusive process in which, by whatever means, a public advisory committee is formed. This committee should be inclusive in membership, encouraging participation of representatives from all stakeholder groups, including involved scientists, representatives of government agencies, and so forth. What is required of this committee is regular participation and an honest effort to understand and solve problems. It is also helpful if the representative stakeholders on the advisory committee can maintain regular communication with their constituencies. This committee must develop trust among its members, try to find common ground with representatives of opposed groups, and – just as importantly – serve an educative function with their constituencies. In this way, it is hoped that an "epistemological community" – a group of people with enough trust and shared vision of what the questions and problems are – can begin cooperating in choosing policies, and in using scientific testing to evaluate policies to respond to the problems faced (Lee, 1993; Gunderson et al., 1995). It is also hoped that the members of this advisory committee will communicate well enough with their constituencies to arrive at policies that will have broad public support. Again, the local and situational nature of the process we describe prohibits detailed description of such a committee or its exact workings. Nevertheless, we assume – in order to have a context for our heuristics – that such a committee has been formed as a part of a public process of adaptive management of an ecosystem, that there is enough commitment on the part of members of the committee so that complex questions can be posed and answered through experiments, and that participants remain involved long enough for social learning to occur. Given such an adaptive system in place in a local community, we are then able to offer two heuristics that may help the community to progress toward shared goals and shared measures of environmental success and failure that they associate with those goals.

## A Process Heuristic

The first heuristic is a way of thinking about the process. The Process Heuristic suggests dividing the ongoing process into two tiers, or "phases," which we can call the "action" phase and the "reflective" phase (Page, 1977; Chapters 12–14, this volume). In the *action phase*, there exist several goals and associated "action rules." These rules will include general evaluative criteria, such as the Cost–Benefit test, the Safe Minimum Standard of Conservation, and the Precautionary Principle, and they will also include more specific goals and associated indicators, such as "minimize impervious surfaces in the watershed," or some other indicators that express more distinctive, place-based aspects of the community's environment. These multiple criteria and indicators, in order to be deployed in real decision situations and according to an evaluative plan, however, must be formulated and weighted in a *reflective phase*. In the reflective phase, a second-order public discourse is thus initiated to design an evaluative procedure employing some combination of the various criteria, or decision rules, in the action tier, according to the appropriateness to a given problem situation. In practice – in active community-based processes – the two phases will of course normally overlap and proceed simultaneously. This two-phase, iterative mechanism is thus simply a heuristic designed to help discussants shift focus from *evaluating* development paths to the reflective task of *choosing an appropriate evaluative model in a given, particular situation, and back again* (see Figure 14.1).

## An Evaluative Heuristic

We turn now, more specifically, to the choice of evaluative criteria for use in this process. As noted above, we evaluate development paths, which are ways that a community could develop in the future, given its current status. There are of course many possible development paths proceeding from any point, but it may be possible to identify a few alternative directions and associate these with policy choices facing a community. If so, it may also be possible to specify a small battery of measurable criteria, or "indicators," that could be used to evaluate proposed development paths (UNCHS, 1994; Alberti, 1996). For this task, we need a system of valuation that encourages articulation of multiple values and goals, coupled with a process of ongoing discussion, debate, information-gathering, and revision of goals as described above. Citizens and stakeholders must be engaged in an ongoing iterative process that builds both trust and an expanding data base, creating an atmosphere conducive to social learning. One important role of stakeholder and citizen participants in management is to help adaptive managers focus attention on problems that are considered important by responsible community members. Since, especially in the beginning stages of a participatory process, we can expect divergent values and concerns to be stated, a multi-criteria approach to valuation allows participants to express their own values in their own terms. Social scientists, as part of the process, can help participants to articulate their varied concerns more clearly and precisely. Ideally, there will emerge a small cluster of measurable environmental indicators, with each of these being advanced by some or all participants as useful measures that are associated with worthy social values. Since stakeholders, arrayed in ongoing participatory groups, can continue conversations about goals and values, adaptive managers can hope that clarification and sharpening of specific values, along with some systematization, consolidation, and simplification of multiple evaluative criteria, will occur. In the process, the choice of goals, values, indicators, and evaluation criteria all become a part of the ongoing experimental approach to management.

We have described a process, consisting (at least implicitly) of two phases, in which an ongoing advisory committee moves back and forth between a reflective phase of goal-setting and discussing associated proposals of particular indicators, or slates of indicators, based on outcomes of the actions taken, and a more action-oriented phase of proposing and choosing policies. This action-oriented phase involves the application of criteria already judged appropriate in a reflection on goals and possible measures associated with them. In it the group proceeds to choose policies by which to pursue those goals, and then attempts implementation and evaluation of those policies. Every cycle through the phases provides further information to feed back into the reflective phase, and this information can either confirm or provoke reconsideration of goals, values, and indicators. This discussion proceeds by focusing primarily on choosing widely acceptable indicators, all the while encouraging people to express their values and suggestions as a part of the ongoing reflection on goals. Our approach offers no ready solutions or decisive algorithms, and certainly no one-size-fits-all criteria or indicators. What we can offer, as the elements of our more systematic, procedural approach to environmental evaluation and decisions, are (a) a general theory based on adaptive management and on democratic practice that supports a public process in a setting designed to encourage experimentation and to induce social learning and (b) some simple heuristics to help particular communities in particular situations to ask the right questions and to gradually move toward agreement regarding goals of environmental management and regarding how to measure success in seeking those goals.

Throughout both phases it is assumed that the process is open and that participants interact with the broader public, both in order to inform and educate the broader public about the process and also to get feedback from the public about proposed management goals and progress in achieving them. Because our theory is community-based and democratic, it is not possible, dealing on the theoretical level of this paper, to be both substantive and specific in defining goals and indicators within the system we describe. This, again, is not surprising in that our theory locates the definition of sustainability and community goals in a local, public process that is expected to have varied outcomes in different communities.

The evaluative heuristic recommends that, in local management situations and with advisory/stakeholder committees fully involved, an iterative process be begun with an exercise in choosing an open-ended slate of indicators that express all participants' values. At first the list of indicators will be inclusive, and there should be discussion of how the various indicators can be associated with policy goals, and how experimental initiatives might be undertaken to establish relationships between various indicators and broad management goals. One requirement of a good indicator will be that it must be measurable, and that measurement must be reasonably efficient and effective; this practical requirement will be, in the course of discussion, balanced against the expected correlation of various indicators and measures with broad social goals of the community. Once a slate of measurable indicators is proposed, the task of gathering baseline data and formulating goals for changing current states of the environment can be undertaken, and a round of policy discussions about options can lead into the action phase, where the proposed multiple criteria are used to rank various proposed management options. Here, adaptive managers will advocate experimental management initiatives, localizable experiments, and pilot projects – with controls – that allow the community to learn about outcomes of policies in a limited locale and also to learn about and assess the performance of the current slate of indicators. Throughout this process, it is expected that stakeholders and laypersons will interact regularly with scientists and technicians, learning about the technical strengths and weaknesses of particular indicators, as well as ascertaining how well the indicators track social values of interest. As the process passes, implicitly or explicitly, through the action and reflective phases time and again, it is hoped that social learning will occur.

One of the key goals of our multi-criteria approach is to allow participants in the process to articulate and gradually agree upon some goals – especially long-term goals – that are expressed in non-economic terms, such as explicit moral commitments to hold open certain options and opportunities that give character and distinctiveness to a place. These are values that participants are not comfortable "trading off" against short-term economic gains; these values, one might say, are privileged within that community because they represent what we will call its "constitutive" values (Ariansen, 1997). Constitutive values are values which, to participants and community members, represent a voluntary self-identification with the peculiarities and charms of a particular place. If constitutive values of a place are threatened, a community member would fear for the special identity of his or her home place. Such fears might be expressed as "If *that* were to happen – if my community were to change in *that* way – I wouldn't even care to live here anymore." This outcome occurs when a place loses its "integrity" and the constitutive link between a community, its environment, and its values is (at least figuratively) severed (Ehrenfeld, 1993). While it may be argued that loss of communities and the values they cherish should, on the Darwinian idea of selection, be considered a natural outcome

537

of the competitive process, our purpose is to ensure that communities can, if they choose to be proactive, articulate policies that maintain a commitment to local natural and cultural history.

In this part we have tried to describe a process of public participation that is rich enough to fulfill the demands of an ongoing project to manage a watershed or an ecosystem according to the principles of adaptive management. We introduced our approach by showing ways in which it differs in important respects from both single-valued criteria and from most of the multi-criteria approaches currently under discussion. We have also shown how our process, if adapted to apply in many different communities, could provide a context for fruitful discussion of environmental goals. Our process endogenizes choices of goals and indicators, and anticipates social learning in the realm of values and community planning. The multi-scalar nature of adaptive management makes multi-scalar monitoring and evaluation possible, and it is a challenge, but hopefully a realistic one, for communities to devise multiple modes of evaluation for impacts that occur at different scales and on differing cycles. We have assumed a multi-criteria system of evaluation embedded in an adaptive management project, with hierarchy theory structuring space–time relations. We believe this pluralistic system, if embedded in an adaptive, participatory process and supplemented with our two heuristics, can be expected to help diverse communities move toward consensus in articulating goals and also in choosing ways to measure attainment of those goals.

PART IV. EVALUATING DEVELOPMENT PATHS IN THE SOUTHERN
APPALACHIANS: AN APPLICATION

In order to give some concreteness to discussions of projects of this sort, we quote the sincere expression, by a local environmental activist from southern Tennessee, of what was to us a convincing "environmental value," and which may be representative of sense of place values. The activist was expressing his frustration at a series of governmental and private decisions, decisions that seemed to make it more and more inevitable that large multi-national corporations would be allowed, even encouraged, to construct mega-mills along the Tennessee River, huge mills for grinding hardwood forests into chips. The extraordinary scale of these mills would ensure that virtually all of the remaining hardwood forests in the Southern Appalachians will be "chipped out." The activist said, "If they let the chip mills in, they'll scour the Southeast, and replant fast-growing pines in straight rows. I grew up in a hardwood forest. We like our hardwoods. I'll fight to stop them, but it seems pretty hopeless, with the government talking 'jobs,' and the big Japanese money behind the mills." The chip mills, and the fast-growing plantation pine forests that will inevitably follow the cutting of hardwoods, will predictably ensure jobs and income for the area for the foreseeable future, so it seems doubtful that the value the activist was expressing was an economic value in any simple sense. What exactly is the value he was expressing?

The value surely has an aesthetic component – the activist was expressing an aesthetic preference for mixed hardwood landscapes over pine plantation landscapes – but this is surely not the whole of the value as experienced (Chapter 19, this volume). Conceptually, it makes sense to think of the additional value, beyond the aesthetic preference, as a value placed on retaining key options or opportunities in the location where the activist lives. Suppose the activist is a hunter; he might have continued his argument[2]: "I love hunting; it makes me come alive each fall; I usually go hunting in the river valley, over in the national forest. My grandpa and my pa used to go there, and they showed me where the deer pass through a little draw early in the morning on their way to the river. Now, my father doesn't usually feel good enough to go along, but my son and I have hunted there every season since he turned twelve years old." These embellishments to the story – which could of course just as easily have included hiking, photography, or bird watching experiences – are important because they begin to show how aesthetic preferences, experiences, and choices all play a role in the individual evaluation process.

Obviously, it is impossible, scientifically, to capture all of this detail in basic measures, so our goal is to offer principles that might guide a process in which this rich fabric of individual experience is fed into – and shaped by – a participatory ecosystem process. Speaking generally about these very specific experiences, we can say that, for the activist, there is a range of experiences or options which are especially important, experiences that are somehow essential to his sense of self and to his sense of family and community. If these options are destroyed as a result of the destruction of the hardwood ecosystem he has grown up with, this outcome would leave him poorer by eliminating options that give meaning to his life, that connect him to his past, and that give him hope for the future. Following Ariansen (1997), we have called the values associated with these options "constitutive values," because, if they are lost, the integrity of a place – its identity as a place – is diminished, as is the sense of self of community members.

Building on this example, we note (a) that the value the activist defends is independent of economic growth issues – there is little doubt that the entry of chip mills will stimulate economic activity in the area – so the values involved are unlikely to be captured in exclusively economic measures; (b) the threatened loss of value can be given context and meaning only in a longer time frame of decades and even generations – it is therefore not easily expressible in "present dollars"; (c) the loss is clearly place-based – the activist is not making a claim that hardwoods are always and everywhere better than pine forests, but rather that hardwoods are naturally and culturally "appropriate" to his home place; (d) the value in question has more to with holding open certain valued options, options which provide meaning and continuity to a community and its culture – the threatened loss that motivated the activist would represent a restriction of the future options open to him and his children; and (e) the value in question seems to refer, not so much to "objects" or "elements" of nature, but to variations in the type of economic development that emerges in the region.

Assuming that our activist is likely to favor at least some economic growth and increasing the standard of living for the region, and given that we have just analyzed the values that motivate his activism as non-economic, we may now have a simple example of how one might use a two-criteria system of value as part of a process to help our activist to integrate two conflicting values. The situation faced by our activist can be characterized as follows. Development interests have proposed to pursue a particular path toward economic development, a path that would positively affect economic activity and likely increase income levels in the area over coming decades. According to economic criteria, then, the chip mill path scores high, perhaps higher than any other development opportunities, if projected over a few years. But our activist also knows that there will be predictable ecological and landscape effects if that development path is pursued. We have interpreted our activist as criticizing the chip mill path as eventually reducing and eliminating certain options which, to him, are highly valued in a non-economic sense having to do with his personal, family, and community identity. If certain options are gradually obliterated as his community pursues the chip mill path, these longer-term and more personal values will be obliterated as well, reducing the continuity he feels with his children and with the communities that evolve in his place in the future. Logically speaking, then, the loss of these valued options, which support important values constitutive of the activist's sense of self and community, can be understood as losses that are not directly compensable in economic terms. The activist and his family, he believes, will be worse off than they would have been if a different path toward development were followed. This can be described as a non-compensable loss because it is attributed even though the chipmill path to development is likely to make them richer. And, as long as both goals can be expressed in terms of more or less, then dialogue can continue. If our activist could live with a 25 or

40 percent reduction in hardwood cover, a variety of more diversified growth paths would open up for discussion.

Our activist's objection to the chipmill path can now be given expression as follows: "While the chip mill path to development scores very high on projections regarding its impact on economic growth in the region, it has unacceptable consequences. I am seeking a development path that scores reasonably well on economic growth measures and is also able to hold open important options that give meaning to my life and to my social interactions; it is important to me, and to my community, that these options be held open for the future – they represent our identity as a family and as a community. I will work to implement such a policy because I simply cannot accept the personal and social costs of destroying options that are so key to our long-term attachment to this place." This set of concerns, if expressed by an activist, is not perspicuously discussed according to a single-criterion system. It seems more like a problem of finding a development path that comes closer to fulfilling two criteria based on independent variables. Since it is impossible to maximize more than one variable in a system, one must find a prudent and efficient trade-off between development goals.

Some readers may be concerned that, by saying that such losses are non-compensable, we imply that intergenerational values will be applied as lexicographically prior, and thus "trump" all economic values. We do not intend to suggest strict lexicography, but rather treat both economic growth and protecting options for the future as important goals. Neither criterion need be given absolute, or lexical, priority, but it might make sense to set *de minimus* standards for each criterion and restrict serious consideration to paths of development that can be expected to achieve minimal levels for each. If our activist were a member of an advisory committee in an adaptive management process, we can imagine him proposing that his community should choose "percentage-of-area-in-mixed-hardwoods" as a useful indicator, and he could explain that the hardwoods are, to him, a useful stand-in for many of his values. Assuming there is also a participant representing local business interests, we can expect her to make a case for setting a goal of consistent and robust economic growth. Discussion and negotiation now becomes a matter of trade-offs, between goals and degrees of achieving them, within a democratic process. While the values advocated by the varied stakeholders are not commensurable, both are at least roughly quantifiable and representable as matters of degree. The value of our evaluative heuristic is now clear: if we maintain multiple criteria throughout the participatory process, it will be easy for the participants to discuss the usefulness and importance of the two indicators, and which one should be emphasized in which situation.

Further, once the goals and indicators are stated, and gradually improved over time, their more-or-less nature will encourage the development of many more alternative paths, or scenarios for development. Relatively little attention has been given to the creation and evaluation of alternatives based on clearly articulated stakeholder values (Gregory et al., 1992; Gregory and Keeney, 1994). Moreover, environmental decision-making methods have focused on the selection of the "best" alternative from a selected set of alternatives, rather than the process by which alternatives can be refined, created, and evaluated (Keeney, 1996). Our approach could, in principle, result in better mixes of economic development and environmental protection, and even the generation of new and creative responses to perceived environmental problems.

We realize that the case we develop here is somewhat idiosyncratic, and perhaps simpler than would be many full-fledged public and community processes. The case we choose for illustration is admittedly favorable for our case because we focused only on one stakeholder, an activist who has already chosen his high-priority issue, and an issue that just happens to be associated with a measurable feature of the landscape. In a real case, there would be a very "noisy" process of getting from many diverse goals and values to a small number of indicators that are candidates to guide management choices. The point of the example is not to draw any generalizations about the nature of environmental values in all situations, but rather to work through one example to show how a systematic, but not monistic or technical, approach to environmental valuation may encourage communication and community-based cooperative management.

At this point, we have helped the activist, his colleagues, and those who favor economic development to express the multiple, and not immediately commensurable, values that affect an important decision. In our approach, participants are able to express their various concerns in a simple conceptual model and, at the same time, our approach offers simple heuristics. The challenge for the activist and the community he lives in is to find a development path that scores high enough on the economic growth criterion, and avoids the unacceptable consequence of creating an ecological and historical or social discontinuity in a single generation. In order to accomplish this, the community must articulate and examine multiple possibilities in search of shared long-term values they can adopt as long-term commitments of their community. This task, undertaken by an advisory committee, will involve weighing risks of various actions and policies, but it will also involve choosing which options – and associated values – are to be privileged as constitutive of the community's commitment to, and cultural connection with, their past and future.

Since we do not consider our activists' values to be fixed, and we assume other members of his community would express different values, all of this will be part of a complex and changing process. If we can encourage our activist into a public participation process, however, then he and his neighbors – some appearing as plain citizens and others as representative of various interests in the community – can begin to articulate which outcomes and risks are unacceptable, and to play off economic criteria against other criteria in search of acceptable compromises. In this sense, we have created a context in which a very simple multi-criteria system with only two incommensurable criteria – pushed forward by the energies of conflicting interest groups and (hopefully) a shared desire to adopt a policy – that can serve as an opportunity for building toward consensus. The problem remaining, of course, is the big one – to identify, to articulate, and then to associate these options and values with measurable features of the environment. This act of choosing appropriate indicators must be undertaken by any community that accepts the challenge of pursuing adaptive and democratic environmental policy formation. If our theory and our speculation about place-based, pluralistic, and dynamic valuation is correct, however, we have perhaps pushed the argument to its limit in this pre-empirical examination of theory and issues. The identification, articulation, and measurement of these important values must be undertaken, we have argued, within a broad-based, participatory, iterative process; a process that must be begun, and pursued continually, within a larger adaptive management process in each particular place that resolves to live sustainably, according to a definition they have actively chosen.

CONCLUSION

Our approach to valuation studies has been consciously shaped by the core principles of adaptive management, which we have taken as representative of an emerging trend in search of a more comprehensive paradigm for environmental management. Our approach to valuation is, accordingly – and in correspondence to the three core principles of adaptive managers – experimental, multi-scalar, and place-based. In this paper we have presented an approach to environmental valuation that is both pluralistic and, to some degree, systematic; it is an approach that involves a political process assisted by heuristics. Our approach differs from technical decision processes such as quantitative risk assessment or multi-attribute utility theory by being openly political and value-laden; it differs from the usual political discourse by encouraging rational discussion of values in the context of a search for shared indicators and management objectives, rather than relying on emotion and differentials in political power. The goal is to embody people's commitments to important values in their choice of appropriate indicators and policy goals.

We believe that a shift to this approach to valuation studies can improve the role of public involvement in environmental decision making. Public involvement is often a discrete event or events before project implementation, rather than an ongoing, adaptive process. In this regard, public involvement methods share the problem of traditional economic valuation methods: they elicit preferences as they exist at a specific time and provide often only a snapshot of pre-project conditions. Our approach recognizes that environmental conditions and individuals' perceptions can and do change, and that ongoing community involvement is central to the evaluation process. We offer two heuristics: a process heuristic that encourages alternation between action and reflection, and an evaluative heuristic that encourages the development of multiple criteria to assist in choosing among various development paths. By applying these heuristics, our approach could, in principle, permit communities to design, and choose among, alternative development paths in order to preserve valued place-based features and to chart a course toward sustainability.

ACKNOWLEDGMENTS. This research received support from the Methodology, Measurement, and Statistics Program of the National Science Foundation (SBR9729229). Any opinions, findings, or conclusions are those of the authors and do not necessarily reflect the views of the National Science Foundation. This work also benefited from collaborations with Bruce Beck and his research team, University of Georgia, under support from the Water and Watersheds Program, Environmental Protection Agency (R825758).

### NOTES

1. The question of preference stability is sometimes conflated with the question of "consumer sovereignty" (see, for example, Stigler and Becker, 1977) – the view that individuals are the best judge of their own well-being – but these are clearly separable issues. Although we have elsewhere expressed concern regarding consumer sovereignty as an assumption in environmental valuation (Chapter 11, this volume), our emphasis here is on the narrower question of changeability of preferences as an important aspect of public involvement in environmental goal-setting.
2. The activist is a real person, and most of the information above was based on a real conversation in which most of these points were either made or implied.

REFERENCES

Agee, J. K., and Johnson, D. R. (1988) *Ecosystem Management for Parks and Wilderness*. Seattle, WA: University of Washington Press.

Alberti, M. (1996) "Measuring Urban Sustainability," *Environmental Impact Assessment Review*, 16(4–6):381–424.

Allen, T. F. H., and Hoekstra, T. W. (1992) *Toward a Unified Ecology: Complexity in Ecological Systems*. New York: Columbia University Press.

Allen, T. F. H., and Starr, T. B. (1982) *Hierarchy: Perspectives for Ecological Complexity*. Chicago: The University of Chicago Press.

Ariansen, Per. (1997) "The Non-Utility Value of Nature. An Investigation into Biodiversity and the Value of Natural Wholes." In *Skogforsk, Communications of the Norwegian Forest Research Institute* (Meddelelser fra Skogforsk) 47. Aas, Norway: Agricultural University of Norway.

Blamey, R. K., Common, M. S., and Norton, T. W. (1993) "Sustainability and Environmental Valuation," *Environmental Values*, 2: 299–334.

Burgess, J, Harrison, C. M., and Filus, P. (1998) "Environmental Communication and the Cultural Politics of Environmental Citizenship," *Environment and Planning A* 30(8): 1445–1460.

Burgess, J., Harrison, C. M., and Limb, M. (1988a) "Exploring Environmental Values Through the Medium of Small Groups. Part One: Theory and Practice," *Environment and Planning A* 20: 309–326.

———. (1988b) "Exploring Environmental Values Through the Medium of Small Groups. Part Two: Illustrations of a Group at Work," *Environment and Planning A* 20: 457–476.

Cortner, H. J., and Moote, M. A. 1994 "Trends and Issues in Land and Water Resources Management: Setting the Agena for Change," *Environmental Management* 18: 167–173.

Daniels, S., and Walker, G. (1996) "Collaborative Learning: Improving Public Deliberation in Ecosystem-Based Management," *Environmental Impact Assessment Review*, 16: 71–102.

Ehrenfeld, D. (1993) *Beginning Again: People and Nature in the New Millennium*. New York: Oxford University Press.

Freeman, A. M. III. (1993) *The Measurement of Environmental and Resource Values: Theory and Practice*. Washington, DC: Resources for the Future.

Glasser, H. (1995) *Towards a Descriptive, Participatory Theory of Environmental Policy Analysis*, Ph.D. Dissertation, Department of Civil and Environmental Engineering, University of California, Davis.

Gregory, R., and Keeney, R. L. (1994) "Creating Policy Alternatives Using Stakeholder Values," *Management Science*, 40(8): 1035–1048.

Gregory, R., Keeney, R. L., and von Winterfeldt, D. (1992) "Adapting the Environmental Impact Statement Process to Inform Decisionmakers," *Journal of Policy Analysis and Management*, 11(1): 58–75.

Gregory, R., Lichtenstein, S., and Slovic, P. (1993) "Valuing Environmental Resources: A Constructive Approach," *Journal of Risk Uncertainty*, 7: 177–197.

Grether, D. M. and Plott, C. R. (1979) "Economic Theory of Choice and the Preference Reversal Phenomenon," *American Economic Review*, 69: 623–638.

545

Grumbine, R. E. (1994) "What Is Ecosystem Management?" *Conservation Biology*, 1: 27–38.

Gundersen, A. G. (1995) *The Environmental Promise of Democratic Deliberation.* Madison, WI: University of Wisconsin Press.

Gunderson, L., Holling, C. S., and Light, S, (1995) "Barriers Broken and Bridges Built: A Synthesis," in L. Gunderson, C. S. Holling, and S. Light (eds.), *Barriers and Bridges.* New York: Columbia University Press.

Harrison, C. M., Burgess, J, and Filius, P. (1996) "Rationalising Environmental Responsibilities: Comparison of Lay Publics in the UK and the Netherlands," *Global Environmental Change*, 6(3): 215–234.

Holling, C. S. (1978) "Adaptive Environmental Assessment and Management," *Wiley IIASA International Series on Applied Systems Analysis.* New York: John Wiley & Sons.

———. (1992) "Cross-Scale Morphology, Geometry and Dynamics of Ecosystems," *Ecological Monographs*, 62(4): 447–502.

———. (1996), "Engineering Resilience versus Ecological Resilience," in C. S. Holling (ed.), *Engineering within Ecological Constraints.* Washington, D.C.: The National Academy Press.

Innes, J. E., and Booher, D. F. 1999a. "Consensus Building as Role Playing and Bricolage: Toward a Theory of Collaborative Planning," *Journal of the American Planning Association*, 65(1): 9–26.

———. 1999b. "Consensus Building and Complex Adaptive Systems: A Framework for Evaluating Collaborative Planning," *Journal of the American Planning Association*, 65(4): 412–423.

Kahneman, D., Slovic, P., and Tversky, A. (eds.) (1982) *Judgement under Uncertainty: Heuristics and Biases*, New York: Cambridge University Press.

Keeney, R. L. (1996) "Value-Focused Thinking: Identifying Decision Opportunities and Creating Alternatives," *European Journal of Operational Research*, 92: 537–549.

Kemmis, D. (1990) *Community and Politics of Place.* Norman, OK: University of Oklahoma.

Kempton, W., Boster, J. S., and Hartley, J. A. (1995) *Environmental Values in American Culture.* Cambridge, MA: The MIT Press.

Kuhn, T. (1996), *Structure of Scientific Revolutions, 3rd edition.* Chicago, IL: University of Chicago Press.

Lee, K. (1993) *Compass and Gyroscope.* Covelo, CA: Island Press.

Leopold, A. (1949) *A Sand County Almanac.* London: Oxford University Press.

McClain, R. J., and Lee, R. G. (1996) "Adaptive Management: Promises and Pitfalls," *Environmental Management*, 20: 437–448.

Morrison, R. (1995) *Ecological Democracy.* Boston, MA: Boston South End Press.

Norton, B. G. (1990), "Context and Hierarchy in Aldo Leopold's Theory of Environmental Management," *Ecological Economics*, 2: 119–127.

———. (1991), *Toward Unity Among Environmentalists*, New York; Oxford University Press.

———. (1999) "Ecology and Opportunity: Intergenerational Equity and Sustainable Options," In Dobson, A. (ed.), *Fairness and Futurity.* Oxford: Oxford University Press.

O'Neil, R. V., DeAngelis, D. L., Waide, J. B., and Allen, T. F. H. (1986) *A Hierarchical Concept of Ecosystems*. Princeton, NJ: Princeton University Press.

Page, T. (1977) *Conservation and Economic Efficiency*. Baltimore, MD: Johns Hopkins University Press.

Reich, R. B. (1998) "Policy Making in a Democracy," in R. B. Reich (eds.), *The Power of Public Ideas*. Cambridge, MA: Harvard University Press.

Sabatier, P. A. (1998) "The Advocacy Coalition Framework: Revisions and Relevance for Europe," *Journal of European Public* Policy 5(1): 98–130.

Sagoff, M. (1988) *The Economy of the Earth*. Cambridge: Cambridge University Press.

_____. (1998) "Aggregation and Deliberation in Valuing Environmental Public Goods: A Look Beyond Contingent Pricing," *Ecological Economics*, 24(2,3): 213–230.

Samson, F. B., and Knopf, F. L. (eds.). (1996) *Ecosystem Management*. New York: Springer-Verlag, Inc.

Shepherd, A. (1998) "Post Project Monitoring and Impact Assessment," in H. Fittipaldi, and A. Porter, (eds.), *Environmental Methods Review: Retooling Impact Assessment for the New Century*. Washington, DC: Army Environmental Policy Institute.

Shepherd, A., and Bowler, C. (1997) "Beyond the Requirements: Improving Public Participation in EIA," *Journal of Environmental Planning and Management*, 40(6): 725–738.

Shrader-Frechette, K. S., and McCoy, E. D. (1994) *Method in Ecology: Strategies in Conservation*. New York: Cambridge University Press.

Slovik, P. (1995) "The Construction of Preference," *American Psychologist*, (50): 364–371.

Slovic, P., Griffin, D., and Tversky, A. (1990) "Compatibility Effects in Judgement and Choice," in R. M. Hogarth (ed.), *Insights in Decision Making: A Tribute to Hillel J. Einhorn*. Chicago, IL: University of Chicago Press.

Stigler, G. J., and Becker, G. S. (1977) "De gustibus non est disputandum," *American Economic Review*, 67: 76–90.

Toulmin, S. (1972) *Human Knowledge*, Vol. 1 Princeton, NJ: Princeton University Press.

Tversky, A., Slovic, P., and Kahneman, D. (1990) "The Causes of Preference Reversal," *American Economic Review*, (80): 204–217.

United Nations Conference on Human Settlements (UNCHS) (1994) *Report of the Expert Group Meeting on Urban Indicators for Country Reporting*. Geneva: UNCHS (Habitat II).

United States Environmental Protection Agency (USEPA), Science Advisory Board (1990) *Reducing Risk: Setting Priorities and Strategies for Environmental Protection*. Washington, DC: Environmental Protection Agency.

Vatn, A., and Bromley, D. W. (1994) "Choices without Prices without Apologies," *Journal of Environmental Economics and Management*, (26): 129–148.

von Winterfeldt D, and Edwards W, (1986) *Decision Analysis and Behavioral Research*. Cambridge: Cambridge University Press.

Walters, C. J. (1986) *Adaptive Management of Natural Resources*. New York: Macmillan.

_____. (1997) "Challenges in Adaptive Management of Riparian and Coastal Ecosystems," *Conservation Ecology* 1(2). http://www.consecol.org/vol1/iss2/art1

Webler, T., Kastenholz, H., and Renn, O. (1995) "Public Participation in Impact Assessment: A Social Learning Perspective," *Environmental Impact Assessment Review*, (15): 443–463.

# Index